P9-DDQ-516

# The Organic Chemistry of Drug Design and Drug Action

# The Organic Chemistry of Drug Design and Drug Action

**Richard B. Silverman**
Department of Chemistry
Northwestern University
Evanston, Illinois

**Academic Press, Inc.**
Harcourt Brace Jovanovich, Publishers
San Diego New York Boston London Sydney Tokyo Toronto

This book is printed on acid-free paper. ∞



Academic Press, Inc.
1250 Sixth Avenue, San Diego, California 92101-4311

*United Kingdom Edition published by*
Academic Press Limited
24–28 Oval Road, London NW1 7DX

Library of Congress Cataloging-in-Publication Data

Silverman, Richard B.
The organic chemistry of drug design and drug action / Richard B. Silverman.
p. cm.
Includes index.
ISBN 0-12-643730-0 (hardcover)
1. Pharmaceutical chemistry. 2. Bioorganic chemistry.
3. Molecular pharmacology. 4. Drugs--Design. I. Title.
[DNLM: 1. Chemistry, Organic. 2. Chemistry, Pharmaceutical.
3. Drug Design. 4. Pharmacokinetics. QV 744 S587o]
RS403.S55 1992
615'.19--dc20
DNLM/DLC
for Library of Congress 91-47041
CIP

PRINTED IN THE UNITED STATES OF AMERICA

92 93 94 95 96 97 HA 9 8 7 6 5 4 3 2 1

To Mom and the memory of Dad,
for their warmth, their humor, their ethics, their inspiration,
but mostly for their genes.

# Contents

## Chapter 3
## Receptors



## Chapter 4
## Enzymes (Catalytic Receptors)

## Chapter 6
## DNA

## Chapter 7
## **Drug Metabolism**

## Chapter 8
## Prodrugs and Drug Delivery Systems

# Preface

From 1985 to 1989 I taught a one-semester course in medicinal chemistry to senior undergraduates and first-year graduate students majoring in chemistry or biochemistry. Standard medicinal chemistry courses are generally organized by classes of drugs with the emphasis on descriptions of their biological and pharmacological effects. I thought that there was a need to teach a course based on the organic chemical aspects of medicinal chemistry. It was apparent then, as it still is today, that there is no text that concentrates exclusively on the organic chemistry of drug design, drug development, and drug action. This book has evolved to fill that important gap and, because of the emphasis on the mechanistic organic chemistry of these biologically important reactions, it also can serve as a text for advanced bioorganic chemistry. (However, if the reader is interested in learning about a specific class of drugs, its biochemistry, pharmacology, and physiology, he or she is advised to look elsewhere for that information.)

Organic chemical principles and reactions vital to drug design and drug action are the emphasis of this text and clinically important drugs are used as examples. Therefore, only one (or at most a few) representative examples of drugs that exemplify a particular principle are given; no attempt is made to be comprehensive in any area. When more than one example is given, it generally is used to demonstrate different chemistry. It is assumed that the reader has taken a one-year course in organic chemistry that included the bioorganic components—amino acids, proteins, and carbohydrates—and is familiar with organic structures and basic organic reaction mechanisms. Only the chemistry and biochemistry background information pertinent to understanding the material in this text is discussed. Related background topics are briefly discussed or are referenced in the general readings section at the end of each chapter.

Depending on the depth of coverage that is desired, this text could be used for a one-semester or a full-year course. The references cited could be ignored in a shorter course or could be assigned for more detailed discussion in an intensive or full-year course. Additionally, not all sections need to be covered, particularly when multiple examples of a particular principle are described. The instructor can select those examples that may be of most interest to the class.

It is my intent that the reader, whether a student or a scientist interested in entering the field of medicinal chemistry, will learn to take a rational physical organic chemical approach to drug design and drug development and to appreciate the chemistry of drug action. This knowledge is of utmost importance to understand how drugs function at the molecular level. The principles are the same regardless of the particular receptor or enzyme involved. Once the fundamentals of drug design and drug action are understood, these concepts can be applied to understand the many classes of drugs described in classical medicinal chemistry texts. This basic understanding can be the foundation for future elucidation of drug action or the rational discovery of new drugs that utilize organic chemical phenomena.

I am very grateful to Carol Slingo for single-handedly typing the entire manuscript and to Cindy Colvin, Eric Lightcap, Katie Bichler, Yury Zelechonok, Cheryl Chamberlain, Ting Su, Zhaozhong Ding, Jon Woo, Xingliang Lu, and Bill Hawe for the computer-generation of all of the structures, schemes, figures, and tables (except where they were reproduced directly with permission). Any errors found in the artwork, however, are the result of my editing oversights.

Evanston, Illinois
November, 1991

Richard B. Silverman

CHAPTER 1

# Introduction



## I. Medicinal Chemistry Folklore

Medicinal chemistry is the science that deals with the discovery or design of new therapeutic chemicals and their development into useful medicines. It may involve synthesis of new compounds, investigations of the relationships between the structure of natural and/or synthetic compounds and their biological activities, elucidations of their interactions with receptors of various kinds, including enzymes and DNA, the determination of their absorption, transport, and distribution properties, and studies of the metabolic transformations of these chemicals into other chemicals.

Medicinal chemistry, in its crudest sense, has been practiced for several thousand years. Man has searched for cures of illnesses by chewing herbs, berries, roots, and barks. Some of these early clinical trials were quite successful; however, not until the last 100–150 years has knowledge of the active constituents of these natural sources been known. The earliest written records of the Chinese, Indian, South American, and Mediterranean cultures described the therapeutic effects of various plant concoctions.[1–3]

Two of the earliest medicines were described about 5100 years ago by the Chinese Emperor Shen Nung in his book of herbs called *Pentsao*.[4] One of these is *Ch'ang Shan*, the root *Dichroa febrifuga*, which was prescribed for fevers. This plant contains alkaloids which are used in the treatment of malaria today. Another plant called *Ma Huang* (now known as *Ephedra sinica*) was used as a heart stimulant, a diaphoretic agent (perspiration producer), and to allay coughing. It contains ephedrine, a drug that raises the blood pressure and relieves bronchial spasms. Theophrastus in the third century B.C. mentioned opium poppy juice as an analgetic, and in the tenth century A.D. Rhazes (Persia) introduced opium pills for coughs, mental disorders, aches, and pains. The opium poppy *Papaver somniferum* contains morphine, a potent analgetic, and codeine, prescribed today as a cough suppressant. The Orientals and the Greeks used henbane, which contains scopolamine (truth serum), as a sleep inducer. Inca mail runners and silver miners in the high Andean mountains chewed coca leaves (cocaine) as a stimulant and euphoric. The antihypertensive drug reserpine was extracted by ancient Hindus from the snakelike root of the *Rauwolfia serpentina* plant and used to treat hypertension, insomnia, and insanity. Alexander of Tralles in the sixth century A.D. recommended the autumn crocus (*Colchicum autumnale*) for relief of pain of the joints, and it was used by Avicenna (eleventh century Persia) and by Baron Anton von Störck (1763) for the treatment of gout. Benjamin Franklin heard about this medicine and brought it to America. The active principle in this plant is the alkaloid colchicine, which is used today to treat gout.

In 1633 a monk named Calancha, who accompanied the Spanish Conquistadors to Central and South America, introduced one of the greatest herbal medicines to Europe upon his return. The South American Indians would extract the bark of *Cinchona* trees and use it for chills and fevers; the Europeans used it for the same and for malaria. In 1820 the active constituent was isolated and later determined to be quinine, an antimalarial drug.

Modern therapeutics is considered to have begun with an extract of the foxglove plant, which was cited by Welsh physicians in 1250, named by Fuchsius in 1542, and introduced for the treatment of dropsy (now congestive heart failure) in 1785 by Withering.[2,5] The active constituents are secondary glycosides from *Digitalis purpurea* (the foxglove plant) and *Digitalis lanata*, namely, digitoxin and digoxin, respectively, both important drugs for the treatment of heart failure. Today, digitalis, which refers to all of the cardiac glycosides, is still manufactured by extraction of foxglove and related plants.

## II. Discovery of New Drugs

If the approach to drug discovery continued as in ancient times, few diseases would be treatable today. Natural products make up a small percentage of drugs on the current market. Typically, when a natural product is found to be active, it is chemically modified in order to improve its properties. As a result of advances made in synthesis and separation methods and in biochemical techniques since the late 1940s, a more rational approach to drug discovery has been possible, namely, one which involves the element of design. In Chapter 2 a discussion is presented regarding how drugs are discovered and chemically modified in order to improve or change their medicinal properties.

## References


1. Bauer, W. W. 1969. "Potions, Remedies and Old Wives' Tales." Doubleday, New York.
2. Withering, W. 1785. "An Account of the Foxglove and Some of Its Medicinal Uses: With Practical Remarks on Dropsy and Other Diseases." C. G. J. Robinson and J. Robinson, London; reprinted in *Med Class.* **2,** 305 (1937).

3a. Sneader, W. 1985. "Drug Discovery: The Evolution of Modern Medicines." Wiley, Chichester.

3b. Margotta, R. 1968. "An Illustrated History of Medicine." Paul Hamlyn, Middlesex, England.

4. Chen, K. K. 1925. *J. Am. Pharm. Assoc.* **14,** 189.
5. Burger, A. 1980. *In* "Burger's Medicinal Chemistry" (Wolff, M. E., ed.), 4th Ed., Part I, Chap. 1. Wiley, New York.


## General References

The following references are excellent sources of material for this entire book.

### Journals


*Annual Reports in Medicinal Chemistry.* Academic Press, San Diego, California.
*Biochemical Pharmacology.* Pergamon, New York.
*Journal of Medicinal Chemistry.* American Chemical Society, Washington, D.C.
*Medicinal Research Reviews.* Wiley, New York.
*Molecular Pharmacology.* Academic Press. San Diego, California.
*Progress in Medicinal Chemistry.* Elsevier/North-Holland, Amsterdam.
*Progress in Drug Research.* Birkäuser, Basel.


### Books


Albert, A. 1985. "Selective Toxicity," 7th Ed. Chapman & Hall, London.
Ariëns, E. J., ed. 1971–1980. "Drug Design," Vols. 1–10. Academic Press, New York.
Budavari, S., ed. 1989. "The Merck Index," 11th Ed. Merck & Co., Rahway, New Jersey.
Burger, A. 1983. "A Guide to the Chemical Basis of Drug Design." Wiley, New York.
Foye, W. O. 1989. "Principles of Medicinal Chemistry," 3rd Ed. Lea & Febiger, Philadelphia, Pennsylvania.
Gilman, A. L., Goodman, L. S., Rall, T. W., and Murad, F. 1985. "Goodman and Gilman's The Pharmacological Basis of Therapeutics," 7th Ed. Macmillan, New York.
Hansch, C., Emmett, J. C., Kennewell, P. D., Ramsden, C. A., Sammes, P. G., and Taylor, J. B., eds. 1990. "Comprehensive Medicinal Chemistry," Vols. 1–6. Pergamon, Oxford.
Korolkovas, A. 1988. "Essentials of Medicinal Chemistry," 2nd Ed. Wiley, New York.
Wolff, M. E., ed. 1979–1981. "Burger's Medicinal Chemistry," 4th Ed., Parts 1–3. Wiley, New York.

CHAPTER 2

# Drug Discovery, Design, and Development



## I. Drug Discovery

In general, clinically used drugs are not discovered. What is more likely discovered is known as a *lead* compound. The lead is a prototype compound that has the desired biological or pharmacological activity, but may have many other undesirable characteristics, for example, high toxicity, other biological activities, insolubility, or metabolism problems. The structure of the lead compound is then modified by synthesis to amplify the desired activity and to minimize or eliminate the unwanted properties. Prior to an elaboration of approaches to lead discovery and lead modification, two of the rare drugs discovered without a lead are discussed.

## A. Drug Discovery without a Lead

### 1. Penicillins

In 1928 Alexander Fleming noticed a green mold growing in a culture of *Staphylococcus aureus*, and where the two had converged, the bacteria were lysed.[1] This led to the discovery of penicillin, which was produced by the mold. It may be thought that this observation was made by other scientists who just ignored it, and, therefore, Fleming was unique for following up on it. However, this is not the case. Fleming tried many times to rediscover this phenomenon without success; it was his colleague, Dr. Ronald Hare,[2,3] who was able to reproduce the observation. It only occurred the first time because a combination of unlikely events all took place simultaneously. Hare found that very special conditions were required to produce the phenomenon initially observed by Fleming. The culture dish inoculated by Fleming must have become accidentally and simultaneously contaminated with the mold spore. Instead of placing the dish in the refrigerator or incubator when he went on vacation as is normally done, Fleming inadvertently left it on his lab bench. When he returned the following month, he noticed the lysed bacteria. Ordinarily, penicillin does not lyse these bacteria; it prevents them from developing, but it has no effect if added after the bacteria have developed. However, while Fleming was on vacation (July to August) the weather was unseasonably cold, and this provided the particular temperature required for the mold and the staphylococci to grow slowly and produce the lysis. Another extraordinary circumstance was that the particular strain of the mold on Fleming's culture was a relatively good penicillin producer, although most strains of that mold (*Penicillium*) produce no penicillin at all. The mold presumably came from the laboratory just below Fleming's where research on molds was going on at the time.

Although Fleming suggested that penicillin could be useful as a topical antiseptic, he was not successful in producing penicillin in a form suitable to treat infections. Nothing more was done until Sir Howard Florey at Oxford University reinvestigated the possibility of producing penicillin in a useful form. In 1940 he succeeded in producing penicillin that could be administered topically and systemically,[4] but the full extent of the value of penicillin was not revealed until the late 1940s.[5] Two reasons for the delay in the universal utilization of penicillin were the emergence of the sulfonamide antibacterials (sulfa drugs, **2.1**; see Chapter 5, Section IV,B,1) in 1935 and the outbreak of World War II. The pharmacology, production, and clinical application of penicillin were not revealed until after the war so that this wonder drug would

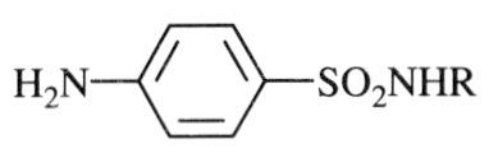

$H_2N-C_6H_4-SO_2NHR$

**2.1**

not be used by the Germans. A team of Allied scientists who were interrogating German scientists involved in chemotherapeutic research were told that the Germans thought the initial report of penicillin was made just for commercial reasons to compete with the sulfa drugs. They did not take the report seriously.

The original mold was *Penicillium notatum*, a strain that gave a relatively low yield of penicillin. It was replaced by *Penicillium chrysogenum*,[6] which had been cultured from a mold growing on a grapefruit in a market in Peoria, Illinois! The correct structure of penicillin (**2.2**) was elucidated in 1943 by Sir Robert Robinson (Oxford) and Karl Folkers (Merck). Several different penicillin analogs (R group varied) were isolated early on; only two of these (**2.2**, R = $PhOCH_2$, penicillin V, and **2.2**, R = $CH_2Ph$, penicillin G) are still in use today.

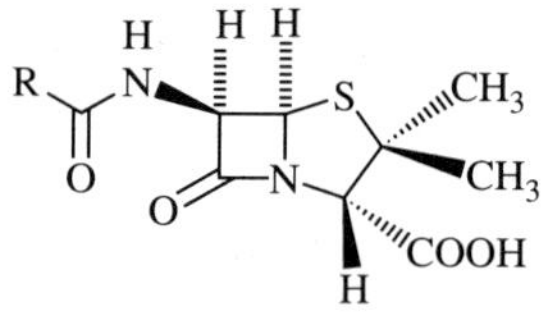

**2.2**

## 2. Librium

The first benzodiazepine tranquilizer drug, Librium [7-chloro-2-(methylamino)-5-phenyl-3*H*-1,4-benzodiazepine 4-oxide, **2.3**], was discovered serendipitously.[7] Dr. Leo Sternbach at Roche was involved in a program to synthesize a new class of tranquilizer drugs. He originally set out to prepare a series of benzheptoxdiazines (**2.4**), but when $R^1$ was $CH_2NR_2$ and $R^2$ was $C_6H_5$, it was found that the actual structure was that of a quinazoline 3-oxide (**2.5**). However, none of these compounds gave any interesting pharmacological results. The program was abandoned in 1955 in order for Sternbach to work on a different project. In 1957 during a general laboratory cleanup a vial containing what was thought to be **2.5** (X = 7-Cl, $R^1$ = $CH_2NHCH_3$, $R^2$ = $C_6H_5$) was found and, as a last effort, was submitted for pharmacological testing. Unlike all the other compounds submitted, this one gave very promising results in six different tests used for preliminary screening of tranquilizers.

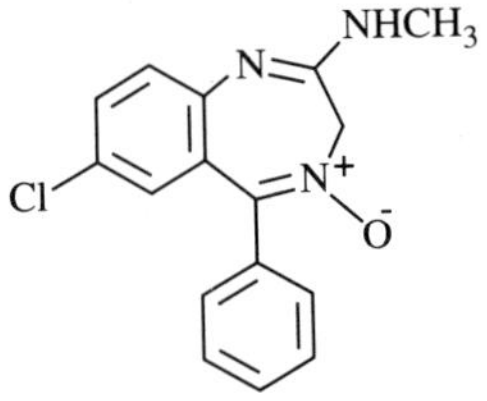

**2.3** 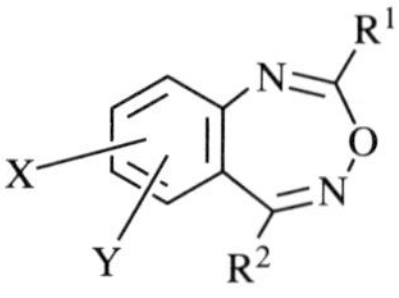 **2.4** 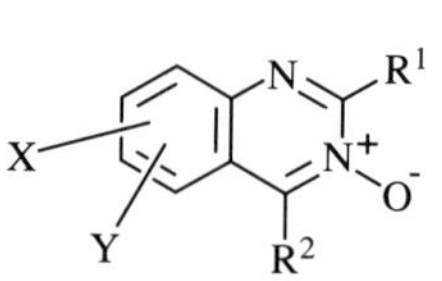 **2.5**

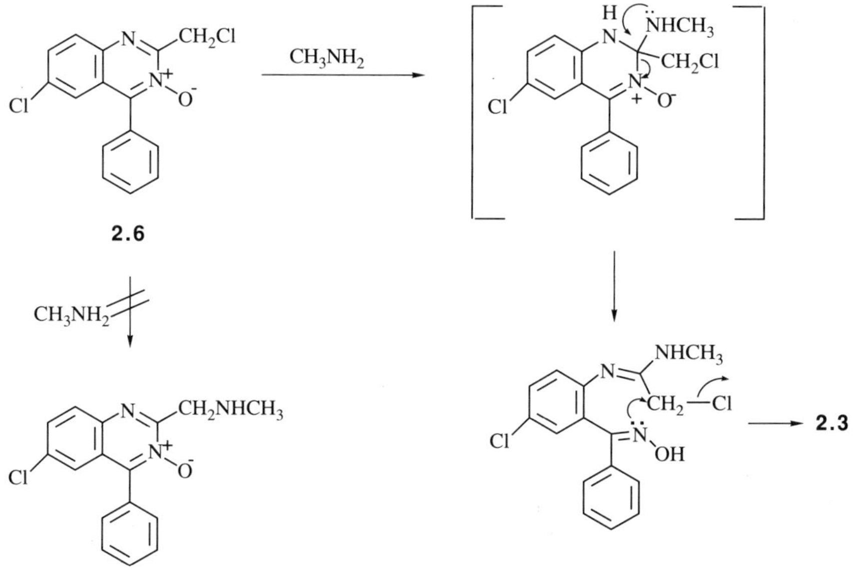

Scheme 2.1. Mechanism for formation of Librium.

Further investigation revealed that the compound was not a quinazoline 3-oxide but, rather, was the benzodiazepine 4-oxide (**2.3**), presumably produced in an unexpected reaction of the corresponding chloromethyl quinazoline 3-oxide (**2.6**) with methylamine (Scheme 2.1). If this compound had not been found in the laboratory cleanup, all of the negative pharmacological results would have been reported for the quinazoline 3-oxide class of compounds, and benzodiazepine 4-oxides may not have been discovered for many years to come.

The examples of drug discovery without a lead are quite few in number. The typical occurrence is that a lead compound is identified and its structure is modified to give, eventually, the drug that goes to the clinic.

## B. Lead Discovery

Penicillin V and Librium are, indeed, two important drugs that were discovered without a lead. Once they were identified, however, they then became lead compounds for future analogs. There are now a myriad of penicillin-derived antibacterials that have been synthesized as the result of the structure elucidation of the earliest penicillins. Valium (diazepam, **2.7**) was synthesized at Roche even before Librium was introduced on to the market; this drug was derived from the lead compound Librium and is almost 10 times more potent than the lead.

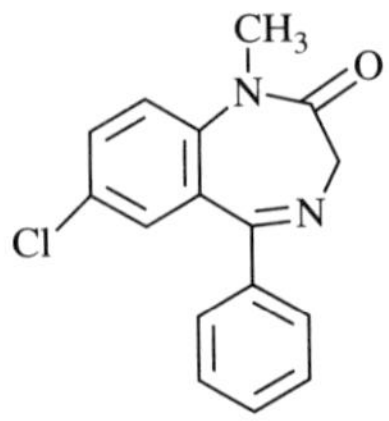

**2.7**

In general, the difficulty arises in the discovery of the lead compound. There are several approaches that can be taken to identify a lead. The first requirement for all of the approaches is to have a means to assay compounds for a particular biological activity, so that it will be known when a compound is active. A *bioassay* (or *screen*) is a means of determining in a biological system, relative to a control compound, whether a compound has the desired activity and, if so, what the relative potency of the compound is. Note the distinction between the terms activity and potency. *Activity* is the particular biological or pharmacological effect (e.g., antibacterial activity or anticonvulsant activity); *potency* is the strength of that effect. Some bioassays (or screens) begin as *in vitro* tests, for example, the inhibition of an enzyme or antagonism of a receptor; others are *in vivo* tests, for example, the ability of the compound to prevent an induced seizure in a mouse. In general, the *in vitro* tests are quicker and less expensive. Once the bioassay is developed, there are a variety of approaches to identify a lead.

### 1. Random Screening

*Random screening* involves no intellectualization; all compounds are tested in the bioassay without regard to their structures. Prior to 1935 (the discovery of sulfa drugs), random screening was essentially the only approach; today this method is used to a lesser degree. However, random screening programs are still very important in order to discover drugs or leads that have unexpected and unusual structures for various targets.

The two major classes of materials screened are synthetic chemicals and natural products (microbial, plant, and marine). An example of a random screen of synthetic and natural compounds is the "war on cancer" declared by Congress and the National Cancer Institute in the early 1970s. Any new compound submitted was screened in a mouse tumor bioassay. Few new anticancer drugs resulted from that screen, but many known anticancer drugs also did not show activity in the screen used. As a result of that observation, multiple bioassay systems are now utilized. In the 1940s and 1950s a random screen by various pharmaceutical companies of soil samples in search of new antibiotics was undertaken. In this case, however, not only were numerous leads uncovered, but two important antibiotics, streptomycin and the tetracyclines, were found.

### 2. Nonrandom Screening

*Nonrandom screening* is a slightly more narrow approach than is random screening. In this case compounds having a vague resemblance to weakly active compounds uncovered in a random screen or compounds containing different functional groups than leads may be tested selectively. By the late 1970s the National Cancer Institute's random screen was modified to a nonrandom screen because of budgetary and manpower restrictions. Also, the single tumor screen was changed to a variety of tumor screens, as it was realized that cancer is not just a single disease.

### 3. Drug Metabolism Studies

During metabolism studies *drug metabolites* (drug degradation products generated *in vivo*) that are isolated are screened in order to determine if the activity observed is derived from the drug candidate or from a metabolite. For example, the anti-inflammatory drug sulindac (**2.8**) is not the active agent; the metabolic reduction product, **2.9**, is responsible for the activity.[8] A classic example of this approach is the discovery of the antibacterial agent sulfanilamide (**2.1**, R = H), which was found to be a metabolite of prontosil (**2.10**) (see Chapter 5, Section IV,B,1 for details).

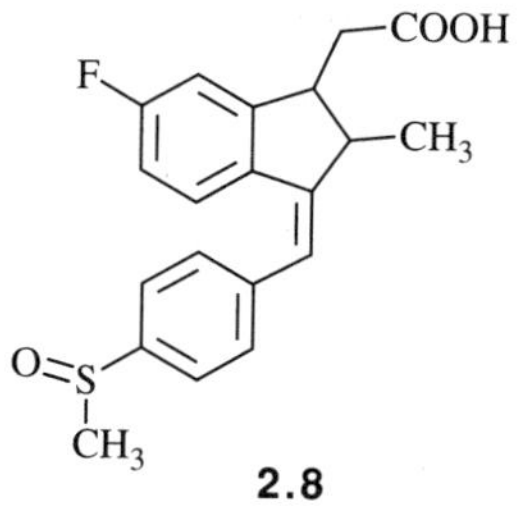

**2.8** 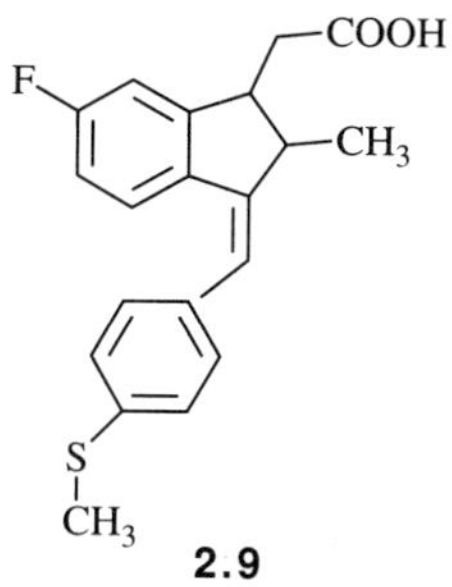 **2.9**

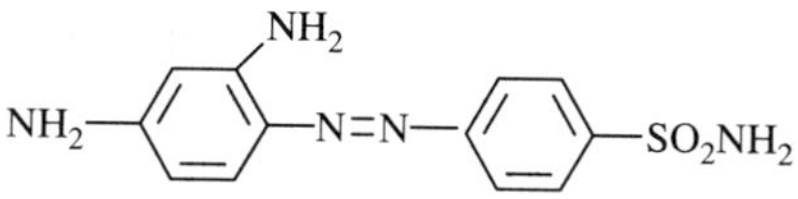

**2.10**

### 4. Clinical Observations

Often a drug candidate during animal testing or clinical trials will exhibit more than one pharmacological activity; that is, it may produce a side effect. This compound, then, can be used as a lead for the secondary activity. In 1947 an antihistamine, dimenhydrinate (**2.11**; Dramamine®) was tested at the allergy clinic at Johns Hopkins University and was found also to be effective in relieving a patient who suffered from car sickness; a further study proved its effectiveness in the treatment of seasickness[9] and airsickness.[10] It is now the most widely used drug for the treatment of all forms of motion sickness.

An antibacterial agent, carbutamide (**2.12**, R = $NH_2$), was found to have an antidiabetic side effect. However, it could not be used as an antidiabetic drug because of its antibacterial activity. Carbutamide, then, was a lead for the discovery of tolbutamide (**2.12**, R = $CH_3$), an antidiabetic drug without antibacterial activity.

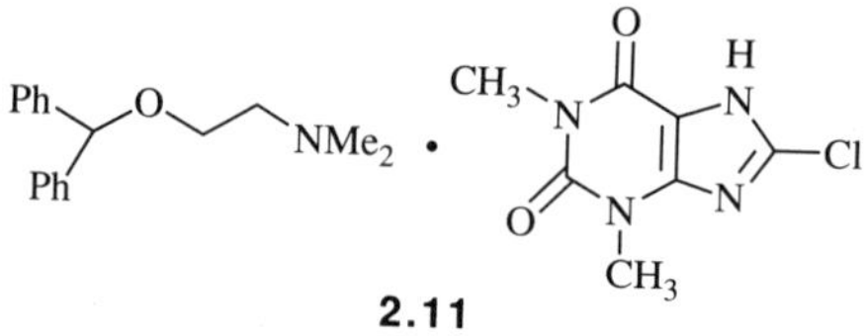

**2.11**

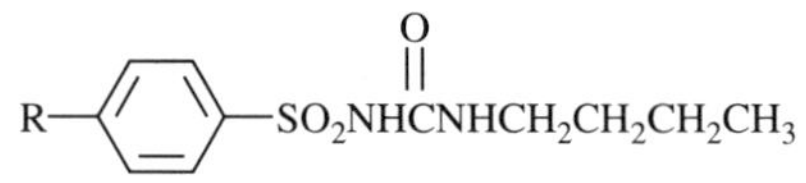

**2.12**

## 5. Rational Approaches to Lead Discovery

None of the above approaches to lead discovery involves a major rational component. The lead is just found by screening techniques, as a by-product of drug metabolism studies, or from whole animal investigations. Is it possible to design a compound having a particular activity? Rational approaches to drug design have now become the major routes to lead discovery. The first step is to identify the cause for the disease state. Most diseases, or at least the symptoms of diseases, arise from an imbalance of particular chemicals in the body, from the invasion of a foreign organism, or from aberrant cell growth. As discussed in later chapters, the effects of the imbalance can be corrected by antagonism or agonism of a receptor (see Chapter 3) or by inhibition of a particular enzyme (see Chapter 5). Foreign organism enzyme inhibition and interference with DNA biosynthesis or function (see Chapter 6) are also important approaches to treat diseases arising from microorganisms and aberrant cell growth.

Once the relevant biochemical system is identified, lead compounds then become the natural receptor agonists or enzyme substrates. For example, lead compounds for the contraceptives (+)-norgestrel (**2.13**) and 17$\alpha$-ethynylestradiol (**2.14**) were the steroidal hormones progesterone (**2.15**) and 17$\beta$-estradiol

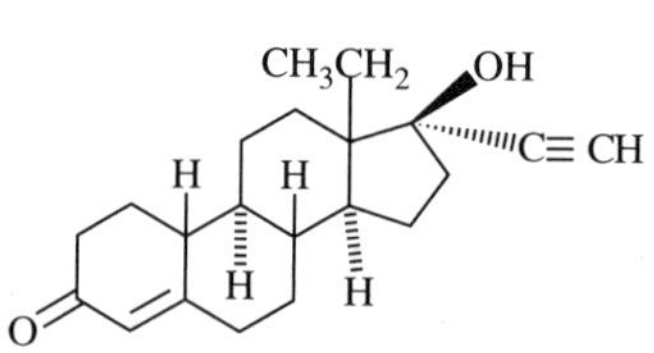

**2.13**

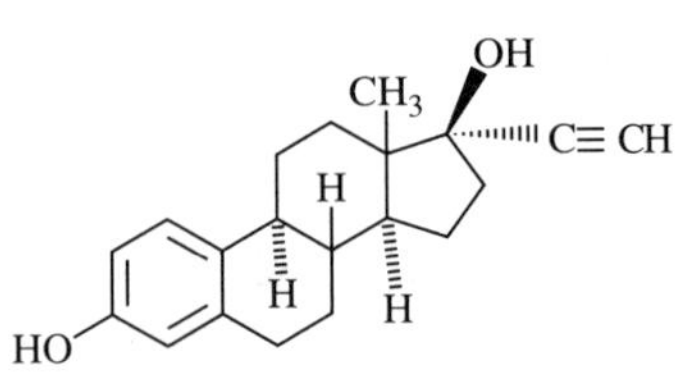

**2.14**

(**2.16**). Whereas the steroid hormones **2.15** and **2.16** show weak and short-lasting effects, the oral contraceptives **2.13** and **2.14** exert strong progestational activity of long duration.

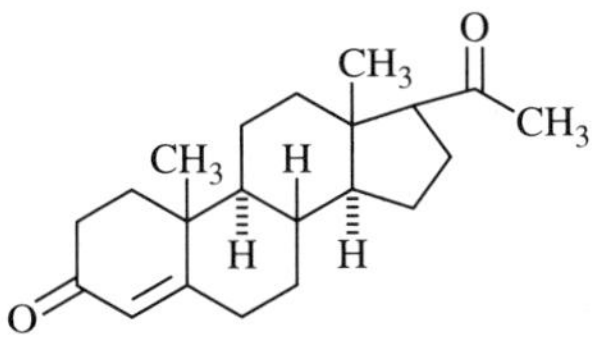

**2.15** 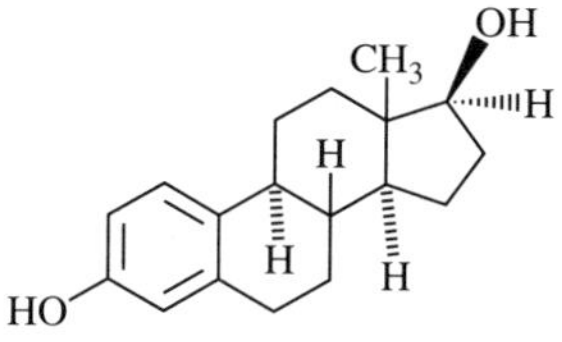 **2.16**

At Merck it was believed that serotonin (**2.17**) was a possible mediator of inflammation. Consequently, serotonin was used as a lead for anti-inflammatory agents, and from this lead the anti-inflammatory drug indomethacin (**2.18**) was developed.[11]

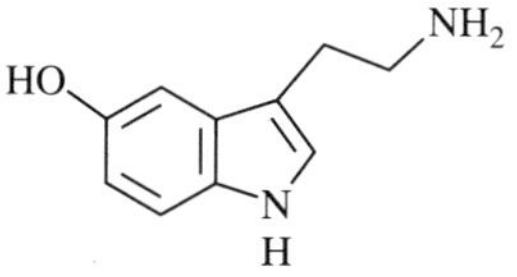

**2.17** 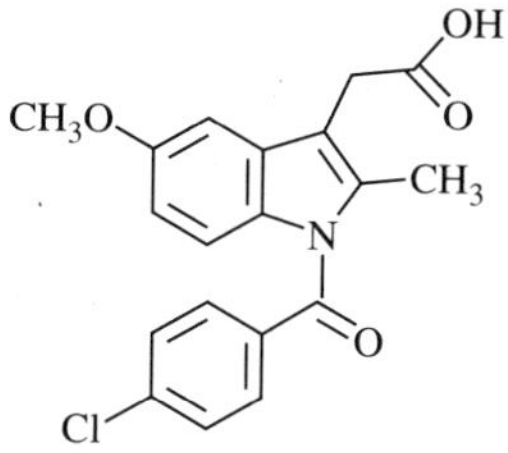 **2.18**

The rational approaches are directed at lead discovery. It is not possible, with much accuracy, to foretell toxicity and side effects, anticipate transport characteristics, or predict the metabolic fate of a drug. Once a lead is identified, its structure can be modified until an effective drug is prepared.

## II. Drug Development: Lead Modification

Once your lead compound is in hand, how do you know what to modify in order to improve the desired pharmacological properties?

### *A. Identification of the Active Part: The Pharmacophore*

Interactions of drugs with receptors are very specific (see Chapter 3). Therefore, only a small part of the lead compound may be involved in the appropriate interactions. The relevant groups on a molecule that interact with a receptor and are responsible for the activity are collectively known as the *pharmacophore*. If the lead compound has additional groups, they may interfere with the appropriate interactions. One approach to lead modification is to cut away sections of the molecule in order to determine what parts are essential and which are superfluous.

As an example of how a molecule can be trimmed and still result in increased potency or modified activity, consider the addictive analgetics morphine (**2.19**, R = R′ = H), codeine (**2.19**, R = $CH_3$, R′ = H), and heroin (R = R′ = $COCH_3$). The pharmacophore is darkened. If the dihydrofuran oxygen is excised, morphinan (**2.20**, R = H)[12] results; the hydroxy analog levorphanol[13] (**2.20**, R = OH) is 3 to 4 times more potent than morphine as an analgetic, but it retains the addictive properties. Removal of half of the cyclohexene ring, leaving only methyl substituents, gives benzomorphan (**2.21**, R = $CH_3$).[14] This compound shows some separation of analgetic and addictive effects; cyclazocine (**2.21**, R = $CH_2$—◁) and pentazocine [**2.21**, R = $CH_2CH{=}C(CH_3)_2$] are analogs with much lower addiction liabilities. Cutting away the cyclohexane fused ring (**2.22**) also has little effect on the analgetic activity in animal tests. Removal of all fused rings, for example, in the case of meperidine (**2.23**, Demerol®), gives an analgetic still possessing 10–12% of the overall potency of morphine.[15] Even acyclic analogs are active. Dextropropoxyphene (**2.24**, Darvon®) is one-half to two-thirds as potent as codeine; its activity can be ascribed to the fact that it can assume a conformation related to that of the morphine pharmacophore. Another acyclic analog is methadone (**2.25**) which is as potent an analgetic as morphine; the (−)-isomer is used in the treatment of opioid abstinence syndromes in heroin abusers.

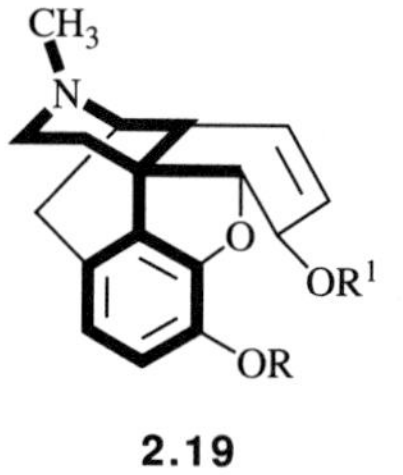

**2.19** 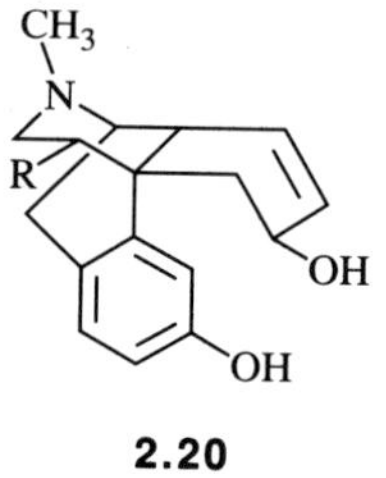 **2.20** 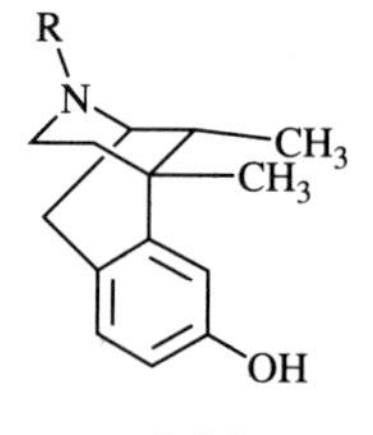 **2.21** 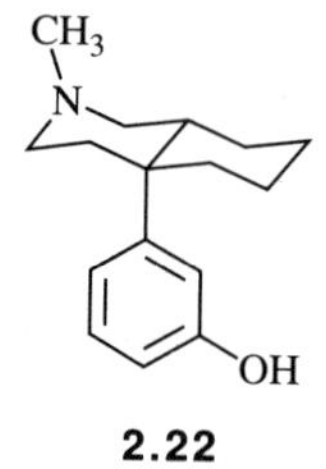 **2.22**

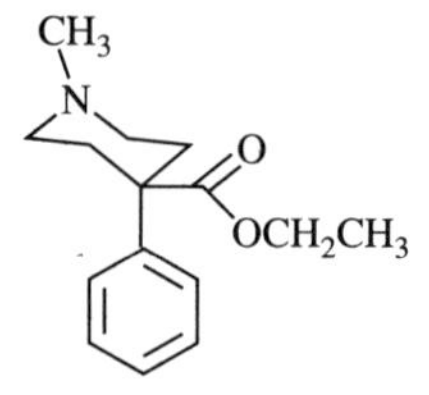

**2.23** 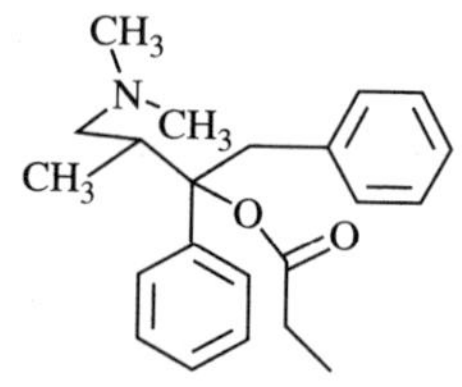 **2.24** 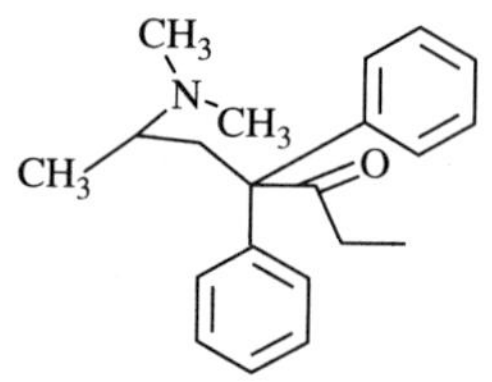 **2.25**

In some cases an increase in structural complexity and/or rigidity can lead to increased potency. For example, an oripavine derivative such as etorphine (**2.26**, R = $CH_3$, R′ = $C_3H_7$), which has a two-carbon bridge and a substituent not in morphine, is about 1000 times more potent than morphine[16] and, therefore, is used in veterinary medicine to immobilize large animals. The related analog, buprenorphine (**2.26**, R = $CH_2-\triangleleft$, R′ = *tert*-Bu, double bond reduced) is 10–20 times more potent than morphine and has a very low level of dependence liability. Apparently, the additional rigidity of the oripavine derivatives increases the appropriate receptor interactions (see Chapter 3).

Once the pharmacophore is identified, manipulation of functional groups becomes consequential.

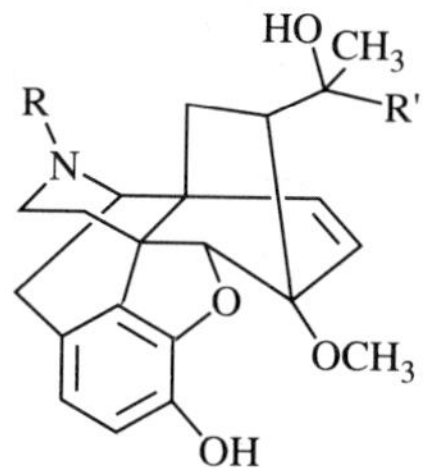

**2.26**

### B. Functional Group Modification

The importance of functional group modification was seen in Section I,B,4; the amino group of carbutamide (**2.12**, R = $NH_2$) was replaced by a methyl group to give tolbutamide (**2.12**, R = $CH_3$), and in so doing the antibacterial activity was separated away from the antidiabetic activity. In some cases an experienced medicinal chemist knows what functional group will elicit a particular effect. Chlorothiazide (**2.27**) is an antihypertensive agent that has a strong diuretic (increased urine excretion) effect as well. It was known from sulfanilamide work that the sulfonamide side chain can give diuretic activity (see Section II,C). Consequently, diazoxide (**2.28**) was prepared as an antihypertensive drug without diuretic activity.

There, obviously, is a relationship between the molecular structure of a compound and its activity. This phenomenon was first realized over 120 years ago.

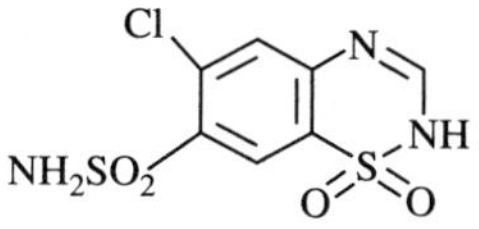

**2.27** 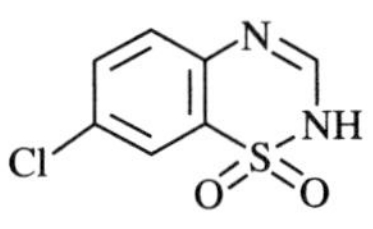 **2.28**

## C. Structure–Activity Relationships

In 1868 Crum-Brown and Fraser,[17] suspecting that the quaternary ammonium character of curare may be responsible for its muscular paralytic properties, examined the neuromuscular blocking effects of a variety of simple quaternary ammonium salts and quaternized alkaloids in animals. From these studies they concluded that the physiological action of a molecule was a function of its chemical constitution. Shortly thereafter, Richardson[18] noted that the hypnotic activity of aliphatic alcohols was a function of their molecular weight. These observations are the basis for future structure–activity relationships (SAR).

Drugs can be classified as being structurally specific or structurally nonspecific. *Structurally specific drugs*, which most drugs are, act at specific sites, such as a receptor or an enzyme. Their activity and potency are very susceptible to small changes in chemical structure; molecules with similar biological activities tend to have common structural features. *Structurally nonspecific drugs* have no specific site of action and usually have lower potency. Similar biological activities may occur with a variety of structures. Examples of these drugs are gaseous anesthetics, sedatives and hypnotics, and many antiseptics and disinfectants.

Even though only a part of the molecule may be associated with the activity, there are a multitude of molecular modifications that could be made. Early SAR studies (prior to the 1960s) simply involved the syntheses of as many analogs as possible of the lead and their testing to determine the effect of structure on activity (or potency). Once enough analogs were prepared and sufficient data accumulated, conclusions could be made regarding structure–activity relationships.

An excellent example of this approach came from the development of the sulfonamide antibacterial agents (sulfa drugs). After a number of analogs of the lead compound sulfanilamide (**2.1**, R = H) were prepared, it was found that compounds of this general structure exhibited diuretic and antidiabetic activities as well as antimicrobial activity. Compounds with each type of activity eventually were shown to possess certain structural features in common. On the basis of the biological results of greater than 10,000 compounds, several SAR generalizations have been made.[19] Antimicrobial agents have structure **2.29** (R = $SO_2NHR'$ or $SO_3H$) where (1) the amino and sulfonyl groups on the benzene ring should be para; (2) the anilino amino group may be unsubstituted (as shown) or may have a substituent that is removed *in vivo*; (3) replacement of the benzene ring by other ring systems, or the introduction of

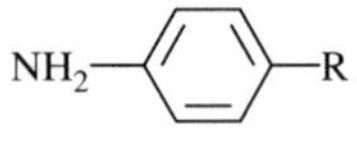

$NH_2$—$C_6H_4$—R

**2.29**

additional substituents on it, decreases the potency or abolishes the activity; (4) R may be

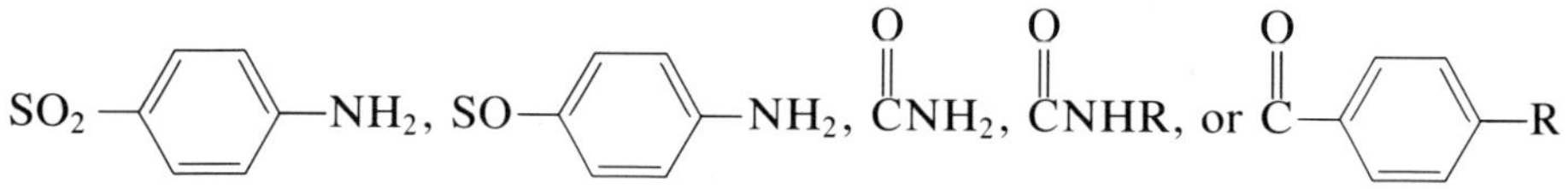

$$SO_2-C_6H_4-NH_2,\; SO-C_6H_4-NH_2,\; \overset{O}{\overset{\|}{C}}NH_2,\; \overset{O}{\overset{\|}{C}}NHR,\; \text{or}\; \overset{O}{\overset{\|}{C}}-C_6H_4-R$$

but the potency is reduced in most cases; (5) N′-monosubstitution ($R = SO_2NHR'$) results in more potent compounds, and the potency increases with heteroaromatic substitution; and (6) N′-disubstitution ($R = SO_2NR'_2$), in general, leads to inactive compounds.

Antidiabetic agents are compounds with structure **2.30**, where X may be O, S, or N incorporated into a heteroaromatic structure such as a thiadiazole or a pyrimidine or in an acyclic structure such as a urea or thiourea. In the case of ureas, the $N^2$ should carry as a substituent a chain of at least two carbon atoms.[20]

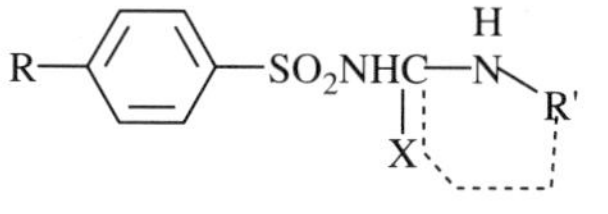

R—$C_6H_4$—$SO_2NHC(=X)$—N(H)—R′

**2.30**

Sulfonamide diuretics are of two general structural types, hydrochlorothiazides (**2.31**) and the high ceiling type (**2.32**). The former compounds have 1,3-disulfamyl groups on the benzene ring, and $R^2$ is an electronegative group such as Cl, $CF_3$, or NHR. The high ceiling compounds contain 1-sulfamyl-3-carboxy groups. Substituent $R^2$ is Cl, Ph, or PhZ, where Z may be O, S, CO, or NH, and X can be at position 2 or 3 and is normally NHR, OR, or SR.[21]

The sulfonamide example is strong evidence to support the notion that a correlation does exist between structure and activity, but how do you know what molecular modifications to make in order to fine-tune the lead compound?

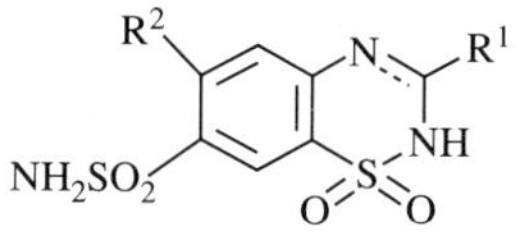

$R^2$, $NH_2SO_2$, N, $R^1$, NH, S, O, O

**2.31**

$R^2$, X, $R^1NHSO_2$, $CO_2H$

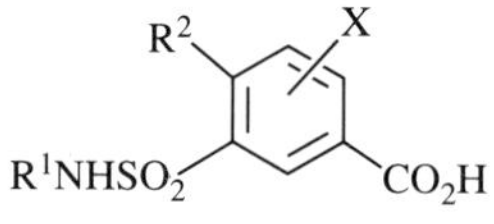

**2.32**

### D. Structure Modifications to Increase Potency and Therapeutic Index

In the preceding section it was made clear that structure modifications were the keys to activity and potency manipulations. After years of structure–activity relationship studies, various standard molecular modification approaches have been developed for the systematic improvement of the *therapeutic index* (also called the *therapeutic ratio*), which is a measure of the ratio of undesirable to desirable drug effects. For *in vivo* systems the therapeutic index could be the ratio of the $LD_{50}$ (the lethal dose for 50% of the test animals) to the $ED_{50}$ (the effective dose that produces the maximum therapeutic effect in 50% of the test animals). The larger the therapeutic index, the greater the margin of safety of the compound. A number of these structural modification methodologies follow.

### 1. Homologation

A *homologous series* is a group of compounds that differ by a constant unit, generally a $CH_2$ group. As will become more apparent in Section II,E, biological properties of homologous compounds show regularities of increase and decrease. For many series of compounds, lengthening of a saturated carbon side chain from one (methyl) to five to nine atoms (pentyl to nonyl) produces an increase in pharmacological effects; further lengthening results in a sudden decrease in potency (Fig. 2.1). In Section II,E,2,b it will be shown that this phenomenon corresponds to increased lipophilicity of the molecule, which permits penetration into cell membranes' until its lowered water solubility becomes problematic in its transport through aqueous media. In the case of aliphatic amines another problem is micelle formation, which begins at about $C_{12}$. This effectively removes the compound from potential interaction with the appropriate receptors. One of, if not the, earliest example of this potency versus chain length phenomenon was reported by Richardson,[18] who was

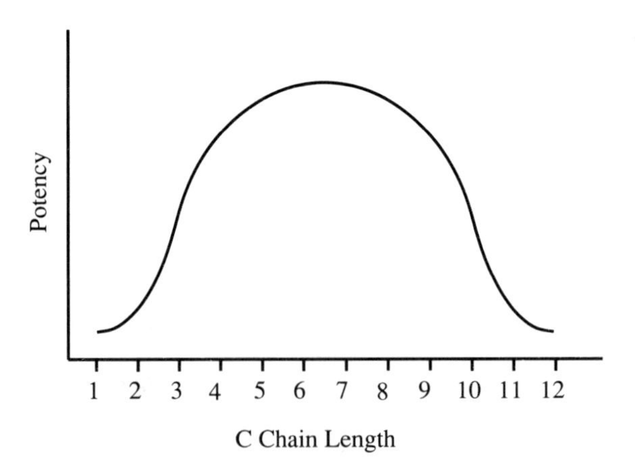


**Figure 2.1.** General effect of carbon chain length on drug potency.

**Table 2.1** Effect of Chain Length on Potency: Antibacterial Activity of 4-*n*-Alkylresorcinols[22a] and Spasmolytic Activity of Mandelate Esters[22b]

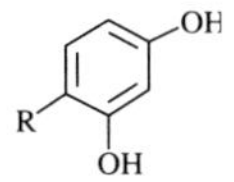

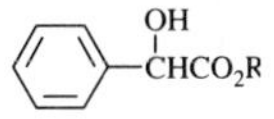

| R | Phenol coefficient | % Spasmolytic activity[a] |
|---|---|---|
| methyl | — | 0.3 |
| ethyl | — | 0.7 |
| *n*-propyl | 5 | 2.4 |
| *n*-butyl | 22 | 9.8 |
| *n*-pentyl | 33 | 28 |
| *n*-hexyl | 51 | 35 |
| *n*-heptyl | 30 | 51 |
| *n*-octyl | 0 | 130 |
| *n*-nonyl | 0 | 190 |
| *n*-decyl | 0 | 37 |
| *n*-undecyl | 0 | 22 |
| *i*-propyl | — | 0.9 |
| *i*-butyl | 15.2 | 8.3 |
| *i*-amyl | 23.8 | 28 |
| *i*-hexyl | 27 | — |

[a] Relative to 3,3,5-trimethylcyclohexanol, set at 100%.

investigating the hypnotic activity of alcohols. The maximum effect occurred for 1-hexanol to 1-octanol; then the potency declined upon chain lengthening until no activity was observed for hexadecanol.

A study by Dohme *et al.*[22a] on 4-alkyl-substituted resorcinol derivatives showed that the peak antibacterial activity occurred with 4-*n*-hexylresorcinol (see Table 2.1), a compound now used as a topical anesthetic in a variety of throat lozenges. Funcke *et al.*[22b] found that the peak spasmolytic activity of a series of mandelate esters occurred with the *n*-nonyl ester (see Table 2.1).

## 2. Chain Branching

When a simple lipophilic relationship is important as described above, then *chain branching* lowers the potency of a compound. This phenomenon is exemplified by the lower potency of the compounds having isoalkyl chains in Table 2.1. Chain branching also can interfere with receptor binding. For example, phenethylamine ($PhCH_2CH_2NH_2$) is an excellent substrate for monoamine oxidase [amine oxidase (flavin-containing)], but $\alpha$-methylphenethylamine (amphetamine) is a poor substrate. Primary amines often are more potent than secondary amines which are more potent than tertiary amines. For example, the antimalarial drug primaquine (**2.33**) is much more potent than its secondary or tertiary amine homologs.

Major pharmacological changes can occur with chain branching and homologation. Consider the 10-aminoalkylphenothiazines (**2.34**, X = H). When R is $CH_2CH(CH_3)N(CH_3)_2$ (promethazine) or $CH_2CH_2N(CH_3)_2$ (diethazine), antispasmodic and antihistaminic activities predominate. However, the homolog **2.34** with R being $CH_2CH_2CH_2N(CH_3)_2$ (promazine) has greatly reduced antispasmodic and antihistaminic activities, but sedative and tranquilizing activities are greatly enhanced. In the case of the branched chain analog **2.34** with R equal to $CH_2CH(CH_3)CH_2N(CH_3)_2$ (trimeprazine), the tranquilizing activity is reduced and antipruritic (anti-itch) activity increases.

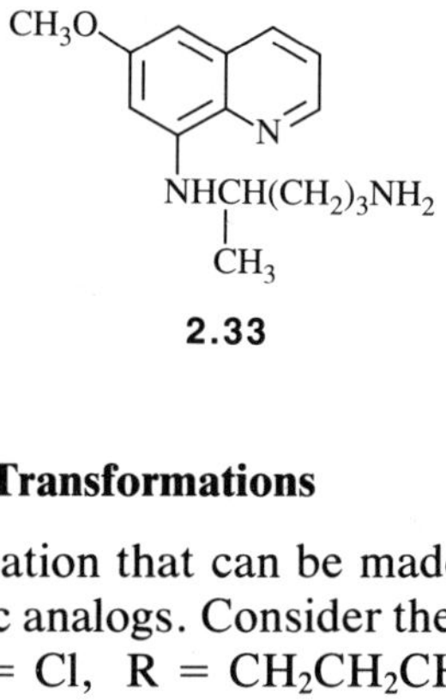

$CH_3O$ … N … $NHCH(CH_2)_3NH_2$ / $CH_3$

**2.33**

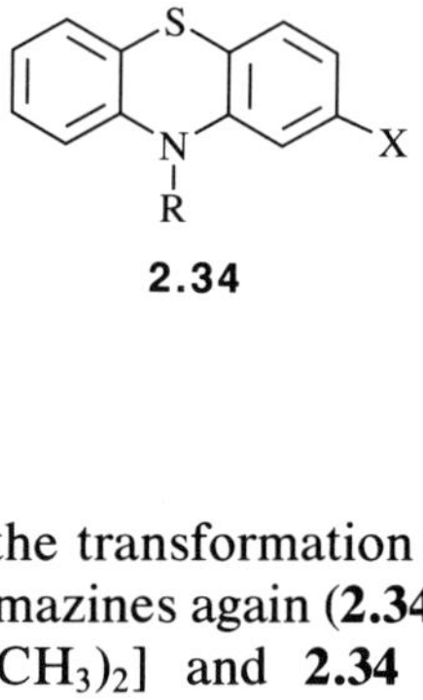

S … N … X / R

**2.34**

## 3. Ring–Chain Transformations

Another modification that can be made is the transformation of alkyl substituents into cyclic analogs. Consider the promazines again (**2.34**). Chlorpromazine [**2.34**, X = Cl, R = $CH_2CH_2CH_2N(CH_3)_2$] and **2.34** (X = Cl, R = $CH_2CH_2CH_2N$(pyrrolidine ring)) are equivalent as tranquilizers in animal tests. Trimeprazine [**2.34**, X = H, R = $CH_2CH(CH_3)CH_2N(CH_3)_2$] and methdilazine [**2.34**, X = H, R = $CH_2$—CH—$CH_2$ (ring: CH—$CH_2$—$CH_2$—N—$CH_3$, with $CH_2$—N)]

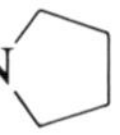

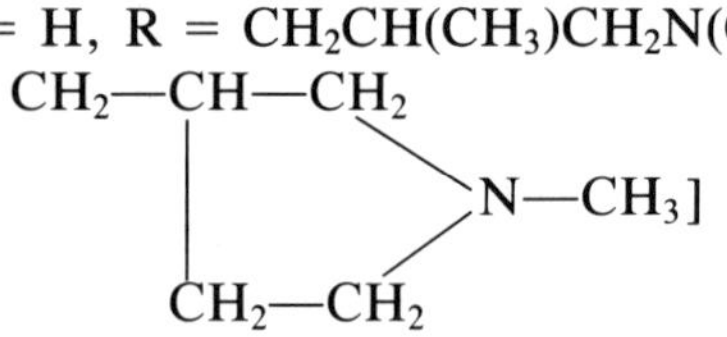

have similar antipruritic activity in man.

Different activities can result from a ring–chain transformation as well. For example, if the dimethylamino group of chlorpromazine is substituted by a methylpiperazine ring (**2.34**, X = Cl, R = $CH_2CH_2CH_2N\langle\rangle NCH_3$; prochlorperazine), the antiemetic (prevents nausea and vomiting) activity is greatly enhanced. In this case, however, an additional amino group is added.

## 4. Bioisosterism

*Bioisosteres* are substituents or groups that have chemical or physical similarities, and which produce broadly similar biological properties.[23] Bioisosterism is a lead modification approach that has been shown to be useful to attenuate toxicity or to modify the activity of a lead, and it may have a significant role in the alteration of metabolism of a lead. There are classical isosteres[24,25] and nonclassical isosteres.[23,26] In 1925 Grimm[27] formulated the *hydride displacement law* to describe similarities between groups that have the same number of valence electrons but may have a different number of atoms. Erlenmeyer[28] later redefined isosteres as atoms, ions, or molecules in which the peripheral layers of electrons can be considered to be identical. These two definitions describe *classical isosteres*; examples are shown in Table 2.2. *Nonclassical*

**Table 2.2** Classical Isosteres[24,25]

| | | | | | |
|---|---|---|---|---|---|
| 1. Univalent atoms and groups | | | | | |
| a. $CH_3$ | $NH_2$ | OH | F | Cl | |
| b. Cl | $PH_2$ | SH | | | |
| c. Br | *i*-Pr | | | | |
| d. I | *t*-Bu | | | | |
| 2. Bivalent atoms and groups | | | | | |
| a. $—CH_2—$ | —NH— | —O— | —S— | —Se— | |
| b. $—COCH_2R$ | —CONHR | $—CO_2R$ | —COSR | | |
| 3. Trivalent atoms and groups | | | | | |
| a. —CH= | —N= | | | | |
| b. —P= | —As= | | | | |
| 4. Tetravalent atoms | | | | | |
| a. $—\overset{\vert}{\underset{\vert}{C}}—$ | $—\overset{\vert}{\underset{\vert}{Si}}—$ | | | | |
| b. =C= | $=\overset{+}{N}=$ | $=\overset{+}{P}=$ | | | |
| 5. Ring equivalents | | | | | |
| a. —CH=CH— | —S— | (e.g., benzene, thiophene) | | | |
| b. —CH= | —N= | (e.g., benzene, pyridine) | | | |
| c. —O— | —S— | $—CH_2—$ | —NH— | (e.g., tetrahydrofuran, tetrahydrothiophene, cyclopentane, pyrrolidine) | |

**Table 2.3** Nonclassical Bioisosteres[23]

1. **Carbonyl group**

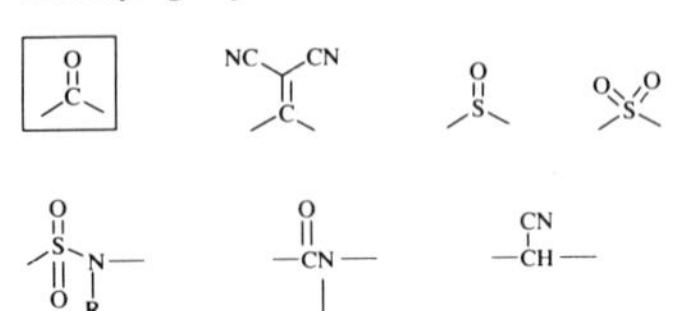


2. **Carboxylic acid group**

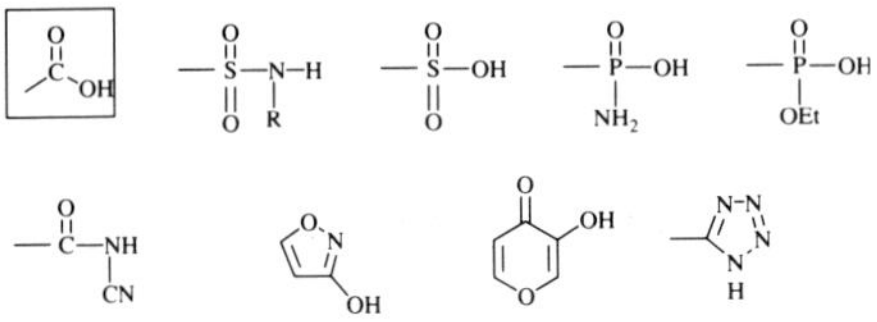

$-C(=O)OH$   $-SO_2-NH-R$   $-SO_2-OH$   $-P(=O)(NH_2)-OH$   $-P(=O)(OEt)-OH$

$-C(=O)-NH-CN$   OH   O OH O   N-N N N H

3. **Hydroxy group**

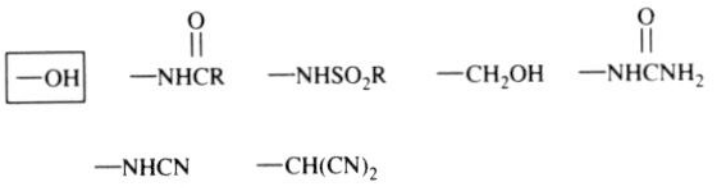

$-OH$   $-NHC(=O)R$   $-NHSO_2R$   $-CH_2OH$   $-NHC(=O)NH_2$

$-NHCN$   $-CH(CN)_2$

4. **Catechol**

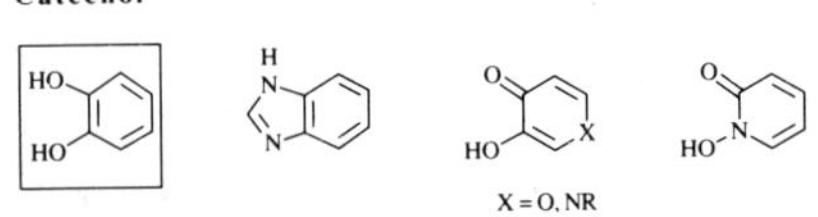


X = O, NR

5. **Halogen**

X   $CF_3$   CN   $N(CN)_2$   $C(CN)_3$

6. **Thioether**

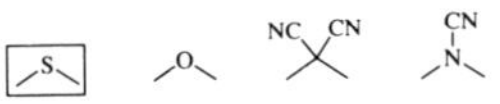

S   O   NC CN   CN N

7. **Thiourea**

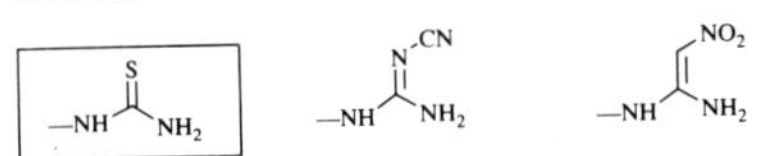


8. **Azomethine**

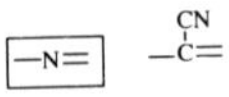

$-N=$   $-C(CN)=$

9. **Pyridine**

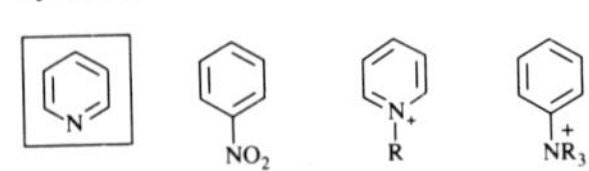


10. **Spacer group**

$-(CH_2)_3-$   $-C_6H_4-$

11. **Hydrogen**

H   F

*bioisosteres* do not have the same number of atoms and do not fit the steric and electronic rules of the classical isosteres, but they do produce a similarity in biological activity. Examples of these are shown in Table 2.3.

Ring–chain transformations also can be considered to be isosteric interchanges. There are hundreds of examples of compounds that differ by a bioisosteric interchange[23,26]; some examples are shown in Table 2.4. Bioisosterism also can lead to changes in activity. If the sulfur atom of the phenothiazine neuroleptic drugs (**2.34**) is replaced by the —CH=CH— or —$CH_2CH_2$— bioisosteres, then dibenzazepine antidepressant drugs (**2.35**) result.

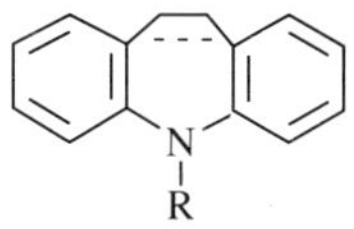

N
R

**2.35**

It is, actually, quite surprising that bioisosterism should be such a successful approach to lead modification. Perusal of Table 2.2, and especially of Table 2.3, makes it clear that in making a bioisosteric replacement, one or more of the following parameters will change: size, shape, electronic distribution, lipid solubility, water solubility, p$K_a$, chemical reactivity, and hydrogen bonding. Because a drug must get to the site of action, then interact with it (see Chapter 3), modifications made to a molecule may have one or more of the following effects:

1. Structural. If the moiety that is replaced by a bioisostere has a structural role in holding other functionalities in a particular geometry, then size, shape, and hydrogen bonding will be important.
2. Receptor interactions. If the moiety replaced is involved in a specific interaction with a receptor or enzyme, then all of the parameters except lipid and water solubility will be important.
3. Pharmacokinetics. If the moiety replaced is necessary for absorption, transport, and excretion (collectively, with metabolism, termed *pharmacokinetics*) of the compound, then lipophilicity, hydrophilicity, p$K_a$, and hydrogen bonding will be important.
4. Metabolism. If the moiety replaced is involved in blocking or aiding metabolism, then the chemical reactivity will be important.

It is because of these subtle changes that bioisosterism is effective. This approach allows the medicinal chemist to tinker with only some of the parameters in order to augment the potency, selectivity, and duration of action and to reduce toxicity. Multiple alterations may be necessary to counterbalance effects. For example, if modification of a functionality involved in binding also decreases the lipophilicity of the molecule, thereby reducing its ability to

**Table 2.4** Examples of Bioisosteric Analogs[23,26]

1. **Neuroleptics (antipsycnotics)**

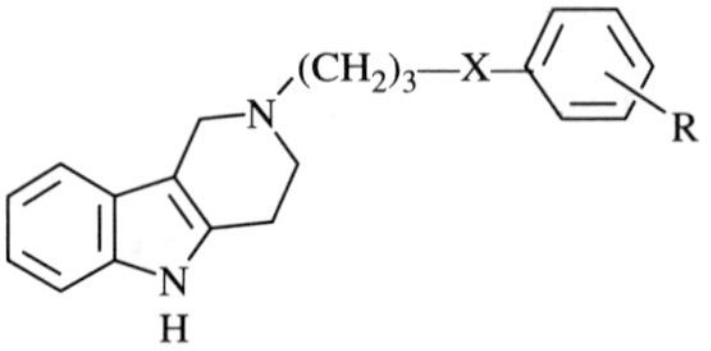

X= 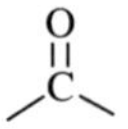 or CHCN

2. **Anti-inflammatory agents**

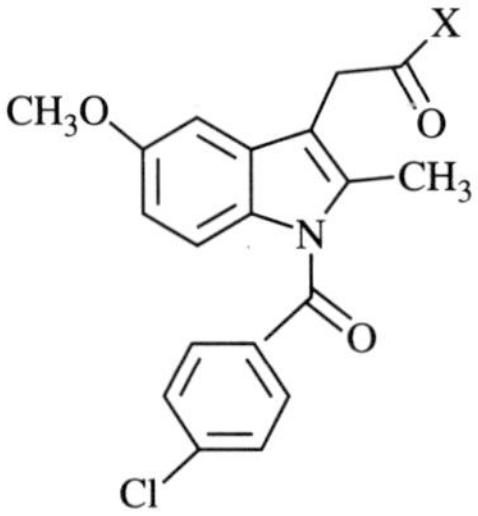

X = OH (indomethacin)

= NHOH

= 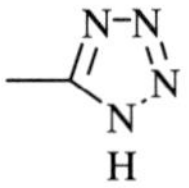

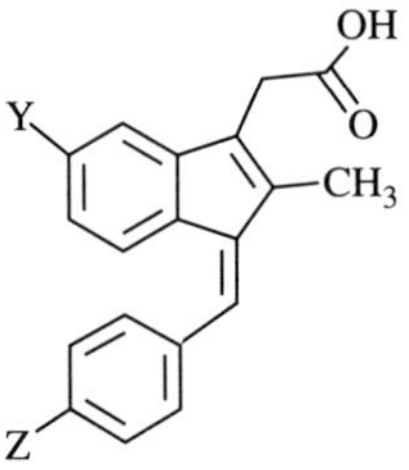

Y = $CH_3O$ Z = Cl

Y = F Z = $SCH_3$ (sulindac)

3. **Antihistamines**

R — X — $(CH_2)_n$ — Y

X = NH, O, $CH_2$

Y = $N(CH_3)_2$ (n = 2)

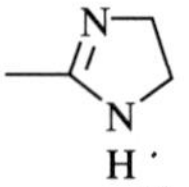 (n = 1)

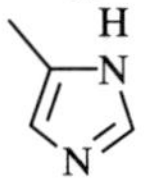 (n = 1, 2)

penetrate cell walls and cross other membranes, the molecule can be substituted with additional lipophilic groups at sites distant from that involved with binding. Modifications of this sort may change the overall molecular shape and result in another activity.

Up to this point we have been discussing more or less random molecular modifications to make qualitative differences in a lead. In 1868 Crum-Brown and Fraser[17] predicted that some day a mathematical relationship between structure and activity would be expressed. It was not until almost 100 years later that this prediction began to be realized and a new era in drug design was born. In 1962 Corwin Hansch attempted to quantify the effects of particular substituent modifications, and from this quantitative structure–activity relationship (QSAR) studies developed.[29]

## E. Quantitative Structure–Activity Relationships

### 1. Historical

The concept of quantitative drug design is based on the fact that the biological properties of a compound are a function of its *physicochemical parameters*, that is, physical properties, such as solubility, lipophilicity, electronic effects, ionization, and stereochemistry, that have a profound influence on the chemistry of the compounds. The first attempt to relate a physicochemical parameter to a pharmacological effect was reported in 1893 by Richet.[30] He observed that the narcotic action of a group of organic compounds was inversely related to their water solubility (Richet's rule). Overton[31] and Meyer[32] related tadpole narcosis induced by a series of nonionized compounds added to the water in which the tadpoles were swimming to the ability of the compounds to partition between oil and water. These early observations regarding the depressant action of structurally nonspecific drugs were rationalized by Ferguson.[33] He reasoned that, for a state of equilibrium, simple thermodynamic principles could be applied to drug activities, and that the important parameter for correlation of narcotic activities was the relative saturation (termed *thermodynamic activity* by Ferguson) of the drug in the external phase or extracellular fluids. This is known as *Ferguson's principle*, which is useful for the classification of the general mode of action of a drug and for predicting the degree of its biological effect. The numerical range of the thermodynamic activity for structurally nonspecific drugs is 0.01 to 1.0, indicating that they are active only at relatively high concentrations. Structurally specific drugs have thermodynamic activities considerably less than 0.01 and normally below 0.001.

In 1951 Hansch *et al.*[34] noted a correlation between the plant growth activity of phenoxyacetic acid derivatives and the electron density at the ortho position (lower electron density gave increased activity). They made an attempt to quantify this relationship by the application of the Hammett $\sigma$ functions (see Section II,E,2,a), but this was unsuccessful.

### 2. Physicochemical Parameters

***a. Electronic Effects: The Hammett Equation.*** In 1940 L. P. Hammett published a book entitled *Physical Organic Chemistry*[35] that marked the beginning of a new era in organic chemistry, namely, quantitative organic chemistry. Hammett's postulate was that the electronic effects (both the inductive and resonance effects) of a set of substituents on different organic reactions should be similar. Therefore, if numbers could be assigned to substituents in a standard organic reaction, these same numbers could be used to estimate rates in a new organic reaction. This was the first approach toward the prediction of reaction rates. Hammett chose benzoic acids as the standard system.

Consider the reaction shown in Scheme 2.2. Intuitively, it seems reasonable that as X becomes electron withdrawing (relative to H), the equilibrium constant ($K_a$) should increase (the reaction to the right is favored) because X is inductively pulling electron density from the carboxylic acid group, making it more acidic (ground state argument); it also is stabilizing the incipient negative charge on the carboxylate group in the transition state (transition state argument). Conversely, when X is electron donating, the equilibrium constant should decrease. A similar relationship should exist for a rate constant ($k$) where charge develops in the transition state. Hammett chose the reaction shown in Scheme 2.3 as the standard system. If $K_a$ is measured from Scheme 2.2 and $k$ from Scheme 2.3 for a series of substituents X, and the data are expressed in a double logarithm plot (Fig. 2.2[36]), then a straight line can be drawn through most of the data points. This is known as a *linear free-energy relationship*. When X is a meta or para substituent, then virtually all of the points fall on the straight line; the ortho substituent points are badly scattered. The Hammett relationship does not hold for ortho substituents because of steric interactions and polar effects.

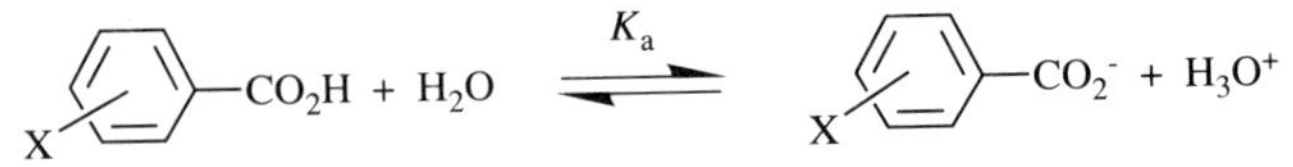

$$X\text{-}C_6H_4\text{-}CO_2H + H_2O \overset{K_a}{\rightleftharpoons} X\text{-}C_6H_4\text{-}CO_2^- + H_3O^+$$

**Scheme 2.2.** Ionization of substituted benzoates.

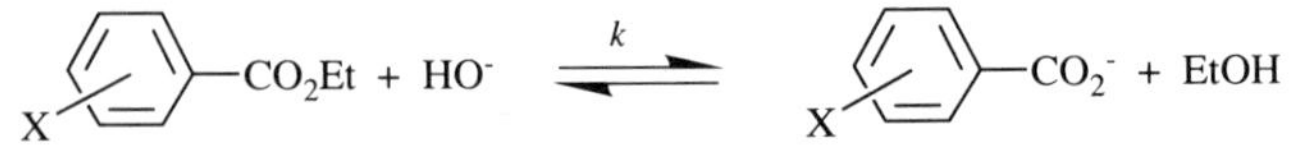

$$X\text{-}C_6H_4\text{-}CO_2Et + HO^- \overset{k}{\rightleftharpoons} X\text{-}C_6H_4\text{-}CO_2^- + EtOH$$

**Scheme 2.3.** Saponification of substituted ethyl benzoates.

The linear correlation for the meta and para substituents is observed for rate or equilibrium constants for a wide variety of organic reactions. The straight

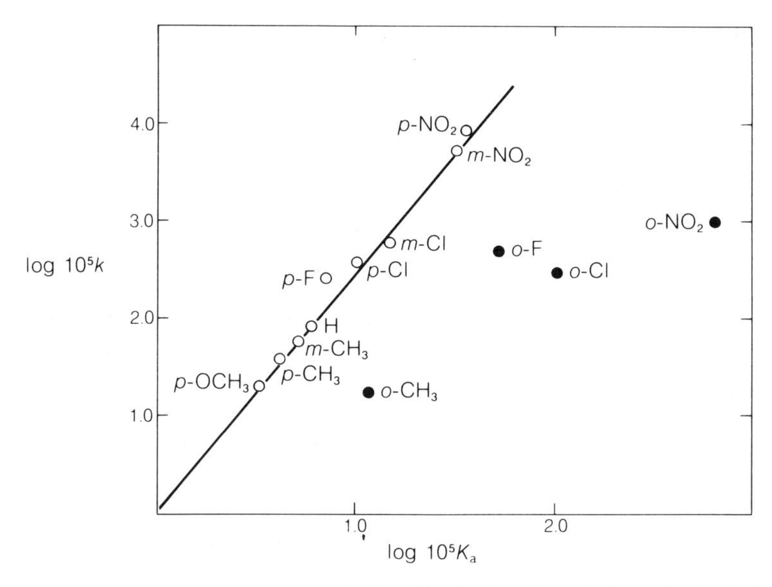


**Figure 2.2.** Linear free energy relationship for the dissociation of substituted benzoic acids in water at 25°C ($K_a$) against the rates of alkaline hydrolysis of substituted ethyl benzoates in 85% ethanol–water at 30°C ($k$). [Reprinted with permission from Roberts, J. D., and Caserio, M. C. (1977). "Basic Principles of Organic Chemistry," 2nd ed., p. 1331. W. A. Benjamin, Menlo Park, CA. Copyright © 1977 Benjamin/Cummings Publishing Company.]

line can be expressed by Eq. (2.1), where the two variables are log $k$ and log $K$. The slope of the line is $\rho$ and the intercept is $C$. When there is no substituent, that is, when X is H, then Eq. (2.2) holds. Subtraction of Eq. (2.2) from Eq. (2.1) gives Eq. (2.3), where $k$ and $K$ are the rate and equilibrium constants, respectively, for compounds with a substituent X, and $k_o$ and $K_o$ are the same for the parent compound (X = H). If log $K/K_o$ is defined as $\sigma$, then Eq. (2.3) reduces to Eq. (2.4), the *Hammett equation*. The *electronic parameter*, $\sigma$, depends on the electronic properties and position of the substituent on the ring and, therefore, is also called the *substituent constant*. The more electron withdrawing a substituent, the more positive is its $\sigma$ value (relative to H, which is set at 0.0); conversely, the more electron donating, the more negative is its $\sigma$ value. The meta $\sigma$ constants result from inductive effects, but the para $\sigma$ constants correspond to the net inductive and resonance effects. Therefore, $\sigma_{\text{meta}}$ and $\sigma_{\text{para}}$ for the same substituent, generally, are not the same.

$$\log k = \rho \log K + C \tag{2.1}$$

$$\log k_o = \rho \log K_o + C \tag{2.2}$$

$$\log \frac{k}{k_o} = \rho \log \frac{K}{K_o} \tag{2.3}$$

$$\log \frac{k}{k_o} = \rho\sigma \tag{2.4}$$

The $\rho$ values (the slope) depend on the particular type of reaction and the reaction conditions (e.g., temperature and solvent) and, therefore, are called *reaction constants*. The importance of $\rho$ is that it is a measure of the sensitivity of the reaction to the electronic effects of the meta and para substituents. A large $\rho$, either positive or negative, indicates great sensitivity to substituent effects. Reactions that are favored by high electron density in the transition state have negative $\rho$ values; reactions that are aided by electron withdrawal have positive $\rho$ values.

As we shall see in Section II,E,3, the substituent constant $\sigma$ will be of major significance to QSAR.

***b. Lipophilicity Effects: The Basis for the Hansch Equation.*** The crucial breakthrough in QSAR came when Hansch and co-workers[29,37] conceptualized the action of a drug as depending on two processes. The first process is the journey of the drug from its point of entry into the body to the site of action (*pharmacokinetics*), and the second process is the interaction of the drug with the specific site (*pharmacodynamics*). Hansch proposed that the first step in the overall process was a random walk, a diffusion process, in which the drug made its way from a dilute solution outside of the cell to a particular site in the cell. This was visualized as being a relatively slow process, the rate of which being highly dependent on the molecular structure of the drug.

For the drug to reach the site of action, it must be able to interact with two different environments, lipophilic (e.g., membranes) and aqueous (the exobiophase, such as the cytoplasm). The cytoplasm of a cell is essentially a dilute solution of salts in water; all living cells are surrounded by a nonaqueous phase, the membrane. The functions of membranes are to protect the cell from water-soluble substances, to form a surface to which enzymes and other proteins can attach in order to produce a localization and structural organization, and to separate solutions of different electrochemical potentials (e.g., in nerve conduction). One of the most important membranes is known as the *blood–brain barrier*, a membrane that surrounds the capillaries of the circulatory system in the brain and protects it from passive diffusion of undesirable chemicals from the bloodstream. This is an important prophylactic boundary, but it also can block the delivery of central nervous system drugs to their site of action.

Although the structure of membranes has not been resolved, the most widely accepted model is the *fluid mosaic model* (Fig. 2.3).[38] In this depiction

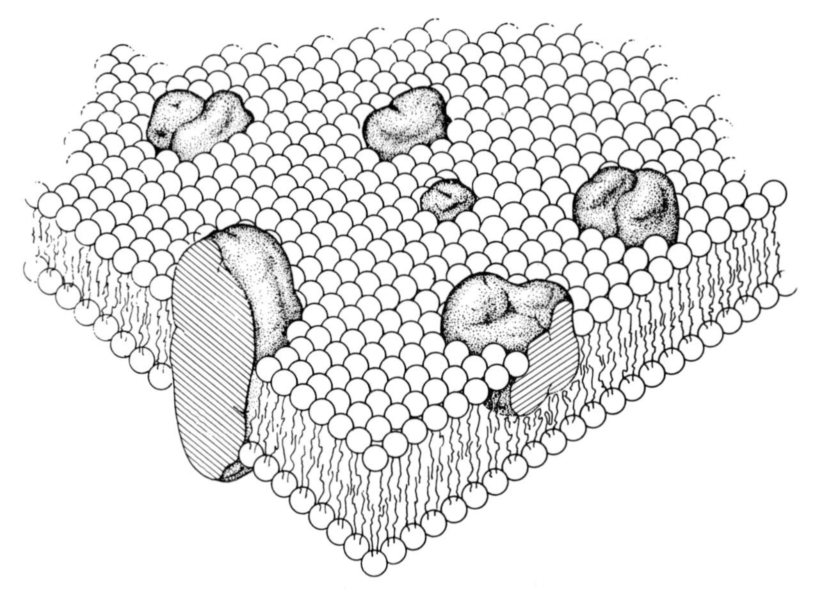

**Figure 2.3.** Fluid mosaic model of a membrane. The balls represent polar end groups, and the wavy lines are the hydrocarbon chains of the lipids. [From Singer, S. J., and Nicolson, G. L. (1972). *Science* **175,** 720. Copyright © 1972 by the AAAS with permission.]

*integral proteins* are embedded in a lipid bilayer; *peripheral proteins* are associated with only one membrane surface. The structure of the membrane is primarily determined by the structure of the lipids of which it is comprised. The principal classes of lipids found in membranes are neutral cholesterol (**2.36**) and the ionic phospholipids, for example, phosphatidylcholine [**2.37**, R = $(CH_3)_3\overset{+}{N}CH_2CH_2$], phosphatidylethanolamine (**2.37**, R = $\overset{+}{N}H_3CH_2CH_2$), phosphatidylserine [**2.37**, R = $^{-}OOC(NH_3^+)CHCH_2$), phosphatidylinositol (**2.37**, R = inositol), and sphingomyelin [**2.38**, R = $(CH_3)_3\overset{+}{N}CH_2CH_2O_2PO^-$]; R′CO and R″CO in **2.37** and **2.38** are derived from fatty acids. Glycolipids (**2.38**, R = sugar) also are important membrane constituents.

**2.36** **2.37** **2.38**

All of these lipids are *amphipathic*, meaning that one end of the molecule is *hydrophilic* (water soluble) and the other is *hydrophobic* or, if you wish, *lipophilic* (water insoluble; soluble in organic solvents). Thus, the hydroxyl group in cholesterol, the ammonium groups in the phospholipids, and the sugar residue in the glycolipids are the polar, hydrophilic ends, and the steroid and hydrocarbon moieties are the lipophilic ends. The hydrocarbon part (R′ and R″) actually can be a mixture of chains from 14 to 24 carbon atoms long; approximately 50% of the chains contain a double bond. The polar groups of the lipid bilayer are in contact with the aqueous phase; the hydrocarbon chains project toward each other in the interior with a space between the layers. The stability of the membrane arises from the stabilization of the ionic charges by ion–dipole interactions with the water (see Chapter 3) and from association of the nonpolar groups. The hydrocarbon chains are relatively free to move; therefore, the core is similar to a liquid hydrocarbon.

It occurred to Hansch that the fluidity of the hydrocarbon region of the membrane may explain the correlation noted by Richet,[30] Overton,[31] and Meyer[32] between lipid solubility and biological activity. The Hansch group[29,39] suggested that a reasonable model for the first step in drug action (transport to the site of action) would be the ability of a compound to partition between 1-octanol, which would simulate a lipid membrane, and water (the aqueous phase). 1-Octanol has a long saturated alkyl chain and a hydroxyl group for hydrogen bonding, and it dissolves water to the extent of 1.7 *M* (saturation). This combination of lipophilic chains, hydrophilic groups, and water molecules gives 1-octanol properties very close to those of natural membranes and macromolecules.

Hansch believed that, just as in the case of the Hammett equation, there should be a linear free energy relationship between lipophilicity and biological activity. As a suitable measure of lipophilicity, the *partition coefficient*, $P$, between 1-octanol and water was proposed,[29,39] and $P$ was determined by Eq. (2.5), where $\alpha$ is the degree of dissociation of the compound in water calculated from ionization constants.

$$P = \frac{[\text{compound}]_{\text{oct}}}{[\text{compound}]_{\text{aq}}(1 - \alpha)} \tag{2.5}$$

The partition coefficient is derived experimentally by placing a compound in a shaking device (such as a separatory funnel) with varying volumes of 1-octanol and water, determining the concentration of the compound in each layer after mixing, and utilizing Eq. (2.5) to calculate $P$. The value of $P$ varies slightly with temperature and concentration of the solute, but with neutral molecules in dilute solutions ($<0.01$ *M*) and small temperature changes ($\pm 5°C$), variations in $P$ are minor.

Collander[40] had shown previously that the rate of movement of a variety of organic compounds through cellular material was approximately proportional to the logarithm of their partition coefficients between an organic solvent and water. Therefore, as a model for a drug traversing through the body to its site of action, the relative potency of the drug, expressed as $\log 1/C$, where $C$ is the concentration of the drug that produces some standard biological effect, was related to its lipophilicity by the parabolic expression shown in Eq. (2.6).[41]

$$\log 1/C = -k(\log P)^2 + k'(\log P) + k'' \tag{2.6}$$

On the basis of Eq. (2.5), it is apparent that if a compound is more soluble in water than in 1-octanol, $P$ is less than 1, and, therefore, $\log P$ is negative. Conversely, a molecule more soluble in 1-octanol has a $P$ value greater than 1, and $\log P$ is positive. The larger the value of $P$, the more there will be an interaction of the drug with the lipid phase (i.e., membranes). As $P$ approaches infinity, the drug interaction will become so great that the drug will not be able to cross the aqueous phase, and it will localize in the first lipophilic phase with which it comes into contact. As $P$ approaches zero, the drug will be so water soluble that it will not be capable of crossing the lipid phase and will localize in the aqueous phase. Somewhere between $P = 0$ and $P = \infty$, there will be a value of $P$ such that drugs having this value will be least hindered in their journey through macromolecules to their site of action. This value is called $\log P_0$, the optimum partition coefficient for biological activity.

This random walk analysis supports the parabolic relationship [Eq. (2.6)] between potency ($\log 1/C$) and $\log P$ (Fig. 2.4). Note the correlation of Fig. 2.4 with the generalization regarding homologous series of compounds (Section II,D,1; Fig. 2.1). An increase in the alkyl chain length increases the lipophilicity of the molecule; apparently, the $\log P_0$ generally occurs in the range of 5–9 carbon atoms. Hansch *et al.*[41] found that a number of series of nonspecific hypnotics had similar $\log P_0$ values, approximately 2, and they suggested that this is the value of $\log P_0$ needed for penetration into the central nervous system (CNS). If a hypnotic agent has a $\log P$ considerably different from 2, then its activity probably is derived from mechanisms other than just lipid

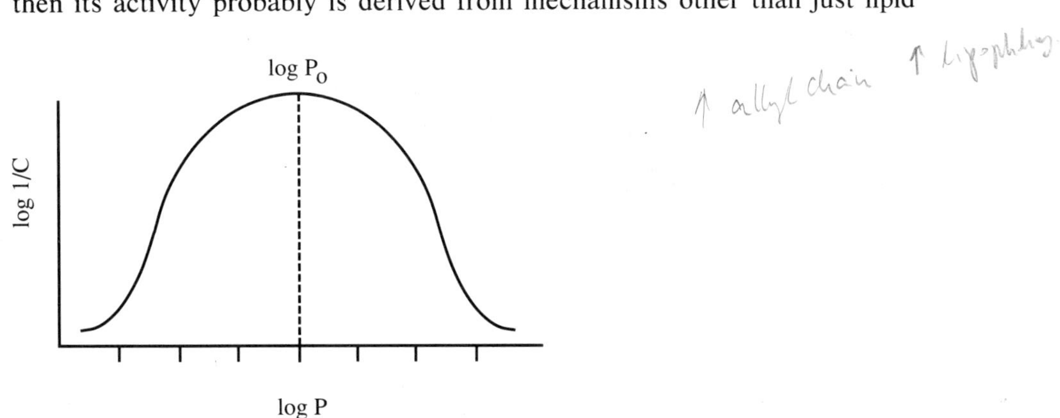


**Figure 2.4.** Effect of log $P$ on biological response. $P$ is the partition coefficient, and $C$ is the concentration of the compound required to produce a standard biological effect.

transport. If a lead compound has modest CNS activity and has a log $P$ value of 0, it would be reasonable to synthesize an analog with a higher log $P$.

Can you predict what analog will have a higher log $P$? In the same way that substituent constants were derived by Hammett for the electronic effects of atoms and groups ($\sigma$ constants), Hansch and co-workers[29,37,39] derived substituent constants for the contribution of individual atoms and groups to the partition coefficient. The *lipophilicity substituent constant*, $\pi$, is defined by Eq. (2.7), which has the same derivation as the Hammett equation. The term $P_X$ is the partition coefficient for the compound with substituent X, and $P_H$ is the partition coefficient for the parent molecule (X = H). As in the case of the Hammett substituent constant $\sigma$, $\pi$ is additive and constitutive. *Additive* means that multiple substituents exert an influence equal to the sum of the individual substituents. *Constitutive* indicates that the effect of a substituent may differ depending on the molecule to which it is attached or on its environment. Alkyl groups are some of the least constitutive. For example, methyl groups attached at the meta or para positions of 15 different benzene derivatives had $\pi_{CH_3}$ values with a mean and standard derivation of $0.50 \pm 0.04$. Because of the additive nature of $\pi$ values, $\pi_{CH_2}$ can be determined as shown in Eq. (2.8), where the log $P$ values are obtained from standard tables.[42] Because, by definition, $\pi_H = 0$, then $\pi_{CH_2} = \pi_{CH_3}$.

$$\pi = \log P_X - \log P_H = \log \frac{P_X}{P_H} \tag{2.7}$$

$$\begin{aligned} \pi_{CH_2} &= \log P_{\text{nitroethane}} - \log P_{\text{nitromethane}} \\ &= 0.18 - (-0.33) = 0.51 \end{aligned} \tag{2.8}$$

As was alluded to in Section II,D,2 on molecular modification, branching in an alkyl chain lowers the log $P$ or $\pi$ as a result of the larger molar volumes and shapes of branched compounds. As a rule of thumb, the value of log $P$ or $\pi$ is lowered by 0.2 unit per branch. For example, the $\pi_{i\text{-Pr}}$ value in 3-isopropylphenoxyacetic acid is 1.30; $\pi_{Pr}$ is 3(0.5) = 1.50. Another case where $\pi$ values are fairly constant is conjugated systems, as exemplified by $\pi_{CH=CHCH=CH}$ in Table 2.5.

Inductive effects are quite important to lipophilicity.[43] In general, electron-withdrawing groups increase $\pi$ when a hydrogen-bonding group is involved. For example $\pi_{CH_2OH}$ varies as a function of the proximity of an electron-withdrawing phenyl group [Eq. (2.9)],[44] and $\pi_{NO_2}$ varies as a function of the inductive effect of the nitro group on the hydroxyl group [Eq. (2.10)].[43] The electron-withdrawing inductive effects of the phenyl group [Eq. (2.9)] and the nitro group [Eq. (2.10)] make the nonbonded electrons on the hydroxyl group less available for hydrogen bonding, thereby reducing the affinity of this functional group for the aqueous phase. This, then, increases the log $P$ or $\pi$. Also

**Table 2.5** Constancy of $\pi$ for —CH=CH—CH=CH—[41,43]

| Log P Difference | | | $\pi_{CH=CHCH=CH}$ |
|---|---|---|---|
| log P | 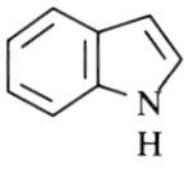 — log P | 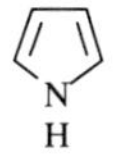 = 2.14 — 0.75 = | 1.39 |
| log P | 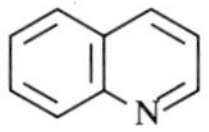 — log P | 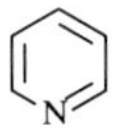 = 2.03 — 0.65 = | 1.38 |
| log P | 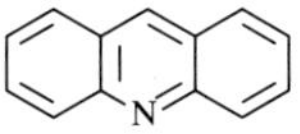 — log P | 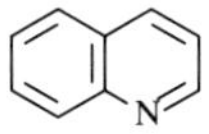 = 3.40 — 2.03 = | 1.37 |
| log P | 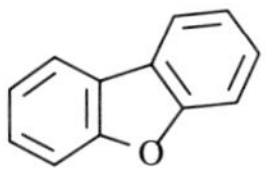 — log P | 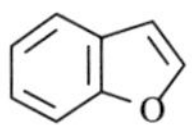 = 4.12 — 2.67 = | 1.45 |
| log P | 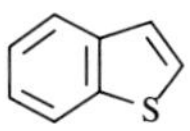 — log P |  = 3.12 — 1.81 = | 1.31 |
| log P | 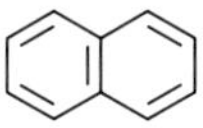 — log P |  = 3.45 — 2.13 = | 1.32 |
| 2/3 log P |  | = 2/3 (2.13) = | 1.42 |
| log P (OH) | — log P (OH) | 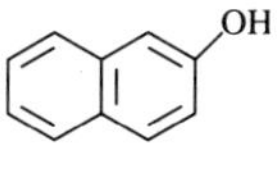 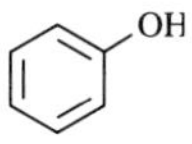 = 2.84 — 1.46 = | 1.38 |
| | | | ave. 1.38 ± 0.046 |

note in Eqs. (2.9) and (2.10) that, because $\pi_H = 0$ by definition, $\log P_{benzene} = \pi_{Ph}$.

$$\pi_{CH_2OH} = \log P_{Ph(CH_2)_2OH} - \log P_{PhCH_3} = -1.33$$
$$\pi_{CH_2OH} = \log P_{PhCH_2OH} - \log P_{PhH} = -1.03 \quad (2.9)$$
$$\pi_{NO_2} = \log P_{pHNO_2} - \log P_{PhH} = -0.28$$
$$\pi_{NO_2} = \log P_{4\text{-}NO_2PhCH_2OH} - \log P_{PhCH_2OH} = 0.11 \quad (2.10)$$

Resonance effects also are important to the lipophilicity much the same way as are inductive effects.[43] Delocalization of nonbonded electrons into aromatic systems decreases their availability for hydrogen bonding with the

**Table 2.6** Effect of Folding of Alkyl Chains on $\pi$[43]

| X | $\pi_X$ (aromatic)[a] | $\pi_X$ (aliphatic)[b] | $\Delta\pi_X$ |
|---|---|---|---|
| OH | −1.80 | −1.16 | 0.64 |
| F | −0.73 | −0.17 | 0.56 |
| Cl | −0.13 | 0.39 | 0.52 |
| Br | 0.04 | 0.60 | 0.56 |
| I | 0.22 | 1.00 | 0.78 |
| COOH | −1.26 | −0.67 | 0.59 |
| $CO_2CH_3$ | −0.91 | −0.27 | 0.64 |
| $COCH_3$ | −1.26 | −0.71 | 0.55 |
| $NH_2$ | −1.85 | −1.19 | 0.66 |
| CN | −1.47 | −0.84 | 0.63 |
| $OCH_3$ | −0.98 | −0.47 | 0.51 |
| $CONH_2$ | −2.28 | −1.71 | 0.57 |
| | | Average | 0.60 ± 0.05 |

[a] Log $P_{Ph(CH_2)_3X}$ − log $P_{Ph(CH_2)_3H}$.
[b] Log $P_{CH_3(CH_2)_3X}$ − log $P_{CH_3(CH_2)_3H}$.

aqueous phase and, therefore, increases the $\pi$. This is supported by the general trend that aromatic $\pi_X$ values are greater than aliphatic $\pi_X$ values, again emphasizing the constitutive nature of $\pi$ and log *P*.

Steric effects are variable.[43] If a group sterically shields nonbonded electrons, then aqueous interactions will decrease, and the $\pi$ value will increase. However, crowding of functional groups involved in hydrophobic interactions (see Chapter 3) will have the opposite effect. Conformational effects also can affect the $\pi$ value.[43] The $\pi_X$ values for $Ph(CH_2)_3X$ are consistently lower (more water soluble) than $\pi_X$ values for $CH_3(CH_2)_3X$ (Table 2.6). This phenomenon is believed to be the result of folding of the side chain onto the phenyl ring (**2.39**), which means a smaller apolar surface for organic solvation. The folding may be caused by the interaction of the $CH_2$–X dipole with the phenyl $\pi$ electrons and by intramolecular hydrophobic interactions.

X, $CH_2$, $CH_2$, $CH_2$

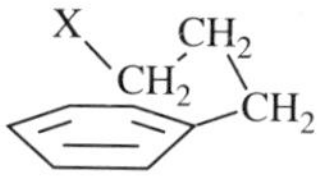

**2.39**

Two examples follow to show the additivity of $\pi$ constants in predicting log *P* values. A calculation of the log *P* for the anticancer drug diethylstilbestrol (**2.40**) is as follows:

$$\begin{aligned} \text{Calc. log } P &= 2\pi_{CH_3} + 2\pi_{CH_2} + \pi_{CH=CH} + 2 \log P_{PhOH} - 0.40 \\ &= 2(0.50) + 2(0.50) + 0.69 + 2(1.46) - 0.40 \\ &= 5.21 \end{aligned} \quad (2.11)$$

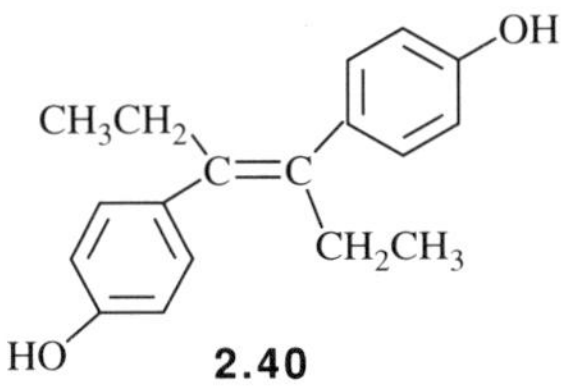

**2.40**

In Eq. (2.11), $\pi_{CH=CH} = \frac{1}{2}(\pi_{CH=CHCH=CH})$, which was shown in Table 2.5 to be $\frac{1}{2}(1.38)$; $-0.40$ is added into the equation to account for two branching points (each end of the alkene). The calculated log $P$ value of 5.21 is quite remarkable considering that the experimental log $P$ value is 5.07.

A calculation of log $P$ for the antihistamine diphenhydramine (**2.41**) is shown in Eq. (2.12). In Eq. (2.12), 2.13 is log $P$ for benzene, which is the same as $\pi_{Ph}$; 0.30 is $\pi_{CH_2}(0.50) - 0.20$ for branching; $-0.73$ was obtained by subtracting 1.50 ($2\pi_{CH_3} + \pi_{CH_2}$) from log $P_{CH_3CH_2OCH_2CH_3}$ ($=0.77$); and $-0.95$ is the value for $\pi_{NMe_2}$ obtained from $Ph(CH_2)_3NMe_2$.[43] The experimental log $P$ value is 3.27.

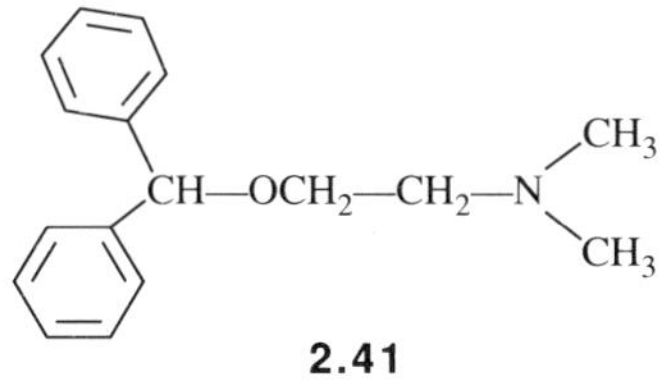

**2.41**

$$\begin{aligned} \text{Calc. log } P &= 2\pi_{Ph} + \pi_{CH} + \pi_{COH_2} + \pi_{CH_2} + \pi_{NMe_2} \\ &= 2(2.13) + 0.30 - 0.73 + 0.50 - 0.95 \\ &= 3.38 \end{aligned} \quad (2.12)$$

The chore of calculating log $P$ values for molecules has been lessened considerably by the computerization of the method.[45] A nonlinear regression model for the estimation of partition coefficients was developed by Bodor *et al.*[46] using the following molecular descriptors: molecular surface, volume, weight, and charge densities. It was shown to have excellent predictive power for the estimation of log $P$ for complex molecules.

Although the log $P$ values determined from 1-octanol/water partitioning are excellent models for *in vivo* lipophilicity, it has been found for a variety of aromatic compounds with log $P$ values exceeding 5.5 (very lipophilic) or molar volumes greater than 230 $cm^3$/mol that there is a breakdown in the correlation of these values with those determined from partitioning between L-$\alpha$-phosphatidylcholine dimyristoyl membrane vesicles and water.[47] Above a log $P$ value of 5.5 the solvent solubility for these molecules is greater than their membrane solubility. As the compound increases in size more energy per unit volume is required to form a cavity in the structured membrane phase. This is consistent with observations that branched molecules have lower log $P$ values than their straight chain counterparts and that this effect is even greater in membranes than in organic solvents.

It should be noted that although log $P$ values are most commonly determined with 1-octanol/water mixtures, this is not universal. For example, Seiler[48] introduced a new additive constitutive substituent constant for solvents other than 1-octanol. Therefore, when using log $P$ values, it is important to be aware of the solvent used to obtain the log $P$ data.

***c. Steric Effects: The Taft Equation.*** Since interaction of a drug with a receptor involves the mutual approach of two molecules, another important parameter for QSAR is the *steric effect*. In much the same way that Hammett derived quantitative electronic effects (see Section II,E,2,a), Taft[49] defined the *steric parameter* $E_s$ [Eq. (2.13)]. Taft used for the reference reaction the relative rates of the acid-catalyzed hydrolysis of $\alpha$-substituted acetates ($XCH_2CO_2Me$). This parameter is normally standardized to the methyl group ($XCH_2 = CH_3$) so that $E_s(CH_3) = 0.0$; it is possible to standardize it to hydrogen by adding 1.24 to every methyl-based $E_s$ value.[50] Hancock *et al.*[51] claimed that this model reaction was under the influence of hyperconjugative effects and, therefore, developed corrected $E_s$ values for the hyperconjugation of $\alpha$-hydrogen atoms [Eq. (2.14)], where $E_s{}^c$ is the corrected $E_s$ value and $n$ is the number of $\alpha$-hydrogen atoms.

$$E_s = \log k_{XCOMe} - \log k_{CH_3CO_2Me} = \log \frac{k_X}{k_o} \tag{2.13}$$

$$E_s{}^c = E_s + 0.306(n - 3) \tag{2.14}$$

Two other steric parameters worth mentioning are molar refractivity ($MR$) and the Verloop parameter. *Molar refractivity*[52] is defined by the Lorentz–Lorenz equation [Eq. (2.15)], where $n$ is the index of refraction at the sodium D line, MW is the molecular weight, and $d$ is the density of the compound. The greater the positive $MR$ value of a substituent, the larger is its steric or bulk effect. This parameter also measures the electronic effect and, therefore, may reflect dipole–dipole interactions at the receptor site.

$$MR = \frac{n^2 - 1}{n^2 + 2} \frac{\text{MW}}{d} \tag{2.15}$$

The *Verloop steric parameters*[53] are used in a program called STERIMOL to calculate the steric substituent values from standard bond angles, van der Waals radii, bond lengths, and user-determined reasonable conformations. Five parameters are involved. One ($L$) is the length of the substituent along the axis of the bond between the substituent and the parent molecule. Four width parameters ($B_1$–$B_4$) are measured perpendicular to the bond axis. These five parameters describe the positions, relative to the point of attachment and the bond axis, of five planes which closely surround the group. In contrast to $E_s$ values which, because of the reaction on which they are based, cannot be determined for many substituents, the Verloop parameters are available for any substituent.

### 3. Methods Used to Correlate Physicochemical Parameters with Biological Activity

Now that we can obtain numerous physicochemical parameters (also called descriptors) for any substituent, how do we use these parameters to gain information regarding what compound to synthesize next in an attempt to optimize the lead compound? First, several (usually, many) compounds related to the lead are synthesized, and the biological activities are determined in some bioassay. These data, then, can be manipulated by a number of QSAR methods. The most popular is Hansch analysis.

***a. Hansch Analysis: A Linear Multiple Regression Analysis.*** With the realization that there are (at least) two considerations for biological activity, namely, lipophilicity (required for the journey of the drug to the site of action) and electronic factors (required for drug interaction with the site of action), and that lipophilicity is a parabolic function, Hansch and Fujita[37] expanded Eq. (2.6) to that shown in either Eq. (2.16a) or (2.16b) known as the *Hansch equation*, where $C$ is the molar concentration (or dose) that elicits a standard biological response (e.g., $ED_{50}$, the dose required for 50% of the maximal effect, $IC_{50}$, the concentration that gives 50% inhibition of an enzyme or antagonism of a receptor; or $LD_{50}$, the lethal dose for 50% of the animal population). The terms $k$, $k'$, $\rho$, and $k''$ are the regression coefficients derived from statistical curve fitting, and $\pi$ and $\sigma$ are the lipophilicity and electronic substituent constants, respectively. The reciprocal of the concentration ($1/C$) reflects the fact that greater potency is associated with a lower dose, and the negative sign for the $\pi^2$ [or $(\log P)^2$] term reflects the expectation of an optimum lipophilicity, that is, the $\pi_o$ or $\log P_o$.

$$\log 1/C = -k\pi^2 + k'\pi + \rho\sigma + k'' \quad (2.16a)$$

$$\log 1/C = -k(\log P)^2 + k'(\log P) + \rho\sigma + k'' \quad (2.16b)$$

Because of the importance of steric effects and other shape factors of molecules for receptor interactions, an $E_s$ term and a variety of other shape, size, or topography terms ($S$) have been added to the Hansch equation [see Eq. (2.17)]. The way these parameters are used is by the application of the method of *linear multiple regression analysis*.[54] The best least squares fit of the dependent variable (the biological activity) to a linear combination of the independent variables (the descriptors) is determined. *Hansch analysis*, also called the *extrathermodynamic method*, then, is a linear free energy approach to drug design in congeneric series in which equations are set up involving different combinations of the physicochemical parameters; the statistical methodology allows the best equation to be selected and the statistical significance of the correlation to be assessed. Once this equation has been established, it can be used to predict the activities of untested compounds. Problems associated with the use of multiple regression analysis in QSAR studies have been discussed by Deardon.[55]

$$\log 1/C = -a\pi^2 + b\pi + \rho\sigma + cE_s + dS + e \qquad (2.17)$$

Several assumptions must be made when the extrathermodynamic method is utilized: conformational changes in receptors can be ignored, metabolism does not interfere, linear free energy terms relevant to receptor affinity are additive, the potency–lipophilicity relationship is parabolic or linear, and correlation implies a causal relationship. According to Martin[56a] and Tute[56b] there is a balance of assets and liabilities to the extrathermodynamic method. The strengths are severalfold: (1) the use of descriptors ($\pi$, $\sigma$, $E_s$, $MR$, and so forth) permits data collected from simple organic chemical model systems to be utilized for the prediction of biological activity in complex systems; (2) the predictions are quantitative with statistical confidence limits; (3) the method is easy to use and is inexpensive; and (4) conclusions that are reached may have application beyond the substituents included in the particular analysis.

The weaknesses of this method are that (1) there must be parameter values available for the substituents in the data set; (2) a large number of compounds must be included in the analysis in order to have confidence in the derived equations; (3) expertise in statistics and computer use is essential; (4) small molecule interactions are imperfect models for biological systems; (5) in contrast to chemical reactions in which one knows the atoms that interact with the reagent, steric effects in biological systems may not be relevant, since it is often not certain which atoms in the drug interact with the receptor; (6) organic reactions used to determine the descriptors usually are studied under acidic or basic conditions when all analogs are fully protonated or deprotonated, whereas in biological systems the drug may be partially protonated; (7) since QSAR study is empirical, it is a retrospective technique that depends on the pharmacological activity of compounds belonging to the same structural type, and, therefore, new types of active compounds are not discovered (i.e., it is a lead optimization technique, not a lead discovery approach); and (8) like other empirical relationships, extrapolations frequently lead to false predictions.

Despite the weaknesses of this approach it is used widely, and several successes in drug design attributable to Hansch analysis have been reported.[57] As pointed out in Section III,F,2 of Chapter 3, however, caution should be used when applying QSAR methods to racemic mixtures if only one enantiomer is active. Although Hansch analysis is the foremost method, there are other important statistical approaches that will be mentioned briefly.

***b. Free and Wilson or de Novo Method.*** Not long after Hansch proposed the extrathermodynamic approach, Free and Wilson[58] reported a general mathematical method for assessing the occurrence of additive substituent effects, and for quantitatively estimating their magnitude. It is a method for the optimization of substituents within a given molecular framework that is based on the assumption that the introduction of a particular substituent at any one position in a molecule always changes the relative potency by the same amount, regardless of what other substituents are present in the molecule. A series of linear equations, constructed of the form shown in Eq. (2.18), where $BA$ is the magnitude of the biological activity, $X_i$ is the $i$th substituent with a value of 1 if present and 0 if not, $a_i$ is the contribution of the $i$th substituent to the $BA$, and $\mu$ is the overall average activity of the parent skeleton, are solved by the method of least squares for the $a_i$ and $\mu$. All activity contributions at each position of substitution must sum to zero. The pros and cons of the Free–Wilson method have been discussed by Blankley.[59] Fujita and Ban[60] suggested two modifications of the Free–Wilson approach on the assumption that the effect on the activity of a certain substituent at a certain position in a compound is constant and additive. First, they suggested that the biological activity should be expressed as log $A/A_0$, where $A$ and $A_0$ represent the magnitude of the activity of the substituted and unsubstituted compounds, respectively, and that $a_i$ is the log activity contribution of the $i$th substituent relative to H. This allows the derived substituent constants to be compared directly with other parameters related to free energy that are additive. Second, they suggested that $\mu$ become analogous to the theoretically predicted (calculated) activity of the parent compound of the series. Both of these modifications have been widely accepted.

$$BA = \sum a_i X_i + \mu \tag{2.18}$$

As an example of the Free–Wilson approach, consider the hypothetical compound **2.42**.[56a] If in one pair of analogs for which $R^1$, $R^2$, $R^3$, and $R^4$ are constant and $R^5$ is Cl or $CH_3$, the methyl compound is one-tenth as potent as the chloro analog, then the Free–Wilson method assumes that every $R^5$ methyl analog (where $R^1$–$R^4$ are varied) will be one-tenth as potent as the corresponding $R^5$ chloro analog. A requirement for this approach, then, is a series of compounds that have changes at more than one position. In addition, each type of substituent must occur more than once at each position in which it is found. The outcome is a table of the contribution to potency of each substituent at each position. If the free energy relationships of the extrathermodynamic method are linear or position specific, then Free–Wilson calculations will be successful.

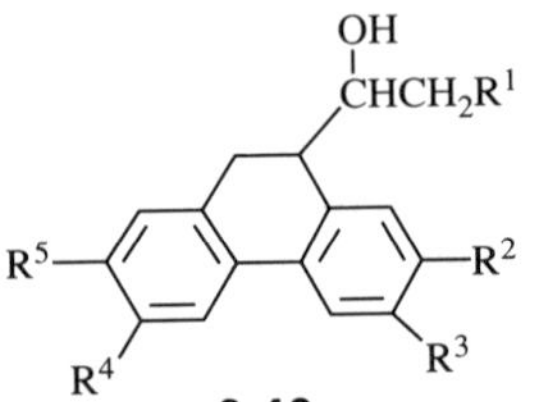

**2.42**

The *interaction model*[61] is a mathematical model similar to that of the Free–Wilson additive model with an additional term ($e_X e_Y$) to account for possible interactions between substituents X and Y.

***c. Enhancement Factor.*** One of the earliest QSAR observations resulted from a retrospective analysis of a large number of synthetic corticosteroids.[62] Examination of the biological properties of steroids prepared by the introduction of halogen, hydroxyl, alkyl, or double bond modifications revealed that each substituent affects the activity of the molecule in a quantitative sense, and almost independently of other groups. The effect (whether positive or negative) of each substituent was assigned a numerical value termed the *enhancement factor*. Multiplication of the enhancement factor for each substituent by the biological activity of the unsubstituted compound gave the potency of the modified steroid.

***d. Manual Stepwise Methods: Topliss Operational Schemes and Others.*** Since organic chemists are, by nature, more likely to be intuitive and less so mathematical, it was not long before Topliss[63] developed a nonmathematical, nonstatistical, and noncomputerized (hence, *manual*) guide to the use of the Hansch principles. This method is most useful when the synthesis of large numbers of compounds is difficult and when biological testing of compounds is readily available. It is an approach for the efficient optimization of the potency of a lead compound with minimization of the number of compounds needed to be synthesized. The only prerequisite for the technique is that the lead compound must contain an unfused benzene ring. However, according to literature surveys at the time that this method was published, 40% of all reported compounds[64] contain an unfused benzene ring and 50% of drug-oriented patents[65] are concerned with substituted benzenes. This approach relies heavily on $\pi$ and $\sigma$ values and to a much lesser degree $E_s$ values. The methodology is outlined here; a more detailed discussion can be found in the Topliss paper.[63]

Consider that the lead compound is benzenesulfonamide (**2.43**, R = H) and its potency has been measured in whatever bioassay is being used. Since many systems are $+\pi$ dependent, that is, the potency increases with increasing $\pi$ values, then a good choice for the first analog would be one with a substituent having a $+\pi$ value. Since $\pi_{4\text{-Cl}} = 0.71$ and $\sigma_{4\text{-Cl}} = 0.23$ (remember, $\pi_H = \sigma_H = 0$), the 4-chloro analog (**2.43**, R = Cl) should be synthesized and tested. There are three possible outcomes of this effort, namely, the 4-chloro analog is more potent, equipotent, or less potent than the parent compound. If it is more potent, then it can be attributed to a $+\pi$ effect, a $+\sigma$ effect, or to both. In this case, the 3,4-dichloro analog ($\pi_{3,4\text{-Cl}_2} = 1.25$, $\sigma_{3,4\text{-Cl}_2} = 0.52$) could be synthesized next and tested. Again, the 3,4-dichloro analog could be more potent, equipotent, or less potent than the 4-chloro compound. If it is more potent, then determination of whether $+\pi$ or $+\sigma$ is more important could be made by selection next of the 4-SPh analog ($\pi_{\text{SPh}} = 2.32$, $\sigma_{\text{SPh}} = 0.18$) or the 3-trifluoromethyl-4-nitro analog ($\pi_{3\text{-CF}_3\text{-4-NO}_2} = 0.60$, $\sigma_{3\text{-CF}_3\text{-4-NO}_2} = 1.21$).

R $SO_2NH_2$

**2.43**

At this point a potency tree, termed a *Topliss decision tree*, could be constructed (Fig. 2.5), and additional analogs could be made. It must be stressed that this analysis was based solely on $\pi$ and $\sigma$ values, and other factors such as steric effects have been neglected.

If the 3,4-dichloro compound was less potent than the 4-chloro analog, it could be that the optimum values of $\pi$ and $\sigma$ were exceeded or that the 3-chloro group has an unfavorable steric effect. The latter hypothesis could be tested by the synthesis of the 4-trifluoromethyl analog ($\pi_{4\text{-CF}_3} = 0.88$, $\sigma_{4\text{-CF}_3} =$

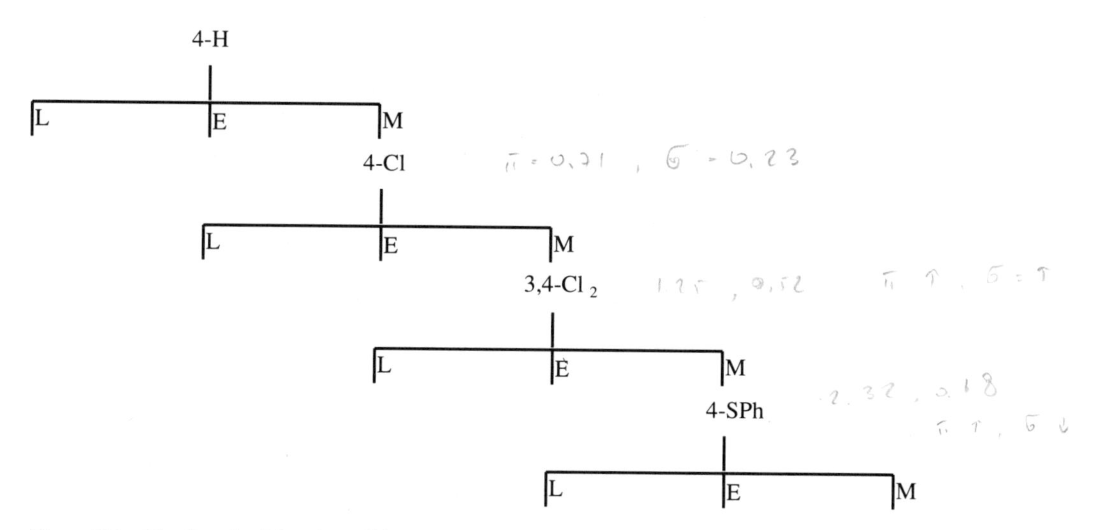


**Figure 2.5.** Topliss decision tree (M, more potent; E, equipotent; L, less potent).

0.54), which has no 3-substituent but has a high $\sigma$ and intermediate $\pi$ value. If this analog is more potent than the 4-chloro analog, the 4-nitro analog ($\pi_{4\text{-NO}_2} = -0.28$, $\sigma_{4\text{-NO}_2} = 0.78$) or the 4-ethyl analog ($\pi_{4\text{-Et}} = 1.02$, $\sigma_{4\text{-Et}} = -0.15$) could be synthesized in order to determine the importance of $\pi$ and $\sigma$ values, respectively.

What if the 4-chloro analog was equipotent with the parent compound? This could result from a favorable $+\pi$ effect counterbalanced by an unfavorable $\sigma$ effect or vice versa. If the former is the case, then the 4-methyl analog ($\pi_{4\text{-Me}} = 0.56$, $\sigma_{4\text{-Me}} = -0.17$) should show enhanced potency. Further enhancement of potency by the 4-methyl analog would suggest that the synthesis of analogs with increasing $\pi$ values and decreasing $\sigma$ values would be propitious. If the 4-methyl analog is worse than the 4-chloro analog, perhaps the equipotency of the 4-chloro compound was the result of a favorable $\sigma$ effect and an unfavorable $\pi$ effect. The 4-nitro analog ($\pi_{4\text{-NO}_2} = -0.28$, $\sigma_{4\text{-NO}_2} = 0.78$) would, then, be a wise next choice.

If the 4-chloro analog was less potent than the lead, then there may be a steric problem at the 4 position, or increased potency may depend on $-\pi$ and $-\sigma$ values. The 3-chloro analog ($\pi_{3\text{-Cl}} = 0.71$, $\sigma_{3\text{-Cl}} = 0.37$) could be synthesized to determine if a steric effect is the problem. Note that the $\sigma$ constant for the 3-Cl substituent is different from that for the 4-Cl one because these descriptors are constitutive. If there is no steric effect, then the 4-methoxy compound ($\pi_{4\text{-OMe}} = -0.04$, $\sigma_{4\text{-OMe}} = -0.27$) could be prepared to investigate the effect of adding a $-\sigma$ substituent. An increased potency of the 4-OMe substituent would suggest that other substituents with more negative $\pi$ and/or $\sigma$ constants be tried.

Topliss[63] extended the operational scheme for side-chain problems when the group is adjacent to a carbonyl, amino, or amide functionality, namely, —COR, —NHR, —CONHR, —NHCOR, where R is the variable substituent. This approach is applicable to a variety of situations other than direct substitution on the aromatic nucleus. In this case, the parent molecule is the one where R is $CH_3$, and $\pi$, $\sigma$, and $E_s$ parameters are used.

Note that in the Topliss operational scheme, as for other methods in this section, the procedure is *stepwise*, that is, the next compound is determined on the basis of the results obtained with the previous one. Three other manual, stepwise methods are mentioned only briefly: Craig plots,[66] Fibonacci search method,[67] and sequential simplex strategy.[68] The Topliss decision tree approach evolved from the work of Craig,[66] who pointed out the utility of a simple graphical plot of $\pi$ versus $\sigma$ (or any two parameters) to guide the choice of a substituent (Fig. 2.6). Once the Hansch equation has been expressed for an initial set of compounds, the sign and magnitude of the $\pi$ and $\sigma$ regression coefficients determine the particular quadrant of the Craig plot that is to be used to direct further synthesis. Thus, if both the $\pi$ and $\sigma$ terms have

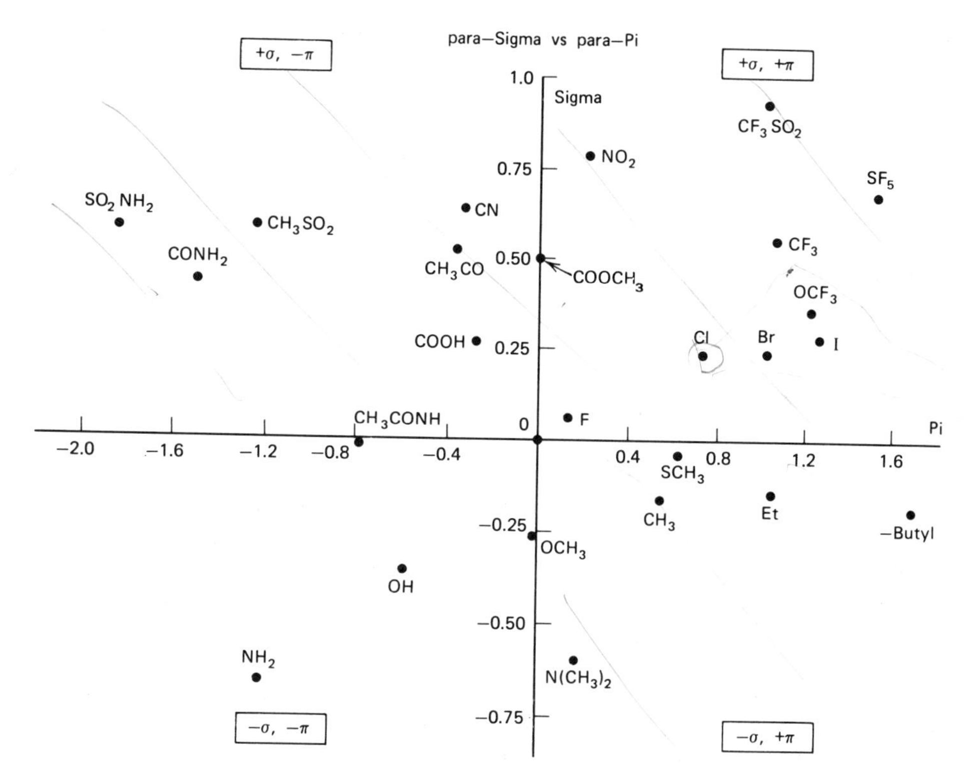


**Figure 2.6.** Craig plot of $\sigma$ constants versus $\pi$ values for aromatic substituents.[66] [Reprinted by permission of John Wiley & Sons, Inc. from Craig, P. N. (1980). *In* "Burger's Medicinal Chemistry," (M. E. Wolff, ed.), 4th ed., Part I, p. 343. Wiley, New York. Copyright © 1980 John Wiley & Sons, Inc.]

positive coefficients, then substituents in the upper right-hand quadrant of the plot (Fig. 2.6) should be selected for future analogs.

The Fibonacci search technique[67] is a manual method to discover the optimum of some parabolic function, such as potency versus log $P$, in a minimum number of steps. Sequential simplex strategy[68] is another stepwise technique suggested when potency depends on two physicochemical parameters such as $\pi$ and $\sigma$.

***e. Batch Selection Methods: Batchwise Topliss Operational Scheme, Cluster Analysis, and Others.*** The inherent problem with the Topliss operational scheme described in Section II,E,3,d is its stepwise nature. Provided that pharmacological results can be obtained quickly, this is probably not much of a problem; however, biological evaluation is often slow. Topliss[69] proposed an alternative scheme that uses *batchwise* analysis of small groups of com-

**Table 2.7** Potency Order for Various Parameter Dependencies [With permission from Topliss, J. G. (1977). *J. Med. Chem* **20,** 463. Copyright © 1977 American Chemical Society.]

| | Parameters | | | | | | | | | |
|---|---|---|---|---|---|---|---|---|---|---|
| Substituent | $\pi$ | $2\pi - \pi^2$ | $\sigma$ | $-\sigma$ | $\pi + \sigma$ | $2\pi - \sigma$ | $\pi - \sigma$ | $\pi - 2\sigma$ | $\pi - 3\sigma$ | $E_4$[a] |
| 3,4-$Cl_2$ | 1 | 1–2 | 1 | 5 | 1 | 1 | 1–2 | 3–4 | 5 | 2–5 |
| 4-Cl | 2 | 1–2 | 2 | 4 | 2 | 2–3 | 3 | 3–4 | 3–4 | 2–5 |
| 4-$CH_3$ | 3 | 3 | 4 | 2 | 3 | 2–3 | 1–2 | 1 | 1 | 2–5 |
| 4-$OCH_3$ | 4–5 | 4–5 | 5 | 1 | 5 | 4 | 4 | 2 | 2 | 2–5 |
| H | 4–5 | 4–5 | 3 | 3 | 4 | 5 | 5 | 5 | 3–4 | 1 |

[a] Unfavorable steric effect from 4-substitution.

pounds. Substituents were grouped by Topliss[69] according to $\pi$, $\sigma$, $\pi^2$, and a variety of $x\pi$ and $y\sigma$ weighted combinations. The approach starts with the synthesis of five derivatives, the unsubstituted (4-H), 4-chloro, 3,4-dichloro, 4-methyl, and 4-methoxy compounds. After these five analogs have been tested in the bioassay, they are ranked in order of decreasing potency. The potency order determined for these analogs is then compared with the rankings in Table 2.7 to determine which parameter or combination of parameters is most dominant. If, for example, the potency order is 4-$OCH_3$ > 4-$CH_3$ > H > 4-Cl > 3,4-$Cl_2$, then $-\sigma$ is the dominant parameter. Once the parameter dependency is determined, Table 2.8 is consulted in order to discover what substituents should be investigated next. In the above example, 4-$N(C_2H_5)_2$, 4-$N(CH_3)_2$, 4-$NH_2$, 4-$NHC_4H_9$, 4-OH, 4-$OCH(CH_3)_2$, 3-$CH_3$, and 4-$OCH_3$ would be suitable choices. The major weakness of this approach is that it is

**Table 2.8** New Substituent Selections [With permission from Topliss, J. G. (1977). *J. Med. Chem.* **20,** 463. Copyright © 1977 American Chemical Society.]

| Probable operative parameters | New substituent selection |
|---|---|
| $\pi$, $\pi + \sigma$, $\sigma$ | 3-$CF_3$, 4-Cl; 3-$CF_3$, 4-$NO_2$; 4-$CF_3$, 2,4-$Cl_2$; 4-$c$-$C_5H_9$; 4-$c$-$C_6H_{11}$ |
| $\pi$, $2\pi - \sigma$, $\pi - \sigma$ | 4-$CH(CH_3)_2$; 4-$C(CH_3)_3$; 3,4-$(CH_3)_2$; 4-$O(CH_2)_3CH_3$; 4-$OCH_2Ph$; 4-$N(C_2H_5)_2$ |
| $\pi - 2\sigma$, $\pi - 3\sigma$, $-\sigma$ | 4-$N(C_2H_5)_2$; 4-$N(CH_3)_2$; 4-$NH_2$; 4-$NHC_4H_9$; 4-OH; 4-$OCH(CH_3)_2$; 3-$CH_3$, 4-$OCH_3$ |
| $2\pi - \pi^2$ | 4-Br; 3-$CF_3$; 3,4-$(CH_3)_2$; 4-$C_2H_5$; 4-$O(CH_2)_2CH_3$; 3-$CH_3$, 4-Cl; 3-Cl; 3-$CH_3$; 3-$OCH_3$; 3-$N(CH_3)_2$; 3-$CF_3$; 3,5-$Cl_2$ |
| Ortho effect | 2-Cl; 2-$CH_3$; 2-$OCH_3$; 2-F |
| Other | 4-F; 4-$NHCOCH_3$; 4-$NHSO_2CH_3$; 4-$NO_2$; 4-$COCH_3$; 4-$SO_2CH_3$; 4-$CONH_2$; 4-$SO_2NH_2$ |

**Table 2.9** Typical Members of Clusters Based on $\alpha$, $\pi^2$, *F*, *R*, *MR*, and *MW* [Reprinted with permission from Martin, Y. C. (1979). *In* "Drug Design" (E. J. Ariens, ed.), Vol. VIII, p. 5. Academic Press, New York. Copyright © 1979 Academic Press, Inc.]

| Cluster number[a] | Typical members |
|---|---|
| 1 | Me, H, 3,4-($OCH_2O$), $CH_2CH_2COOH$, CH=$CH_2$, Et, $CH_2OH$ |
| 2 | CH=CHCOOH |
| 3a | CN, $NO_2$, CHO, COOH, COMe |
| 3b | C≡CH, $CH_2Cl$, Cl, NNN, SH, SMe, CH=NOH, $CH_2CN$, OCOMe, SCOMe, COOMe, SCN |
| 4a | $CONH_2$, CONHMe, $SO_2NH_2$, $SO_2Me$, SOMe |
| 4b | NHCHO, NHCOMe, $NHCONH_2$, $NHCSNH_2$, $NHSO_2Me$ |
| 5 | F, OMe, $NH_2$, $NHNH_2$, OH, NHMe, NHEt, $NMe_2$ |
| 6 | Br, $OCF_3$, $CF_3$, NCS, I, $SF_5$, $SO_2F$ |
| 7 | $CH_2Br$, SeMe, $NHCO_2Et$, $SO_2Ph$, $OSO_2Me$ |
| 8 | NHCOPh, $NHSO_2Ph$, $OSO_2Ph$, COPh, N=NPh, OCOPh, $PO_2Ph$ |
| 9 | 3,4-$(CH_2)_3$, 3,4-$(CH_2)_4$, Pr, *i*-Pr, 3,4-$(CH)_4$, NHBu, Ph, $CH_2Ph$, *t*-Bu, OPh |
| 10 | Ferrocenyl, adamantyl |

[a] Clusters 3 and 4 contain many of the common substituents used in medicinal chemistry; hence, these clusters are further subdivided according to their cluster membership when 20 clusters have been made.

difficult to extend the method to additional parameters unless computers are utilized.

A computer-based batch selection method, known as *cluster analysis*, was introduced by Hansch *et al.*[70] Substituents were grouped into clusters with similar properties according to their $\sigma$, $\pi$, $\pi^2$, $E_s$, *F* (field constant), *R* (resonance constant), *MR* (molar refractivity), and MW (molecular weight) values. Some of the clusters are shown in Table 2.9.[71] One member of each cluster would be selected for substitution into the lead compound, and the compounds would be synthesized and tested. If a substituent showed dominant potency, then other substituents from that cluster would be selected for further investigation. The important advantage of the batch selection methods is that the initial batch of analogs prepared is derived from the widest range of parameters possible so that the dominant physicochemical property can be revealed early in the lead modification process.

There are other selection methods and statistical analysis techniques, but they lie outside the scope of this book (see General References, this chapter).

## 4. Computer-Based Methods of QSAR Related to Receptor Binding

There are a variety of computer-based methods that have been used to correlate molecular structure with receptor binding, and, therefore, activity. Some are mentioned here. Crippen and co-workers[72a–c] devised a linear free energy model, termed the *distance geometry approach*, for calculating QSAR from receptor binding data. The distances between various atoms in the molecule, compiled into a table called the *distance matrix*, define the conformation of the molecule. Rotations about single bonds change the molecular conformation and, therefore, these distances; consequently, an upper and lower distance limit is set on each distance. Experimentally determined free energies of binding of a series of compounds to the receptor are used with the distance matrix of each molecule in a computerized method to deduce possible binding sites in terms of geometry and chemical character of the site. Although this approach requires more computational effort and adjustable parameters than Hansch analysis, it is suggested[72b] to give good results on more difficult data sets.

The distance geometry approach was extended by Sheridan *et al.*[73] to treat two or more molecules as a single ensemble. The *ensemble approach to distance geometry* can be used to find a common pharmacophore for a receptor with unknown structure from a small set of biologically active molecules.

Hopfinger[74] has developed a set of computational procedures termed *molecular shape analysis* for the determination of the active conformations and, thereby, molecular shapes during receptor binding. Common pairwise overlap steric volumes calculated from low-energy conformations of molecules are used to obtain three-dimensional molecular shape descriptors which can be treated quantitatively and used with other physicochemical parameter descriptors.

Two other descriptors for substructure representation, the *atom pair*[75] and the *topological torsion*,[76] have been described by Venkataraghavan and coworkers. These descriptors characterize molecules in fundamental ways that are useful for the selection of potentially active compounds from hundreds of thousands of structures in a database. The atom pair method can select compounds from diverse structural classes that have atoms within the entire molecule similar to those of a particular active structure. The topological torsion descriptor is complementary to the atom pair descriptor, and it focuses on a local environment of a molecule for comparison with active structures.

### F. Molecular Graphics-Based Drug Design

Quantitative structure–activity relationship studies have relied heavily on the use of computers from the beginning for statistical calculations involving multiparameter equations. It was soon realized that drug design could be aided significantly if structures of receptors and drugs could be displayed on a terminal and molecular processes could be visualized. *Molecular graphics* is the use of computer graphics to represent and manipulate molecular structures. The origins of molecular graphics has been traced by Hassall[77] to the project MAC (Multiple Access Computer),[78] which produced computer graphics models of macromolecules for the first time. The potential to apply this technology to protein crystallography was quickly realized, and by the early 1970s electron density data from X-ray diffraction studies could be presented and manipulated in stick or space-filling multicolored representations on a computer terminal.[79]

Medicinal chemists saw the potential of this approach in drug design as well. Stick (Dreiding) and space-filling (CPK) molecular models have been used extensively by organic chemists for years, but these hand-held models have major disadvantages.[80] Space-filling models often obscure the structure of the molecule, and wire or plastic models can give false impressions of molecular flexibility and often tend to change into unfavorable conformations at inopportune moments. A three-dimensional computer graphics representation that can be manipulated in three dimensions allows the operator to visualize the interactions of small molecules with biologically important macromolecules. Superimposition of structures, which is cumbersome at best with manual models, can be performed easily by molecular graphics. Also, some systems have the capability to synthesize graphically new structures by the assemblage of appropriate molecular fragments from a fragment file.

There are numerous molecular graphics systems available,[80,81a,b] but the typical system, utilized by every major pharmaceutical company in the United States, Western Europe, and Japan, consists of a mainframe or supermini computer linked to a high-resolution graphics terminal with local intelligence. The graphics terminal may be equipped with a variety of peripheral devices such as graphic tablets, light pens, function keys, and dials to effect the molecular display and three-dimensional manipulations. The mainframe or minicomputer executes all of the molecular calculations, such as calculations of bond lengths, bond angles, and quantum chemical or force field calculations.

Once the computer graphics system is set up, there are a variety of approaches that can be taken to utilize it for drug design. The basic premise in the utilization of molecular graphics is that the better the complementary fit of the drug to the receptor, the more potent the drug will be. This is the *lock-and-key hypothesis* of Fischer[82] in which the receptor is the lock into which the key (i.e., the drug) fits. In order to apply this concept it would appear that the structure of the receptor would have to be known, then different drug analogs could be docked into the receptor. *Docking* is a molecular graphics term for the computer-assisted movement of a terminal-displayed molecule into its receptor. However, as indicated in Chapter 3, the structures of very few receptors, except for enzymes, are known. This methodology, then, would appear to be quite limited; however, the ability to accomplish this would be a simple and useful drug design approach. Kuntz *et al.*[83] reported an algorithm designed to fit small molecules into their macromolecular receptors. This shape-matching method, which is restricted to rigid *ligands* (receptor-bound molecules) and receptors, was modified[84] for flexible ligands where a ligand is approximated as a small set of rigid fragments. The drawbacks of this approach are the assumptions that binding is determined primarily by shape complementarity and that only small changes in the shape of the receptor occur upon ligand binding. An important advantage, though, is that this method is not limited to docking of known ligands. A library of molecular shapes can be scanned to determine which shapes best fit a particular receptor binding site.

Since the energetics, as well as the shape complementarity, of a drug–receptor complex are vital to its stability, Goodford[85] described a method that simultaneously displays the energy contour surfaces and the macromolecular structure on the computer graphics system. This allows both the energy and shape to be considered together when considering the design of molecules that have the optimal fit into the receptor.

There are relatively few receptors with known X-ray structures. Consequently, it would appear that the approach of *receptor fitting* is of little importance to drug design. However, because of the great advances in computer technology and software development,[81b] it is possible to use molecular graphics to obtain information about the ligand binding site of an unknown receptor. One approach is to deduce the topography of an unknown receptor site from related known receptor structures.[86] Another approach is to use molecular graphics visualization of an electron density map for a known drug–receptor complex obtained by X-ray crystallography. This may reveal empty pockets in the complex that could be filled by appropriate modification, an approach taken by Blaney *et al.*[87] in the design of new thyroid hormone analogs.

A third important molecular graphics technique useful for identification of the pharmacophore geometry is called *receptor mapping*. This method utilizes data from known ligand binding studies to an unknown receptor. With this technique, which also is founded on the premise that receptor topography is complementary to that of drugs, the structure of the lock is deduced from the shape of the keys that fit it. A variety of receptor mapping techniques have been described. An approach termed *steric mapping*[88] uses molecular graphics to combine the volumes of compounds known to bind to the desired receptor. This composite volume generates an *enzyme-excluded volume map* which defines that region of the binding site available for binding by drug analogs and, therefore, not occupied by the receptor itself. The same procedure is then carried out for similar molecules that are inactive. The composite volume is inspected for regions of volume overlap common to all of the inactive analogs. These are the *enzyme-essential regions*, sites required by the receptor itself and unavailable for occupancy by ligands. Any other molecule that overlaps with these regions should be inactive. Drug design, then, would involve the synthesis of compounds with the appropriate pharmacophore that filled the enzyme-excluded regions and that avoided the enzyme-essential regions.

In the modified method for docking flexible ligands into a receptor[84] described above, an X-ray structure of the receptor is not necessary to characterize the shape of the receptor binding site. Rather, the receptor binding site can be deduced from the shapes of active ligands.

Although molecular graphics approaches are widely used, it is not clear if any drugs have yet been designed *de novo* by this method, although leads have been discovered and lead optimization has been assisted by these techniques.[77,89] Several problems with this approach may contribute to its less than optimal effectiveness. The major problem derives from the fact that pharmacokinetics are ignored by this method. Prior to the drug candidate interacting with a receptor, it must be properly absorbed, it must reach the receptor without metabolic or chemical degradation (unless it is a prodrug; see Chapter 8), excretion must be appropriate, and the drug candidate and metabolites must not be toxic nor lead to undesirable side effects. Another problem with molecular graphics approaches is that energy minimization programs[90] are used to determine the lowest energy conformers of molecules, and the calculations are carried out for ground state molecules; however, as will be discussed in Chapters 3 and 5, the conformation of a molecule during the receptor binding process is not necessarily the one having the lowest energy, and it can be quite different from the ground state conformation. Also, these calculations generally are performed on molecules in the absence of solvent effects. Finally, programs written for drug–receptor interactions assume that the receptor is reasonably rigid during binding, which may not be the case.

## G. Epilogue

On the basis of what was discussed in this chapter, it appears that even if one uncovers a lead, it may be a fairly random process to optimize its potency. In fact, less than 1 in 10,000 compounds synthesized in drug companies makes it to the drug market, and, in so doing, it takes about 10 years of research at a cost of $200–250 million. However, more rational approaches to lead discovery and lead optimization, based on chemical and biochemical principles, can be used. Some of these approaches are discussed in Chapters 3, 5, and 6.

## References


1. Fleming, A. 1929. *Br. J. Exp. Pathol.* **10,** 226.
2. Hare, R. 1970. "The Birth of Penicillin." Allen & Unwin, London.
3. Beveridge, W. I. B. 1981. "Seeds of Discovery." Norton, New York.

4. Abraham, E. P., Chain, E., Fletcher, C. M., Gardner, A. D., Heatley, N. G., Jennings, M. A., and Florey, H. W. 1941. *Lancet 2,* 177.
5. Florey, H. W., Chain, E., Heatley, N. G., Jennings, M. A., Sanders, A. G., Abraham, E. P., and Florey, M. E. 1949. "Antibiotics," Vol. 2. Oxford Univ. Press, London.
6. Moyer, A. J., and Coghill, R. D. 1946. *J. Bacteriol.* **51,** 79.
7. Sternbach, L. H. 1979. *J. Med. Chem.* **22,** 1.
8. Shen, T. Y. 1976. *In* "Clinoril in the Treatment of Rheumatic Disorders" (Huskisson, E. C., and Franchimont, P., eds.), Raven, New York.
9. Gay, L. N., and Carliner, P. E. 1949. *Science* **109,** 359.
10. Strickland, B. A., Jr., and Hahn, G. L. 1949. *Science* **109,** 359.
11. Shen, T. Y., and Winter, C. A. 1977. *Adv. Drug Res.* **12,** 89.
12. Grewe, R. 1946. *Naturwissenschaften* **33,** 333.
13. Schnider, O., and Grüssner, A. 1949. *Helv. Chim. Acta* **32,** 821.
14. May, E. L., and Murphy, J. G. 1955. *J. Org. Chem.* **20,** 257.
15. Schaumann, O. 1940. *Naunyn-Schmiedeberg's Arch. Pharmacol. Exp. Pathol.* **196,** 109.
16. Bentley, K. W., and Hardy, D. G. 1967. *J. Am. Chem. Soc.* **89,** 3267; Bentley, K. W., Hardy, D. G., and Meek, B. 1967. *J. Am. Chem. Soc.* **89,** 3273.
17. Crum-Brown, A., and Fraser, T. R. 1868–1869. *Trans. R. Soc. Edinburgh* **25,** 151 and 693; Crum-Brown, A., and Fraser, T. R. 1869. *Proc. R. Soc. Edinburgh* **6,** 556; Crum-Brown, A., and Fraser, T. R. 1872. *Proc. R. Soc. Edinburgh* **7,** 663.
18. Richardson, B. W. 1869. *Med. Times Gaz.* **18,** 703.
19. Northey, E. H. 1948. "The Sulfonamides and Allied Compounds." American Chemical Society Monograph Series, Reinhold, New York.
20. Loubatieres, A. 1969. *In* "Oral Hypoglycemic Agents" (Campbell, G. D., ed.). Academic Press, New York.
21. Sprague, J. M. 1968. *In* "Topics in Medicinal Chemistry" (Robinowitz, J. L., and Myerson, R. M., eds.), Vol. 2. Wiley, New York.
22a. Dohme, A. R. L., Cox, E. H., and Miller, E. 1926. *J. Am. Chem. Soc.* **48,** 1688.
22b. Funcke, A. B. H., Ernsting, M. J. E., Rekker, R. F., and Nauta, W. T. 1953. *Arzneim.-Forsch.* **3,** 503.
23. Thornber, C. W. 1979. *Chem. Soc. Rev.* **8,** 563.
24. Burger, A. 1970. *In* "Medicinal Chemistry" (Burger, A., ed.), 3rd Ed., Wiley, New York.
25. Korolkovas, A. 1970. "Essentials of Molecular Pharmacology: Background for Drug Design," p. 54. Wiley, New York.
26. Lipinski, C. A. 1986. *Annu. Rep. Med. Chem.* **21,** 283.
27. Grimm, H. G. 1925. *Z. Elektrochem.* **31,** 474; Grimm, H. G. 1928. *Z. Elektrochem.* **34,** 430.
28. Erlenmeyer, H. 1948. *Bull. Soc. Chim. Biol.* **30,** 792.
29. Hansch, C., Maloney, P. P., Fujita, T., and Muir, R. M. 1962. *Nature (London)* **194,** 178.
30. Richet, M. C. 1893. *C. R. Seances Soc. Biol. Ses Fil.* **45,** 775.
31. Overton, E. 1897. *Z. Phys. Chem.* **22,** 189.
32. Meyer, H. 1899. *Arch. Exp. Pathol. Pharmacol.* **42,** 109.
33. Ferguson, J. 1939. *Proc. R. Soc. London, Ser. B* **127,** 387.
34. Hansch, C., Muir, R. M., and Metzenberg, R. L., Jr. 1951. *Plant Physiol.* **26,** 812.
35. Hammett, L. P. 1940. "Physical Organic Chemistry." McGraw-Hill, New York.
36. Roberts, J. D., and Caserio, M. C. 1977. "Basic Principles of Organic Chemistry," 2nd Ed., p. 1331. Benjamin, Menlo Park, California.
37. Hansch, C., and Fujita, T. 1964. *J. Am. Chem. Soc.* **86,** 1616.
38. Singer, S. J., and Nicolson, G. L. 1972. *Science* **175,** 720.
39. Fujita, T., Iwasa, J., and Hansch, C. 1964. *J. Am. Chem. Soc.* **86,** 5175.
40. Collander, R. 1954. *Physiol. Plant* **7,** 420; Collander, R. 1951. *Acta Chem. Scand.* **5,** 774.
41. Hansch, C., Steward, A. R., Anderson, S. M., and Bentley, D. 1968. *J. Med. Chem.* **11,** 1.

42. Hansch, C., and Leo, A. 1979. "Substituent Constants for Correlation Analysis in Chemistry and Biology." Wiley, New York.
43. Leo, A., Hansch, C., and Elkins, D. 1971. *Chem. Rev.* **71,** 525.
44. Iwasa, J., Fujita, T., and Hansch, C. 1965. *J. Med. Chem.* **8,** 150.
45. Chou, J. T., and Jurs, P. C. 1979. *J. Chem. Inf. Comput. Sci.* **19,** 171; Pomona College Medicinal Chemistry Project; see Hansch, C., Björkroth, J. P., and Leo, A. 1987. *J. Pharm. Sci.* **76,** 663.
46. Bodor, N., Gabanyi, Z., and Wong, C.-K. 1989. *J. Am. Chem. Soc.* **111,** 3783 and 8062.
47. Gobas, F. A. P. C., Lahittete, J. M., Garofalo, G., Shiu, W. Y., and Mackay, D. 1988. *J. Pharm. Sci.* **77,** 265.
48. Seiler, P. 1974. *Eur. J. Med. Chem.* **9,** 473.
49. Taft, R. W. 1956. *In* "Steric Effects in Organic Chemistry" (Neuman, M. S., ed.), p. 556. Wiley, New York.
50. Unger, S. H., and Hansch, C. 1976. *Prog. Phys. Org. Chem.* **12,** 91.
51. Hancock, C. K., Meyers, E. A., and Yager, B. J. 1961. *J. Am. Chem. Soc.* **83,** 4211.
52. Hansch, C., Leo, A., Unger, S. H., Kim, K. H., Nikaitani, D., and Lien, E. J. 1973. *J. Med. Chem.* **16,** 1207.
53. Verloop, A., Hoogenstraaten, W., and Tipker, J. 1976. *In* "Drug Design" (Ariens, E. J., ed.), Vol. 7, p. 165. Academic Press, New York.
54. Daniel, C., and Wood, F. S. 1971. "Fitting Equations to Data." Wiley, New York; Draper, N. R., and Smith, H. 1966. "Applied Regression Analysis." Wiley, New York; Snedecor, G. W., and Cochran, W. G. 1967. "Statistical Methods." Iowa State Univ. Press, Ames, Iowa.
55. Deardon, J. C. 1987. *In* "Trends in Medicinal Chemistry" (Mutschler, E., and Winterfeldt, E., eds.), p. 109. VCH, Weinheim.
56a. Martin, Y. C. 1978. "Quantitative Drug Design: A Critical Introduction," Chap. 2. Dekker, New York.
56b. Tute, M. S. 1980. *In* "Physical Chemical Properties of Drugs" (Yalkowsky, S. H., Sinkula, A. A., and Valvani, S. C., eds.), p. 141. Dekker, New York.
57. Unger, S. H. 1984. *In* "QSAR in Design of Bioactive Compounds" (Kuchar, M., ed.), p. 1. Prous, Barcelona; Hopfinger, A. J. 1985. *J. Med. Chem.* **28,** 1133; Fujita, T. 1984. *In* "Drug Design: Fact or Fantasy?" (Jolles, G., and Wooldridge, K. R. H., eds.), Chap. 2. Academic Press, London.
58. Free, S. M., Jr., and Wilson, J. W. 1964. *J. Med. Chem.* **7,** 395.
59. Blankley, C. J. 1983. *In* "Quantitative Structure–Activity Relationships of Drugs" (Topliss, J. G., ed.), Chap. 1. Academic Press, New York.
60. Fujita, T., and Ban, T. 1971. *J. Med. Chem.* **14,** 148.
61. Bocek, K., Kopecký, J., Krivucová, M., and Vlachová, D. 1964. *Experientia* **20,** 667; Kopecký, J., Bocek, K., and Vlachová, D. 1965. *Nature* (*London*) **207,** 981.
62. Fried, J., and Borman, A. 1958. *Vitam. Horm.* (*N.Y.*) **16,** 303.
63. Topliss, J. G. 1972. *J. Med. Chem.* **15,** 1006.
64. Granito, C. E., Becker, G. T., Roberts, S., Wiswesser, W. J., and Windlinz, K. J. 1971. *J. Chem. Soc.* **11,** 106.
65. Goodford, P. J. 1973. *Adv. Pharmacol. Chemother.* **11,** 51.
66. Craig, P. N. 1971. *J. Med. Chem.* **14,** 680; Craig, P. N. 1980. *In* "Burger's Medicinal Chemistry" (Wolff, M. E., ed.), 4th Ed., Part 1, Chap. 8. Wiley, New York.
67. Bustard, T. M. 1974. *J. Med. Chem.* **17,** 777; Santora, N. J., and Auyang, K. 1975. *J. Med. Chem.* **18,** 959; Deming, S. N. 1976. *J. Med. Chem.* **19,** 977.
68. Darvas, F. 1974. *J. Med. Chem.* **17,** 799.
69. Topliss, J. G. 1977. *J. Med. Chem.* **20,** 463.
70. Hansch, C., Unger, S. H., and Forsythe, A. B. 1973. *J. Med. Chem.* **16,** 1217.

71. Martin, Y. C. 1979. *In* "Drug Design" (Ariens, E. J., ed.), Vol. 8, p. 5. Academic Press, New York.
72a. Crippen, G. M. 1979. *J. Med. Chem.* **22,** 988.
72b. Ghose, A. K., and Crippen, G. M. 1982. *J. Med. Chem.* **25,** 892.
72c. Crippen, G. M. 1981. "Distance Geometry and Conformational Calculations." Research Studies Press, New York.
73. Sheridan, R. P., Nilakantan, R., Dixon, J. S., and Venkataraghavan, R. 1986. *J. Med. Chem.* **29,** 899; Sheridan, R. P., and Venkataraghaven, R. 1987. *Acc. Chem. Res.* **20,** 322.
74. Hopfinger, A. J. 1980. *J. Am. Chem. Soc.* **102,** 7196; Hopfinger, A. J. 1981. *J. Med. Chem.* **24,** 818; Hopfinger, A. J. 1983. *J. Med. Chem.* **26,** 990.
75. Carhart, R. E., Smith, D. H., and Venkataraghavan, R. 1985. *J. Chem. Inf. Comput. Sci.* **25,** 64.
76. Nilakantan, R., Bauman, N., Dixon, J. S., and Venkataraghavan, R. 1987. *J. Chem. Inf. Comput. Sci.* **27,** 82.
77. Hassall, C. H. 1985. *Chem. Brit.* **21,** 39.
78. Levinthal, C. 1966. *Sci. Am.* **214(6),** 42.
79. Barry, C. D., and North, A. C. T. 1971. *Cold Spring Harbor Symp. Quant. Biol.* **36,** 577; Barry, C. D. 1971. *Nature (London)* **232,** 236.
80. Tollenaere, J. P., and Janssen, P. A. 1988. *J. Med. Res. Rev.* **8,** 1.
81a. Gund, P., Halgren, T. A., and Smith, G. M. 1987. *Annu. Rep. Med. Chem.* **22,** 269.
81b. Gund, T., and Gund, P. 1987. *In* "Molecular Structure and Energetics" (Liebman, J. F., and Greenberg, A., eds.), Vol. 4, Chap. 10. VCH, Weinheim.
82. Fischer, E. 1894. *Ber. Dtsch. Chem. Ges.* **27,** 2985.
83. Kuntz, I. D., Blaney, J. M., Oatley, S. J., Langridge, R., and Ferrin, T. E. 1982. *J. Mol. Biol.* **161,** 269.
84. DesJarlais, R. L., Sheridan, R. P., Dixon, J. S., Kuntz, I. D., and Venkatarghavan, R. 1986. *J. Med. Chem.* **29,** 2149; DesJarlais, R. L., Sheridan, R. P., Seibel, G. L., Dixon, J. S., Kuntz, I. D., and Venkataraghavan, R. 1988. *J. Med. Chem.* **31,** 722.
85. Goodford, P. J. 1985. *J. Med. Chem.* **28,** 849.
86. Carlson, G. M., MacDonald, R. J., and Meyer, E. F., Jr. 1986. *J. Theor. Biol.* **119,** 107.
87. Blaney, J. M., Jorgensen, E. C., Connolly, M. L., Ferrin, T. E., Langridge, R., Oatley, S. J., Burridge, J. M., and Blake, C. C. F. 1982. *J. Med. Chem.* **25,** 785.
88. Sufrin, J. R., Dunn, D. A., and Marshall, G. R. 1981. *Mol. Pharmacol.* **19,** 307; Humblet, C., and Marshall, G. R. 1980. *Annu. Rep. Med. Chem.* **15,** 267; Marshall, G. R. 1985. *Ann. N.Y. Acad. Sci.* **439,** 162.
89. Hopfinger, A. J. 1985. *J. Med. Chem.* **28,** 1133.
90. Burkert, U., and Allinger, N. L. 1982. "Molecular Mechanics." American Chemical Society, Washington, D.C.; Boyd, D. B., and Lipkowitz, K. B. 1982. *J. Chem. Educ.* **59,** 269.

## General References

### Membranes and Membrane Biochemistry

Gennis, R. B. 1989. "Biomembranes: Molecular Structure and Function." Springer-Verlag, New York.
Jain, M. K., Wagner, R. C. 1980. "Introduction to Biological Membranes." Wiley, New York.
Vance, D. E., and Vance, J. E., eds. 1985. "Biochemistry of Lipids and Membranes." Benjamin/Cummings, Menlo Park, California.
Yeagle, P. 1987. "The Membranes of Cells." Academic Press, Orlando, Florida.

## QSAR

Blankley, C. J. 1983. *In* "Quantiative Structure–Activity Relationships of Drugs" (Topliss, J. G., ed.). Academic Press, New York.

Chu, K. C. 1980. *In* "Burger's Medicinal Chemistry" (Wolff, M. E., ed.), Part 1, Chap. 10. Wiley, New York.

Martin, Y. C. 1978. "Quantitative Drug Design: A Critical Introduction." Dekker, New York.

Martin, Y. C. 1981. *J. Med. Chem.* **24,** 229.

## Partition Coefficients

Hansch, C., and Leo, A. 1979. "Substituent Constants for Correlation Analysis in Chemistry and Biology." Wiley, New York.

Leo, A., Hansch, C., and Elkins, D. 1971. *Chem. Rev.* **71,** 525.

## Molecular Graphics

Cohen, N. C., Blaney, J. M., Humblet, C., Gund, P., and Barry, D. C. 1990. *J. Med. Chem.* **33,** 883.

Gund, P., Halgren, T. A., and Smith, G. M. 1987. *Annu. Rep. Med. Chem.* **22,** 269.

Hassall, C. H. 1985. *Chem. Brit.* **21,** 39.

Hopfinger, A. J. 1985. *J. Med. Chem.* **28,** 1133.

Perun, T. J., and Propst, C. L., eds. 1989. "Computer-Aided Drug Design: Methods and Applications." Dekker, New York.

Tollenaere, J. P., and Janssen, P. A. J. 1988. *Med. Res. Rev.* **8,** 1.

Venkataraghavan, R., and Feldmann, R. J., eds. 1985. "Macromolecular Structure and Specificity: Computer-Assisted Modeling and Applications," *Ann. N.Y. Acad. Sci.* **439.**

*Journal of Computer-Aided Molecular Design*. ESCOM Science Publishers, Leiden, The Netherlands.

*Journal of Molecular Graphics*. Butterworth-Heinemann Ltd., Stoneham, MA.

CHAPTER 3

# Receptors



## I. Introduction

Up to this point in our discussion it appears that a drug is taken, and by some kind of magic it travels through the body and elicits a pharmaceutical effect. *Pharmacokinetics* (absorption, distribution, metabolism, and excretion) was mentioned in Chapter 2, but no discussion was presented regarding what produces the pharmaceutical effect. The site of drug action, which is ultimately responsible for the pharmaceutical effect, is called a *receptor*. The interaction of the drug with the receptor constitutes *pharmacodynamics*. In this chapter the emphasis is placed on pharmacodynamics of general noncatalytic receptors, in Chapter 4 a special class of receptors that have catalytic properties, called enzymes, will be discussed, and in Chapter 6 another receptor, DNA, will be the topic of discussion. The drug–receptor properties described in this chapter also apply to drug–enzyme and drug–DNA complexes.

## II. Receptor Structure

### A. Historical

In 1878 John N. Langley,[1] a physiology student at Cambridge University, while studying the mutually antagonistic action of the alkaloids atropine (**3.1**; now used as an antisecretory agent) and pilocarpine (**3.2**; used in the treatment of glaucoma, but causes sweating and salivation) on cat salivary flow, suggested that both of these chemicals interacted with some substance in the nerve endings of the gland cells. Langley, however, did not follow up this notion for over 25 years.

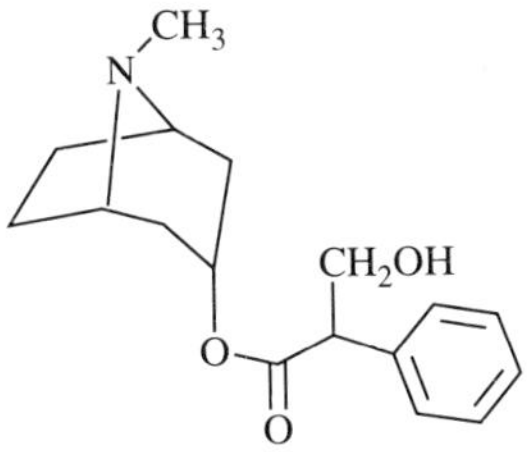

**3.1** 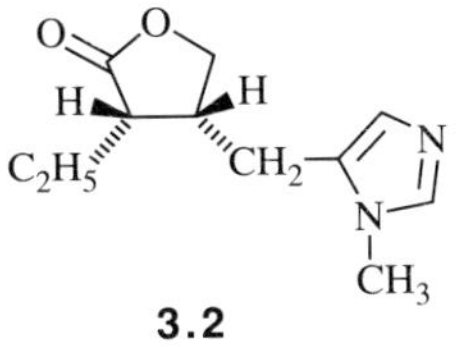 **3.2**

Paul Ehrlich[2] suggested his *side chain theory* in 1897. According to this hypothesis, cells have side chains attached to them that contain specific groups capable of combining with a particular group of a toxin. Ehrlich termed these side chains receptors. Another facet of this hypothesis was that when toxins combined with the side chains, excess side chains were produced and released into the bloodstream. In today's biochemical vernacular these excess side chains would be called *antibodies*, and they combine with toxins stoichiometrically.

In 1905 and 1906 Langley[3] studied the antagonistic effects of curare (a generic term for a variety of South American quaternary alkaloid poisons that cause muscular paralysis) on nicotine stimulation of skeletal muscle. He concluded that there was a receptive substance that received the stimulus and, by transmitting it, caused muscle contraction. This was really the first time that attention was drawn to the two fundamental characteristics of a receptor, namely, a *recognition capacity* for specific ligands and an *amplification component*, the ability of the ligand–receptor complex to initiate a biological response.

### B. What Is a Receptor?

In general, receptors are integral proteins (i.e., polypeptide macromolecules) that are embedded in the phospholipid bilayer of cell membranes (see Fig. 2.3). They, typically, function in the membrane environment; consequently, their properties and mechanisms of action depend on the phospholipid milieu. Vigorous treatment of cells with detergents is required to dissociate these proteins from the membrane. Once they become dissociated, however, they can lose their integrity. Since they generally exist in minute quantities and can be unstable, few receptors have been purified, and little structural information is known about them. Advances in molecular biology more recently have permitted the isolation, cloning, and sequencing of receptors,[4] and this is leading to further approaches to molecular characterization of these proteins. However, these receptors, unlike many enzymes, are still typically characterized in terms of their function rather than by their structural properties. The two functional components of receptors, the recognition component and the amplification component, may represent the same or different sites on the same protein. Various hypotheses regarding the mechanism by which drugs may initiate a biological response are discussed in Section III,E.

## III. Drug–Receptor Interactions

### A. General Considerations

In order to appreciate mechanisms of drug action it is important to understand the forces of interaction that bind drugs to their receptors. Because of the low concentration of drugs and receptors in the bloodstream and other biological fluids, the law of mass action alone cannot account for the ability of small doses of structurally specific drugs to elicit a total response by combination with all, or practically all, of the appropriate receptors. One of my all-time favorite calculations, shown below, supports the notion that something more than mass action is required to get the desired drug–receptor interaction.[5] One mole of a drug contains $6.02 \times 10^{23}$ molecules (Avogadro's number). If the molecular weight of an average drug is 200 g/mol, then 1 mg (often an effective dose) will contain $6.02 \times 10^{23}(10^{-3})/200 = 3 \times 10^{18}$ molecules of drug. The human organism is composed of about $3 \times 10^{13}$ cells. Therefore, each cell will be acted upon by $3 \times 10^{18}/3 \times 10^{13} = 10^{5}$ drug molecules. One erythrocyte cell contains about $10^{10}$ molecules. On the assumption that the same number of molecules is found in all cells, then for each drug molecule, there are $10^{10}/10^{5} = 10^{5}$ molecules of the human body! With this ratio of human molecules to drug molecules, Le Chatelier would have a difficult time explaining how the drug could interact and form a stable complex with the desired receptor.

The driving force for the drug–receptor interaction can be considered as a low-energy state of the drug–receptor complex [Eq. (3.1)], where $k_{on}$ is the rate constant for formation of the drug–receptor complex, which depends on the concentrations of the drug and the receptor, and $k_{off}$ is the rate constant for breakdown of the complex, which depends on the concentration of the drug–receptor complex as well as other forces. The biological activity of a drug is related to its affinity for the receptor, which is measured by its $K_D$, the dissociation constant at equilibrium [Eq. (3.2)]. Note that $K_D$ is a *dissociation* constant, so that the smaller the $K_D$, the larger the concentration of the drug–receptor complex, and the greater is the affinity of the drug for the receptor.

$$\text{Drug} + \text{receptor} \underset{k_{off}}{\overset{k_{on}}{\rightleftharpoons}} \text{drug–receptor complex} \quad (3.1)$$

$$K_D = \frac{[\text{drug}][\text{receptor}]}{[\text{drug–receptor complex}]} \quad (3.2)$$

### *B. Forces Involved in the Drug–Receptor Complex*

The forces involved in the drug–receptor complex are the same forces experienced by all interacting organic molecules and include covalent bonding, ionic (electrostatic) interactions, ion–dipole and dipole–dipole interactions, hydrogen bonding, charge-transfer interactions, hydrophobic interactions, and van der Waals interactions. Weak interactions usually are possible only when molecular surfaces are close and complementary, that is, bond strength is distance dependent. The spontaneous formation of a bond between atoms occurs with a decrease in free energy, that is, $\Delta G$ is negative. The change in free energy is related to the binding equilibrium constant ($K_{eq}$) by Eq. (3.3). Therefore, at physiological temperature (37°C) changes in free energy of $-2$ to $-3$ kcal/mol can have a major effect on the establishment of good secondary interactions. In fact, a decrease in $\Delta G°$ of $-2.7$ kcal/mol changes the binding equilibrium constant from 1 to 100. If the $K_{eq}$ were only 0.01 (i.e., 1% of the equilibrium mixture in the form of the drug–receptor complex), then a $\Delta G°$ of interaction of $-5.45$ kcal/mol would shift the binding equilibrium constant to 100 (i.e., 99% in the form of the drug–receptor complex).

$$\Delta G° = -RT \ln K_{eq} \quad (3.3)$$

In general, the bonds formed between a drug and a receptor are weak noncovalent interactions; consequently, the effects produced are reversible. Because of this, a drug becomes inactive as soon as its concentration in the extracellular fluids decreases. Often it is desirable for the drug effect to last only a limited time so that the pharmacological action can be terminated. In the case of CNS stimulants and depressants, for example, a prolonged action could be harmful. Sometimes, however, the effect produced by a drug should persist, and even be irreversible. For example, it is most desirable for a *chemotherapeutic agent*, a drug that acts selectively on a foreign organism or tumor cell, to form an irreversible complex with its receptor so that the drug can exert its toxic action for a prolonged period.[6] In this case, a covalent bond would be desirable.

In the following subsections the various types of possible drug–receptor interactions are discussed briefly. These interactions are applicable to all types of receptors, including enzymes and DNA, that are described in this book.

### 1. Covalent Bonds

The *covalent bond* is the strongest bond, generally worth anywhere from −40 to −110 kcal/mol in stability. It is seldom formed by a drug–receptor interaction, except with enzymes and DNA. These bonds will be discussed further in Chapters 5 and 6.

### 2. Ionic (or Electrostatic) Interactions

For protein receptors at *physiological pH* (generally taken to mean pH 7.4), basic groups such as the amino side chains of arginine, lysine, and, to a much lesser extent, histidine are protonated and, therefore, provide a cationic environment. Acidic groups, such as the carboxylic acid side chains of aspartic acid and glutamic acid, are deprotonated to give anionic groups.

Drug and receptor groups will be mutually attracted provided they have opposite charges. This *ionic interaction* can be effective at distances farther than those required for other types of interactions, and they can persist longer. A simple ionic interaction can provide a $\Delta G^\circ = -5$ kcal/mol which declines by the square of the distance between the charges. If this interaction is reinforced by other simultaneous interactions, the ionic interaction becomes stronger ($\Delta G^\circ = -10$ kcal/mol) and persists longer. Acetylcholine is used as an example of a molecule that can undergo an ionic interaction (Fig. 3.1).

### 3. Ion–Dipole and Dipole–Dipole Interactions

As a result of the greater electronegativity of atoms such as oxygen, nitrogen, sulfur, and halogens relative to that of carbon, C—X bonds in drugs and receptors, where X is an electronegative atom, will have an asymmetric distribution of electrons; this produces electronic dipoles. The dipoles in a drug molecule can be attracted by ions (*ion–dipole interaction*) or by other dipoles (*dipole–dipole interaction*) in the receptor, provided charges of opposite sign are properly aligned. Since the charge of a dipole is less than that of an ion, a

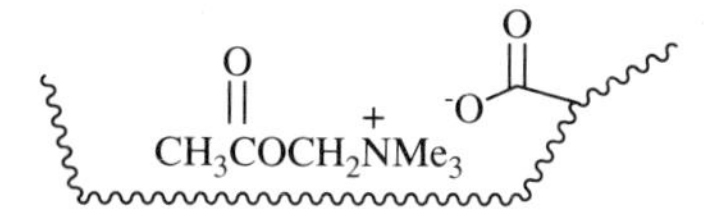


**Figure 3.1.** Example of a simple ionic interaction. The wavy line represents the receptor surface.

dipole–dipole interaction is weaker than an ion–dipole interaction. In Fig. 3.2 acetylcholine is used to demonstrate these interactions, which can provide a $\Delta G°$ of $-1$ to $-7$ kcal/mol.

### 4. Hydrogen Bonds

*Hydrogen bonds* are a type of dipole–dipole interaction formed between the proton of a group X—H, where X is an electronegative atom, and other electronegative atoms (Y) containing a pair of nonbonded electrons. The only significant hydrogen bonds occur in molecules where X and Y are N, O, or F. X removes electron density from the hydrogen so it has a partial positive charge, which is strongly attracted to nonbonded electrons of Y. The interaction is denoted as a dotted line, —X—H····Y—, to indicate that a covalent bond between X and H still exists, but that an interaction between H and Y also occurs. When X and Y are equivalent in electronegativity and degree of ionization, the proton can be shared equally between the two groups, that is, —X····H····Y—.

The hydrogen bond is unique to hydrogen because it is the only atom that can carry a positive charge at physiological pH while remaining covalently bonded in a molecule, and hydrogen also is small enough to allow close approach of a second electronegative atom. The strength of the hydrogen bond is related to the Hammett $\sigma$ constants.[7]

There are *intramolecular* and *intermolecular* hydrogen bonds; the former are stronger (see Fig. 3.3). Hydrogen bonding can be quite important for biological activity. For example, methyl salicylate (**3.3**), an active ingredient

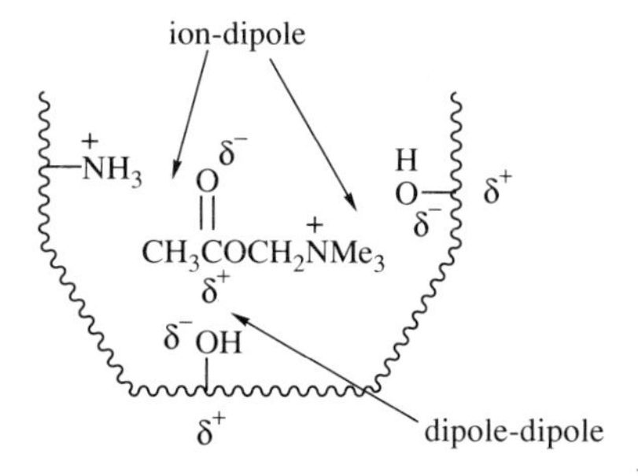


**Figure 3.2.** Examples of ion–dipole and dipole–dipole interactions. The wavy line represents the receptor surface.

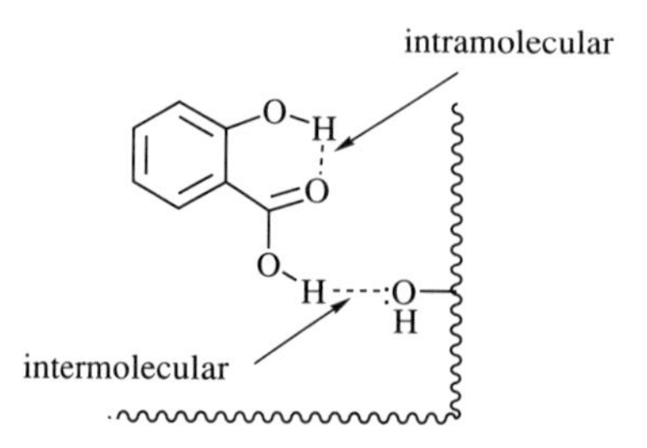


**Figure 3.3.** Examples of hydrogen bonds. The wavy lines represents the receptor surface.

in many muscle pain remedies and at least one antiseptic, is a weak antibacterial agent. The corresponding para isomer, methyl *p*-hydroxybenzoate (**3.4**), however, is considerably more active as an antibacterial agent and is used as a food preservative. It is believed that the antibacterial activity of **3.4** is derived from the phenolic hydroxyl group. In **3.3** this group is masked by intramolecular hydrogen bonding.[8]

**3.3** **3.4**

Hydrogen bonds are essential in maintaining the structural integrity of $\alpha$-helix and $\beta$-sheet conformations of peptides and proteins (**3.5**)[1] and the double helix of DNA (**3.6**).[2] As discussed in Chapter 6, many antitumor agents act by intercalation into the DNA base pairs or by alkylation of the DNA bases, thereby preventing hydrogen bonding. This disrupts the double helix and destroys the DNA.

Another instance where hydrogen bonding is suggested to be important arises when the potency of various oxygen-containing drugs becomes reduced by substitution of a sulfur atom for the oxygen atom in the drug. Sulfur, which is very poor at hydrogen bonding relative to oxygen, presumably cannot interact with the receptor group that hydrogen bonds to the oxygen, and drug–receptor complex stability becomes diminished.

The $\Delta G°$ for hydrogen bonding can be between $-1$ and $-7$ kcal/mol but usually is in the range of $-3$ to $-5$ kcal/mol.

[1] From B. Alberts, D. Bray, J. Lewis, M. Raff, K. Roberts, and J. D. Watson, "Molecular Biology of the Cell," 2nd Ed., pp. 110 and 109, respectively, Garland Publishing, New York, 1989, with permission. Copyright © 1989 Garland Publishing.

[2] From B. Alberts, D. Bray, J. Lewis, M. Raff, K. Roberts, and J. D. Watson, "Molecular Biology of the Cell," 2nd Ed., p. 99. Garland Publishing, New York, 1989, with permission. Copyright © 1989 Garland Publishing.

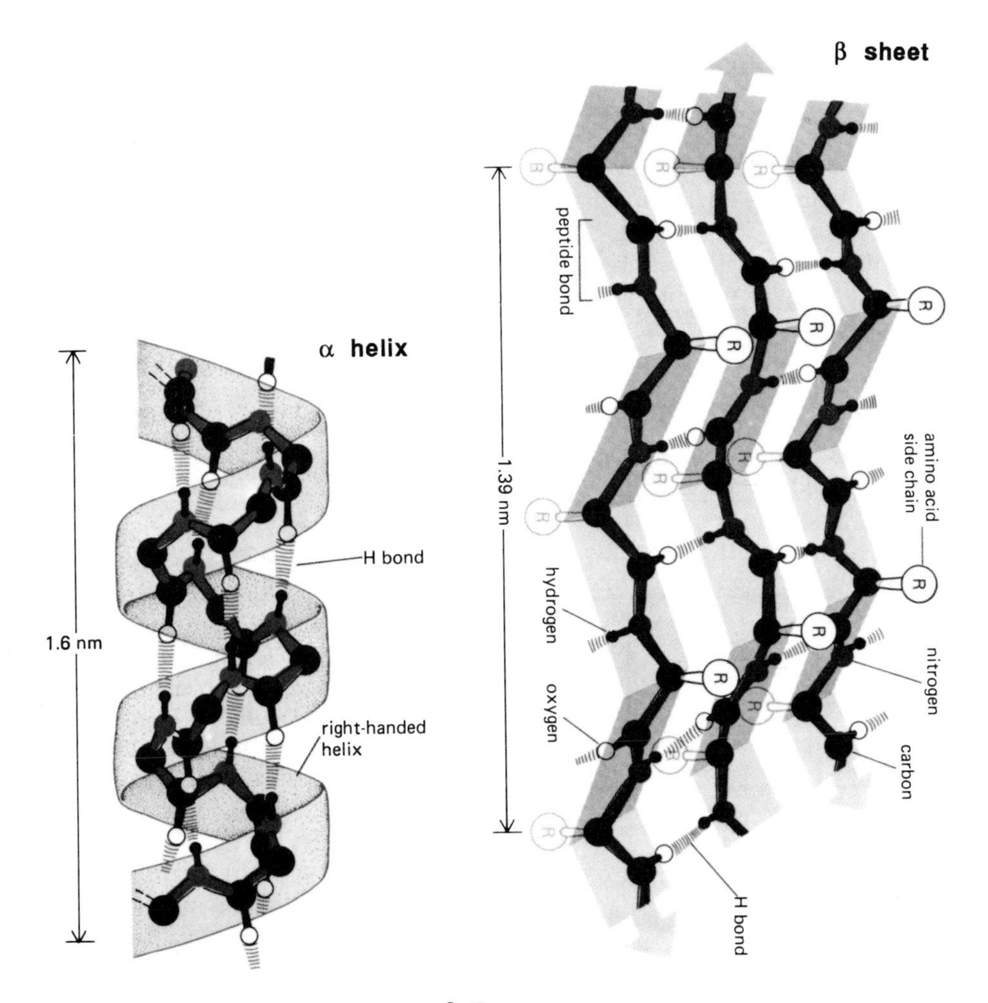
α helix
1.6 nm
H bond
right-handed
helix
β sheet
1.39 nm
peptide bond
amino acid
side chain
hydrogen
nitrogen
oxygen
carbon
H bond
R

**3.5**

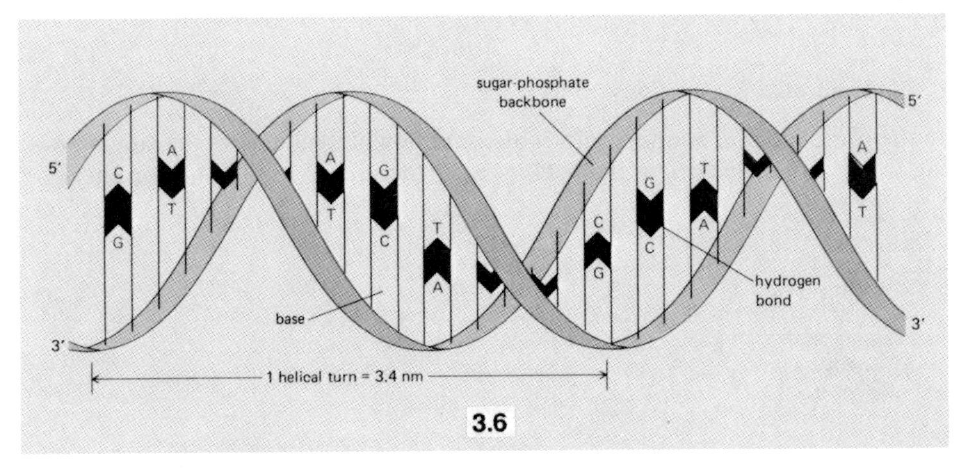
sugar-phosphate
backbone
5′
5′
3′
3′
C
G
A
T
A
T
G
C
T
A
C
G
G
C
T
A
A
T
base
hydrogen
bond
1 helical turn = 3.4 nm

**3.6**

## 5. Charge-Transfer Complexes

When a molecule (or group) that is a good electron donor comes into contact with a molecule (or group) that is a good electron acceptor, the donor may transfer some of its charge to the acceptor. This forms a *charge-transfer complex*, which, in effect, is a molecular dipole–dipole interaction. The potential energy of this interaction is proportional to the difference between the ionization potential of the donor and the electron affinity of the acceptor.

Electron donors contain $\pi$-electrons, for example, alkenes, alkynes, and aromatic moieties with electron-donating substituents, or groups that have a pair of nonbonded electrons, such as oxygen, nitrogen, and sulfur moieties. Acceptor groups contain electron-deficient $\pi$ orbitals, for example, alkenes, alkynes, and aromatic moieties having electron-withdrawing substituents, or weakly acidic protons. There are groups on receptors that can act as electron donors, such as the aromatic ring of tyrosine or the carboxylate group of aspartate, as electron acceptors, such as cysteine, and electron donors and acceptors, such as histidine, tryptophan, and asparagine.

Charge-transfer interactions are believed to provide the energy for intercalation of certain planar aromatic antimalarial drugs, such as chloroquine (**3.7**), into parasitic DNA (see Chapter 6). The fungicide, chlorothalonil, is shown in Fig. 3.4 as a hypothetical example for a charge-transfer interaction with a tyrosine.

Cl N

$NH{-}CH(CH_2)_3N(C_2H_5)_2$

$CH_3$

**3.7**

The $\Delta G°$ for charge-transfer interactions also can range from $-1$ to $-7$ kcal/mol.

## 6. Hydrophobic Interactions

In the presence of a nonpolar molecule or region of a molecule, the surrounding water molecules orient themselves and, therefore, are in a higher energy

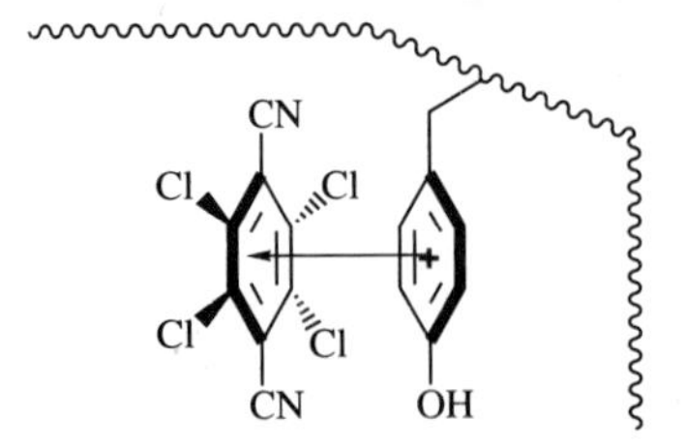


**Figure 3.4.** Example of a charge-transfer interaction. The wavy line is the receptor surface.

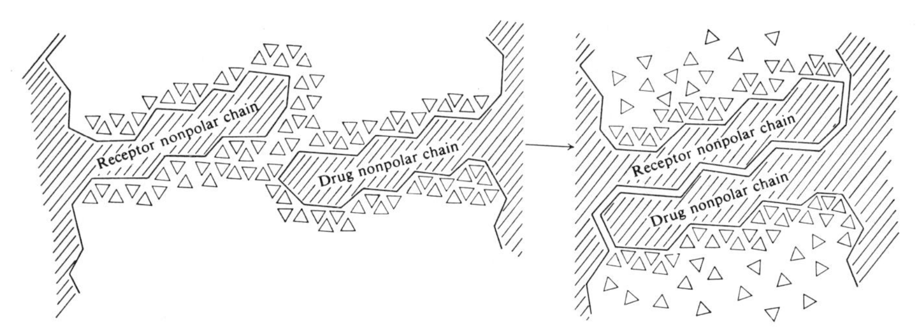


**Figure 3.5.** Formation of hydrophobic interactions. (Reprinted with permission of John Wiley & Sons, Inc. from Korolkovas, A. 1970. "Essentials of Molecular Pharmacology," p. 172. Wiley, New York. Copyright © 1970. John Wiley & Sons, Inc. and by permission of Kopple, K. D. 1966. "Peptides and Amino Acids." Addison-Wesley, Reading, Massachusetts.)

state than when only other water molecules are around. When two nonpolar groups, such as a lipophilic group on a drug and a nonpolar receptor group, each surrounded by ordered water molecules, approach each other, these water molecules become disordered in an attempt to associate with each other. This increase in entropy, therefore, results in a decrease in the free energy that stabilizes the drug–receptor complex. This stabilization is known as a *hydrophobic interaction* (see Fig. 3.5). Consequently, this is not an attractive force of two nonpolar groups "dissolving" in one another but, rather, is the decreased free energy of the nonpolar group because of the increased entropy of the surrounding water molecules. Jencks[9] has suggested that hydrophobic forces may be the most important single factor responsible for noncovalent intermolecular interactions in aqueous solution. Hildebrand,[10] on the other hand, is convinced that hydrophobic effects do not exist. Every methylene-methylene interaction (which actually may be a van der Waals interaction; see Section III,B,7) liberates 0.7 kcal/mol of free energy. In Fig. 3.6 the topical anesthetic butamben is depicted in a hypothetical hydrophobic interaction with an isoleucine group.

### 7. Van der Waals or London Dispersion Forces

Atoms in nonpolar molecules may have a temporary nonsymmetrical distribution of electron density which results in the generation of a temporary dipole.

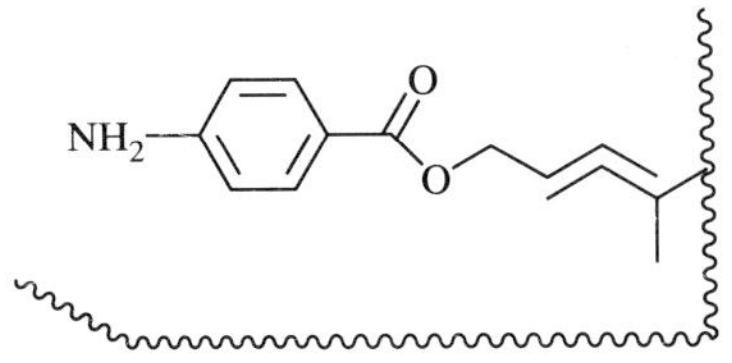

**Figure 3.6.** Example of hydrophobic interactions. The wavy line represents the receptor surface.

As atoms from different molecules (such as a drug and a receptor) approach each other, the temporary dipoles of one molecule induce opposite dipoles in the approaching molecule. Consequently, an intermolecular attraction, known as *van der Waals forces*, results. These weak universal forces only become significant when there is a close surface contact of the atoms; however, when there is molecular complementarity, numerous atomic interactions (each contributing about $-0.5$ kcal/mol to the $\Delta G^\circ$) result, which can add up to a significant overall drug–receptor binding component.

### 8. Conclusion

Since noncovalent interactions are generally weak, cooperativity by several types of interactions is critical. To a first approximation, enthalpy terms will be additive. Once the first interaction has taken place, translational entropy is lost. This results in a much lower entropy loss in the formation of the second interaction. The effect of this cooperativity is that several rather weak interactions may combine to produce a strong interaction. Since several different types of interactions are involved, selectivity in drug–receptor interactions can result. In Fig. 3.7 the local anesthetic dibucaine is used as an example to show the variety of interactions that are possible.

## C. Ionization

At physiological pH (pH 7.4), even mildly acidic groups, such as carboxylic acid groups, will be essentially completely in the carboxylate anionic form; phenolic hydroxyl groups may be partially ionized. Likewise, basic groups, such as amines, will be partially or completely protonated to give the cationic form. The ionization state of a drug will have a profound effect not only on its drug–receptor interaction, but also on its partition coefficient (log $P$; see Section II,E,2,b of Chapter 2).

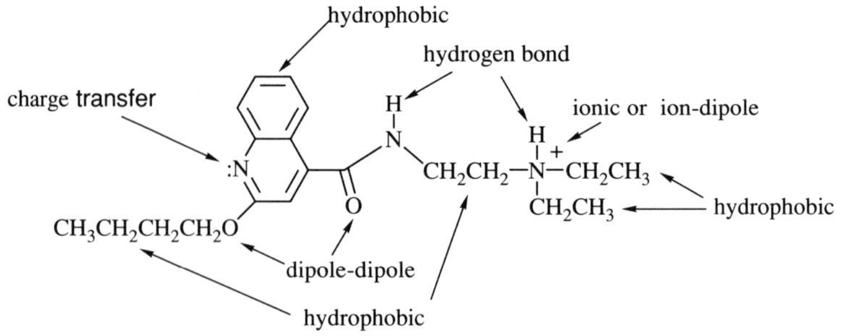

**Figure 3.7.** Examples of potential multiple drug–receptor interactions. The van der Waals interactions are excluded.

The importance of ionization was recognized in 1924 when Stearn and Stearn[11] suggested that the antibacterial activity of stabilized triphenylmethane cationic dyes was related to an interaction of the cation with some anionic group in the bacterium. Increasing the pH of the medium also increased the antibacterial effect, presumably by increasing the ionization of the receptors in the bacterium. Albert and co-workers[12] made the first rigorous proof that a correlation between ionization and biological activity existed. A series of 101 aminoacridines, including the antibacterial drug, 9-aminoacridine or aminacrine (**3.8**), all having a variety of $pK_a$ values, was tested against 22 species of bacteria. A direct correlation was observed between ionization (formation of the cation) of the aminoacridines and antibacterial activity. However, at lower pH values, protons can compete with these cations for the receptor, and antibacterial activity is diminished. When this was realized, Albert[13] notes, the Australian Army during World War II was advised to pretreat wounds with sodium bicarbonate to neutralize any acidity prior to treatment with aminacrine. This, apparently, was quite effective in increasing the potency of the drug. The mechanism of action of aminoacridines is discussed in Chapter 6.

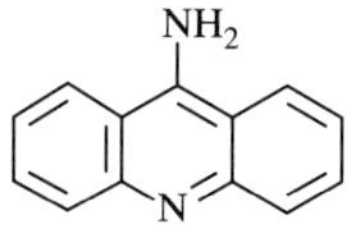

$NH_2$

N

**3.8**

The great majority of alkaloids which act as neuroleptics, local anesthetics, and barbiturates have $pK_a$ values between 6 and 8; consequently both neutral and cationic forms are present at physiological pH.[13] This may allow them to penetrate membranes in the neutral form and exert their biological action in the ionic form. Antihistamines and antidepressants tend to have $pK_a$ values of about 9. The uricosuric (increases urinary excretion of uric acid) drug phenylbutazone [**3.9**, R = $(CH_2)_3CH_3$] has a $pK_a$ of 4.5 and is active as the anion (the OH proton is acidic). However, since the pH of urine is 4.8 or higher, suboptimal concentrations of the anion were found in the urinary system. Sulfinpyrazone (**3.9**, R = $CH_2CH_2SOPh$) has a lower $pK_a$ of 2.8 and is about 20 times more potent than phenylbutazone; the anionic form blocks reabsorption of uric acid by renal tubule cells.[14]

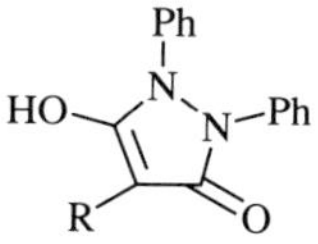

Ph

HO N N–Ph

R O

**3.9**

The antimalarial drug pyrimethamine (**3.10**) has a p$K_a$ of 7.2 and is best absorbed from solutions of sufficient alkalinity that it has a high proportion of molecules in the neutral form (to cross membranes). Its mode of action, the inhibition of the parasitic enzyme dihydrofolate reductase, however, requires that it be in the protonated cationic form.

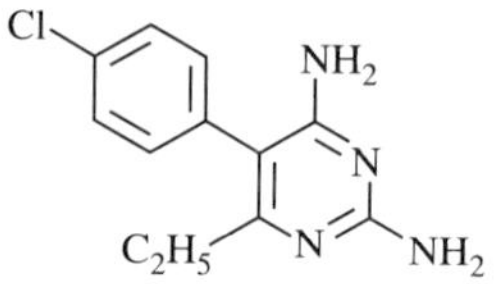

Cl $NH_2$ N $C_2H_5$ N $NH_2$

**3.10**

Similarly, there are drugs such as the anti-inflammatory agent indomethacin (**2.18**) and the antibacterial agent sulfamethoxazole (**3.11**) whose pharmacokinetics (migration to site of action) depend on their nonionized form, but whose pharmacodynamics (interaction with the receptor) depend on the anionic form (carboxylate and sulfonamido ions, respectively). In a cell-free system the antibacterial activity of **3.11** and other sulfonamides was directly proportional to the degree of ionization, but in intact cells, where the drug must cross a membrane to get to the site of action, the antibacterial activity also was dependent on lipophilicity (the neutral form).[15]

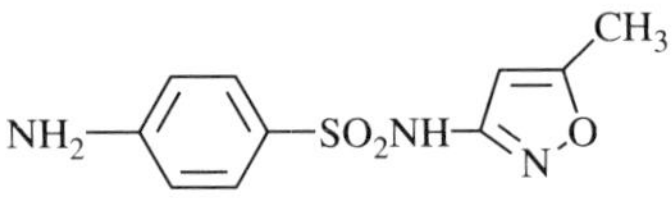

$CH_3$ $NH_2$ $SO_2NH$ O N

**3.11**

Up to this point only the ionization of the drug has been considered. As indicated in Section III,B,2, there are a variety of acidic and basic groups on receptors. Anionic groups in DNA include phosphoric acid groups (p$K_a$ 1.5 or 6.5) and purines and pyrimidines (p$K_a$ ~9); anionic groups in proteins are carboxylic acids (aspartic and glutamic acids; p$K_a$ 3.5–5), phenols (tyrosine; p$K_a$ 9.5–11), sulfhydryls (cysteine; p$K_a$ 8.5), and hydroxyls (serine and threonine; p$K_a$ ~13.5). Cationic groups in DNA include amines (adenine and cytidine; p$K_a$ 3.5–4) and in proteins include imidazole (histidine, p$K_a$ 6.5–7), amino (lysine, p$K_a$ ~10), and guanidino (arginine, p$K_a$ ~13) groups. Therefore, the structure and function of a receptor can be strongly dependent on the pH of the medium, especially if an *in vitro* assay is being used. The p$K_a$ values of various groups embedded in a receptor, however, can be quite variable, and will depend on the microenvironment. If a carboxyl group is in a nonpolar region, its p$K_a$ will be raised because the anionic form is destabilized. Glutamate-35 in lysozyme and the lysozyme–glycolchitin complex has a p$K_a$ of 6.5 and 8.2, respectively.[16] If the carboxylate forms a salt bridge, it will be stabilized and its $pK_a$ will be lower. Likewise, an amino group buried in a nonpolar microenvironment will have a lower $pK_a$ because protonation will be disfavored; the ε-amino group of the active site lysine residue in acetoacetate decarboxylase has a $pK_a$ of 5.9.[17] If the ammonium group forms a salt bridge, it will be stabilized, deprotonation will be inhibited, and the $pK_a$ will be raised.

Now that the importance of drug–receptor interactions has been emphasized, we turn our attention to the principal method for the determination of these interactions.

### D. Determination of Drug–Receptor Interactions

Hormones and neurotransmitters are important natural compounds that are responsible for the regulation of a myriad of physiological functions. These molecules interact with a specific receptor in a tissue and elicit a specific characteristic response. For example, the activation of a muscle by the central nervous system is mediated by release of the neurotransmitter acetylcholine (ACh; the molecule in Figs. 3.1 and 3.2). If a plot is made of the logarithm of the concentration of the acetylcholine added to a muscle tissue preparation versus the percentage of total muscle contraction, the graph shown in Fig. 3.8 may result. This is known as a *dose–response* or *concentration–response curve*. The low concentration part of the curve results from too few neurotransmitter molecules available for collision with the receptor. As the concentration increases, it reaches a point where a linear relationship is observed between the logarithm of the neurotransmitter concentration and the biological response. As most of the receptors become occupied, the probability of a drug and receptor molecule interacting diminishes, and the curve deviates

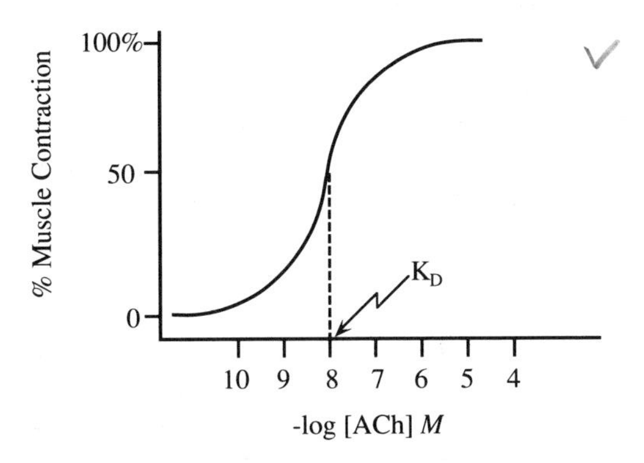


**Figure 3.8.** Effect of increasing the concentration of a neurotransmitter on muscle contraction.

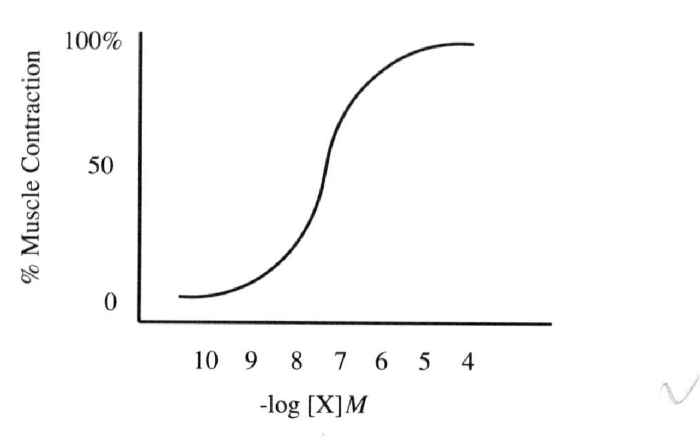


**Figure 3.9.** Dose–response curve for an agonist.

from linearity (the high concentration end). Dose–response curves are a means of measuring drug–receptor interactions and are the standard method for comparing the potencies of various compounds that interact with a particular receptor. Any measure of a response can be plotted on the ordinate, such as $LD_{50}$, $ED_{50}$, or percentage of a physiological effect.

If another compound (X) is added in increasing amounts to the same tissue preparation and the curve shown in Fig. 3.9 results, the compound, which produces the same maximal response as the neurotransmitter, is called an *agonist*. A second compound (Y) added to the tissue preparation shows no response at all (Fig. 3.10A); however, if it is added to the neurotransmitter, the effect of the neurotransmitter is blocked until a higher concentration of the neurotransmitter is added (Fig. 3.10B). Compound Y is called a *competitive antagonist*. There are two general types of antagonists, competitive antagonists and noncompetitive antagonists. The former, which is the larger category, is one in which the degree of antagonism is dependent on the relative concentrations of the agonist and the antagonist; both bind to the same site on the receptor, or, at least, the antagonist directly interferes with the binding of the agonist. The degree of blocking of a *noncompetitive antagonist* (Y′) is independent of the amount of agonist present; two different binding sites may be involved (Fig. 3.10C). Only competitive antagonists will be discussed further in this text.

If a compound Z is added to the tissue preparation and some response is elicited, but not a full response, regardless of how high the concentration of Z used, then Z is called a *partial agonist* (see Fig. 3.11A). A partial agonist has properties of both an agonist and an antagonist. When Z is added to low concentrations of a neurotransmitter sufficient to give a response less than the maximal response of the partial agonist (e.g., 20% as shown in Fig. 3.11B), additive effects are observed as Z is increased, but the maximum response

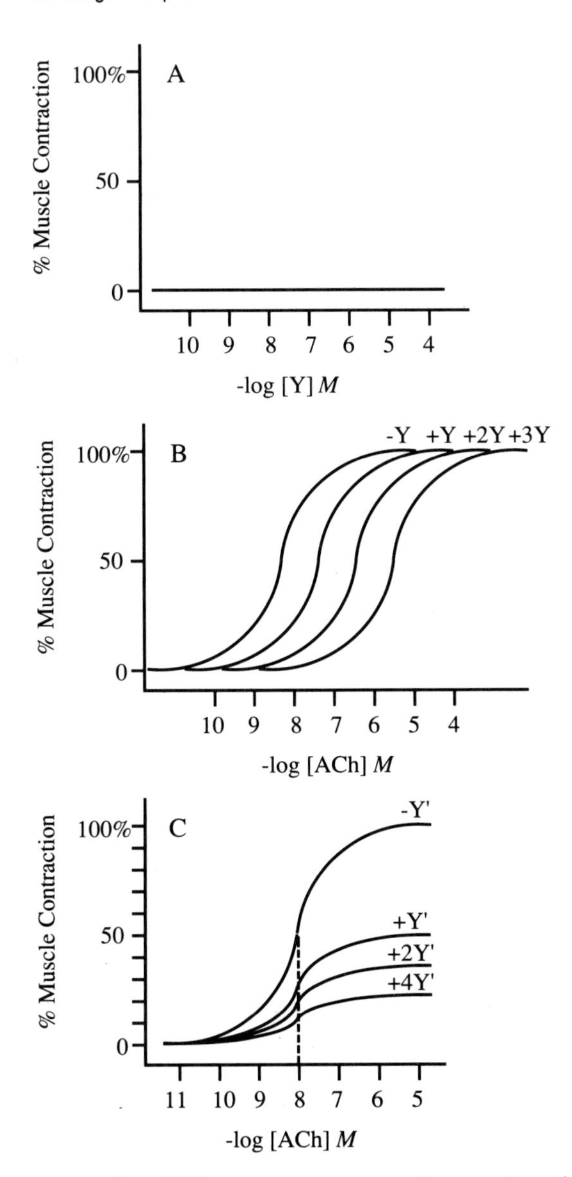


**Figure 3.10.** (A) Dose–response curve for an antagonist; (B) effect of a competitive antagonist (Y) on the response of a neurotransmitter; and (C) effect of a noncompetitive antagonist (Y′) on the response of the neurotransmitter.

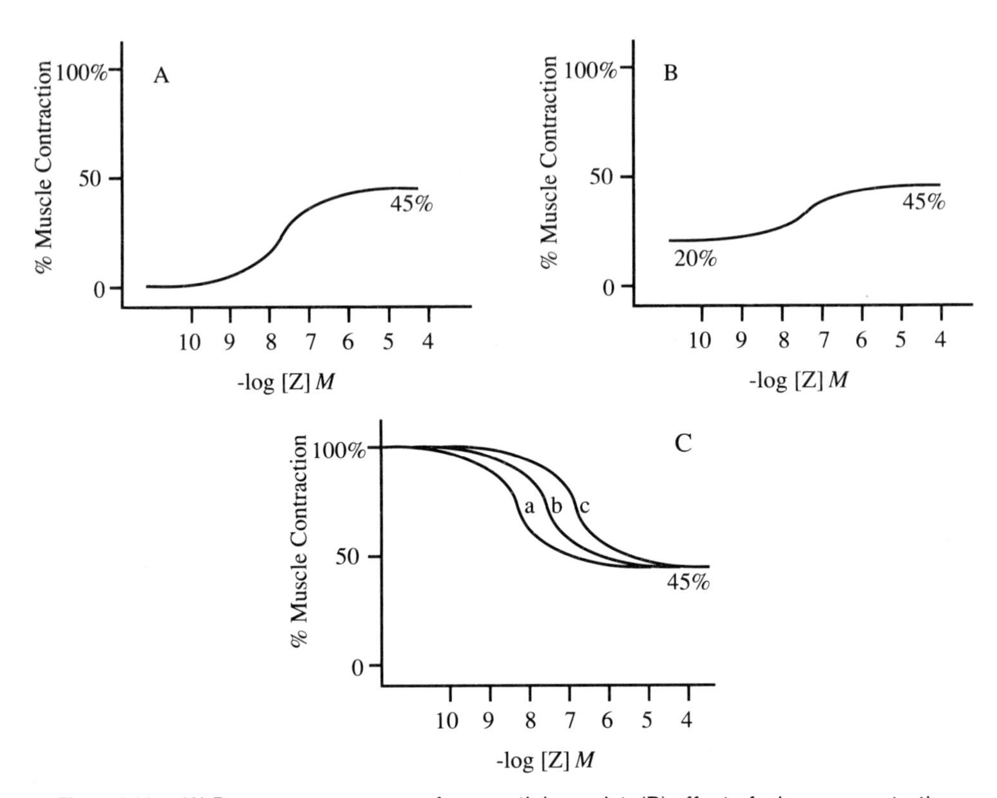


**Figure 3.11.** (A) Dose–response curve for a partial agonist; (B) effect of a low concentration of neurotransmitter on the response of a partial agonist; and (C) effect of a high concentration of neurotransmitter on the response of a partial agonist. In (C) the concentration of the neurotransmitter is c > b > a.

does not exceed that produced by Z alone. Under these conditions, the partial agonist is having an agonistic effect. However, if Z is added to high concentrations of a neurotransmitter sufficient to give full response of the neurotransmitter, then antagonistic effects are observed; as Z increases, the response decreases to the point of maximum response of the partial agonist (Fig. 3.11C). If this same experiment is done starting with higher concentrations of the neurotransmitter, the same results are obtained except that the dose–response curves shift to the right, resembling the situation of adding an antagonist to the neurotransmitter.

On the basis of the above discussion, if you wish to design a drug to effect a certain response, an agonist would be desired; if you wish to design a drug to prevent a particular response of a neurotransmitter or hormone, an antagonist would be required.

In general, there are great structural similarities among a series of agonists, but little structural similarity exists in a series of competitive antagonists. For example, Table 3.1 shows some agonists and antagonists for histamine and epinephrine; a more detailed list of agonists and antagonists for specific receptors has been reported.[18] The differences in the structures of the antagonists are not surprising because a receptor can be blocked by an antagonist simply by its binding to a site near enough to the binding site for the neurotransmitter that it physically blocks the neurotransmitter from reaching its binding site.

**Table 3.1** Agonists and Antagonists

| Neurotransmitter | Agonists | Antagonists |
|---|---|---|
| 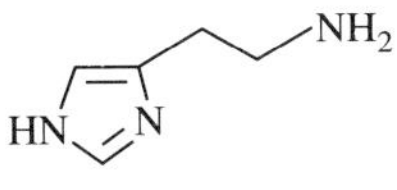 Histamine | 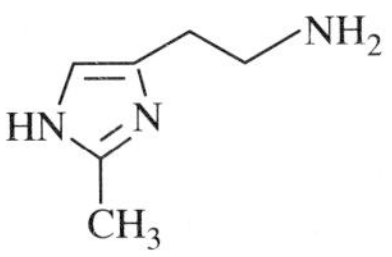 | 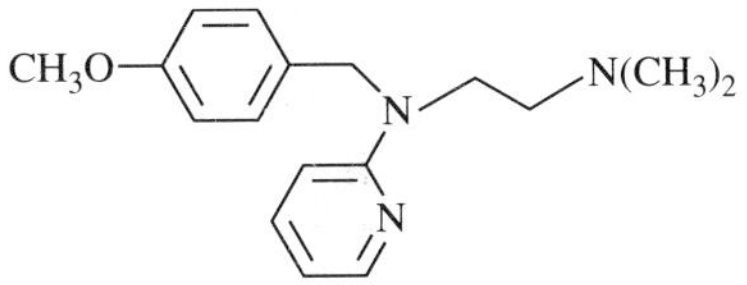 Pyrilamine |
| | 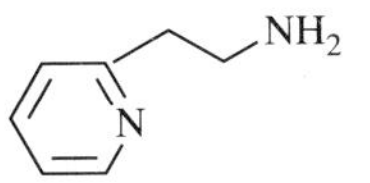 | 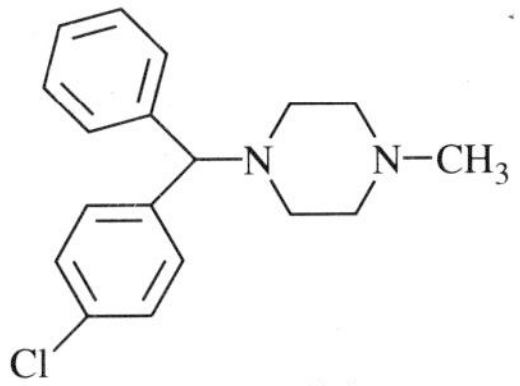 Chlorcyclizine |
| 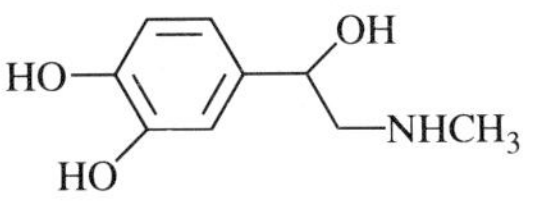 Epinephrine | 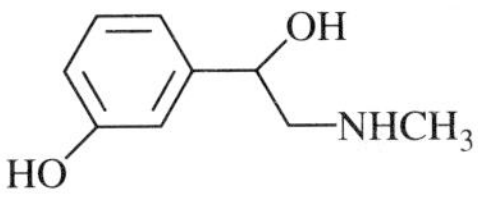 | 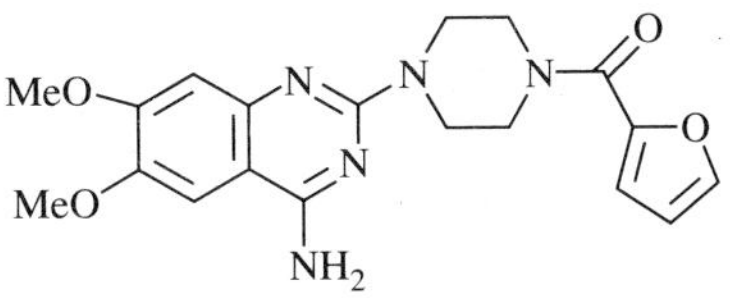 Prazosin |
| | 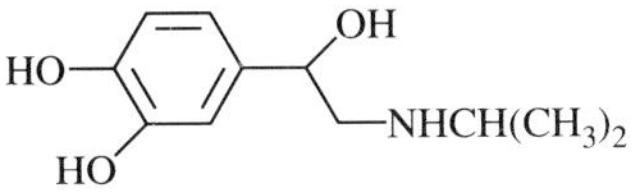 | 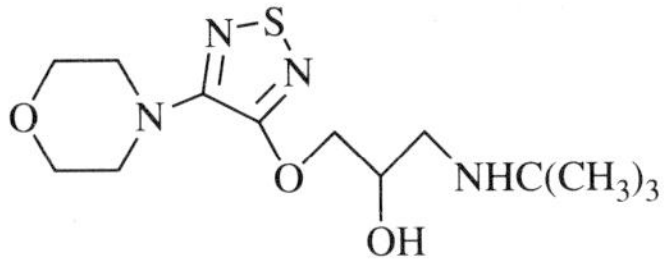 Timolol |

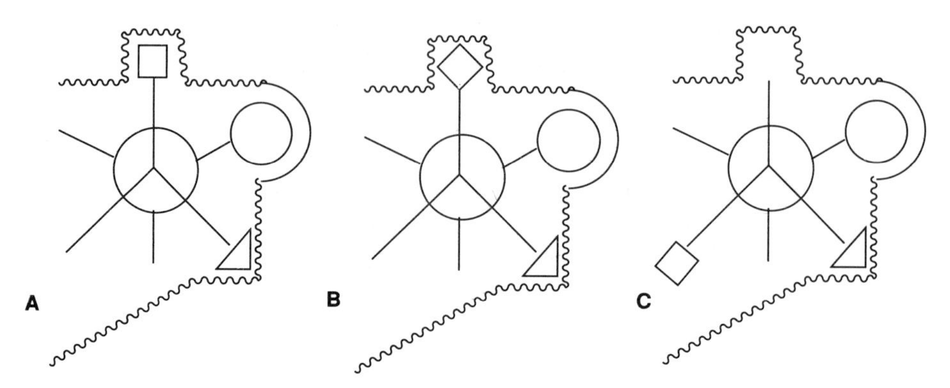


**Figure 3.12.** Inability of an antagonist to elicit a biological response. The wavy line is the receptor surface. (Adapted with permission from W. O. Foye, ed. 1989 "Principles of Medicinal Chemistry," 3rd Ed., p. 63. Copyright © 1989 Lea & Febiger, Philadelphia, Pennsylvania.)

This may explain why antagonists are frequently much more bulky than the corresponding agonists. It is easier to design a molecule that blocks a receptor site than one that interacts with it in the specific way required to elicit a response. An agonist can be transformed into an antagonist by appropriate structural modifications (see Section III,H).

How is it possible for an antagonist to bind to the same site as an agonist and not elicit a biological response? There are several ways that this may occur. Figure 3.12A shows an agonist with appropriate groups interacting with three receptor binding sites and eliciting a response. In Fig. 3.12B the compound has two groups that can interact with the receptor, but one essential group is missing. In the case of optical isomers (Fig. 3.12C), only two groups are able to interact with the proper receptor sites. If appropriate groups must interact with all three binding sites in order for a response to be elicited, then the compounds depicted in Fig. 3.12B and C would be antagonists.

There are two general categories of compounds that interact with receptors: (1) compounds that occur naturally within the body, such as hormones, neurotransmitters, and other agents that modify cellular activity (*autocoids*), and (2) *xenobiotics*, compounds that are foreign to the body. All chemicals naturally occurring in the body are known to act as agonists, but most xenobiotics that interact with receptors are antagonists.

Receptor selectivity is very important but often difficult to attain because receptor structures are generally unknown. Many current drugs are pharmacologically active at multiple receptors, some of which are not associated with the illness that is being treated. This can lead to side effects. For example, the clinical effect of neuroleptics is believed to result from their antagonism of dopamine receptors.[19] In general, this class of drugs also blocks cholinergic and $\alpha$-adrenergic receptors, and this results in side effects such as sedation and hypotension.

### E. Drug–Receptor Theories

Over the years a number of theories have been proposed to account for the ability of a drug to interact with a receptor and elicit a biological response. Several of the more important suggestions are discussed here.

#### 1. Occupancy Theory

The *occupancy theory* of Gaddum[20] and Clark[21] states that the intensity of the pharmacological effect is directly proportional to the number of receptors occupied by the drug. The response ceases when the drug–receptor complex dissociates. However, as discussed in Section III,D, not all agonists produce a maximal response. Therefore, this theory does not rationalize partial agonists.

Ariëns[22] and Stephenson[23] modified the occupancy theory to account for partial agonists, a term coined by Stephenson. These authors utilized the original Langley[3] concept of a receptor that drug–receptor interactions involve two stages: first, there is a complexation of the drug with the receptor, which they both termed the *affinity*; second, there is the initiation of the biological effect which Ariëns termed the *intrinsic activity* and Stephenson called the *efficacy*. Affinity, then, is a measure of the capacity of a drug to bind to the receptor and is dependent on the molecular complementarity of the drug and the receptor. Intrinsic activity ($\alpha$) is a measure of the ability of the drug–receptor complex to initiate the response. In the original theory the latter property was considered to be constant. Examples of affinity and intrinsic activity are given in Fig. 3.13. Figure 3.13A shows the theoretical dose–response curves for five drugs with the same affinity for the receptor ($pK_D$ = 8) but having intrinsic activities varying from 100% of the maximum ($\alpha$ = 1.0)

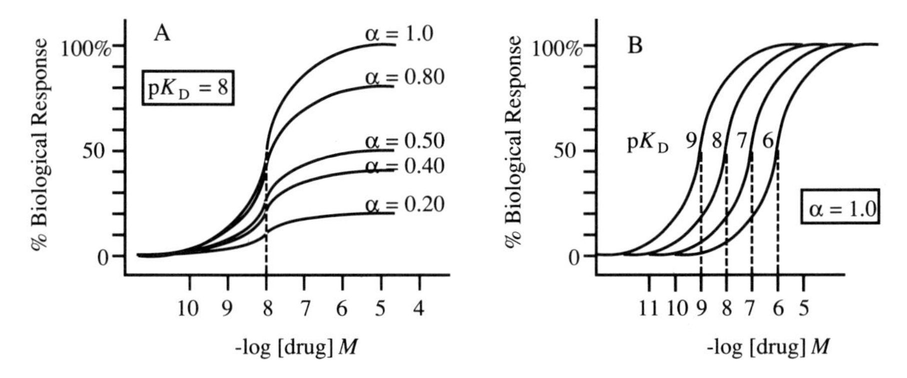


**Figure 3.13.** Theoretical dose–response curves to illustrate (A) drugs with equal affinities and different intrinsic activities and (B) drugs with equal intrinsic activities but different affinities.

to 20% of the maximum ($\alpha = 0.20$). The drug with $\alpha$ equal to 1.0 is a full agonist; the ones with $\alpha$ less than 1.0 are partial agonists. Figure 3.13B shows the dose–response curves for four drugs with the same intrinsic activity ($\alpha = 1.0$) but having different affinities varying from a $pK_D$ of 9 to 6.

In general, antagonists bind tightly to a receptor (great affinity) but are devoid of activity (no efficacy). Potent agonists may have less affinity for their receptors than partial agonists or antagonists. The modified occupancy theory accounts for the existence of partial agonists and antagonists, but it does not account for why two drugs that can occupy the same receptor can act differently, namely, one as an agonist, the other as an antagonist.

### 2. Rate Theory

As an alternative to the occupancy theory, Paton[24] proposed that the activation of receptors is proportional to the total number of encounters of the drug with its receptor per unit time. Therefore, the *rate theory* suggests that the pharmacological activity is a function of the rate of association and dissociation of the drug with the receptor, and not the number of occupied receptors. Each association would produce a quantum of stimulus. In the case of agonists, the rates of both association and dissociation would be fast (the latter faster than the former). The rate of association of an antagonist with a receptor would be fast, but the dissociation would be slow. Partial agonists would have intermediate drug–receptor complex dissociation rates. At equilibrium, the occupancy and rate theories are mathematically equivalent. As in the case of the occupancy theory, the rate theory does not rationalize why the different types of compounds exhibit the characteristics that they do.

### 3. Induced-Fit Theory

The *induced-fit theory* of Koshland[25a–c] was originally proposed for the action of substrates and enzymes, but it could apply to drug–receptor interactions as well. According to this theory the receptor (enzyme) need not necessarily exist in the appropriate conformation required to bind the drug (substrate). As the drug (substrate) approaches the receptor (enzyme), a conformational change is induced which orients the essential binding (catalytic) sites (Fig. 3.14). The conformational change in the receptor could be responsible for the initiation of the biological response. The receptor (enzyme) was suggested to be elastic, and it could return to its original conformation after the drug (substrate) was released. The conformational change need not occur only in the receptor (enzyme); the drug (substrate) also could undergo deformation, even if this resulted in strain in the drug (substrate).

According to the induced-fit theory, an agonist would induce a conformational change and elicit a response, but an antagonist would bind without a

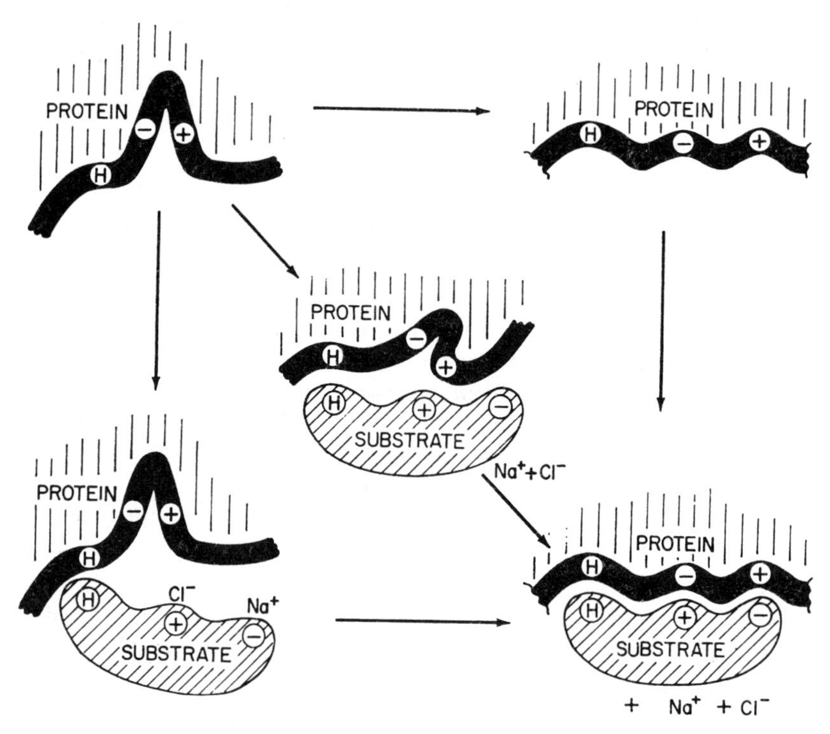


**Figure 3.14.** Schematic of the induced-fit theory. [Reproduced with permission from Koshland, Jr., D. E., and Neet K. E., Annual Review of Biochemistry, Vol. 37, © 1968 by Annual Reviews, Inc.]

conformational change. This theory also can be adapted to the rate theory. An agonist would induce a conformational change in the receptor, resulting in a conformation to which the agonist binds less tightly and from which it can dissociate more easily. If drug–receptor complexation does not cause a conformational change in the receptor, then the drug–receptor complex will be stable, and an antagonist will result. Two other theories evolved from the induced-fit theory, namely, the macromolecular perturbation theory and the activation–aggregation theory.

## 4. Macromolecular Perturbation Theory

Having considered the conformational flexibility of receptors, Belleau[26] suggested that in the interaction of a drug with a receptor two general types of *macromolecular perturbations* could result: *specific conformational perturbation* makes possible the binding of certain molecules that produce a biological response (agonist); *nonspecific conformational perturbation* accommodates other types of molecules that do not elicit a response (antagonist). If the drug contributes to both macromolecular perturbations, a mixture of two complexes will result (partial agonist). This theory offers a physicochemical basis for the rationalization of molecular phenomena that involve receptors.

### 5. Activation–Aggregation Theory

An extension of the macromolecular perturbation theory (which is based on the induced-fit theory) is the *activation–aggregation theory* of Changeux and co-workers[27] and Karlin.[28] According to this theory, even in the absence of drugs, a receptor is in a state of dynamic equilibrium between an activated form ($R_0$), which is responsible for the biological response, and an inactive form ($T_0$). Agonists shift the equilibrium to the activated form, antagonists bind to the inactive form, and partial agonists bind to both conformations. In this model the agonist binding site in the $R_0$ conformation can be different from the antagonist binding site in the $T_0$ conformation. If there are two different binding sites and conformations, then this could account for the structural differences in these classes of compounds and could rationalize why an agonist elicits a biological response but an antagonist does not. This theory can explain the ability of partial agonists to possess both the agonistic and antagonistic properties as depicted in Fig. 3.11. In Fig. 3.11B as the partial agonist interacts with the remaining unoccupied receptors, there is an increase in the response up to the maximal response for the partial agonist interaction. In Fig. 3.11C the partial agonist competes with the neurotransmitter for the receptor sites. As the partial agonist displaces the neurotransmitter, it changes the amount of $R_0$ and $T_0$ receptor forms ($T_0$ increases and, therefore, the response decreases) until all of the receptors have the partial agonist bound.

It is generally accepted in the field of enzymology that conformational changes are quite important to enzyme function. Although noncatalytic receptors are far less characterized, it is reasonable to extrapolate what is known about enzymes to all types of receptors and to assume an important role for conformational changes in drug–receptor interactions in general.

## F. Topographical and Stereochemical Considerations

Up to this point in our discussion of drug–receptor interactions we have been concerned with what stabilizes a drug–receptor complex, how drug–receptor interactions are measured, and possible ways that the drug–receptor complex may form. In this section we turn our attention to molecular aspects and examine the topography and stereochemistry of drug–receptor complexes.

### 1. Spatial Arrangement of Atoms

It was indicated in the discussion of bioisosterism (Chapter 2, Section II,D,4) that many antihistamines have a common structural feature (Fig. 3.15).[29] In Fig. 3.15 $Ar^1$ is aryl, such as phenyl, substituted phenyl, or heteroaryl (2-pyridyl or thienyl); $Ar^2$ is aryl or arylmethyl. The two aryl groups also can be connected through a bridge (as in phenothiazines, **2.34**), and the $CH_2CH_2N$ moiety can be part of another ring (as in chlorcyclizine, Table 3.1). X is CH—O—, N—, or CH—; C—C is a short carbon chain (2 or 3 atoms) which may be saturated, branched, contain a double bond, or be part of a ring system. These compounds are called antihistamines because they are antagonists of a histamine receptor known as the *$H_1$ receptor*. When a sensitized person is exposed to an allergen, an antibody is produced, an antigen–antibody reaction occurs, and histamine is released. Histamine binding to the $H_1$ receptor can cause stimulation of smooth muscle and produce allergic and hypersensitivity reactions such as hay fever, pruritus (itching), contact and atopic dermatitis, drug rashes, urticaria (edematous patches of skin), and anaphylactic shock. Antihistamines are used widely to treat these symptoms. Unlike histamine (see Table 3.1 for structure), most $H_1$ blockers contain tertiary amino groups, usually dimethylamino or pyrrolidino. At physiological pH, then, this group will be protonated, and it is believed that an ionic interaction with the receptor is a key binding contributor.

The commonality of structures of antihistamines suggests that there are specific binding sites on the histamine $H_1$ receptor that have an appropriate topography for interaction with certain groups on the antihistamine which are arranged in a similar configuration (see Section III,B). Those parts of the drug molecule that interact with the receptor are known as the *pharmacophore* of the compound; this is the key interaction that is responsible for the biological response. It must be cautioned, however, that although the antihistamines are competitive antagonists of histamine for the $H_1$ receptor, the same set of atoms on the receptor need not interact with both histamine and the antagonists.[30] Consequently, it is difficult to make conclusions regarding the receptor structure on the basis of antihistamine structure–activity relationships. Because of the essentiality of various parts of antihistamine molecules, it is likely that the minimum binding requirements include a negative charge on the receptor to interact with the ammonium cation and hydrophobic (van der

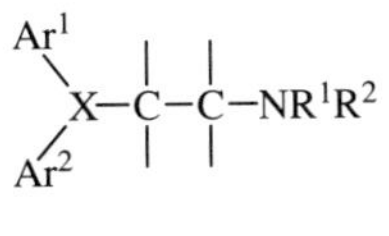

$$(Ar^1)(Ar^2)X-\overset{|}{\underset{|}{C}}-\overset{|}{\underset{|}{C}}-NR^1R^2$$

**Figure 3.15.** General structure of antihistamines.

Waals) interactions with the aryl group. Obviously, many other interactions are involved.

From this very simplistic view of drug–receptor interactions it is not possible to rationalize the fact that enantiomers, that is, mirror image compounds that are identical in all physical and chemical properties except for their effect on the direction of rotation of the plane of polarized light, can have quite different binding properties to receptors. This phenomenon is discussed in more detail in the next section.

## 2. Drug and Receptor Chirality

Histamine is an achiral molecule, and most of the $H_1$ receptor antagonists are achiral molecules as well. However, proteins are polyamino acid macromolecules, and amino acids are chiral molecules (in the case of mammalian proteins, they are all L-isomers); consequently, proteins (receptors) are chiral substances. The two complexes formed between a receptor and two enantiomers are diastereomers and, as a result, have different energies and chemical properties. This suggests that dissociation constants for drug–receptor complexes of enantiomeric drugs may differ, and may even involve different binding sites. Even though histamine is achiral, the chiral antihistamine dexchlorpheniramine (**3.12**) is highly *stereoselective* (one stereoisomer is more active than the other); the (*S*)-(+)-isomer is about 200 times more potent than the (*R*)-(−)-isomer.[31] According to the nomenclature of Ariëns,[32a,b] when there is isomeric stereoselectivity, the more active isomer is termed the *eutomer*; the less active isomer is the *distomer*. The ratio of the potencies (or affinities) of enantiomers is termed the *eudismic ratio*.

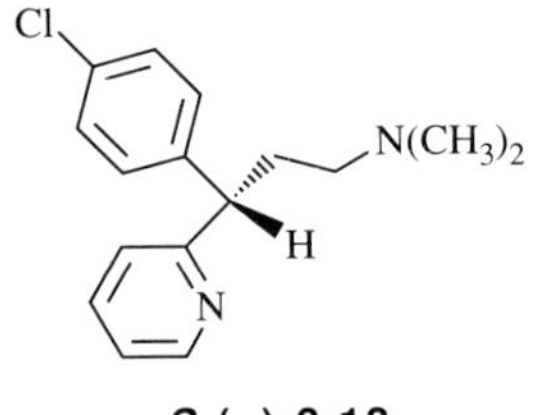

Cl

$N(CH_3)_2$

H

N

**S-(+)-3.12**

High-potency antagonists are those having a high degree of complementarity with the receptor. When the antagonist contains an asymmetric center in the pharmacophore, a high eudismic ratio is usually observed for the stereoisomers because the receptor complementarity would not be retained for the distomer. This increase in eudismic ratio with an increase in potency of the eutomer is *Pfeiffer's rule*.[32b,33] Small eudismic ratios are observed when the eutomer has low affinity for the receptor (poor molecular complementarity) or, in the case of chiral compounds, when the center of asymmetry lies outside of the region critically involved in receptor binding, that is, the pharmacophore.

The distomer actually should be considered as an impurity in the mixture or, in the terminology of Ariëns,[32a] the *isomeric ballast*. It, however, may contribute to undesirable side effects and toxicity; in that case, the distomer for the biological activity may be the eutomer for the side effects. For example, *d*-ketamine (**3.13**; the asterisk marks the chiral carbon) is a hypnotic and analgetic agent; the *l*-isomer is responsible for the undesired side effects[34] [note that *d* is synonymous with (+) and *l* is synonymous with (−)]. It also is possible that both isomers are biologically active, but only one contributes to the toxicity, such as the local anesthetic prilocaine (**3.14**).[35]

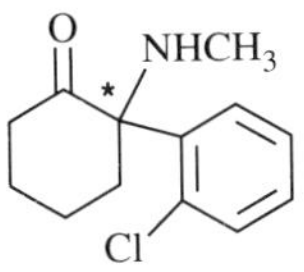

**3.13** 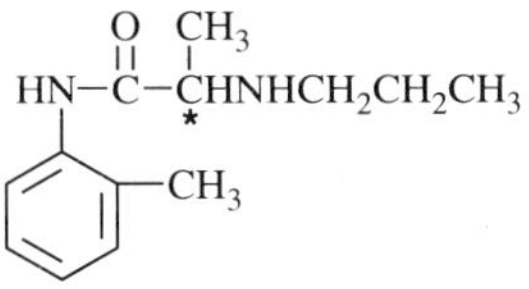 **3.14**

In some cases it is desirable to have both isomers present.[32] Both isomers of bupivacaine (**3.15**) are local anesthetics, but only the *l*-isomer shows vasoconstrictive activity.[36] The experimental diuretic (increases water excretion) drug indacrinone (**3.16**) has a uric acid retention side effect. The *d*-isomer of **3.16** is responsible (i.e., the eutomer) for both the diuretic activity and the side effect. Interestingly, however, the *l*-isomer acts as a uricosuric agent (reduces uric acid levels). Unfortunately, the ratio that gives the optimal therapeutic index (see Chapter 2, Section II,D) is 1*d*:8*l*, not 1:1 as is present in the racemic mixture.[37]

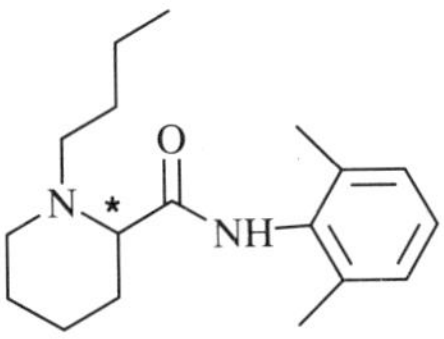

**3.15** 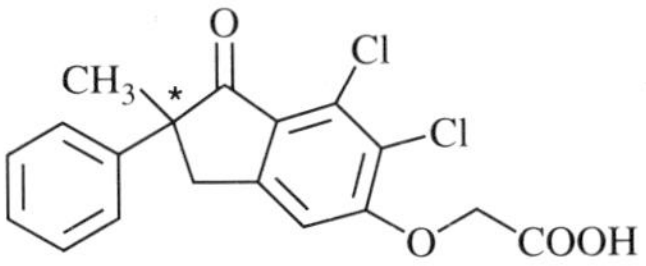 **3.16**

Enantiomers may have different therapeutic activities as well.[38] Darvon® (**3.17**), (2*S*,3*R*)[38a]-(+)-dextropropoxyphene, is an analgetic drug, and its enantiomer Novrad® (**3.18**), (−)-levopropoxyphene, is an antitussive (anticough) agent, an activity that is not compatible with analgetic action. Consequently, these enantiomers are marketed separately. You may have noticed that the trade names are enantiomeric as well!

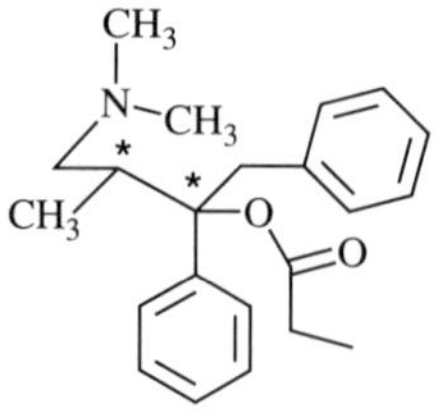

**3.17** 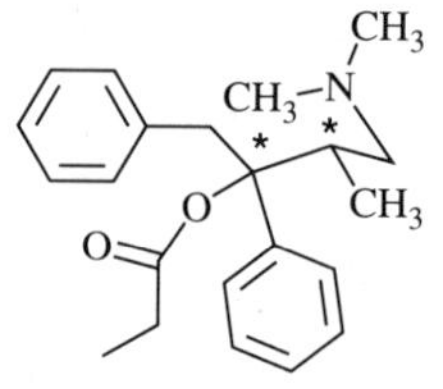 **3.18**

It, also, is possible for the enantiomers to have opposite effects. The *l*-isomers of some barbiturates exhibit depressant activity and the *d*-isomers have convulsant activity; the *l*-isomers can antagonize the *d*-isomers.[39] The *d*-isomer of the experimental narcotic analgetic picenadol (**3.19**) is an opiate agonist, the *l*-isomer is a narcotic antagonist, and the racemate is a partial agonist.[40]

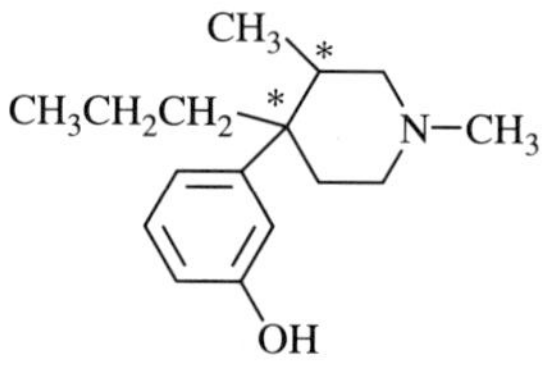

**3.19**

It is quite common for chiral compounds to show stereoselectivity with receptor action, and the stereoselectivity of one compound can vary for different receptors. For example, (+)-butaclamol (**3.20**) is a potent antipsychotic, but the (−)-isomer is essentially inactive; the eudismic ratio (+/−) is 1250 for the $D_2$-dopaminergic, 160 for the $D_1$-dopaminergic, and 73 for the $\alpha$-adrenergic receptors.[32b] (−)-Baclofen (**3.21**) is a muscle relaxant that binds to the $\gamma$-aminobutyric acid-B ($GABA_B$) receptor; the eudismic ratio (−/+) is 800.[41]

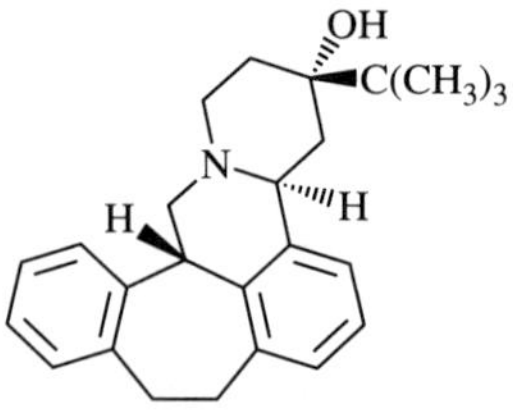

**(+)-3.20** 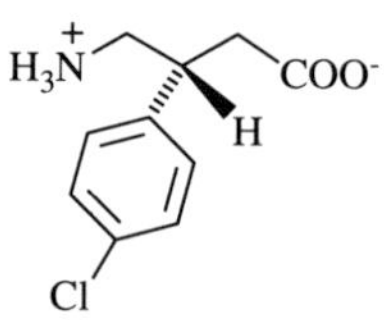 **(-)-3.21**

It should be remembered that the (+) and (−) nomenclature refers to the effect of the compound on the direction of rotation of the plane of polarized light, and it has nothing to do with the stereochemical configuration of the molecule. The stereochemistry about a chiral carbon atom is noted by the

(*R*,*S*) convention of Cahn *et al.*[42] Since the (*R*,*S*) convention is determined by the atomic numbers of the substituents about the chiral center, two compounds having the same stereochemistry, but a different substituent can have opposite chiral nomenclatures. For example, the eutomer of the antihypertensive agent propranolol is the (*S*)-(−)-isomer [**3.22**, X = $NHCH(CH_3)_2$].[43] If X is varied so that the attached atom has an atomic number greater than that of oxygen, such as F, Cl, Br, or SR, then the nomenclature rules[42] dictate that the molecule is designated as an (*R*)-isomer, even though there is no change in the stereochemistry. Note, however, that even though the absolute configuration about the chiral carbon remains unchanged after variation of the X group in **3.22**, the effect on plane polarized light cannot necessarily be predicted; the compound with a different substituent X can be either + or −. The most common examples of this phenomenon in nature are some of the amino acids. (*S*)-Alanine, for example, is the (+)-isomer and (*S*)-serine (same absolute stereochemistry) is the (−)-isomer; the only difference is a $CH_3$ group for alanine and a $CH_2OH$ group for serine.

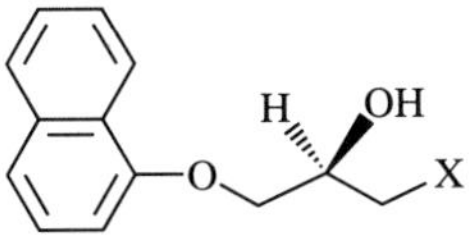

H OH O X

**(-)-3.22**

Propranolol [**3.22**, X = $NHCH(CH_3)_2$] is an antagonist of the β-adrenergic receptor, which triggers vasodilation; the $\beta_1$- and $\beta_2$-adrenergic receptors are important to cardiac and bronchial vasodilation, respectively. The eudismic ratio (*l*/*d*) for propranolol is about 100; however, propranolol also exhibits local anesthetic activity for which the eudesmic ratio is 1. The latter activity apparently is derived from some other mechanism than β-adrenergic blockage. A compound of this type that has two separate mechanisms of action and, therefore, different therapeutic activities, has been called a *hybrid drug* by Ariëns.[44] (+)-Butaclamol (**3.20**), which interacts with a variety of receptors, is another hybrid drug. However, butaclamol has three chiral centers and, therefore, has eight possible isomeric forms. When multiple isomeric forms are involved in the biological activity, the drug is called a *pseudo hybrid drug*.[44] Another important example of this type of drug is the antihypertensive agent, labetalol (Fig. 3.16), which, as a result of two asymmetric carbon atoms, exists in four stereoisomeric forms, having the stereochemistries (*RR*), (*SS*), (*RS*), and (*SR*). This drug has α- and β-adrenergic blocking properties. The (*RR*)-isomer is predominantly the β-blocker (the eutomer for β-adrenergic blocking action), and the (*SR*)-isomer is mostly the α-blocker (the eutomer for α-adrenergic blocking); the other 50% of the isomers, the (*SS*)- and (*RS*)-isomers, are almost inactive (the isomeric ballast). Labetalol,

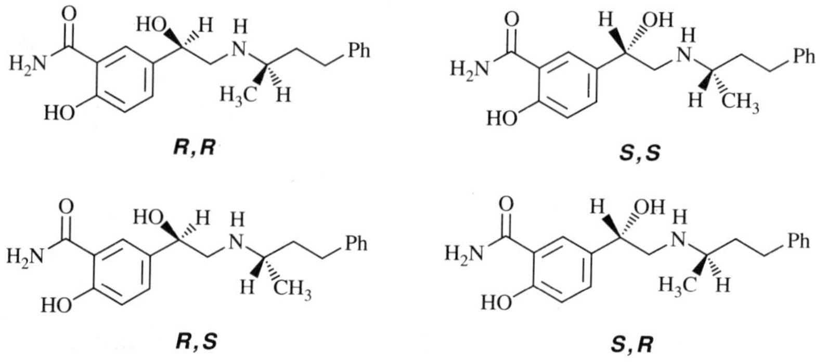

R,R S,S

R,S S,R

**Figure 3.16.** Four stereoisomers of labetalol.

then, is a pseudo hybrid, a mixture of isomers having different receptor-binding properties.

Labetalol also is an example of how relatively minor structural modifications of an agonist can lead to transformation into an antagonist. *l*-Epinephrine (**3.23**) is a natural hybrid molecule that induces both $\alpha$- and $\beta$-adrenergic effects. Introduction of the phenylalkyl substituent on the nitrogen transforms the $\alpha$-adrenergic activity of the agonist *l*-epinephrine into the $\alpha$-adrenergic antagonist labetalol. The modification of one of the catechol hydroxyl groups of *l*-epinephrine to a carbamyl group of labetalol changes the $\beta$-adrenergic action (agonist) to a $\beta$-adrenergic blocking action (antagonist).

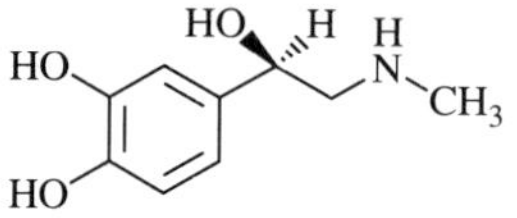

**3.23**

As pointed out by Ariëns[32a,b,44] and by Simonyi,[45] it is quite common for mixtures of isomers, particularly racemates, to be marketed as a single drug, even though at least half of the mixture not only may be inactive for the desired biological activity, but may, in fact, be responsible for various side effects. In the case of $\beta$-adrenergic blockers, antiepileptics, and oral anticoagulants, about 90% of the drugs on the market are racemic mixtures, and for antihistamines, anticholinergics, and local anesthetics about 50% are racemic. In general, about 25% of drugs are sold as racemic mixtures.[45] The isomeric ballast, typically, is not removed for economic reasons; it can be quite expensive to separate the enantiomeric impurity. Keep in mind, however, that because of vast differences in activities of two enantiomers, caution should be used when applying QSAR methods such as Hansch analyses (see Section

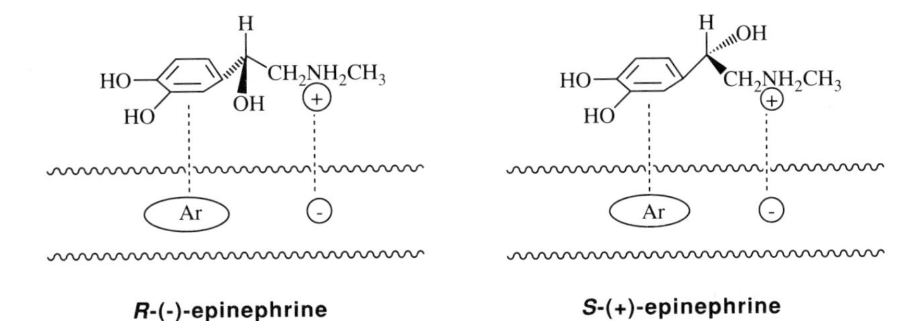


**Figure 3.17.** Binding of epinephrine enantiomers to a two-site receptor. The wavy lines are the receptor surfaces.

II,E,3,a of Chapter 2) to racemic mixtures. These methods really should be applied to the separate isomers.[46]

It is quite apparent from the above discussion that receptors are capable of recognizing and selectively binding optical isomers. Cushny[47] was the first to suggest that enantiomers could have different biological activities because one isomer could fit into a receptor much better than the other. How are they able to accomplish this?

If you consider two enantiomers, such as (*R*)-(−)- and (*S*)-(+)-epinephrine, interacting with a receptor that has only two binding sites (Fig. 3.17), it becomes apparent that the receptor cannot distinguish between them. However, if there are at least three binding sites (Fig. 3.18), the receptor easily can differentiate them. The (*R*)-(−)-isomer has three points of interaction and is held in the conformation shown to maximize molecular complementarity. The (*S*)-(+)-isomer can have only two sites of interaction (the hydroxyl group cannot interact with the hydroxyl binding site, and may even have an adverse steric interaction); consequently it has a lower binding energy. Easson and

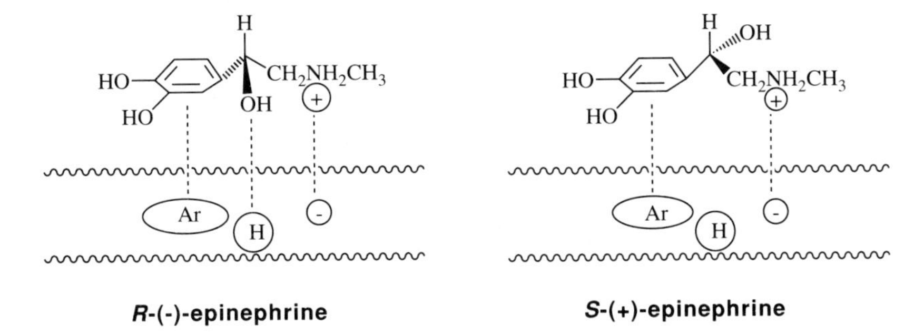


**Figure 3.18.** Binding of epinephrine enantiomers to a three-site receptor. The wavy lines are the receptor surfaces.

Stedman[48] were the first to recognize this "*three-point attachment*" concept: a receptor can differentiate enantiomers if there are as few as three binding sites. As in the case of the β-adrenergic receptors discussed above, the structure of α-adrenergic receptors to which epinephrine binds is unknown. α-Adrenergic receptors appear to mediate vasoconstrictive effects of catecholamines in bronchial, intestinal, and uterine smooth muscle. The eudismic ratio (*R*/*S*) for vasoconstrictor activity of epinephrine is only 12–20,[49a] indicating that there is relatively little difference in binding energy for the two isomers to the α-adrenergic receptor. Although the above discussion was directed at the enantioselectivity of receptor interactions, it should be noted that there also is enantioselectivity with respect to pharmacokinetics, namely, absorption, distribution, metabolism, and excretion.[49b]

### 3. Geometric Isomers

Geometric isomers (*E*- and *Z*-isomers[50] and epimers) are diastereomers, stereoisomers having different spatial arrangements of atoms; consequently, they are different compounds. As a result of their different configurations, receptor interactions will be different. For example, the antipsychotic activity of a series of *Z*-2-substituted doxepin analogs (**3.24a**) was found to be significantly greater than the corresponding *E*-isomers (**3.24b**).[51] Likewise, the neuroleptic potency of the *Z*-isomer of the antipsychotic drug chlorprothixene (**3.25a**) is more than 12 times greater than that of the corresponding *E*-isomer (**3.25b**).[51] On the other hand, the *E*-isomer of the anticancer drug diethylstilbestrol (**3.26a**) has 14 times greater estrogenic activity than the *Z*-isomer (**3.26b**), possibly because its overall structure and the interatomic distance between the two hydroxyls in the *E*-isomer are similar to that of estradiol (**3.27**).

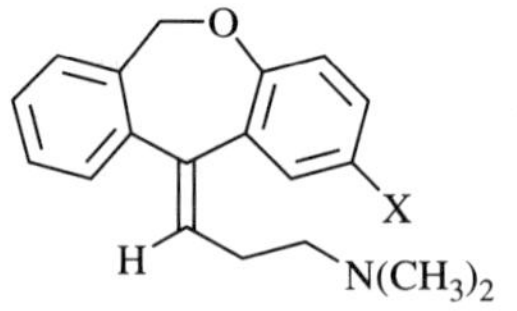

O X H $N(CH_3)_2$ — **3.24a**

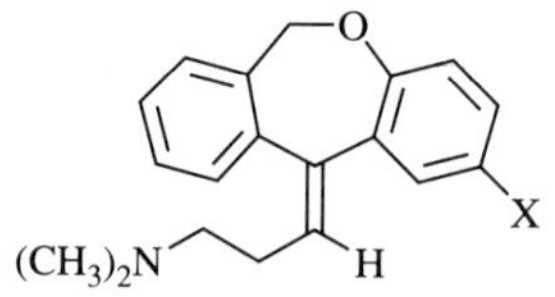

O X $(CH_3)_2N$ H — **3.24b**

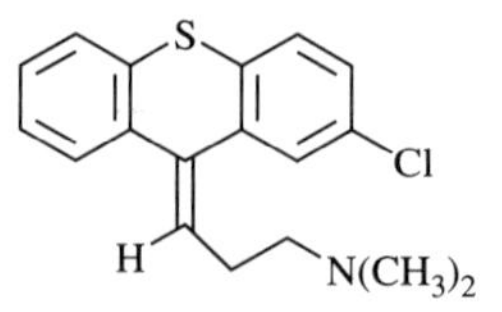

S Cl H $N(CH_3)_2$ — **3.25a**

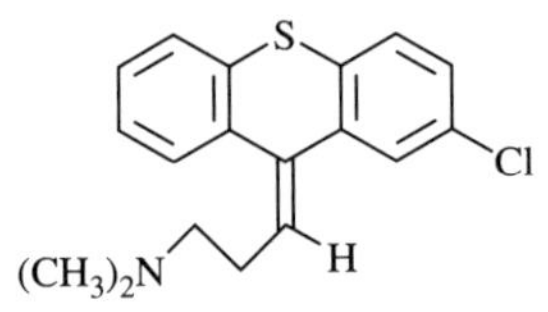

S Cl $(CH_3)_2N$ H — **3.25b**

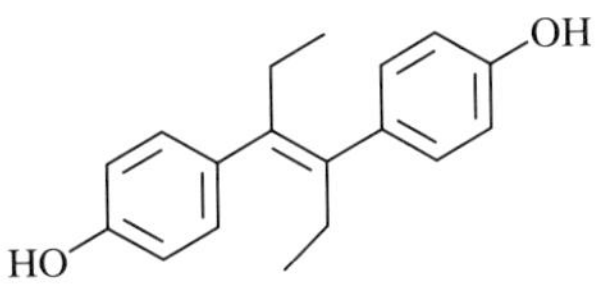

**3.26a** 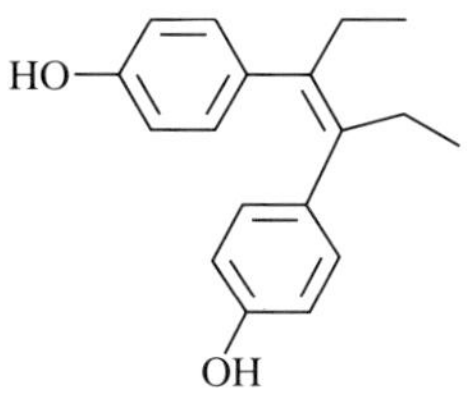 **3.26b** 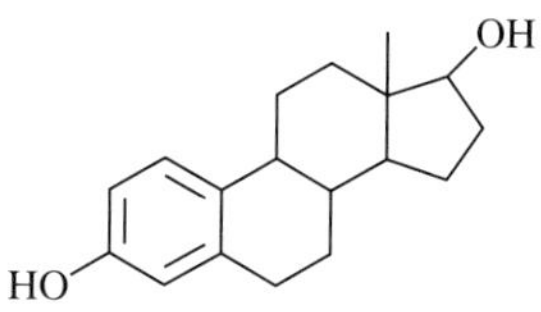 **3.27**

Although in some cases the cis and trans nomenclature does correspond with *Z* and *E*, respectively, it should be kept in mind that these terminologies are based on different conventions,[50] so there may be confusion. The *Z*,*E* nomenclature is unambiguous and should be used.

### 4. Conformational Isomers

As a result of free rotation about single bonds in acyclic molecules and conformational flexibility in many cyclic compounds, a drug molecule can assume a variety of *conformations*, namely, locations of the atoms in space. The pharmacophore of a molecule is defined not only by the configuration of a set of atoms, but also by their conformation in relation to the receptor binding site. A receptor may bind only one of these *conformers* (isomers generated by a change in conformation); the conformer that binds need not necessarily be the lowest energy conformer observed in the crystalline state, as determined by X-ray crystallography, or found in solution, as determined by nuclear magnetic resonance (NMR) spectrometry, or determined theoretically by molecular mechanics calculation. The binding energy to the receptor may overcome the barrier to the formation of an unstable conformer. As was pointed out in Section II,F of Chapter 2, the assumption that a drug–receptor interaction involves the lowest energy conformer is an important problem in much of molecular graphics drug design. In order for drug design to be efficient, it is essential to know the active conformation in the drug–receptor complex. If the lead compound has low potency, it may only be because the population of the active conformer in solution is low (higher in energy).

A unique approach has been taken to determine, with some degree of certainty, the active conformation of a drug molecule in the drug–receptor complex. This approach involves the synthesis of *conformationally rigid analogs* of flexible drug molecules. The potential pharmacophore becomes locked into various configurations by judicious incorporation of cyclic or unsaturated moieties into the drug molecule. The conformationally rigid analogs are, then, tested, and the analog with the optimal activity (or potency) can be used as the prototype for further structural modification. Conformationally rigid analogs are propitious because key functional groups, presumably part of the pharmacophore, are constrained in one position, thereby permitting the determination of the *pharmacophoric conformation*. The major drawback to this approach is that in order to construct a rigid analog of a flexible molecule, additional atoms and/or bonds must be attached to the original compound, and these can affect the chemical and physical properties. Consequently, it is imperative that the conformationally rigid analog and the drug molecule be as similar as possible in size, shape, and molecular weight.

An example of the use of conformationally rigid analogs for the elucidation of receptor binding site topography is the studies of the interaction of the neurotransmitter acetylcholine (ACh) with the muscarinic receptor. There are at least two important receptors for ACh, one activated by the alkaloid muscarine (**3.28**) and the other by the alkaloid nicotine (**3.29**; presumably in the protonated pyrrolidine form). Acetylcholine has a myriad of conformations; four of the more stable possible conformers (group staggered) are **3.30a–3.30d**. There are also conformers with groups eclipsed that are higher in energy. Four different *trans*-decalin stereoisomers were synthesized[52] (**3.31a–3.31d**) corresponding to the four ACh conformers shown in **3.30a–3.30d**. All four isomers exhibited low muscarinic receptor activity; however, **3.31a** was the most potent (0.06 times the potency of ACh). The low potency of **3.31a** is believed to be the result of the unfavorable steric effect of the *trans*-decalin moiety.[52]

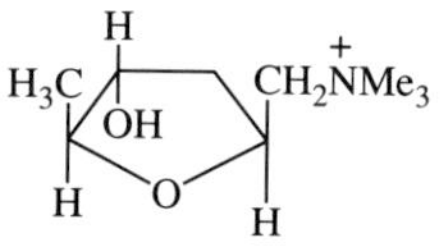

**3.28** 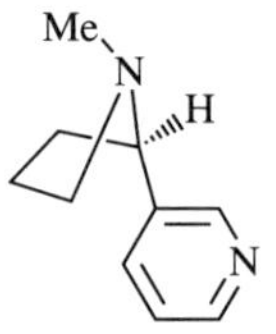 **3.29**

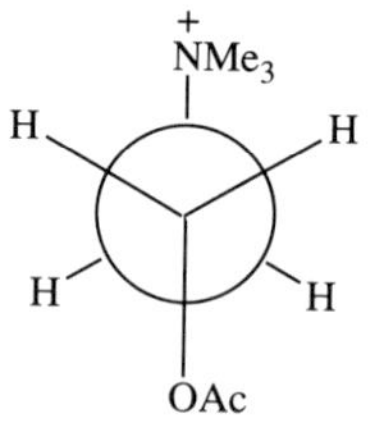

**3.30a** 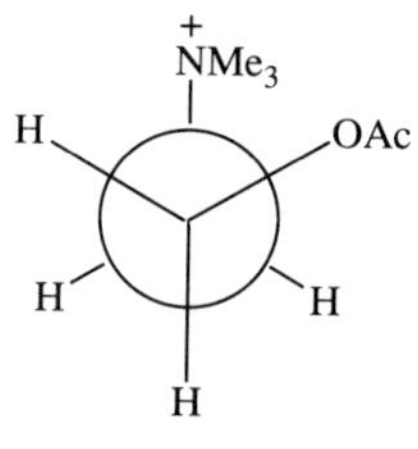 **3.30b**

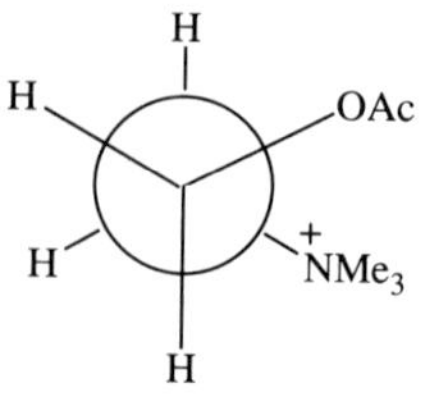

**3.30c** 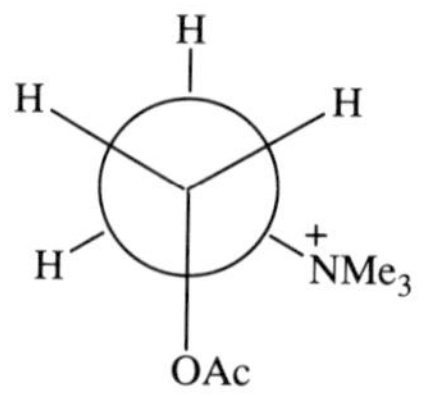 **3.30d**

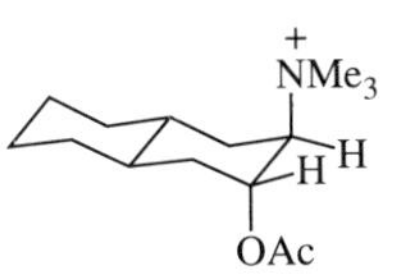

3.31a

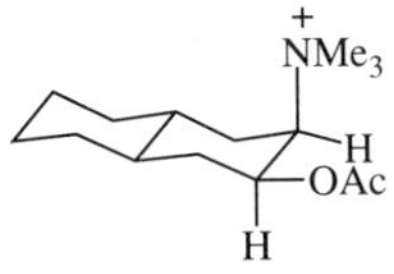

3.31b

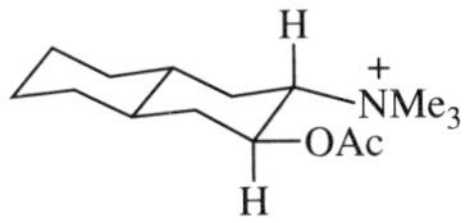

3.31c

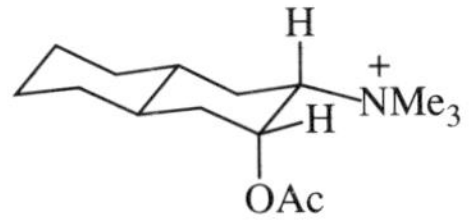

3.31d

A comparison of *erythro*- (**3.32**) and *threo*-2,3-dimethylacetylcholine (**3.33**) gave the startling result that **3.32** was 14 times more potent than ACh and **3.33** was 0.036 times as potent as ACh. Compound **3.31a** corresponds to the threo isomer **3.33** and, therefore, is expected to have low potency. The corresponding erythro analog does not have a *trans*-decalin analogy, so it could not be tested. To minimize the number of extra atoms added to ACh, *trans*- (**3.34**) and *cis*-1-acetoxy-2-trimethylammoniocyclopropanes (**3.35**) were synthesized and tested[53] for *cholinomimetic properties*, that is, production of a response resembling that of ACh. The (+)-trans isomer (shown in **3.34**)[54] has about the same muscarinic activity as does ACh, thus indicating the importance of minimizing additional atoms; the (−)-trans isomer has about 1/500th the potency of ACh. The racemic cis isomer has negligible activity. The (+)-trans isomer was shown to have the same absolute configuration as the active enantiomers of the two muscarinic receptor agonists muscarine and acetyl β-methylcholine.[54] These results suggest that ACh binds in an extended form (**3.30a**). Both the cis and the trans isomers, as well as all of the *trans*-decalin stereoisomers (**3.31a–3.31d**) were weakly active with the nicotinic cholinergic receptor.

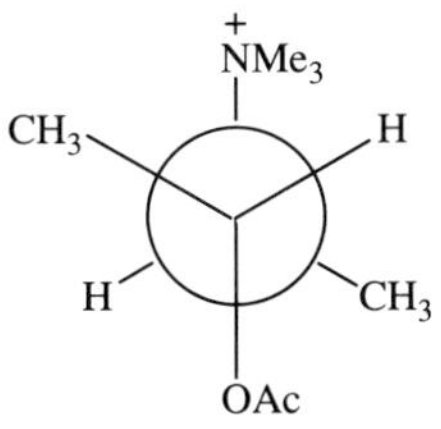

3.32

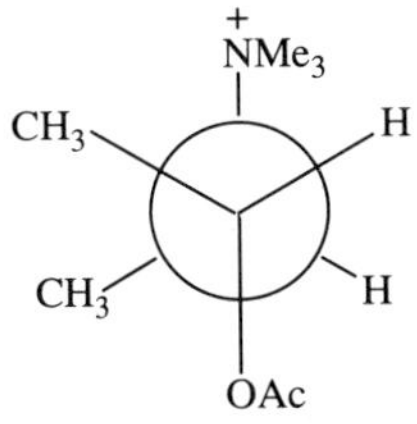

3.33

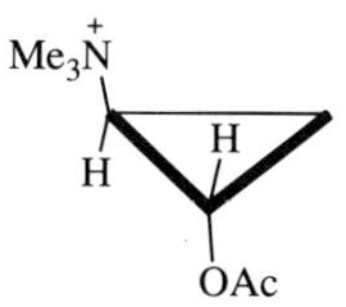

3.34

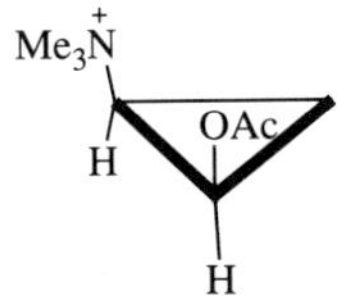

3.35

An example of the use of conformationally rigid analogs in drug design was reported by Li and Biel.[55] 4-(4-Hydroxypiperidino)-4′-fluorobutyrophenone (**3.36**) was found to have moderate tranquilizing activity in lower animals and man; however, unlike the majority of antipsychotic butyrophenone-type compounds, it only has minimal antiemetic (prevents vomiting) activity. The piperidino ring can exist in various conformations [**3.37a–3.37d**, R = F—$C_6H_4CO(CH_2)_3$—], two chair forms (**3.37a** and **3.37d**) and two twist–boat forms (**3.37b** and **3.37c**). The difference in free energy between the axial and equatorial hydroxy conformers of the related compound *N*-methyl-4-piperidinol (**3.37**, R = Me) is 0.94 ± 0.05 kcal/mol at 40°C (the equatorial conformer is favored by a factor of 4.56 over the axial conformers).[56] Energies for the twist–boat conformers are about 6 kcal/mol higher, but because of hydrogen bonding, **3.37b** should be more stable than **3.37c**. On the assumption that the chair conformers are more likely, three conformationally rigid chair analogs, **3.38–3.40**, were synthesized to determine the effect on receptor binding of the hydroxyl in the equatorial (**3.38**), axial (**3.39**), and both (**3.40**) positions. Of course, there will be no hydroxyl group hydrogen-bonding effects with **3.40**. When subjected to muscle relaxation tests, the order of potency was **3.39** > **3.40** > **3.38**, indicating that the conformationally less stable compound with the axial hydroxyl group has better molecular complementarity with the receptor than does the more stable compound with the equatorial hydroxyl group. This suggests that future analogs should be prepared where the axial hydroxyl is the more stable conformer or where it can be held in that configuration.

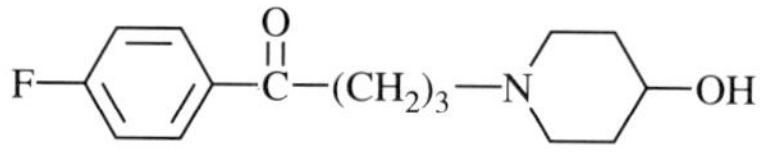

**3.36**

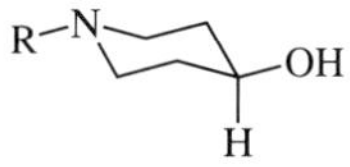

**3.37a** 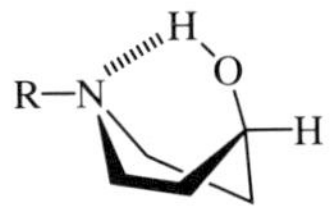 **3.37b** 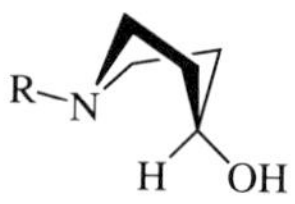 **3.37c** 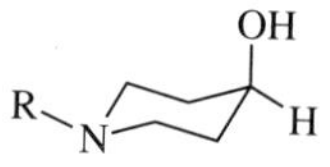 **3.37d**

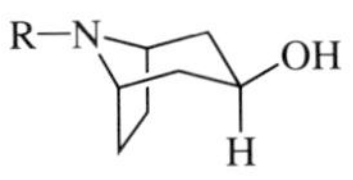

**3.38** 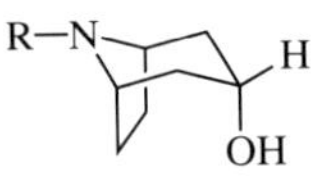 **3.39** 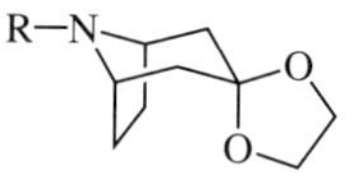 **3.40**

Another use of conformationally rigid analogs is to prepare compounds that have conformational features common to potent analogs which cannot be adopted by inactive analogs. This is the strategy of drug design that can be used in conjunction with the molecular graphics approach known as steric mapping (see Section II,F of Chapter 2).

### 5. Ring Topology

Tricyclic psychomimetic drugs show an almost continuous transition of activity in going from structures such as the tranquilizer chlorpromazine (**3.41**) through the antidepressant amitriptyline (**3.42**), which has a tranquilizing side effect, to the pure antidepressant agent imipramine (**3.43**).[57] Stereoelectronic effects seem to be the key factor, even though tranquilizers and antidepressants have different molecular mechanisms. Three angles can be drawn to define the positions of the two aromatic rings in these compounds (Fig. 3.19). The angle $\alpha$ (**3.44**) describes the bending of the ring planes; $\beta$ (**3.45**) is the annellation angle of the ring axes that pass through carbon 1 and 4 of each aromatic ring; $\gamma$ (**3.46**) is the torsional angle of the aromatic rings as viewed from the side of the molecule. In general, the tranquilizers have only a bending angle $\alpha$ and no $\beta$ and $\gamma$ angles. The mixed tranquilizer–antidepressants have both a bending ($\alpha$) and annellation angle ($\beta$), but no $\gamma$ angle. The pure antidepressants exhibit all three angles.

S, N, Cl, $(CH_2)_3NMe_2$ — **3.41**

$HC{-}(CH_2)_2NMe_2$ — **3.42**

N, $(CH_2)_3NMe_2$ — **3.43**

## G. *Ion Channel Blockers*[58]

A receptor was defined in Section II,A as having two basic characteristics, recognition of a substance and ability to initiate a biological response. Ion channels, then, fulfill the definition of receptors: they selectively bind ions and they mediate a response, namely, ion transport. An *ion channel* is a transmembrane pore that is composed of three elements, a *pore* responsible

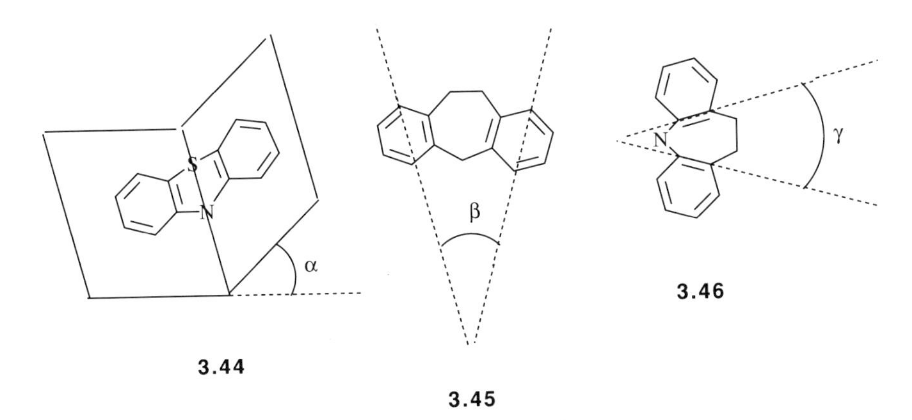


**Figure 3.19.** Ring topology of tricyclic psychomimetic drugs. [Reproduced with permission from Nogrady, T. (1985). *In* "Medicinal Chemistry: A Biochemical Approach," p. 29. Oxford University Press, New York. Copyright © 1985 Oxford University Press.]

for the transit of the ion and one or more *gates* that open and close in response to specific stimuli that are received by the *sensors*. Conformational mobility is an integral component of the function of ion channels; the three states of a channel, closed, open, and activated, are all believed to be regulated by conformational changes. Ligands may gain access to the channel either by membrane permeation or through an open channel state.

The movement of calcium ions into cells is vital to the excitation and contraction of the heart muscle. When a cardiac cell potential reaches a threshold, a sodium ion channel allows rapid influx of sodium ions through the cell membrane. This is followed by a slower movement of calcium ions through a calcium ion channel; the calcium ions maintain the plateau phase of the cardiac action potential. *Calcium ion channel blockers* prevent the influx of calcium ions, which then alters the plateau phase and, therefore, the coronary blood flow. Consequently, calcium channel blockers such as verapamil (**3.47**), nifedipine (**3.48**), and diltiazem (**3.49**) are valuable drugs in the treatment of angina (resulting from reduced oxygen), cardiac arrhythmias, and hypertension.

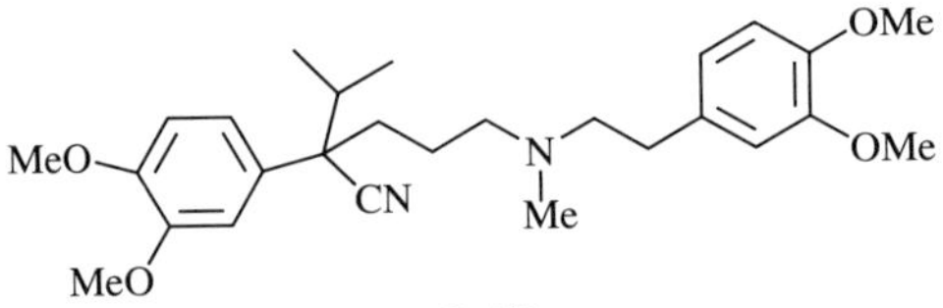

MeO, MeO, CN, N, Me, OMe, OMe

**3.47**

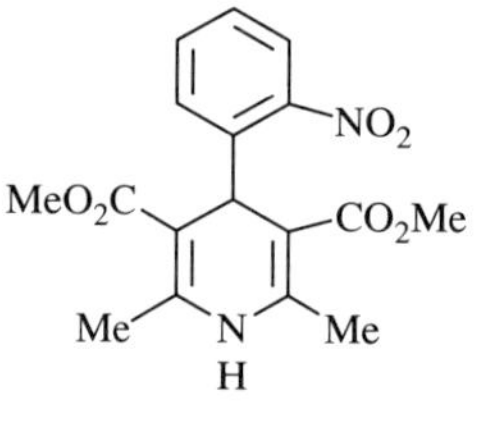

$MeO_2C$, $NO_2$, $CO_2Me$, Me, N, H, Me

**3.48**

OMe, H, S, H, $O-CCH_3$, O, N, O, $CH_2CH_2N(CH_3)_2$

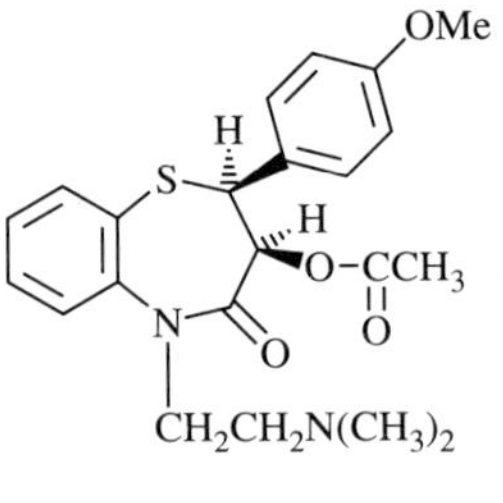

**3.49**

### H. Example of Rational Drug Design of a Receptor Antagonist: Cimetidine

The antiulcer drug cimetidine is a truly elegant example of lead discovery and the use of physical organic chemical principles, coupled with the various lead modification approaches discussed in Chapter 2, to uncover the first histamine $H_2$ receptor antagonist and an entirely new class of drugs. This is a case, however, where neither QSAR nor molecular graphics approaches were utilized. As described in Section III,F,1, histamine binds to the $H_1$ receptor and causes allergic and hypersensitivity reactions, which antihistamines antagonize. It is now known that another action of histamine is the stimulation of gastric acid secretion. However, antihistamines have no effect on this activity; consequently, it was suggested that there was a second histamine receptor, which was termed the $H_2$ receptor. The $H_1$ and $H_2$ receptors can be differentiated by agonists and antagonists. 2-Methylhistamine (**3.50**) preferentially elicits $H_1$ receptor responses, and 4-methylhistamine (**3.51**) has the corresponding preferential effect on $H_2$ receptors. An antagonist of the histamine $H_2$ receptor would be beneficial to the treatment of hypersecretory conditions such as duodenal and gastric ulcers. Consequently, in 1964 Smith, Kline & French Laboratories in England initiated a search for a lead compound that would antagonize the $H_2$ receptor.[29,59]

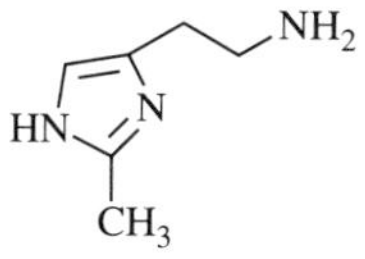

**3.50** 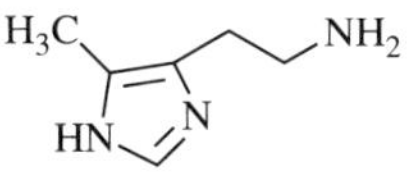 **3.51**

The first requirement for initiation of a lead discovery program is an efficient bioassay. Histamine was infused into anesthetized rats to stimulate gastric acid secretion, then the pH of the perfusate from the lumen of the stomach was measured before and after administration of the compound.

The lead discovery approach that was taken involved a biochemical rationale. Since a histamine receptor antagonist was sought, histamine analogs were synthesized on the assumption that the receptor would recognize that general backbone structure. However, the structure had to be sufficiently different so as not to stimulate a response and defeat the purpose. It took the group at Smith, Kline & French four years and the synthesis of about 200 compounds until the lead compound, $N^{\alpha}$-guanylhistamine (**3.52**), was discovered. This compound was only very weakly active as an inhibitor of histamine stimulation; later it was determined to be a partial agonist, not an antagonist. The isosteric isothiourea (**3.53**) was found to be more active. The corresponding conformationally rigid analog **3.54** was less potent than **3.53**; consequently, it was thought that flexibility in the side chain was important. Many additional compounds were synthesized, but they acted as partial agonists. They could block histamine binding, but they could not inhibit acid secretion.

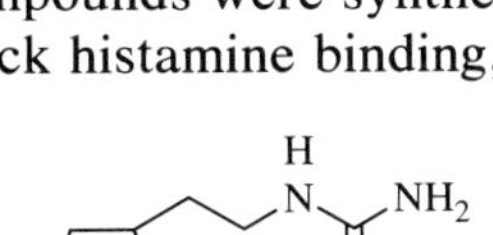

**3.52** 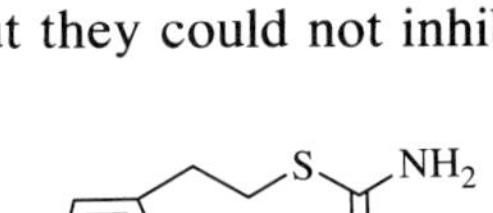 **3.53** 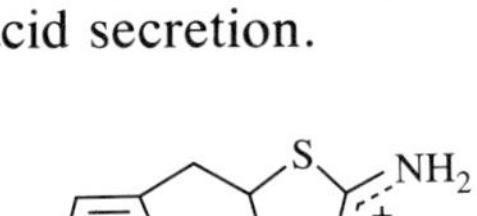 **3.54**

It, therefore, became necessary to separate the agonist and antagonist activities. The reason for the agonistic activity, apparently, was the structural similarity to histamine. Not only were these compounds imidazoles, but at physiological pH the side chains were protonated and positively charged, just like histamine. Consequently, it was reasoned that the imidazole ring should be retained for receptor recognition, but the side chain could be modified to eliminate the positive charge. After numerous substitutions, a thiourea analog (**3.55**) was prepared having weak antagonistic activity without stimulatory activity. Homologation of the side chain gave a purely competitive antagonist (**3.56**, R = H); no agonist effects were observed. The *N*-methyl analog (**3.56**, R = $CH_3$), called burimamide, was found to be highly specific as a competitive antagonist of histamine at the $H_2$ receptor. It was shown to be effective in the inhibition of histamine-stimulated gastric acid secretion in rat, cat, dog, and man. Burimamide was the first $H_2$ receptor antagonist tested in humans, but it lacked adequate oral activity, so the search for more potent analogs was continued.

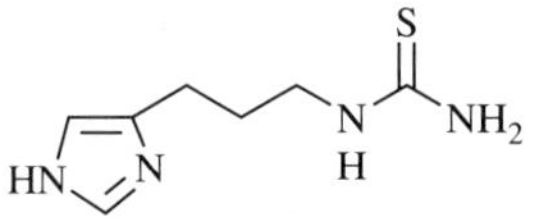

**3.55** 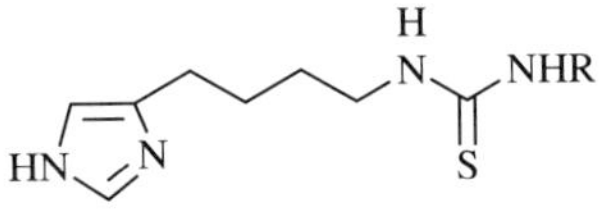 **3.56**

The poor oral potency of burimamide could be a pharmacokinetics problem or a pharmacodynamics problem. Let's consider the latter. In aqueous solution at physiological pH the imidazole ring can exist in three main forms (**3.57a–3.57c**, Fig. 3.20; R is the rest of burimamide). The thioureido group can exist as four conformers (**3.58a–3.58d**, Fig. 3.21; R is the remainder of burimamide). The side chain can exist in a myriad of conformations. Therefore, it is possible that only a very small fraction of the molecules in equilibrium would have the active structure, and this could account for the low potency.

One approach taken to increase the potency of burimamide was to compare the population of the imidazole form in burimamide at physiological pH to that in histamine.[60] The population can be estimated from the electronic influence of the side chain, which alters the electron densities at the ring nitrogen atoms and, therefore, affects the proton acidity. This effect is more important at the nearer nitrogen atom, so if R is electron releasing, **3.57c** (Fig. 3.20) should predominate; if R is electron withdrawing, **3.57a** should be favored.

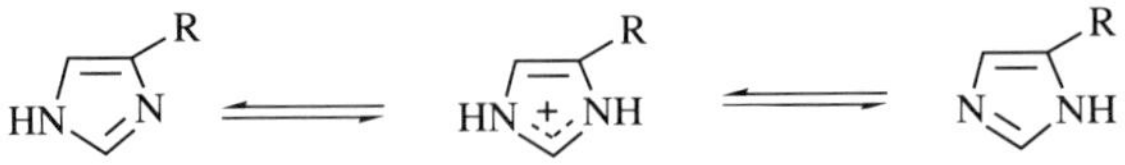

**3.57a** **3.57b** **3.57c**

**Figure 3.20.** Three principal forms of 5-substituted imidazoles at physiological pH.

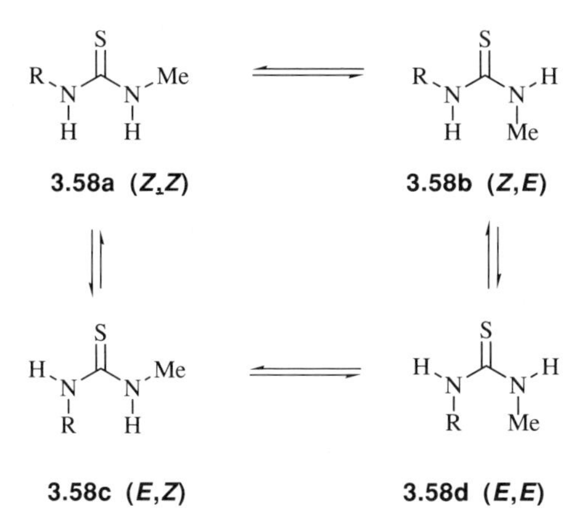


**Figure 3.21.** Four conformers of the thioureido group.

The fraction present as **3.57b** can be determined from the ring p$K_a$ and the pH of the solution. The electronic effect of R can be calculated from the measured ring p$K_a$ with the use of the Hammett equation [Eq. (3.4)], where $\mathrm{p}K_a^R$ is the p$K_a$ of the substituted imidazole, $\mathrm{p}K_a^H$ is that of imidazole (R = H), $\sigma_m$ is the meta electronic substituent constant, and $\rho$ is the reaction constant (see Section II,E,2,a of Chapter 2). Imidazole has a p$K_a$ of 6.80, and at physiological

$$\mathrm{p}K_a^R = \mathrm{p}K_a^H + \rho\sigma_m \quad (3.4)$$

temperature and pH, 20% of the molecules are in the protonated form. The imidazole in histamine under these conditions has a p$K_a$ of 5.90. This indicates that the side chain is electron withdrawing, thus favoring tautomer **3.57a** (to the extent of 80%), and only 3% of the molecules are in the cationic form (**3.57b**). The p$K_a$ of the imidazole in burimamide, however, is 7.25, indicating an electron-donating side chain which favors tautomer **3.57c**. The cation is one of the principal species, about 40% of the molecules. Therefore, even though the side chains in histamine and burimamide appear to be similar, they have opposite electronic effects on the imidazole ring.

On the assumption that the desired form of the imidazole should resemble that in histamine, the Smith, Kline & French group decided to convert the burimamide side chain to an electron-withdrawing group; however, they did not want to make a major structural modification. Incorporation of an electron-withdrawing atom into the side chain near the imidazole ring was contemplated, and the isosteric replacement of a methylene by a sulfur atom to give thiaburimamide (**3.59**, R = H) was carried out. A comparison of the physical properties of the two compounds (**3.56**, R = $CH_3$, and **3.59**, R = H) shows that they have similar van der Waals radii and bond angles, although the C—S bond is slightly longer than the C—C bond and is more flexible. A

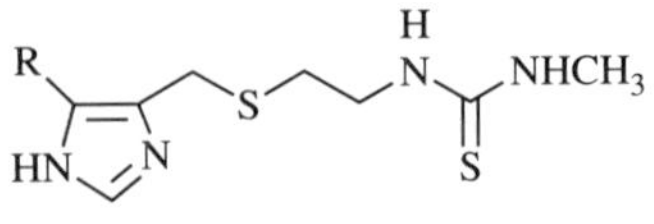

R
H
N NHCH$_3$
S
HN N S

**3.59**

sulfur atom also is more hydrophilic than a methylene group; the log *P* for thiaburimamide is 0.16 and that for burimamide is 0.39. The p$K_a$ of the imidazole in thiaburimamide was determined to be 6.25, indicating that the side chain was electron withdrawing and the favored tautomeric form was the same as in histamine (**3.57a**). Thiaburimamide is about three times more potent as a histamine $H_2$ receptor antagonist *in vitro* than burimamide.

A second way to increase the population of tautomer **3.57a** would be to introduce an electron-donating substituent at the 4-position of the ring, because electron-donating groups favor the form with the hydrogen on the adjacent nitrogen. Since 4-methylhistamine (**3.51**) is a known $H_2$ receptor agonist, there should be no steric problem with a 4-methyl group. However, the addition of an electron-donating group should increase the p$K_a$ of the ring, thereby increasing the population of the cation (**3.57b**). Although the increase in tautomer **3.57a** is somewhat offset by the decrease in the total uncharged population, the overall effect was favorable. Metiamide (**3.59**, R = $CH_3$) has a p$K_a$ identical with that of imidazole, indicating that the effect of the electron-withdrawing side chain exactly balanced the effect of the electron-donating 4-methyl group; the percentage of molecules in the charged form was 20%. The important result, however, is that metiamide is 8 to 9 times more potent than burimamide.

As an aside, it is interesting that the oxygen analog of burimamide also was synthesized in order to increase the electron-withdrawing effect of the side chain even further (oxygen is more electronegative than sulfur); however, oxaburimamide is less potent than burimamide. An explanation for this result is that intramolecular hydrogen bonding produces an unfavorable *conformationally restricted analog*, that is, a conformer stabilized by noncovalent phenomena (**3.60**).

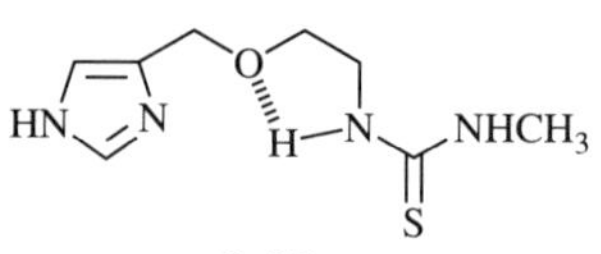

O
HN N H—N NHCH$_3$
S

**3.60**

Metiamide was tested on 700 patients with duodenal ulcers, and it was found to produce significant increases in the healing rate with marked symptomatic relief. However, a few cases of granulocytopenia (deficiency of blood granulocytes and reduced bone marrow) developed. Even though this was a reversible side effect, it was undesirable, and it halted further clinical work with this compound.

The Smith, Kline & French group conjectured that the granulocytopenia that was associated with metiamide use was caused by the thiourea group; consequently, alternative substituents were sought. An isosteric replacement approach was taken. The corresponding urea (**3.61**, X = O) and guanidino (**3.61**, X = NH) analogs were synthesized and found to be 20 times less potent than metiamide. Of course, the guanidino analog would be positively charged at physiological pH, which could be the cause for the lower potency. Charton[61] found a Hammett relationship between the $\sigma$ and $pK_a$ values for N-substituted guanidines; consequently, if guanidino basicity were the problem, then substitution of the guanidino nitrogen with electron-withdrawing groups could lower the $pK_a$. In fact, cyanoguanidine and nitroguanidine have $pK_a$ values of −0.4 and −0.9, respectively (compared with −1.2 for thiourea), a drop of about 14 $pK_a$ units from that of guanidine. The corresponding cyanoguanidine (**3.61**, X = NCN; cimetidine) and nitroguanidine (**3.61**, X = $NNO_2$) were synthesized, and both were potent $H_2$ antagonists, comparable in potency to that of metiamide (cimetidine was slightly more potent than **3.61**, X = $NNO_2$).

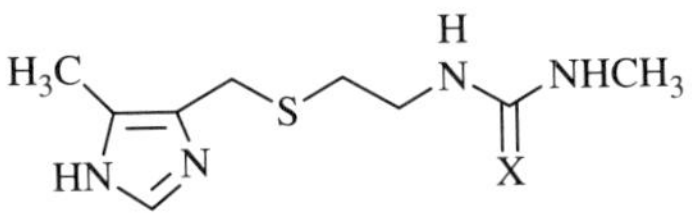

H₃C, HN, N, S, H, N, NHCH₃, X

**3.61**

Since strong electron-withdrawing substituents on the guanidino group favor the imino tautomer, the cyanoguanidino and nitroguanidino groups correspond to the thiourea structure (**3.61**, X = NCN, $NNO_2$, and S, respectively). These three groups are actually bioisosteres; they are all planar structures of similar geometries, are weakly amphoteric (weakly basic and acidic), being un-ionized in the pH range 4–11, are very polar, and are hydrophilic. The crystal structures of metiamide (**3.59**, R = $CH_3$) and cimetidine (**3.61**, X = NCN) are almost identical. The major difference in the two groups is that, whereas N,N′-disubstituted thioureas assume three stable conformers (Fig. 3.21; *Z,Z*, *Z,E*, and *E,Z*), N,N′-disubstituted cyanoguanidines appear to assume only two stable conformers (*Z,E* and *E,Z*). This suggests that the most stable conformer, the *Z,Z* conformer, is not the pharmacologically active form. An isocytosine analog (**3.62**) also was prepared ($pK_a$ 4.0), which can

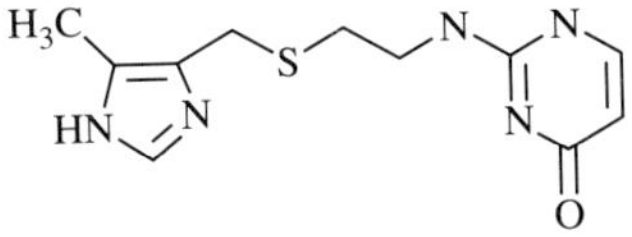

**3.62**

exist only in the *Z,Z* and *E,Z* conformations. It was about one-sixth as potent as cimetidine. However, the isocytosino group has a lower log *P* value (more

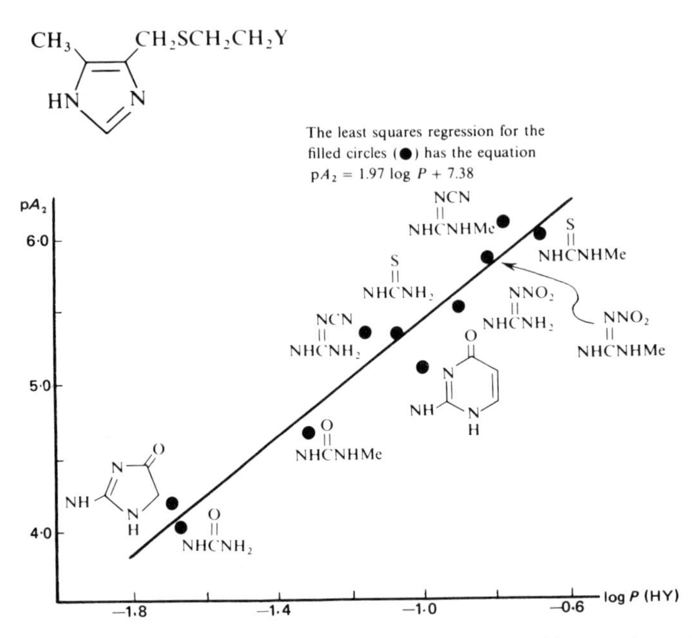


**Figure 3.22.** Linear free energy relationship between $H_2$ receptor antagonist activity and the partition coefficient. The least squares regression for the filled circles (●) has the equation $pA_2 = 1.97 \log P + 7.38$. [Reproduced with permission from Ganellin, C. R., and Parsons, M. E. (1982). *In* "Pharmacology of Histamine Receptors," p. 83. Wright-PSG, Bristol.]

hydrophilic) than that of the *N*-methylcyanoguanidino group, and it was thought that lipophilicity may be an important physicochemical parameter. There was, indeed, a correlation found between the $H_2$ receptor antagonist activity *in vitro* and the octanol–water partition coefficient of the corresponding acid of the substituent Y (Fig. 3.22). Although increased potency correlates with increased lipophilicity, all of these compounds are fairly hydrophilic. Since the correlation was determined in an *in vitro* assay, membrane transport is not a concern; consequently, these results probably reflect a property involved with receptor interaction, not with transport. Therefore, it is not clear if the lower potency of the isocytosine analog is structure or hydrophilicity dependent.

Cimetidine was first marketed in the United Kingdom in 1976; therefore, it took only 12 years from initiation of the $H_2$ receptor antagonist program to commercialization. Subsequent to the introduction of cimetidine onto the U.S. drug market, two other $H_2$ receptor antagonists were approved, ranitidine (Glaxo Laboratories, **3.63**), which rapidly became the largest selling drug worldwide, and famotidine (Merck, Sharp & Dohme, **3.64**). It is obvious that an imidazole ring is not essential for $H_2$ receptor activity and that a positive charge near the heterocyclic ring (the $Me_2N$— and guanidino groups of **3.63** and **3.64**, respectively, will be protonated at physiological pH) is not unfavorable.

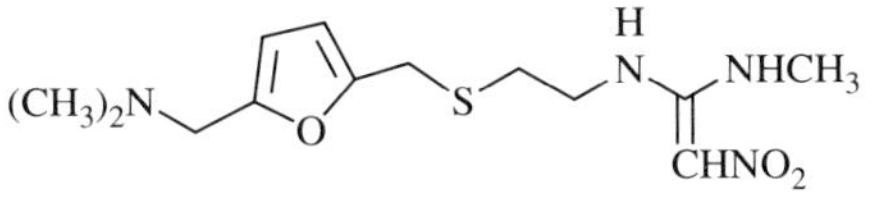

**3.63** 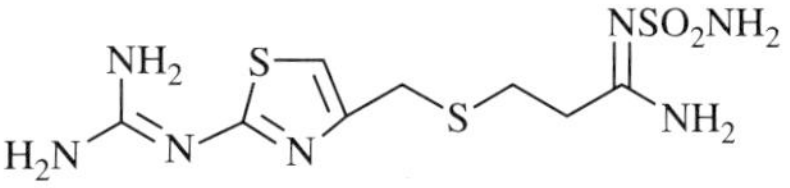 **3.64**

The discovery of cimetidine is one of many examples of how the judicious use of physical organic chemistry can result in, at least, lead discovery, if not in drug discovery. Next, we turn our attention to a special class of receptors called enzymes, which are very important targets for drug design and drug action.

## References


1. Langley, J. N. 1878. *J. Physiol.* (*London*) **1,** 367.
2. Ehrlich, P. 1897. *Klin. Jahr.* **6,** 299.
3. Langley, J. N. 1905. *J. Physiol.* (*London*) **33,** 374; Langley, J. N. 1906. *J. Physiol.* (*London*) **B78,** 170.
4. Lindstrom, J. 1985. *In* "Neurotransmitter Receptor Binding" (Yamamura, H. I., Enna, S. J., and Kuhar, M. J., eds.), p. 123. Raven, New York, 1985; Douglass, J., Civelli, O., and Herbert, E. 1984. *Annu. Rev. Biochem.* **53,** 665.
5. Litter, M. 1961. "Farmacologia," 2nd Ed. El Ateneo, Buenos Aires.
6. Albert, A. 1985. "Selective Toxicity," 7th Ed., p. 206. Chapman & Hall, London.
7. Jencks, W. P. 1969. "Catalysis in Chemistry and Enzymology," p. 340. McGraw-Hill, New York.
8. Korolkovas, A. 1970. "Essentials of Molecular Pharmacology," p. 159. Wiley, New York.
9. Jencks, W. P. 1969. "Catalysis in Chemistry and Enzymology," p. 393. McGraw-Hill, New York.
10. Hildebrand, J. H. 1979. *Proc. Natl. Acad. Sci. U.S.A.* **76,** 194.
11. Stearn, A., and Stearn, E. 1924. *J. Bacteriol.* **9,** 491.
12. Albert, A., Rubbo, S., and Goldacre, R. 1941. *Nature* (*London*) **147,** 332; Albert, A., Rubbo, S., Goldacre, R., Davey, M., and Stone, J. 1945. *Br. J. Exp. Pathol.* **26,** 160; Albert, A., and Goldacre, R. 1948. *Nature* (*London*) **161,** 95; Albert, A. 1966. "The Acridines, Their Preparation, Properties, and Uses," 2nd Ed. Arnold, London.
13. Albert, A. 1985. "Selective Toxicity," 7th Ed., p. 398. Chapman & Hall, London.
14. Burns, J., Yü, T., Dayton, P., Gutman, A., and Brodie, B. 1960. *Ann. N.Y. Acad. Sci.* **86,** 253.
15. Miller, G., Doukos, P., and Seydel, J. 1972. *J. Med. Chem.* **15,** 700.
16. Parsons, S. M., and Raftery, M. A. 1972. *Biochemistry* **11,** 1623, 1630, and 1633.
17. Schmidt, D. E., and Westheimer, F. H. 1971. *Biochemistry* **10,** 1249.
18. Williams, M., and Enna, S. J. 1986. *Annu. Rep. Med. Chem.* **21,** 211.
19. Costall, B., and Naylor, R. J. 1981. *Life Sci.* **28,** 215.
20. Gaddum, J. H. 1926. *J. Physiol.* (*London*) **61,** 141.
21. Clark, A. J. 1926. *J. Physiol.* (*London*) **61,** 530.

22. Ariëns, E. J. 1954. *Arch. Intern. Pharmacodyn. Thér.* **99,** 32; van Rossum, J. M., and Ariëns, E. J. 1962. *Arch. Int. Pharmacodyn. Thér.* **136,** 385; van Rossum, J. M. 1963. *J. Pharm. Pharmacol.* **15,** 285.
23. Stephenson, R. P. 1956. *Br. J. Pharmacol. Chemother.* **11,** 379.
24. Paton, W. D. M. 1961. *Proc. R. Soc. London, Ser. B* **154,** 21.
25a. Koshland, D. E., Jr. 1958. *Proc. Natl. Acad. Sci. U.S.A.* **44,** 98.
25b. Koshland, D. E., Jr. 1961. *Biochem. Pharmacol.* **8,** 57.
25c. Koshland, D. E., Jr., and Neet, K. E. 1968. *Annu. Rev. Biochem.* **37,** 359.
26. Belleau, B. 1964. *J. Med. Chem.* **7,** 776; Belleau, B. 1965. *Adv. Drug Res.* **2,** 89.
27. Monad, J., Wyman, J., and Changeux, J.-P. 1965. *J. Mol. Biol.* **12,** 88.
28. Karlin, A. 1967. *J. Theor. Biol.* **16,** 306.
29. Ganellin, C. R. 1982. *In* "Pharmacology of Histamine Receptors" (Ganellin, C. R., and Parsons, M. E., eds.), Chap. 2. Wright-PSG, Bristol, England.
30. Ariëns, E. J., Simonis, A. M., and van Rossum, J. M. 1964. *In* "Molecular Pharmacology" (Ariëns, E. J., ed.), Vol. 1, pp. 212 and 225. Academic Press, New York.
31. Roth, F. E., and Govier, W. M. 1958. *J. Pharmacol. Exp. Ther.* **124,** 347.
32a. Ariëns, E. J. 1986. *Med. Res. Rev.* **6,** 451.
32b. Ariëns, E. J. 1987. *Med. Res. Rev.* **7,** 367.
33. Pfeiffer, C. 1956. *Science* **124,** 29.
34. White, P., Ham, J., Way, W., and Trevor, A. 1980. *Anesthesiology* **52,** 231.
35. Takada, T., Tada, M., and Kiyomoto, A. 1966. *Nippon Yakurigaku Zasshi* **62,** 64; 1967. *Chem. Abstr.* **67,** 72326s.
36. Aps, C., and Reynolds, F. 1978. *Br. J. Clin. Pharmacol.* **6,** 63.
37. Tobert, J., Cirillo, V., Hitzenberger, G., James, I., Pryor, J., Cook. T, Buntinx, A., Holmes, I., and Lutterbeck, P. 1981. *Clin. Pharmacol. Ther. (St. Louis)* **29,** 344.
38. Drayer, D. E. 1986. *Clin. Pharmacol. Ther. (St. Louis)* **40,** 125.
38a. Sullivan, H. R., Beck, J. R., and Pohland, A. 1963. *J. Org. Chem.* **28,** 2381.
39. Ho, I. K. 1981. *Annu. Rev. Pharmacol. Toxicol.* **21,** 83.
40. Zimmerman, D., and Gesellchen, P. 1982. *Annu. Rep. Med. Chem.* **17,** 21.
41. Hill, D. R., and Bowery, N. G. 1981. *Nature (London)* **290,** 149.
42. Cahn, R. S., Ingold, C. K., and Prelog, V. 1966. *Angew. Chem., Int. Ed. Engl.* **5,** 385.
43. Dukes, M., and Smith, L. H. 1971. *J. Med. Chem.* **14,** 326.
44. Ariëns, E. J. 1988. *Med. Res. Rev.* **8,** 309.
45. Simonyi, M. 1984. *Med. Res. Rev.* **4,** 359.
46. Lien, E. J., Rodrigues de Miranda, J. F., and Ariëns, E. J. 1976. *Mol. Pharmacol.* **12,** 598.
47. Cushny, A. 1926. "Biological Relations of Optically Isomeric Substances." Williams & Wilkins, Baltimore, Maryland.
48. Easson, L. H., and Stedman, E. 1933. *Biochem. J.* **27,** 1257.
49a. Blaschko, H. 1950. *Proc. R. Soc. London Ser B* **137,** 307.
49b. Jamali, F., Mehvar, R., and Pasutto, F. M. 1989. *J. Pharm. Sci.* **78,** 695.
50. Cross, L. C., and Klyne, W. 1976. *Pure Appl. Chem.* **45,** 11.
51. Kaiser, C., and Setler, P. E. 1981. *In* "Burger's Medicinal Chemistry" (Wolff, M. E., ed.), 4th Ed., Part 3, Chap. 56. Wiley, New York.
52. Smissman, E. E., Nelson, W. L., LaPidus, J. B., and Day, J. L. 1966. *J. Med. Chem.* **9,** 458.
53. Armstrong, P. D., Cannon, J. G., and Long, J. P. 1968. *Nature (London)* **220,** 65; Chiou, C. Y., Long, J. P., Cannon, J. G., and Armstrong, P. D. 1969. *J. Pharmacol. Exp. Ther.* **166,** 243.
54. Armstrong, P. D., and Cannon, J. G. 1970. *J. Med. Chem.* **13,** 1037.
55. Li, J. P., and Biel, J. H. 1969. *J. Med. Chem.* **12,** 917.
56. Chen, C.-Y., and LeFèvre, R. J. W. 1965. *Tetrahedron Lett.* 4057.
57. Nogrady, T. 1985. "Medicinal Chemistry," p. 28. Oxford Univ. Press, New York.

58. Triggle, D. J. 1987. *In* "Trends in Medicinal Chemistry" (Mutschler, E., and Winterfeldt, E., eds.), p. 57. VCH, Weinheim; Triggle, D. J., and Janis, R. A. 1987. *Annu. Rev. Pharmacol. Toxicol.* **27,** 347.
59. Ganellin, R. 1981. *J. Med. Chem.* **24,** 913; Ganellin, C. R., and Durant, G. J. *In* "Burger's Medicinal Chemistry" (Wolff, M. E., ed.), 4th Ed., Part 3, Chap. 48. Wiley, New York.
60. Black, J. W., Durant, G. J., Emmett, J. C., and Ganellin, C. R. 1974. *Nature (London)* **248,** 65.
61. Charton, M. 1965. *J. Org. Chem.* **30,** 969.

## General References

### Drug–Receptor Interactions

Albert, A. 1985. "Selective Toxicity," 7th Ed. Chapman & Hall, London.
Korolkovas, A. 1970. "Essentials of Molecular Pharmacology." Wiley, New York.

### Drug–Receptor Theories

O'Brien, R. D., ed. 1979. "The Receptors." Plenum, New York.
Smithies, J. R., and Bradley, R. J., eds. 1978. "Receptors in Pharmacology." Dekker, New York.

### Stereochemical Considerations

Smith, D. F., ed. 1989. "CRC Handbook of Stereoisomers: Therapeutic Drugs." CRC Press, Boca Raton, Florida.

### Ion Channels

Hille, B. 1984. "Ionic Channels of Excitable Membranes." Sinauer, Sunderland, Massachusetts.
Triggle, D. J. 1987. *In* "Trends in Medicinal Chemistry" (Mutschler, E., and Winterfeldt, E., eds.), VCH, Weinheim.
Triggle, D. J., and Janis, R. A. 1987. *Annu. Rev. Pharmacol. Toxicol.* **27,** 347.

### Histamine $H_1$ and $H_2$ Receptors

Ganellin, C. R., and Parsons, M. E., eds. 1982. "Pharmacology of Histamine Receptors." Wright-PSG, Bristol, England.

CHAPTER 4

# Enzymes (Catalytic Receptors)



## I. Enzymes as Catalysts

Enzymes are special types of receptors. The receptors discussed in Chapter 3 interact with agonists to form complexes which then elicit a biological response. Subsequent to the response, the agonist is released intact. Enzymes interact with substrates to form complexes, but, unlike other receptors, it is from these *enzyme–substrate complexes* that enzymes catalyze reactions, thereby transforming the substrates into products that are released. Therefore, the two characteristics of enzymes are their ability to recognize a substrate and to catalyze a reaction of it.

### A. What Are Enzymes?

*Enzymes* are proteins that catalyze reactions in a biological system. The first enzyme to be recognized as a protein was jack bean urease,[1] which catalyzes the hydrolysis of urea to $CO_2$ and $NH_3$. Enzymes can have molecular weights of several thousand to several million, yet they catalyze transformations on molecules as small as $CO_2$ and $N_2$. Carbonic anhydrase (carbonate dehydratase) from bovine erythrocytes, for example, has a molecular weight of 31,000, and each enzyme molecule can catalyze the hydration of 200,000 molecules of $CO_2$ to $H_2CO_3$ per second! This is $10^7$ times faster than the uncatalyzed reaction.

### B. How Do Enzymes Work?

There are a wide variety of rationalizations for enzyme catalysis. In fact, Page[2] has compiled 21 hypotheses, all of which may provide a composite explanation for this phenomenon.

According to Kraut,[3] modern theories of enzyme catalysis originated with Haldane,[4] who introduced the concept that an enzyme–substrate complex requires additional activation energy prior to reaction, which presumably is obtained by substrate strain energy on the enzyme. Eyring[5] then developed *transition-state theory*, which is the basis for the hypothesis of Pauling[6] whose vision of an enzyme (which is amazingly accurate considering how little was known about enzymes at that time) was that it was a flexible template designed by evolution to be complementary to the structure of its substrates at the transition state of the reaction rather than at the ground state. As the reaction proceeds toward the transition state, the enzyme interacts more effectively (increased binding energy) with the transition state geometry and electronic environment, which consequently accelerates the reaction. This is referred to as *transition-state stabilization*. It has been suggested[7] that nearly all of the above-mentioned 21 hypotheses of enzyme catalysis are just alternative expressions of transition-state stabilization or are related factors.

Similar to the case of noncatalytic receptors in which the pharmacophore of the drug interacts with a relatively small part of the total receptor, the substrate likewise binds to only a small part of the enzyme known as the *active site* of the enzyme. So why then are enzymes so large? The interactions between substrate and enzyme (see Chapter 3, Section III,B) must be quite specific in order to attain the catalytic effect. It has been suggested[8] that the large binding energies between substrate and enzyme result from close packing of atoms within the protein, and that the large size may be required to achieve this. Therefore, the entire protein outside of the active site may be functioning to hold the active site in the proper geometry for catalysis. Another function of the protein outside of the active site may be to channel the substrate into the active site.

The two key features of enzyme catalysis are *specificity* and *rate acceleration*. The active site contains moieties responsible for both of these properties.

## 1. Specificity of Enzyme-Catalyzed Reactions

Specificity refers to both *specificity of binding* and *specificity of reaction*. Enzyme catalysis is initiated by a prior interaction between the substrate and the enzyme, known as the *E · S complex* or the *Michaelis complex*. Certain active site constituents are involved in these binding interactions which are responsible for the binding specificity. These interactions are the same as those discussed in Chapter 3 (Section III,B) for the interaction of an agonist with a noncatalytic receptor and include covalent, electrostatic, ion–dipole, dipole–dipole, hydrogen-bonding, charge-transfer, hydrophobic, hydrophilic, and van der Waals interactions.

As indicated above, maximum binding interactions at the active site occur at the transition state of the reaction. Therefore, it is important that an enzyme does not bind to intermediate states excessively, or this will increase the free energy difference between the intermediate and transition state. The binding interactions set up the substrate for the reaction that the enzyme catalyzes (Scheme 4.1).

$$\mathrm{E + S} \overset{K_s}{\rightleftharpoons} \mathrm{E{\bullet}S} \overset{k_{cat}}{\rightleftharpoons} \mathrm{E{\bullet}P} \rightleftharpoons \mathrm{E + P}$$

**Scheme 4.1.** Generalized enzyme-catalyzed reaction.

Binding specificity can be absolute, that is, essentially only one substrate forms an E · S complex with a particular enzyme, which then leads to product formation; or binding specificity can be very broad, in which case many molecules of related structure can bind and be converted to product. Because enzymes are chiral (mammalian enzymes are comprised of only L-amino acids), interactions of an enzyme with a racemic mixture results in two diastereomeric complexes. This is analogous to the principle behind the resolution of racemic mixtures with chiral reagents. If, for example (Scheme 4.2), a pure (*R*)-isomer of a chiral amine such as (*R*)-2-methylbenzylamine (**4.1**) is mixed with a racemic mixture of a carboxylic acid such as the nonsteroidal anti-inflammatory drug ibuprofen (**4.2**), then two diastereomeric salts (*R* · *R*) and (*R* · *S*) will be formed. Because the salts are no longer enantiomers, they will have different properties and can be separated by physical means. This is what happens when an enzyme is exposed to a racemic mixture of a substrate. E·S complex formation may or may not be possible with both enantiomers. The binding energy for E·S complex formation with one enantiomer may be much higher than that with the other enantiomer either because of differential binding interactions as noted above or for steric reasons. If the binding energies for the two complexes are significantly different, then only one E·S complex may form. Alternatively, both E·S complexes may form, but only one E·S complex may lead to product formation. The enantiomer that forms the E·S complex but is not *turned over* (i.e., converted by the enzyme to product) is said to undergo *nonproductive binding* to the enzyme. Enzymes also can demonstrate complete stereospecificity with geometric isomers as these are diastereomers.

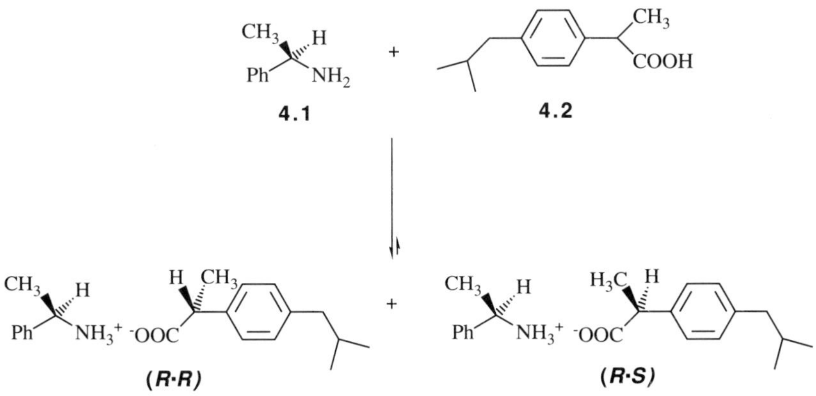

**4.1** + **4.2**

**(R·R)** + **(R·S)**

**Scheme 4.2.** Resolution of a racemic mixture.

Reaction specificity arises from other constituents of the active site, namely, specific acid, base, and nucleophilic functional groups of amino acids (see Section II) and specific organic molecules or transition metal ions called *coenzymes* or *cofactors* (see Section III). Enzymes also can show specificity for chemically identical protons (Fig. 4.1). If there are specific binding sites for R and R′ at the active site of the enzyme, and a base (B) of an amino acid

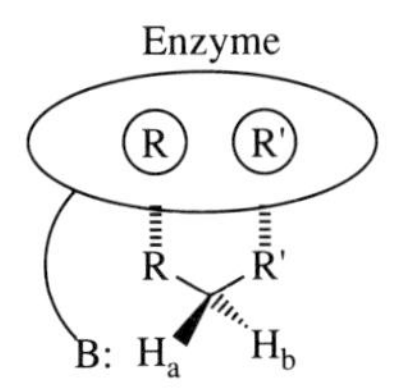


**Figure 4.1.** Enzyme specificity for chemically identical protons. R and R′ on the enzyme are groups that interact specifically with R and R′, respectively, on the substrate.

side chain is juxtaposed so that it can only reach proton $H_a$, then abstraction of $H_a$ will occur stereospecifically, even though in a nonenzymatic reaction $H_a$ and $H_b$ would be chemically equivalent and, therefore, have equal probability to be abstracted.

## 2. Rate Acceleration

In general, catalysts stabilize the transition-state energy relative to the ground state, and this decrease in $\Delta G^{\ddagger}$ is responsible for the rate acceleration that results (Fig. 4.2A). An enzyme has various opportunities to invoke catalysis, for example, by destabilization of the E·S complex, by stabilization of the transition states, by destabilization of intermediates, and during product release. Consequently, multiple steps, each having small $\Delta G^{\ddagger}$ values, may be involved (Fig. 4.2B). As a result of these multiple catalytic steps, rate accelerations of $10^{10}$–$10^{12}$ over the corresponding nonenzymatic reactions are possible. Enzyme catalysis, however, does not alter the equilibrium of a reversible reaction. If an enzyme accelerates the rate of the forward reaction, it must accelerate the rate of the corresponding back reaction as well; its effect is to accelerate the attainment of the equilibrium, but not the relative concentrations of substrates and products at equilibrium.

Typically, enzymes have *turnover numbers*, that is, the number of molecules of substrate converted to product per unit of time per molecule of enzyme active site, of the order of $10^3$ min$^{-1}$. $\Delta^5$-3-Ketosteroid isomerase from *Pseudomonas testosteroni* is one of the most efficient enzymes,[9] having a turnover number of $10^6$ sec$^{-1}$. As there are two other important steps to enzyme catalysis, namely, substrate binding and product release, high turnover numbers are only useful if these two physical steps occur at faster rates. This is not always the case.

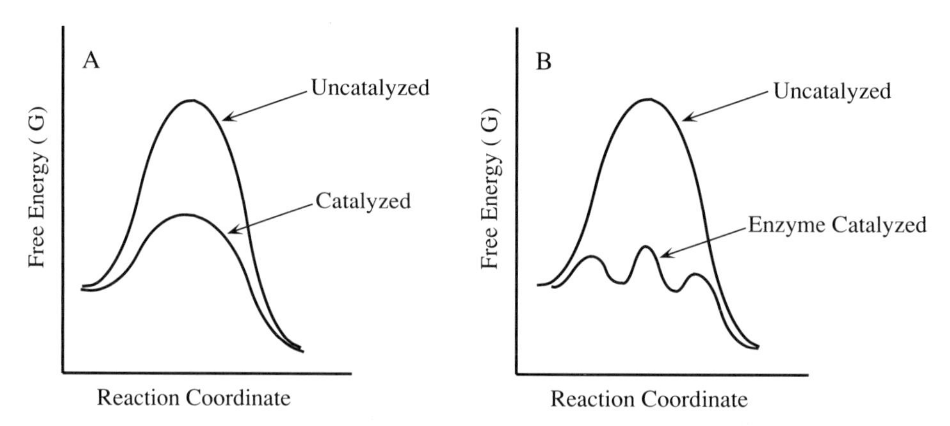


**Figure 4.2.** Effect of (A) a Chemical catalyst and (B) an enzyme on activation energy.

## II. Mechanisms of Enzyme Catalysis

Once the substrate binds to the active site of the enzyme via the interactions noted in Chapter 3 (Section III,B), there are various mechanisms that enzymes utilize to catalyze the conversion of substrate to product.[10–12] The most common mechanisms are approximation, covalent catalysis, general acid–base catalysis, electrostatic catalysis, desolvation, and strain or distortion.

### A. Approximation

*Approximation* is rate enhancement by proximity, that is, the enzyme serves as a template to bind the substrate(s) so that it is (they are) close to the reaction center. This results in a loss of rotational and translational entropies of the substrate upon binding to the enzyme; however, because the catalytic groups are now an integral part of the same molecule, the reaction becomes first order rather than second order when free in solution. Holding the reaction centers in close proximity is equivalent to increasing the concentration of the reacting groups. This phenomenon can be exemplified with nonenzymatic model studies. For example, consider the second-order reaction of acetate with an aryl acetate (Scheme 4.3). If the rate constant $k$ for this reaction is set equal to 1.0 $M^{-1}$ $sec^{-1}$, and then the effect of decreasing the rotational and translational entropy is determined by measuring the corresponding first-order rate constants for related intramolecular reactions, it is apparent from Table 4.1 that forcing the reacting groups to be closer to each other increases the reaction rate.[13,14]

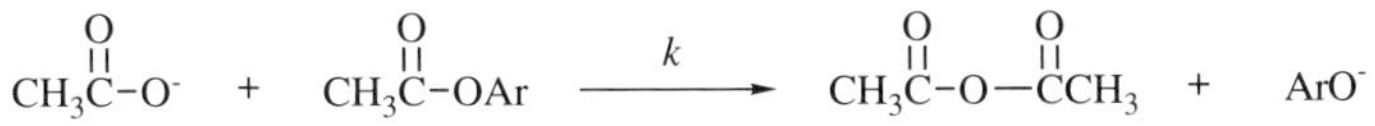

$$CH_3\overset{O}{\overset{\|}{C}}{-}O^- + CH_3\overset{O}{\overset{\|}{C}}{-}OAr \xrightarrow{k} CH_3\overset{O}{\overset{\|}{C}}{-}O{-}\overset{O}{\overset{\|}{C}}CH_3 + ArO^-$$

**Scheme 4.3.** Second-order reaction of acetate with aryl acetate.

Although first- and second-order rate constants cannot be compared directly, the efficiency of an intramolecular reaction can be defined in terms of its effective molarity (EM), the concentration of the catalytic group required to cause the intermolecular reaction to proceed at the observed rate of the intramolecular reaction. The EM is calculated by dividing the first-order rate constant for the intramolecular reaction by the second-order rate constant for the corresponding intermolecular reaction (see Table 4.1). What this indicates is that acetate ion would have to be at a concentration of, for example, 220 *M* in order for the intermolecular reaction to proceed at a rate comparable to that of the glutarate monoester reaction. Of course, 220 *M* acetate ion is an impossibility (pure water is only 55 *M*), so the effect of decreasing the entropy is significant. Effective molarities for a wide range of intramolecular reactions

**Table 4.1** Effect of Approximation on Reaction Rates[13,14]

| Reactants | $k_{obs}$ (30°C) | Relative rate, *k* | EM[a] |
|---|---|---|---|
| 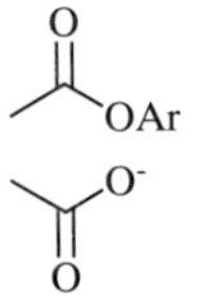 | $3.36 \times 10^{-7}$ $M^{-1}s^{-1}$ | 1.0 $M^{-1}s^{-1}$ | |
| 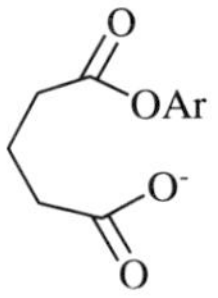 | $7.39 \times 10^{-5}$ $s^{-1}$ | 220 $s^{-1}$ | 220 *M* |
| 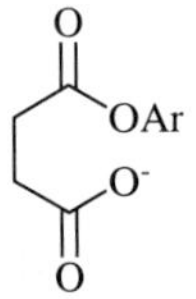 | $1.71 \times 10^{-2}$ $s^{-1}$ | $5.1 \times 10^{4}$ $s^{-1}$ | $5.1 \times 10^{4}$ *M* |
| 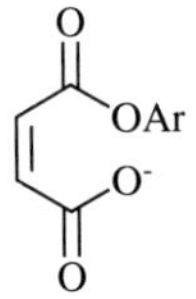 | $7.61 \times 10^{-1}$ $s^{-1}$ | $2.3 \times 10^{6}$ $s^{-1}$ | $2.3 \times 10^{6}$ *M* |
| 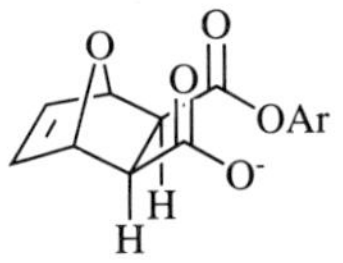 | 3.93 $s^{-1}$ | $1.2 \times 10^{7}$ $s^{-1}$ | $1.2 \times 10^{7}$ *M* |

[a] Effective molarity.

have been measured,[14] and the conclusion is that the efficiency of intramolecular catalysis varies with structure and can be as high as $10^{16}$ *M* for reactive systems. Therefore, holding groups proximal to each other in an enzyme–substrate complex can be an important contributor to catalysis.

## B. Covalent Catalysis

Some enzymes can use nucleophilic amino acid side chains in the active site to form covalent bonds to the substrate; a second substrate then can react with this enzyme-substrate intermediate to generate the product. This is

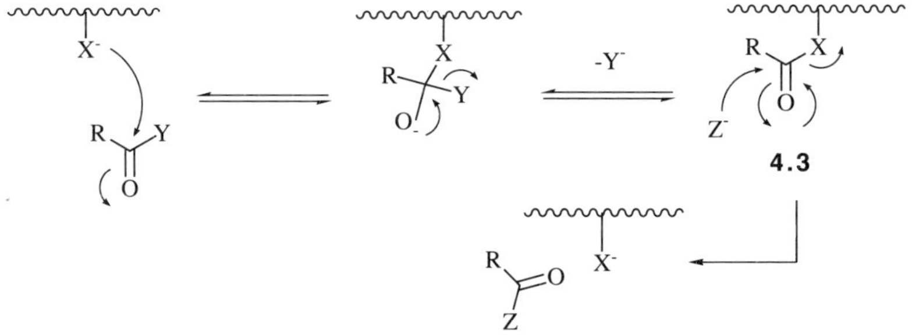

Scheme 4.4. Nucleophilic catalysis.

known as *nucleophilic catalysis* (Scheme 4.4), a subclass of *covalent catalysis* that involves covalent bond formation as a result of attack by an enzyme nucleophile at an electrophilic site on the substrate. For example, if Y in Scheme 4.4 is an amino acid or peptide and $Z^-$ is hydroxide ion, then the enzyme would be a peptidase (or protease). For nucleophilic catalysis to be most effective, Y must be converted to a better leaving group than X, and the covalent intermediate (**4.3**, Scheme 4.4) should be more reactive than the substrate.

Nucleophilic catalysis is the enzymatic analogy to anchimeric assistance by neighboring groups in organic reaction mechanisms. *Anchimeric assistance* is the process by which a neighboring functional group assists in the expulsion of a leaving group by intermediate covalent bond formation.[15] This results in accelerated reaction rates. Scheme 4.5 shows how the sulfur atom of a sulfur mustard nerve gas makes the displacement of the $\beta$-chlorine a much more facile reaction than it would be without the sulfur atom.

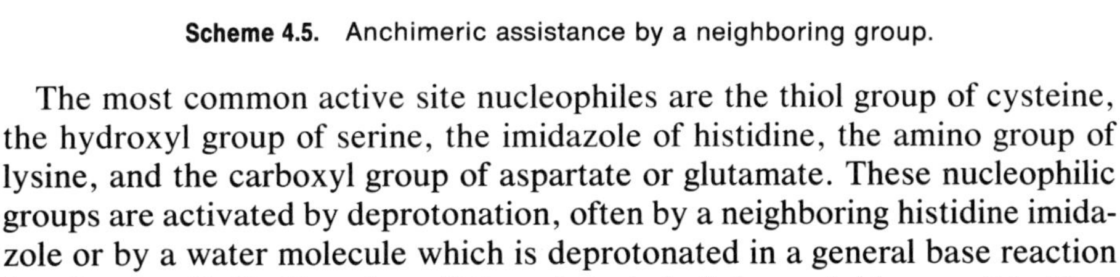

Scheme 4.5. Anchimeric assistance by a neighboring group.

The most common active site nucleophiles are the thiol group of cysteine, the hydroxyl group of serine, the imidazole of histidine, the amino group of lysine, and the carboxyl group of aspartate or glutamate. These nucleophilic groups are activated by deprotonation, often by a neighboring histidine imidazole or by a water molecule which is deprotonated in a general base reaction (see Section II,C). Therefore, if the substrate in Scheme 4.4 is a peptide [R = $NH_2CH$ (R'); Y = amino acid or peptide], then a peptidase would convert the relatively unreactive amide linkage to a covalent intermediate having a much more reactive ester linkage (if the serine hydroxyl group were the nucleophile) or thioester linkage (if a cysteine thiolate were involved), either of which could be rapidly hydrolyzed ($Z^- = HO^-$). The principal catalytic advantage of using an active site residue instead of water directly is that the former is a unimolecular reaction (since the substrate is bound to the enzyme, attack by the serine residue is equivalent to an intramolecular reaction), which is entropically favored over the bimolecular reaction with water. Also, alkoxides (serine) and thiolates (cysteine) are better nucleophiles than hydroxide.

Classic examples where nucleophilic catalysis is important are many of the proteolytic enzymes, for instance, the serine proteases such as elastase (degrades necrotic lung tissue) or plasmin (lyses blood clots) and the cysteine proteases such as papain (used in digestion).

### C. General Acid–Base Catalysis

In any reaction where proton transfer occurs *general acid catalysis* and/or *general base catalysis* is effective. Consider the hydrolysis of ethyl acetate (Scheme 4.6). This is an exceedingly slow reaction at neutral pH because both the nucleophile ($H_2O$) and the electrophile (the carbonyl of ethyl acetate) are unreactive. The reaction rate could be accelerated, however, if either the nucleophile or the electrophile could be activated. An increase in the pH increases the concentration of hydroxide ion, which is a much better nucleophile than is water, and, in fact, the rate of hydrolysis at higher pH increases (Scheme 4.7). Likewise, a decrease in the pH increases the hydronium ion concentration which can protonate the ester carbonyl, thereby increasing its electrophilicity, and this also increases the hydrolysis rate (Scheme 4.8). This being the case, then the hydrolysis rate should be doubly increased if base and acid are added together, right? Of course not. Addition of an acid to a base would only lead to neutralization and loss of any catalytic effect.

$$H_3C-C(=O)-OC_2H_5 \;+\; H_2O \;\rightleftharpoons\; H_3C-C(=O)-OH \;+\; C_2H_5OH$$

**Scheme 4.6.** Hydrolysis of ethyl acetate.

$$CH_3C(=O)-OC_2H_5 \;\xrightarrow{HO^-}\; CH_3C(=O)-OH \;+\; C_2H_5O^- \;\rightleftharpoons\; H_3C-C(=O)O^- \;+\; C_2H_5OH$$

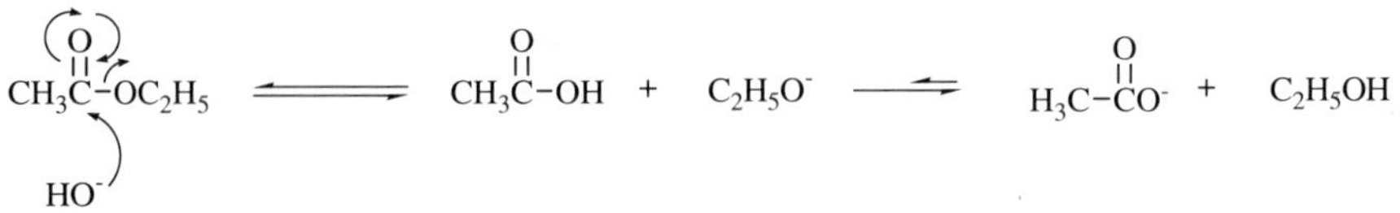

**Scheme 4.7.** Alkaline hydrolysis of ethyl acetate.

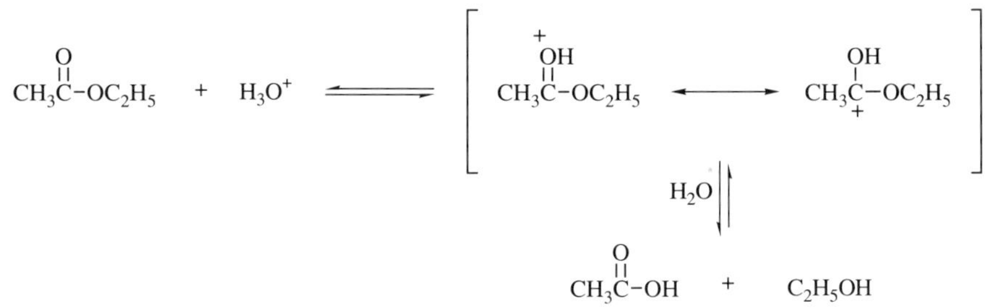

$$CH_3C(=O){-}OC_2H_5 + H_3O^+ \rightleftharpoons \left[ CH_3C(=\overset{+}{O}H){-}OC_2H_5 \longleftrightarrow CH_3\overset{+}{C}(OH){-}OC_2H_5 \right] \xrightleftharpoons{H_2O} CH_3C(=O){-}OH + C_2H_5OH$$

**Scheme 4.8.** Acid hydrolysis of ethyl acetate.

However, unlike reactions in solution, an enzyme can utilize acid and base catalysis simultaneously (Scheme 4.9). The protonated base in Scheme 4.9 is an acidic amino acid side chain and the free base is a basic residue, as was discussed in Section III,C of Chapter 3 for receptors. It is important to appreciate the fact that the $pK_a$ values of amino acid side-chain groups within the active site of enzymes are not necessarily the same as those measured in solution. If several bases are near the essential active site base, its $pK_a$ will be raised, and if acidic groups are adjacent to the essential active site acid, its $pK_a$ will be lowered. Also, the simultaneous donation of a proton to a carbonyl and removal of an $\alpha$-proton could account for the ability of an enzyme to deprotonate relatively high $pK_a$ carbon acids and make the corresponding enols.[15a] Therefore, removal of seemingly higher $pK_a$ protons from substrates by active site bases may not be as unreasonable as would appear if only solution chemistry were taken into consideration.

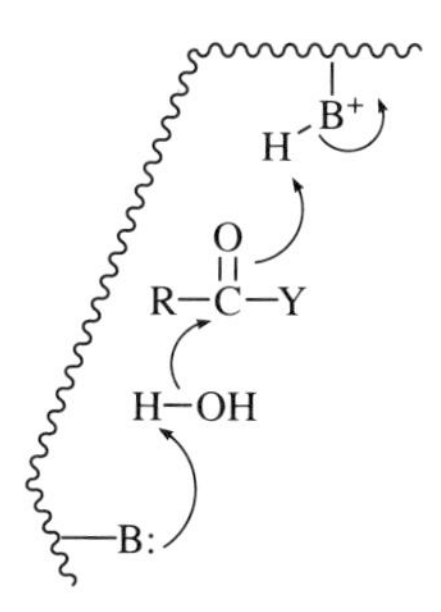


**Scheme 4.9.** Simultaneous acid and base enzyme catalysis.

As an example of general acid–base catalysis, consider the enzyme $\alpha$-chymotrypsin, a serine protease; it utilizes an active site serine residue in a covalent catalytic cleavage of peptide bonds. Since the nucleophilic group of serine is a hydroxyl, it should be a poor nucleophile. However, aspartic acid and histidine residues nearby have been implicated in the microscopic conversion of the serine to an alkoxide by a mechanism called the *charge relay system* (Scheme 4.10). This *catalytic triad* involves the aspartate carboxyl ($pK_a$ 3.9 in solution) removing a proton from the histidine imidazole ($pK_a$ 6.1

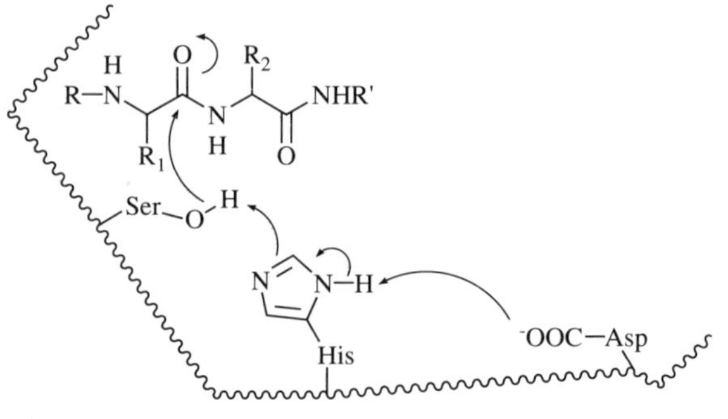

Scheme 4.10. Charge relay system for activation of an active site serine residue.

in solution) which, in turn, removes a proton from the serine hydroxyl group ($pK_a$ 14 in solution). On the basis of the solution $pK_a$ values this would be expected to be a high-energy process; presumably the $pK_a$ values at the active site are different from those in solution.

## D. Electrostatic Catalysis and Desolvation

An enzyme catalyzes a reaction by destabilization of the ground state and by stabilization of the transition state. Ground state destabilization could occur by desolvation of charged groups (removal of water molecules) at the active site upon substrate binding, thereby exposing the substrate to a lower dielectric constant (possibly hydrophobic) environment which would destabilize a charged group on the substrate. Desolvation also could expose a water-bonded charged group at the active site so that it can more effectively participate in stabilization of a charge generated in the transition state. Because these *electrostatic interactions* are much stronger in low dielectric media than in water, the charged groups within the low dielectric environment more strongly stabilize developing charges at the transition state of a reaction (Scheme 4.11). In the case of the tetrahedral intermediate shown in Scheme 4.11 the site in the enzyme that leads to this stabilization is referred to as the

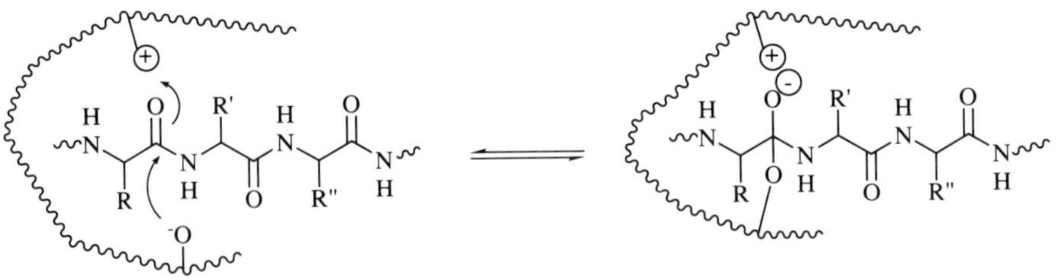

Scheme 4.11. Electrostatic stabilization of the transition state.

*oxyanion hole*.[16] This is analogous to the acceleration of the rate of solvolysis of *tert*-butyl bromide by catalytic silver salts. The silver ion coordinates to the bromine, thereby destabilizing the ground state and stabilizing the transition state.

The electrostatic interaction may not be as obvious as is shown in Scheme 4.11; instead of a full positive charge on the enzyme there may be one or more local dipoles having partial positive charges directed at the incipient transition state anion or a protonated group available for hydrogen bonding. In the case of the serine protease subtilisin it has been suggested that the lowering of the free energy of the activated complex is due largely to hydrogen bonding of the developing oxyanion with protein residues. When the suspected active site proton donor was replaced by a leucine residue using site-directed mutagenesis techniques, the $k_{cat}$ greatly diminished but the $K_m$ remained the same, indicating the importance of the hydrogen bonding to catalysis.[17] Furthermore, a mutant of subtilisin in which all three of the catalytic triad residues (serine 221, histidine 64, and aspartate 32) were replaced by alanine residues using site-directed mutagenesis still was able to hydrolyze amides $10^3$ times faster than the uncatalyzed hydrolysis rate (albeit $2 \times 10^6$ times slower than the wild-type enzyme).[18] This suggests that factors other than nucleophilic and general base catalysis must be important. Sophisticated free energy calculations on serine protease active sites indicated that the primary mechanism for rate acceleration is derived from electrostatic stabilization in the transition state.[19]

## E. Strain or Distortion

*Strain* and *distortion* play an important role in the reactivity of molecules in organic chemistry. The much higher reactivity of epoxides relative to other ethers demonstrates this phenomenon. Cyclic phosphate ester hydrolysis is another example. Considerable ring strain in **4.4** (Scheme 4.12) is released upon alkaline hydrolysis; the rate of hydrolysis of **4.4** is $10^8$ times greater than that for the corresponding acyclic phosphodiester (**4.5**).[20] Therefore, if strain or distortion could be induced during enzyme catalysis, then the reaction rate could be enhanced. This effect could be induced in the enzyme, thereby converting it to a high-activity state, or in the substrate, thereby raising the ground state energy and making it more reactive. In Section III,E,3 of Chap-

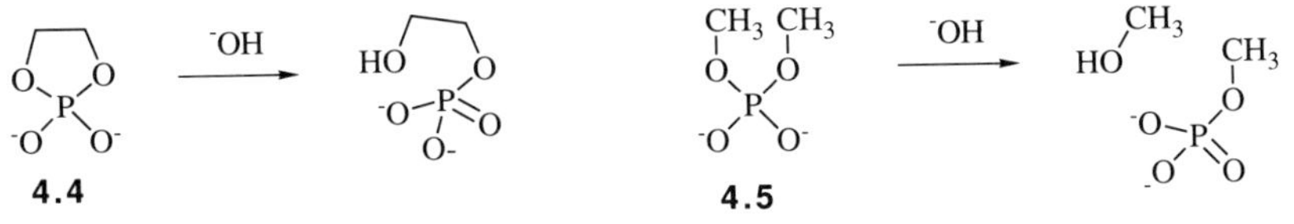

**4.4** $\xrightarrow{^-OH}$ **4.5** $\xrightarrow{^-OH}$

**Scheme 4.12.** Alkaline hydrolysis of phosphodiesters.

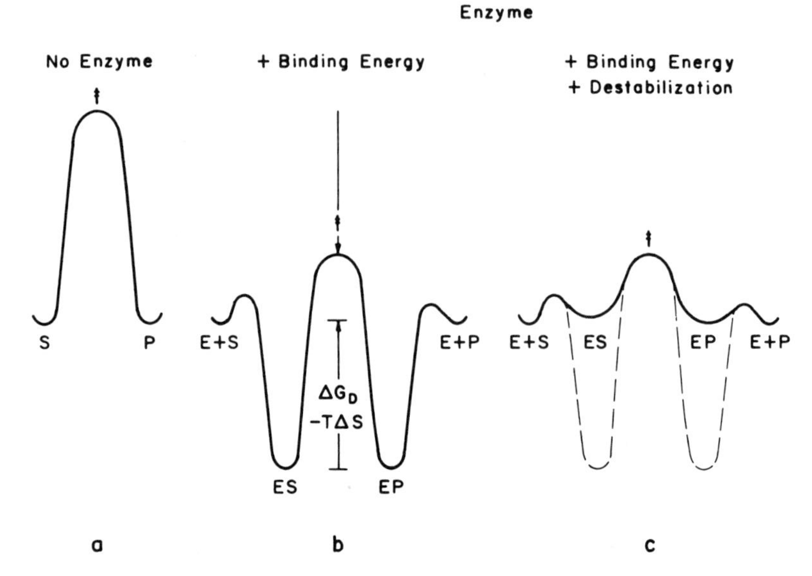


**Figure 4.3.** Energetic effect of enzyme catalysis. [Reproduced with permission from Jencks, W. P. (1987) Cold Spring Harbor Symp. Quant. Biol. **52,** 65. Copyright © 1987 Cold Spring Harbor Laboratory Press.]

ter 3 the induced-fit theory for drug–receptor interactions was discussed. This theory, originally proposed by Koshland and co-workers[21] to explain enzyme catalysis, suggests that when a substrate begins to bind to an enzyme, various groups on the substrate interact with particular active site functional groups and this mutual interaction induces a *conformational change* in the active site of the enzyme. This can result in a change of the enzyme from a low-catalytic form to a high-catalytic form by destabilization of the enzyme (strain or distortion) or by inducing proper alignment of active site groups involved in catalysis.

According to Jencks,[22] strain or distortion of the bound substrate is essential for catalysis. Since ground state stabilization of the substrate occurs concomitant with transition state stabilization, the $\Delta G^{\ddagger}$ is no different from that of the uncatalyzed reaction, only displaced downward (Fig. 4.3b). In order to lower the $\Delta G^{\ddagger}$ for the catalytic reaction, the E·S complex must be destabilized by strain, desolvation, or loss of entropy upon binding, thereby raising the $\Delta G$ of the E·S and E·P complexes (Fig. 4.3c). As the reaction proceeds, the $\Delta G^{\ddagger}$ can be lowered by release of strain energy or by other mechanisms described above.

### F. Example of the Mechanisms of Enzyme Catalysis

A very important bacterial enzyme in medicinal chemistry is the peptidoglycan transpeptidase (peptidoglycan endopeptidase),[l] the enzyme that catalyzes the cross-linking of peptidoglycan strands to make the bacterial cell wall. This enzyme has not been purified, but studies have been conducted with another penicillin-binding protein that is believed to be the *in vivo* transpeptidase which becomes uncoupled during its purification, namely, D-alanine carboxypeptidase. The carboxypeptidase has been purified and has been useful in the elucidation of the mechanism of penicillin action.[23]

Although a detailed mechanistic study of the transpeptidase has not yet been carried out, on the basis of the principles discussed above the hypothetical mechanism shown in Scheme 4.13 can be proposed. The E·S complex formed from the two strands of peptidoglycan with transpeptidase would be stabilized by the appropriate noncovalent binding interactions (see Section III,B of Chapter 3). These interactions could place the peptide carbonyl, which is ultimately the site of transpeptidation, very close (approximation) to the serine residue that is involved in covalent catalysis. Base catalysis could be utilized to activate the active site serine, and electrostatic catalysis (e.g., by a positively charged arginine residue) could stabilize the oxyanion intermediate. Alternatively, an active site acidic group could donate a proton (acid catalysis) to the incipient oxyanion to lower the transition state energy for the formation of the tetrahedral intermediate. Breakdown of this intermediate could be facilitated by proton donation (acid catalysis) to the leaving D-alanine residue in addition to appropriate strain energy in the $sp^3$ tetrahedral carbon produced by a conformational change that favors $sp^2$ hybridization of the product (the covalent intermediate). If a proton-donating mechanism instead of electrostatic mechanism were used to activate the initial covalent reaction, then the active site conjugate base could remove the proton to facilitate tetrahedral intermediate breakdown.

The third step, cross-linking of the second peptidoglycan strand with the newly formed ester linkage of the activated initial peptidoglycan strand, could be catalyzed by approximation of the two reacting centers, by electrostatic catalysis as in the first step, and by another conformational change in the enzyme to distort the $sp^2$ ester carbonyl toward the second $sp^3$ tetrahedral intermediate. Breakdown of that tetrahedral intermediate could be catalyzed again by strain energy ($sp^3$ back to $sp^2$) as well as by base catalysis. Product release may occur more readily with the enzyme in the charged form shown.

[l] For those unfamiliar with enzyme nomenclature, it is, in most cases, a simpleminded proposition. The suffix "-ase" is generally added to the type of reaction that the enzyme catalyzes. Therefore, a transpeptidase catalyzes the conversion of one peptide to another.

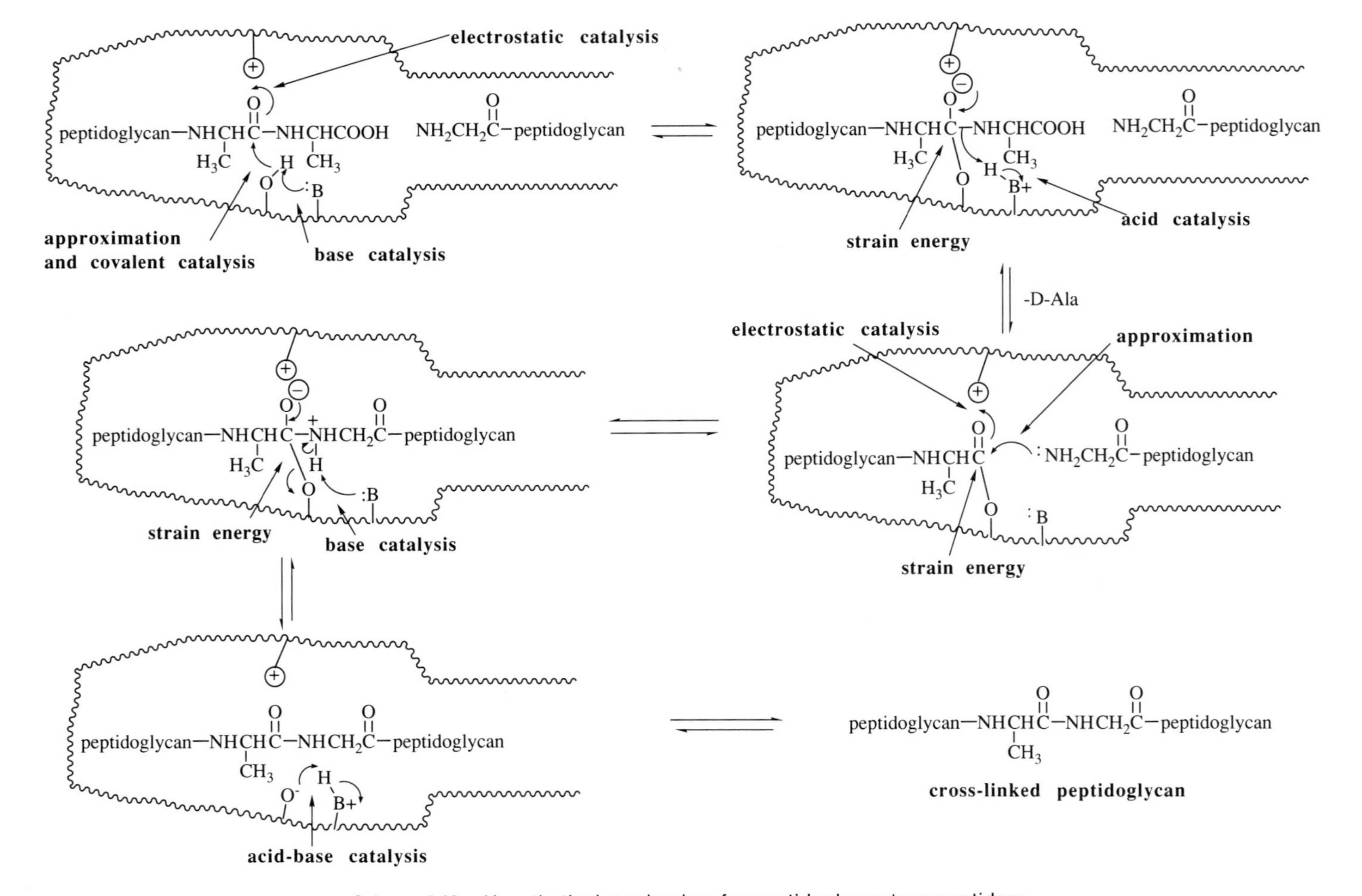


**Scheme 4.13.** Hypothetical mechanism for peptidoglycan transpeptidase.

Proton transfer after product release would return the enzyme to its normal energy state.

It must be emphasized that the mechanism shown in Scheme 4.13 is hypothetical. However, on the basis of the mechanism of enzyme catalysis described it is not an unreasonable hypothesis.

# III. Coenzyme Catalysis

A *coenzyme*, or *cofactor*, is any organic molecule or metal ion that is essential for the catalytic action of the enzyme. The usual organic coenzymes are generally derived as products of the metabolism of vitamins that we consume (Table 4.2). Other organic molecules that are involved in essential enzyme

**Table 4.2** Coenzymes Derived from Vitamins

| Vitamin | Structure | Coenzyme form | Structure | Coenzyme acronym |
|---|---|---|---|---|
| Vitamin $B_1$ (thiamin) | **4.6** (R = H) | Thiamin pyrophosphate | **4.6** (R = $^{-}O_2POPO_3^{2-}$) | TPP |
| Vitamin $B_2$ (riboflavin) | **4.7** (R = H) | Flavin mononucleotide | **4.7** (R = $PO_3^{2-}$) | FMN |
| | | Flavin adenine dinucleotide | **4.7** (R = $^{-}O_2POPO_2$-–$O^{-}$–5′-Ado) | FAD |
| Vitamin $B_3$ (niacinamide) | **4.8a** (no R) | Nicotinamide adenine dinucleotide | **4.8a** (R′ = H) | $NAD^+$ |
| (niacin) | Corresponding carboxylic acid | Nicotinamide dinucleotide phosphate | **4.8a** (R′ = $PO_3^{2-}$) | $NADP^+$ |
| | | Reduced nicotinamide adenine dinucleotide | **4.8b** (R′ = H) | NADH |
| | | Reduced nicotinamide adenine dinucleotide phosphate | **4.8b** (R′ = $PO_3^{2-}$) | NADPH |
| Vitamin $B_6$ (pyridoxine) | **4.9** (R = $CH_2OH$, R′ = H) | Pyridoxal 5′-phosphate | **4.9** (R = CHO, R′ = $PO_3^{2-}$) | PLP |
| | | Pyridoxamine 5′-phosphate | **4.9** (R = $CH_2NH_2$, R′ = $PO_3^{2-}$) | PMP |
| Vitamin $B_{12}$ (cyanocobalamin) | **4.10** (R = CN) | Adenosylcobalamin | **4.10** (R = 5′-Ado) | $CoB_{12}$ |
| Folic acid | **4.11a** (R = OH or poly-γ-glutamyl) | Tetrahydrofolate | **4.11b** (R = OH or poly-γ-glutamyl) | THF |
| Biotin | **4.12** (R = OH) | Covalently bound to enzyme as amide | **4.12** (R = lysine residue of enzyme) | — |
| Pantothenic acid | **4.13a** | Coenzyme A (no relation to vitamin A) | **4.13b** | CoASH |

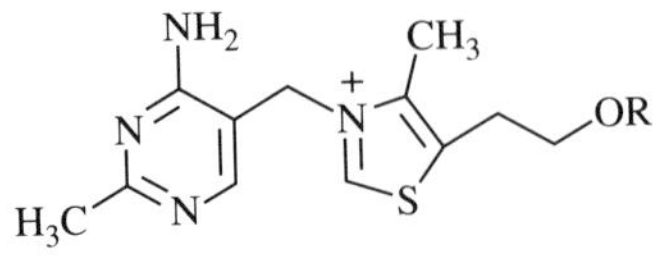

**4.6**

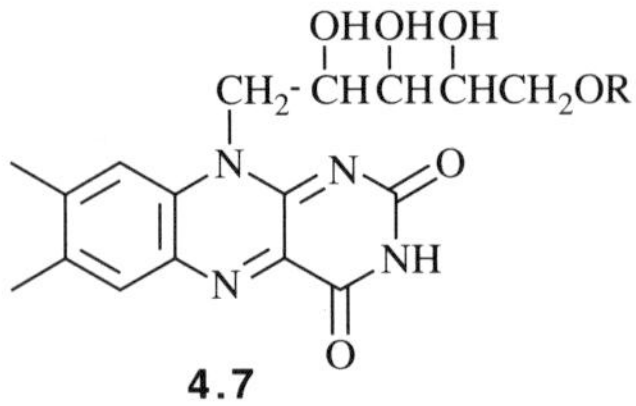

**4.7**

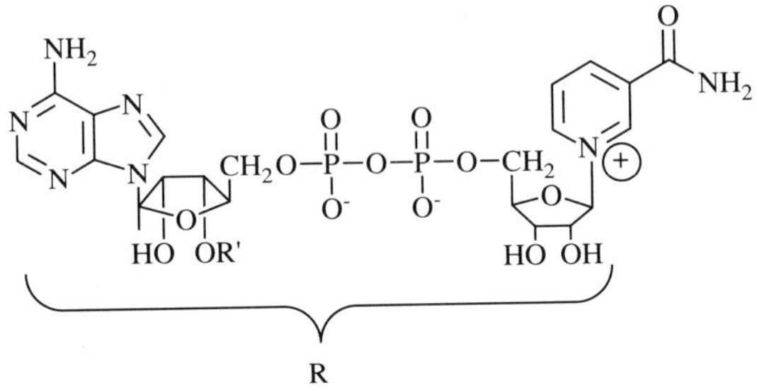

**4.8a**

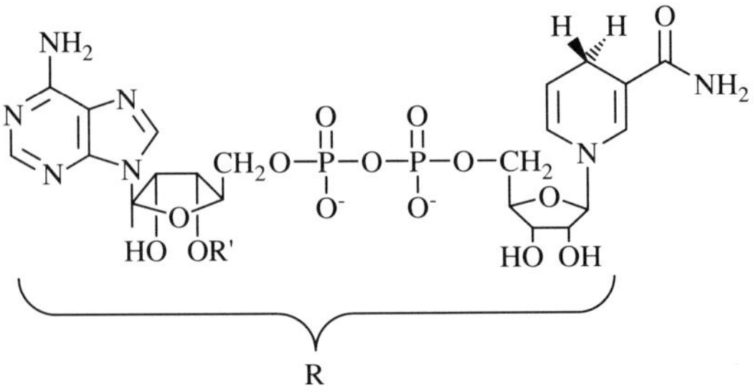

**4.8b**

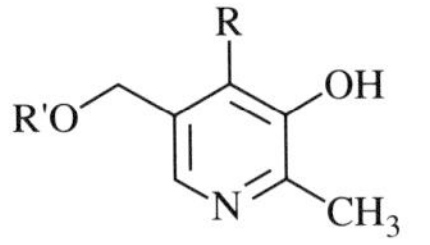

**4.9**

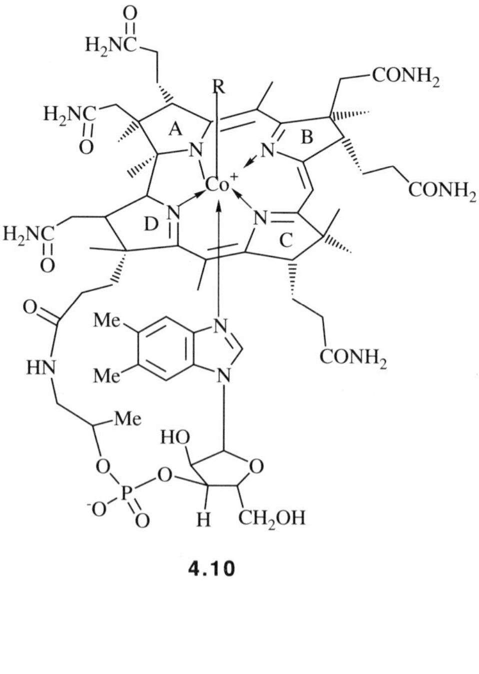

**4.10**

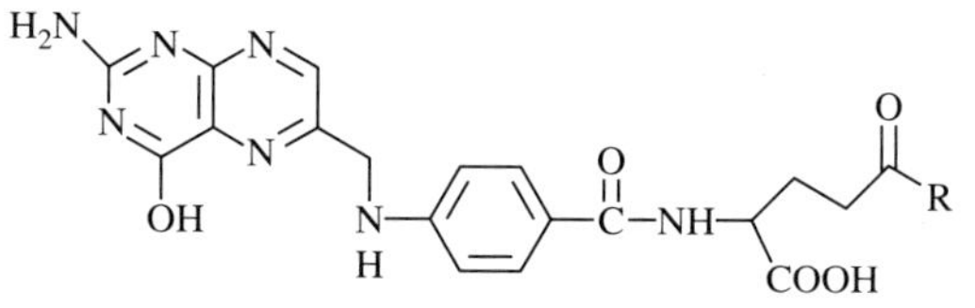

**4.11a**

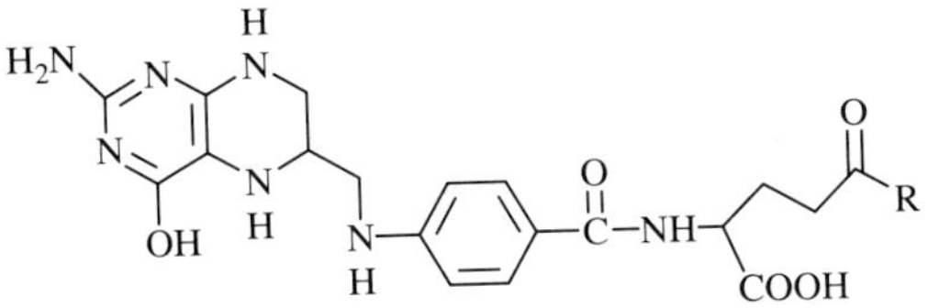

**4.11b**

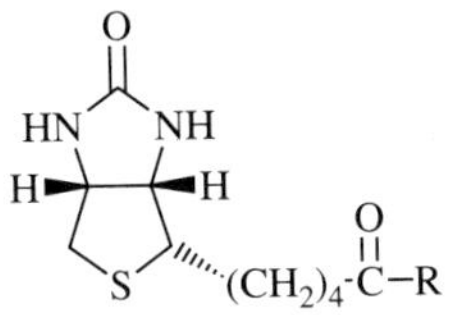

**4.12**

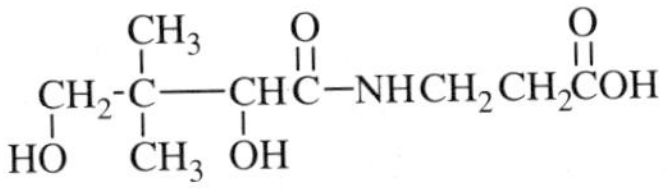

**4.13a**

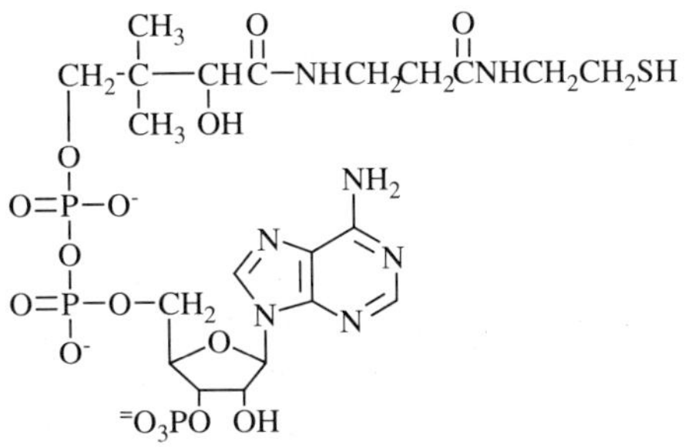

**4.13b**

functioning, but which are not derived from vitamins, include heme (**4.14**; protoporphyrin IX) and the tripeptide glutathione (**4.15**; GSH), which are very important to enzymes involved in drug metabolism (Chapter 8), adenosine triphosphate (**4.16**; ATP), which supplies the energy required to activate certain substrates during enzyme-catalyzed reactions, and coenzyme A (**4.13b**; CoA), which is used to make the CoA thioesters of carboxylic acid substrates.

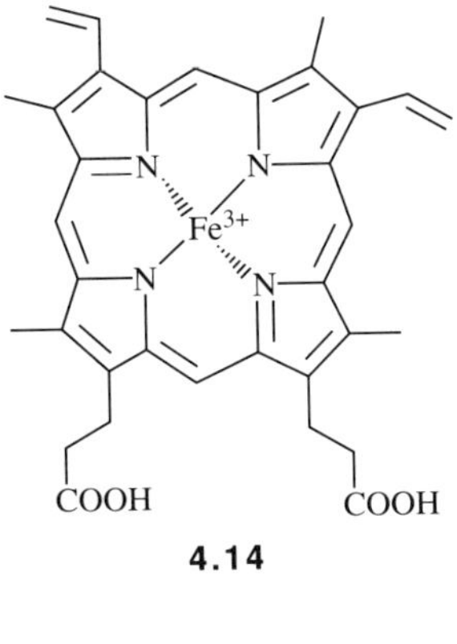

**4.14**

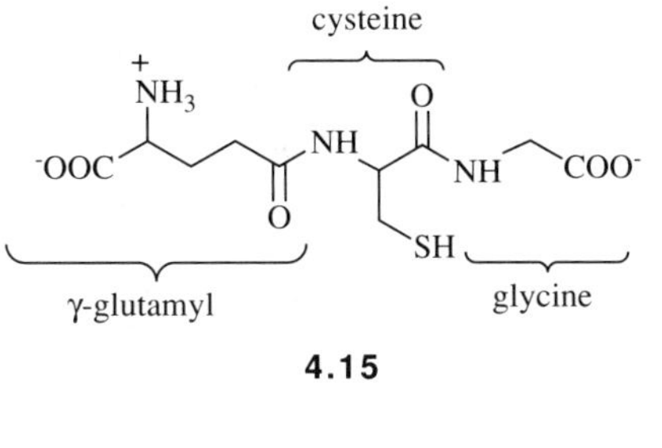

cysteine
γ-glutamyl
glycine

**4.15**

**Table 4.3** Symptoms of Vitamin Deficiency

| Deficient vitamin | Disease state |
|---|---|
| Thiamin | Beriberi |
| Riboflavin | Dermatitis, anemia |
| Niacin | Pellagra |
| Pyridoxine | Dermatitis, convulsions |
| Cyanocobalamin | Pernicious anemia |
| Folic acid | Pernicious anemia |
| Biotin | Dermatitis |
| Pantothenic acid | Neuromuscular effects |

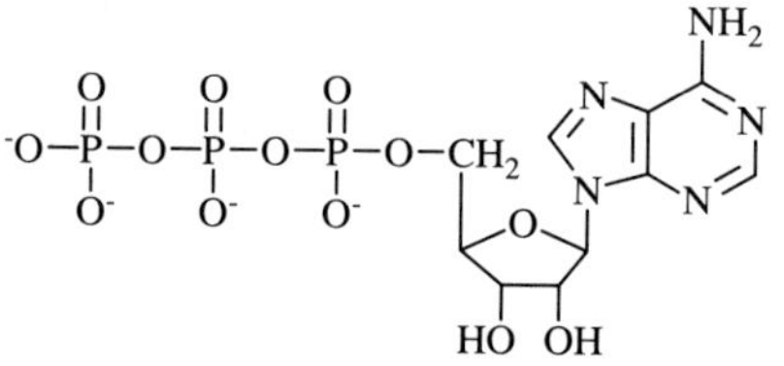

**4.16**

Vitamins are, by definition, essential nutrients; human metabolism is incapable of producing them. Deficiency in a vitamin results in the shutting down of the catalytic activity of various enzymes that require the coenzyme made from the vitamin. This leads to certain disease states (Table 4.3).[24]

The remainder of this chapter is devoted to the chemistry of coenzyme catalysis. The discussion is limited to only those coenzymes whose chemistry of action will be important to the mechanisms of drug action that are described in later chapters.

## A. Pyridoxal 5′-Phosphate

Enzymes dependent on pyridoxal 5′-phosphate (PLP) catalyze several different reactions of amino acids which, at first glance, appear to be unrelated; however, when the mechanisms are discussed, the relationships will become apparent. The overall reactions are summarized in Table 4.4. PLP is the most versatile coenzyme, but all of the reactions are very specific, depending on the enzyme to which the PLP is bound. Although the reactions can be catalyzed by PLP nonenzymatically, typically several of the possible reactions take place simultaneously.[25,26] For a given enzyme, only one reaction occurs almost exclusively.

Although there are several noncovalent binding interactions responsible for holding the PLP in the active site, the major interaction is a covalent one

**Table 4.4** Reactions Catalyzed by Pyridoxal 5′-Phosphate–Dependent Enzymes

| Substrate | Product | Reaction |
|---|---|---|
| $R\text{–}CH(NH_2)\text{–}COOH$ | $R\text{–}CH(NH_2)\text{–}COOH$ (opposite configuration) | racemization |
| $R\text{–}CH(NH_2)\text{–}COOH$ | $R\text{–}CH_2\text{–}NH_2$ + $CO_2$ | decarboxylation |
| $R\text{–}CH(NH_2)\text{–}COOH$ | $R\text{–}C(=O)\text{–}COOH$ + $NH_4^+$ | transamination |
| $R\text{–}CH(NH_2)\text{–}COOH$ | $H_2N\text{–}CH_2\text{–}COOH$ + "$R^+$" | α-cleavage |
| $X\text{–}CH_2\text{–}CH(NH_2)\text{–}COOH$ | $CH_3\text{–}C(=O)\text{–}COOH$ + $NH_4^+$ + $X^-$ | β-elimination |
| $X\text{–}CH_2\text{–}CH(NH_2)\text{–}COOH$ | $Y\text{–}CH_2\text{–}CH(NH_2)\text{–}COOH$ + $X^-$ | β-replacement |
| $X\text{–}CH_2\text{–}CH_2\text{–}CH(NH_2)\text{–}COOH$ | $CH_3\text{–}CH_2\text{–}C(=O)\text{–}COOH$ + $NH_4^+$ + $X^-$ | γ-elimination |

(**4.17a**, Fig. 4.4). Because the pyridine ring substituents appear to be involved in noncovalent binding interactions and not in the chemistry of the enzyme-catalyzed reactions, the structure of PLP will be abbreviated as shown in **4.17b** when the chemistry is discussed. The aldehyde group of the PLP is held tightly at the active site by a Schiff base (immonium) linkage to a lysine residue. In addition to securing the PLP in the optimal position at the active site, Schiff base formation activates the carbonyl for nucleophilic attack. This is very important to the catalysis because the first step in all PLP-dependent

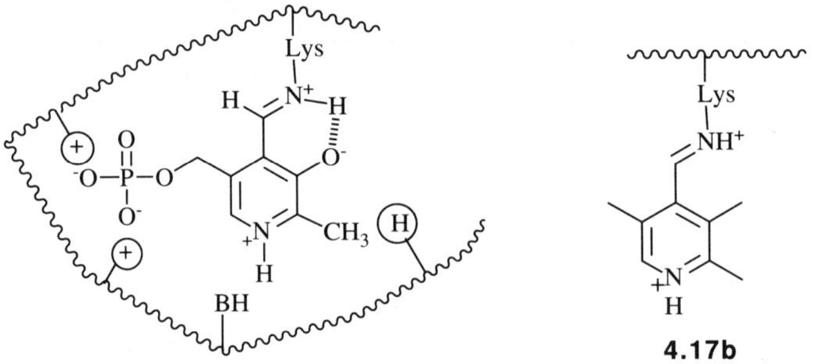

4.17a

4.17b

**Figure 4.4.** Pyridoxal 5′-phosphate covalently bound to the active site of an enzyme.

enzyme transformations of amino acids (see Table 4.4) is a transimination reaction, namely, the conversion of the lysine–PLP imine (**4.18**, Scheme 4.14) to the substrate–PLP imine (**4.19**). In Scheme 4.14 two different bases are shown to be involved in acid–base catalysis; a similar mechanism could be drawn with a single base. It is from **4.19** that all of the reactions shown in Table 4.4 occur. The property of **4.19** that links all of these reactions is that the pyridinium group can act as an electron sink to stabilize electrons by

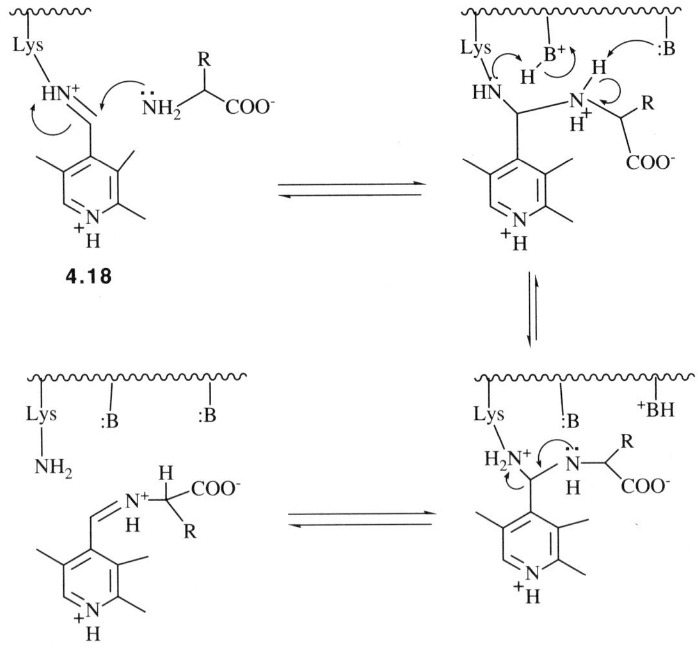

4.18

4.19

**Scheme 4.14.** First step in all pyridoxal 5′-phosphate–dependent enzyme reactions.

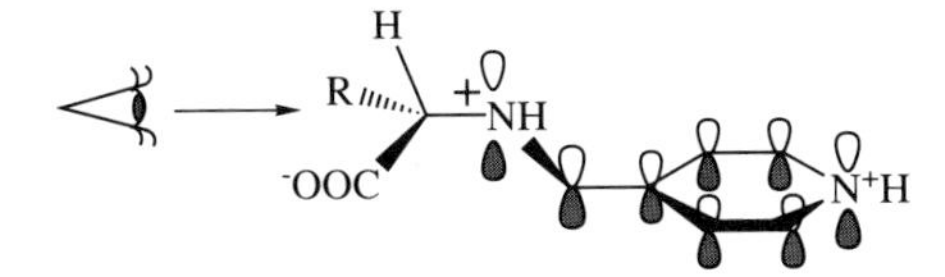


**Figure 4.5.** $\pi$-Electron system of the PLP–imine.

resonance from the C—H, C—$COO^-$, or C—R bonds. This could account for why all three of these bonds can be broken nonenzymatically. The important question is how can an enzyme catalyze the *regiospecific* cleavage of only one of these three bonds?

The bond that breaks must lie in a plane perpendicular to the plane of the PLP–imine $\pi$-electron system. In Fig. 4.5 the C—H bond is the one perpendicular to the plane of the $\pi$ system, that is, parallel with the *p* orbitals. This configuration results in maximum $\sigma$–$\pi$ electron overlap (the $sp^3$ $\sigma$ orbital of the C—H bond and *p* orbital of the aromatic system) and, therefore, minimizes the transition state energy for bond breakage of the C—H bond.[27] The problem for the enzyme to solve, then, is how to control the conformation about the $C_\alpha$—N bond so that only the bond that is to be cleaved is perpendicular to the plane of the $\pi$ system at the active site of the enzyme. The *Dunathan hypothesis*[27] gives a rational explanation for how an enzyme could control the $C_\alpha$—N bond rotation (Fig. 4.6). A positively charged residue at the active site could form a salt bridge with the carboxylate group of the amino acid bound to the PLP. This would make it possible for an enzyme to restrict rotation about the $C_\alpha$–N bond and hold the H (**A**), the $COO^-$ (**B**), or the R (**C**) group perpendicular to the plane of the aromatic system (the rectangles in Fig. 4.6). If the Dunathan hypothesis is accepted, then all of the PLP-dependent enzyme reactions can be readily understood.

In the next four subsections mechanisms for only the classes of PLP-dependent enzymes that will be relevant to later chapters are described. The mechanism for $\alpha$-cleavage is discussed in Section III,B.

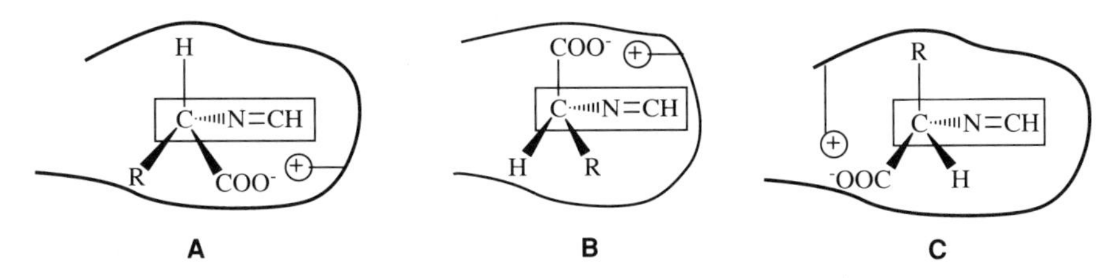


**Figure 4.6.** Dunathan hypothesis[27] for PLP activation of the $C_\alpha$–N bond. The rectangle represents the plane of the pyridine ring of the PLP. The angle of sight is that shown by the eye in Fig. 4.5. (Adapted by permission of John Wiley & Sons, Inc. From Dunathan, H. C. (1971) *In* "Advances in Enzymology," Vol. 35, p. 79, Meister, A., Ed. Copyright © 1971 John Wiley & Sons, Inc.)

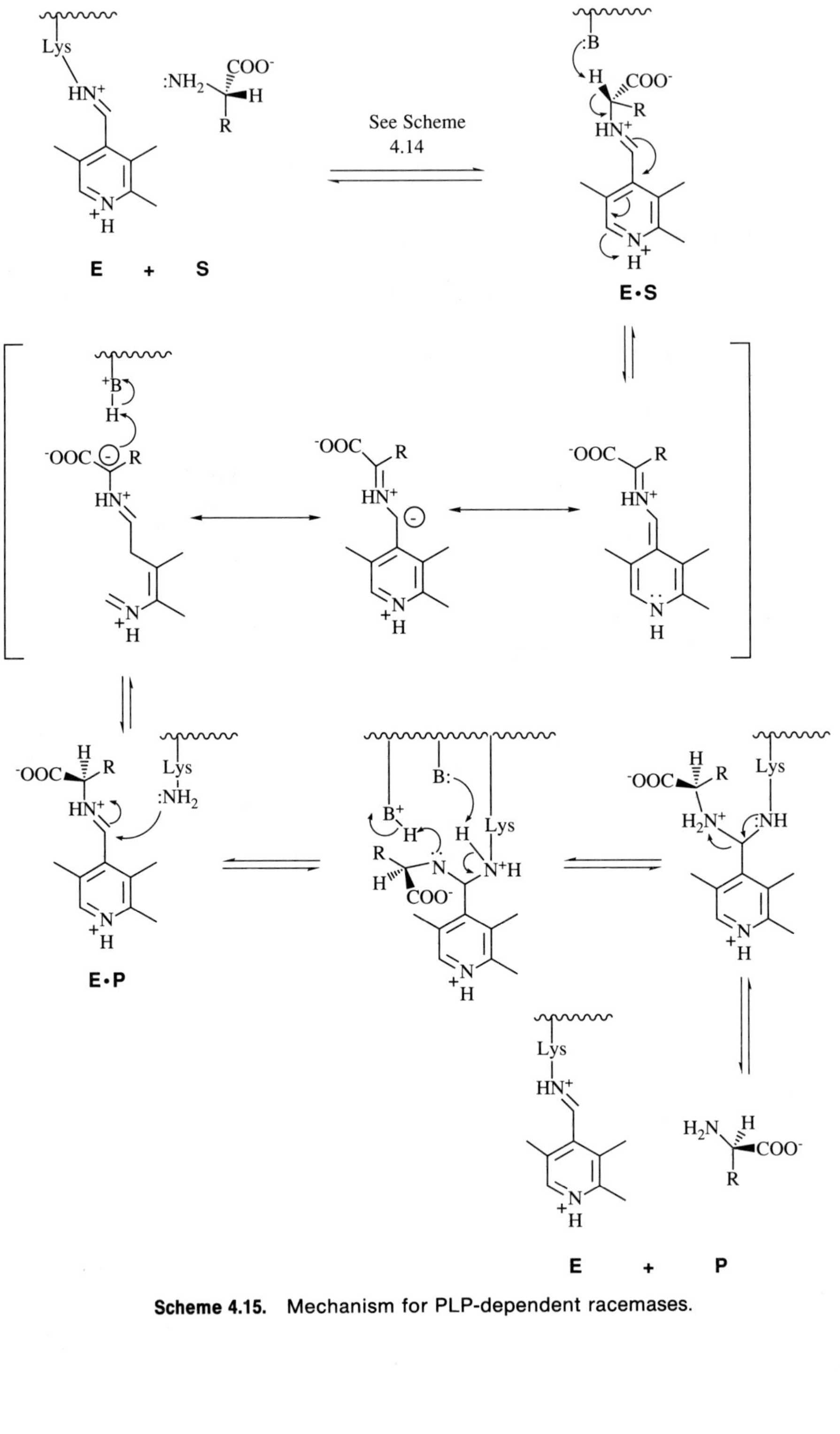

See Scheme 4.14

**E + S**

**E•S**

**E•P**

**E + P**

**Scheme 4.15.** Mechanism for PLP-dependent racemases.

### 1. Racemases

Unlike mammalian cells, bacteria utilize certain D-amino acids in the assembly of their cell wall.[28] Since natural amino acids have the L-conformation, bacteria require enzymes that convert the L-amino acids to their enantiomers. These enzymes are part of the family of enzymes known as *racemases*. The mechanism shown in Scheme 4.15 is typical for PLP-dependent racemases. Because the carbanion produced by $\alpha$-proton removal from the PLP–imine is so highly delocalized, the $pK_a$ of the $\alpha$-proton is much lower than that in the parent amino acid. Some enzymes catalyze the racemization with a single active site base, and others remove the $\alpha$-proton with one base and replace the proton on the other side with a different protonated base at the active site.[29] Note that all of the steps are reversible; therefore, these enzymes can accept either the D- or L-isomer, and the equilibrium mixture will be obtained. Generally, the equilibrium constant is about 1.

### 2. Decarboxylases

*Decarboxylases* catalyze the conversion of an amino acid to an amine (Scheme 4.16).[30] Since the loss of $CO_2$ is irreversible, the amine cannot be converted to the amino acid. However, the last step in Scheme 4.16 is reversible, so it is possible to catalyze the exchange of the $\alpha$-proton of the amine with solvent protons, as demonstrated, for example, when the isolated enzyme reaction is carried out in $^2H_2O$.

### 3. Aminotransferases (Transaminases)

As the name implies, *aminotransferases* catalyze the transfer of the amino group of the substrate amino acid to another molecule, an $\alpha$-keto acid (Scheme 4.17). The labeling pattern given in Scheme 4.17 indicates that the amino group of **4.20** is transferred to the $\alpha$-keto acid (**4.21**) which loses its oxygen as a molecule of water. The amino acid (**4.20**) is converted to an $\alpha$-keto acid (**4.22**), and the starting $\alpha$-keto acid (**4.21**) is converted to an amino acid (**4.23**). A mechanism consistent with these observations is given in Schemes 4.18 and 4.19. In Scheme 4.18 the $^{15}N$ substrate is transferred to the coenzyme, which is converted to $^{15}N$-containing pyridoxamine 5′-phosphate (PMP). This is an internal redox reaction: the amino acid substrate is oxidized to an $\alpha$-keto acid at the expense of the reduction of the PLP to PMP. Note that the first two steps (up to the three resonance structures) are identical to the mechanism for PLP-dependent racemization (Scheme 4.15). So how does the enzyme alter this racemization pathway in favor of transamination? It does so simply by the placement of an appropriate acidic residue closer to the PMP carbon (**4.24b**, Scheme 4.18) than the substrate $\alpha$-carbon (**4.24c**). This leads to

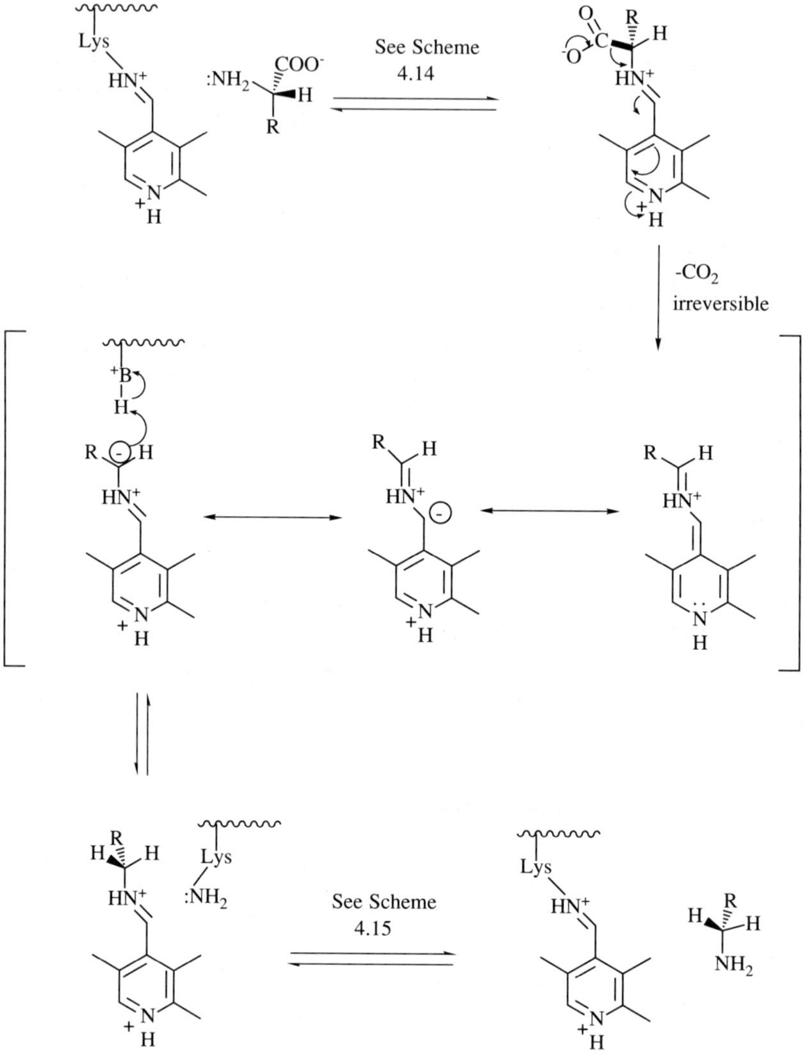

See Scheme 4.14

$-CO_2$ irreversible

See Scheme 4.15

**Scheme 4.16.** Mechanism for PLP-dependent decarboxylases.

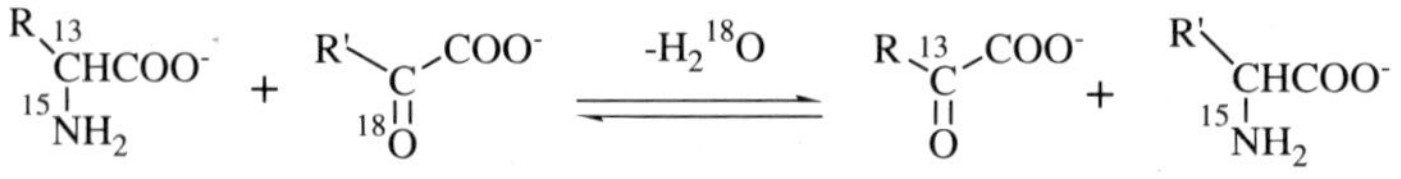

$-H_2{}^{18}O$

4.20 4.21 4.22 4.23

**Scheme 4.17.** Overall reaction catalyzed by PLP-dependent aminotransferases.

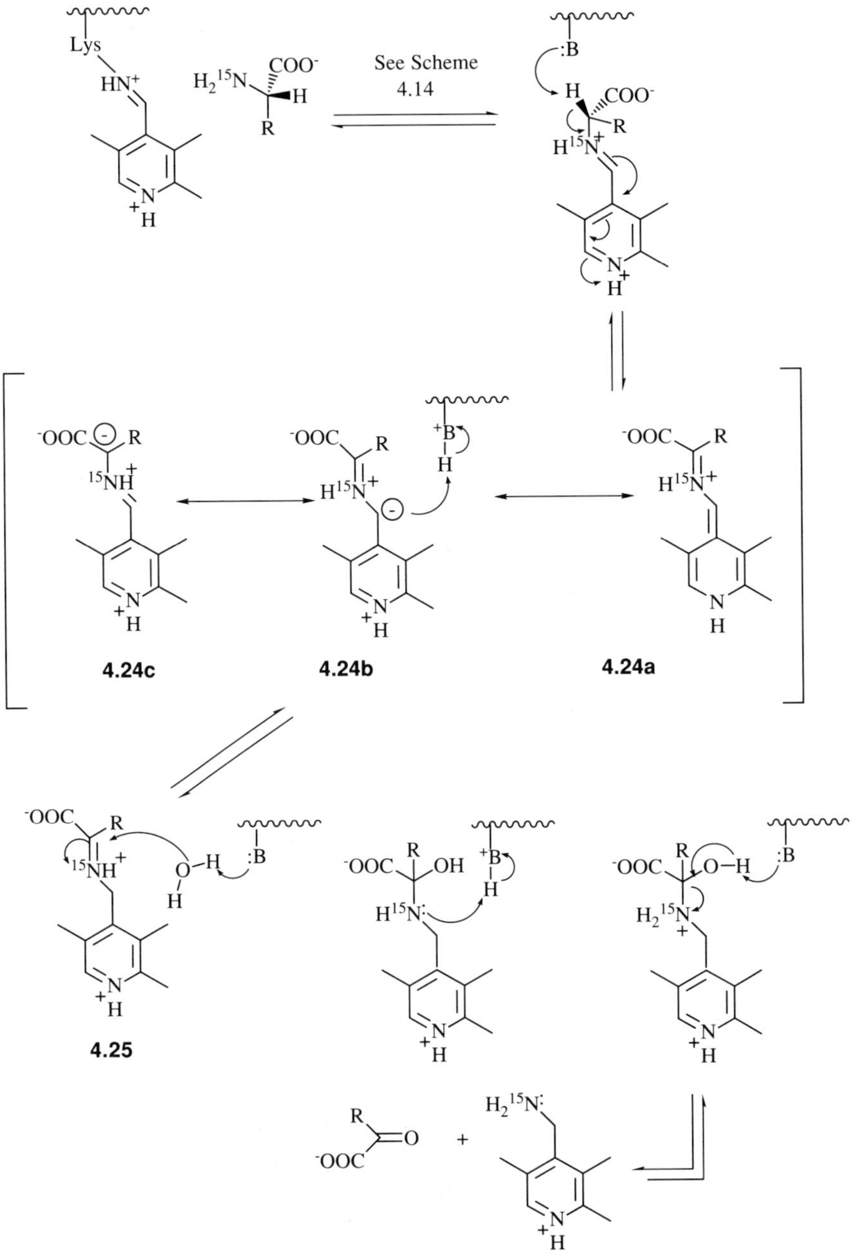

4.24c 4.24b 4.24a

4.25

Scheme 4.18. First half-reaction for the mechanism of PLP-dependent aminotransferases.

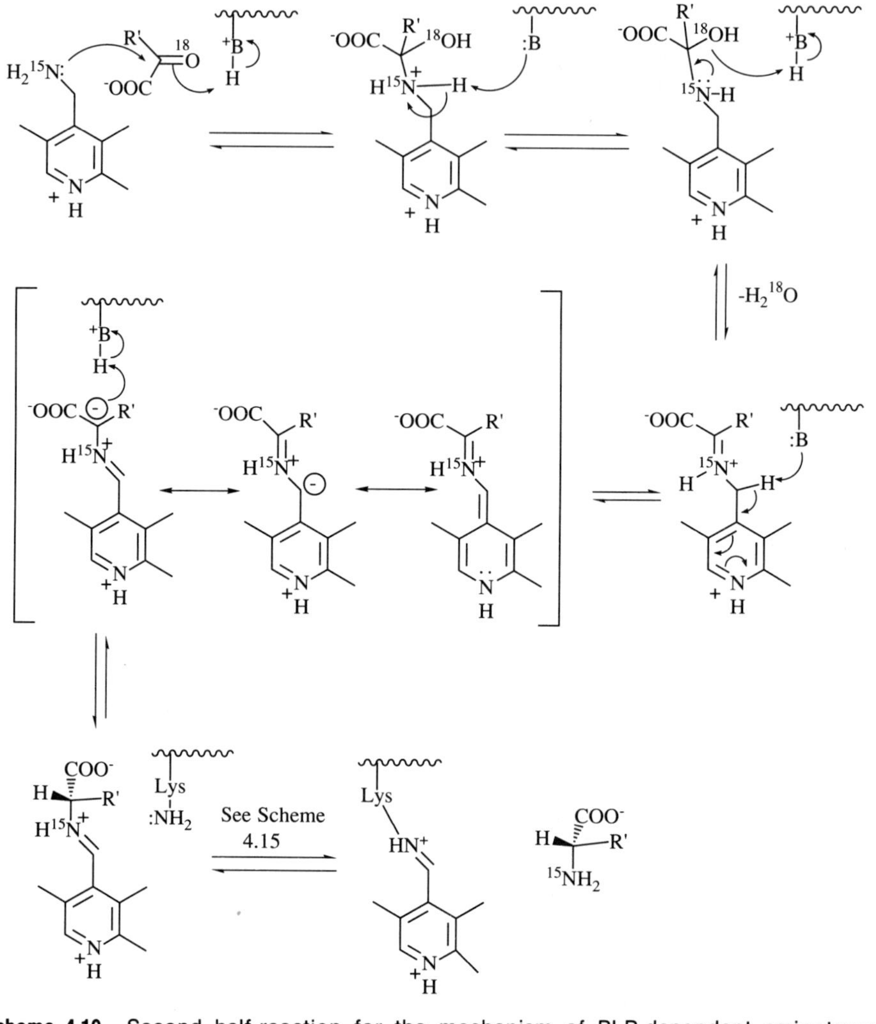

**Scheme 4.19.** Second half-reaction for the mechanism of PLP-dependent aminotransferases.

the PMP–imine Schiff base (**4.25**), which can be hydrolyzed to PMP and the α-keto acid. At this point the coenzyme is no longer in the proper oxidation state (i.e., an imine) for reaction with another amino acid substrate molecule (remember, the first step is Schiff base formation with an amino acid). Consequently, the enzyme is inactive for conversion of a second amino acid molecule to product.

If this were the end of the enzyme reaction, then the enzyme would be a reagent, not a catalyst, since only one turnover has taken place and the enzyme is inactive. Mother Nature handles this problem in a straightforward way. A second substrate, another α-keto acid (not the product α-keto acid), binds to the active site of the enzyme and undergoes the exact reverse of the reaction shown in Scheme 4.18. As depicted in Scheme 4.19, the amino group on the PMP, which was derived from the substrate amino acid, is then transferred to the second substrate, thereby producing a new amino acid. In the process the coenzyme (PMP) is oxidized back to PLP with concomitant reduction of the second substrate (the α-keto acid) to the amino acid. In the overall reaction (Schemes 4.18 and 4.19) the second half-reaction (Scheme 4.19) is only utilized to convert the enzyme back to the active form. There are some aminotransferases, however, in which the converse is true, namely, that the first half-reaction is to get the enzyme in the PMP form so that the intended substrate (an α-keto acid) can be converted to the product amino acid.

### 4. PLP–Dependent β-Elimination

Some PLP-dependent enzymes catalyze the elimination of H—X from substrate molecules. Although these enzymes are not directly relevant to later discussions, this sort of mechanism will be an important alternative pathway for certain inactivators of other PLP-dependent enzymes (see Section V,C,3,b of Chapter 5). The mechanism is shown in Scheme 4.20.

## B. Tetrahydrofolate and Pyridine Nucleotides

Tetrahydrofolate is produced from folic acid by a double reduction that involves the reduced form of the pyridine nucleotide coenzyme, reduced nicotinamide adenine dinucleotide (NADH). This and reduced nicotinamide adenine dinucleotide phosphate (NADPH) can be thought of as Mother Nature's sodium borohydride, a reagent used in organic chemistry to reduce active carbonyl and imine functional groups. As in the case of sodium borohydride, the reduced forms of the pyridine nucleotide coenzymes are believed to transfer their reducing equivalents as hydride ions[31] (Scheme 4.21). Because the sugar phosphate part of the pyridine nucleotides is not involved in the chemistry, it is abbreviated as R; likewise, the part of folic acid not involved in the chemistry is abbreviated by R′.

The carbon atom at the C-4 position of NADH and NADPH is prochiral. An atom is *prochiral* if, by changing one of its substituents, it is converted from achiral to chiral. The C-4 carbon of NADH has two hydrogens attached to it; consequently, it is achiral. If one of the hydrogens is replaced by a deuterium, then the carbon becomes chiral. If the chiral center that is generated by replacement of deuterium for hydrogen has the (*S*)-configuration, then the hydrogen that was replaced is called the pro-(*S*) hydrogen; if the (*R*)-configu-

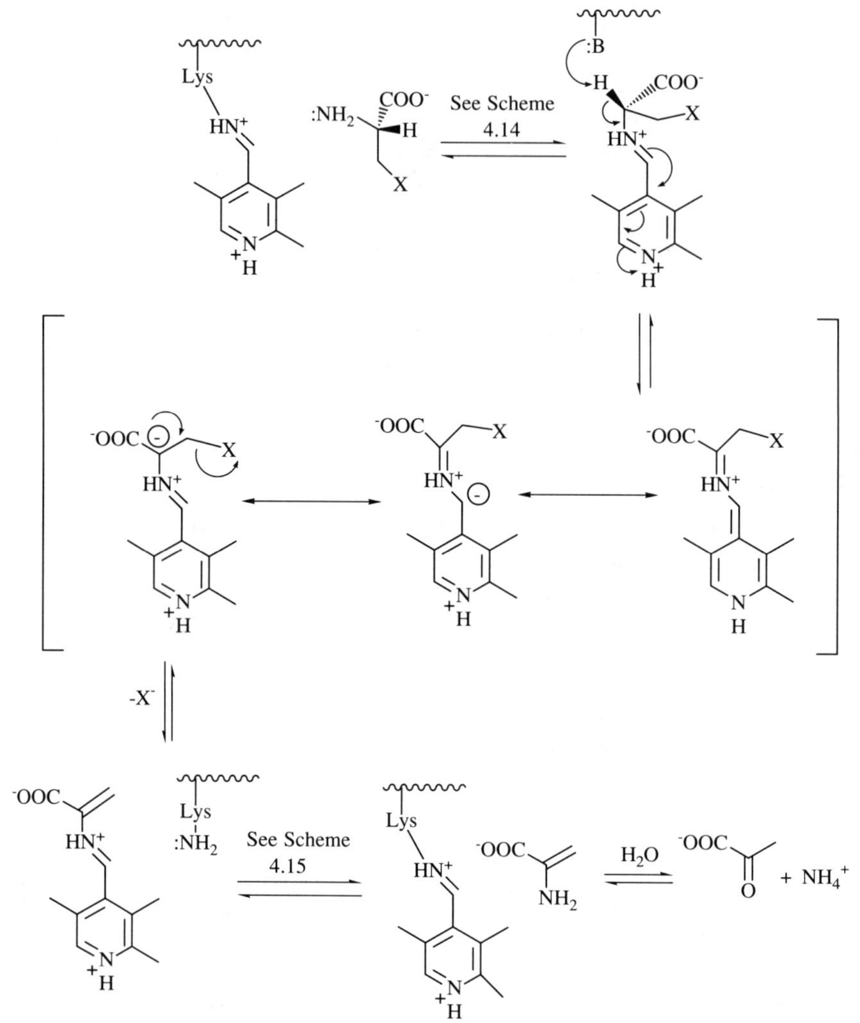

**Scheme 4.20.** Mechanism for PLP-dependent $\beta$-elimination reactions.

ration is produced, then the hydrogen replaced is the pro-($R$) hydrogen (Fig. 4.7). The pro-($R$) and pro-($S$) hydrogens of the pyridine nucleotides are noted in **4.26**. As indicated in Section I,B,1, because enzymes bind molecules in specific orientations, any compound bound to an enzyme can be prochiral, that is, its hydrogens can be differentiated by the enzyme. Some enzymes, such as dihydrofolate reductase, utilize the pro-($R$) hydrogen (also called the A-side hydrogen) of the reduced pyridine nucleotides; others use the pro-($S$), or B-side, hydrogen.[31] However, the reaction always is stereospecific; if an

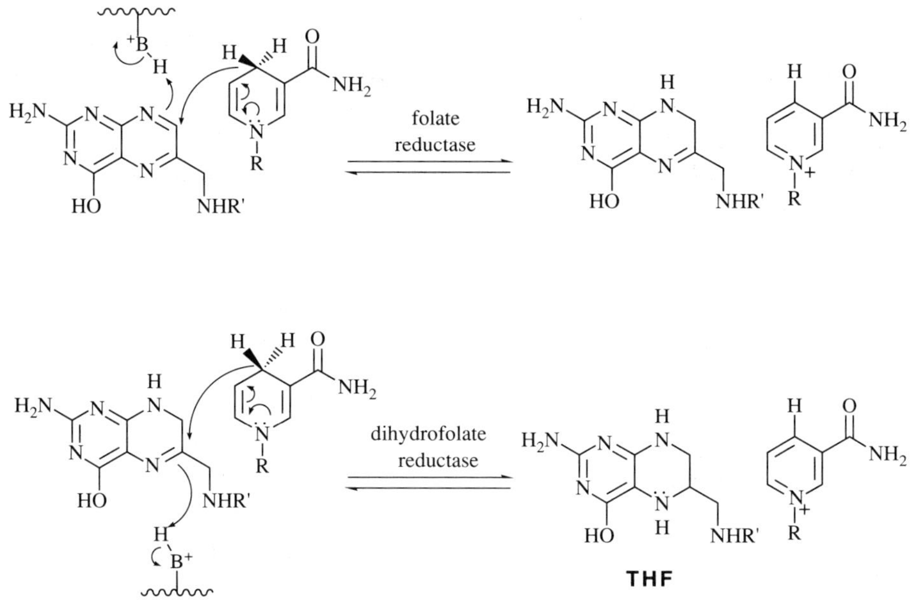

folate reductase

dihydrofolate reductase

THF

Scheme 4.21. Pyridine nucleotide–dependent reduction of folic acid.

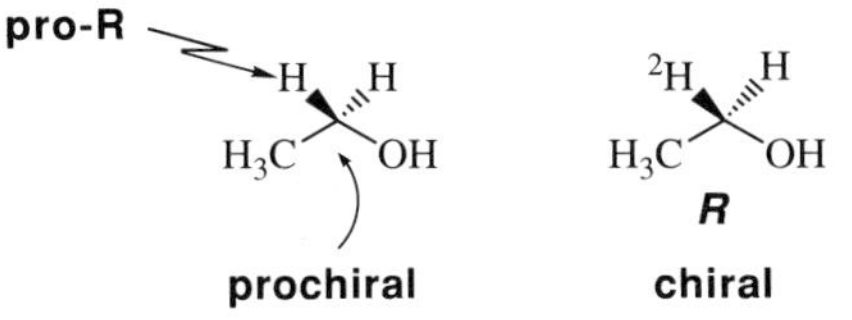

pro-R

prochiral

*R*

chiral

Figure 4.7. Determination of prochirality.

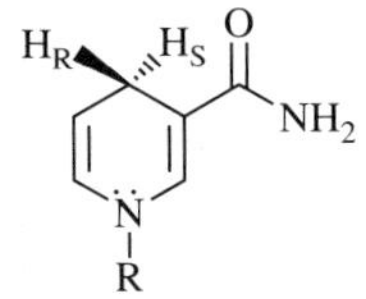

4.26

enzyme uses the pro-(*R*) hydrogen of NADH, then this is the only hydrogen that it transfers.

Although folate is converted to a reduced form (tetrahydrofolate), this coenzyme is generally not involved in redox reactions (except for one case, thymidylate synthase, which will be discussed in Section V,C,3,e of Chapter 5). Tetrahydrofolate-dependent enzymes catalyze the transfer of one-carbon units from one substrate to another. The carbon atom utilized in this transfer is derived from serine (marked with an asterisk in Scheme 4.22) in a PLP-dependent $\alpha$-cleavage reaction (Schemes 4.22 and 4.23). The enzyme that catalyzes the transfer of the carbon unit from serine to tetrahydrofolate is serine hydroxymethylase (glycine hydroxymethyltransferase). As shown in Scheme 4.22 the hydroxymethyl side chain of serine is held perpendicular to the plane of the PLP aromatic system, and it is cleaved in an enzyme-assisted elimination reaction. This $\alpha$-cleavage is an uncommon PLP-dependent reac-

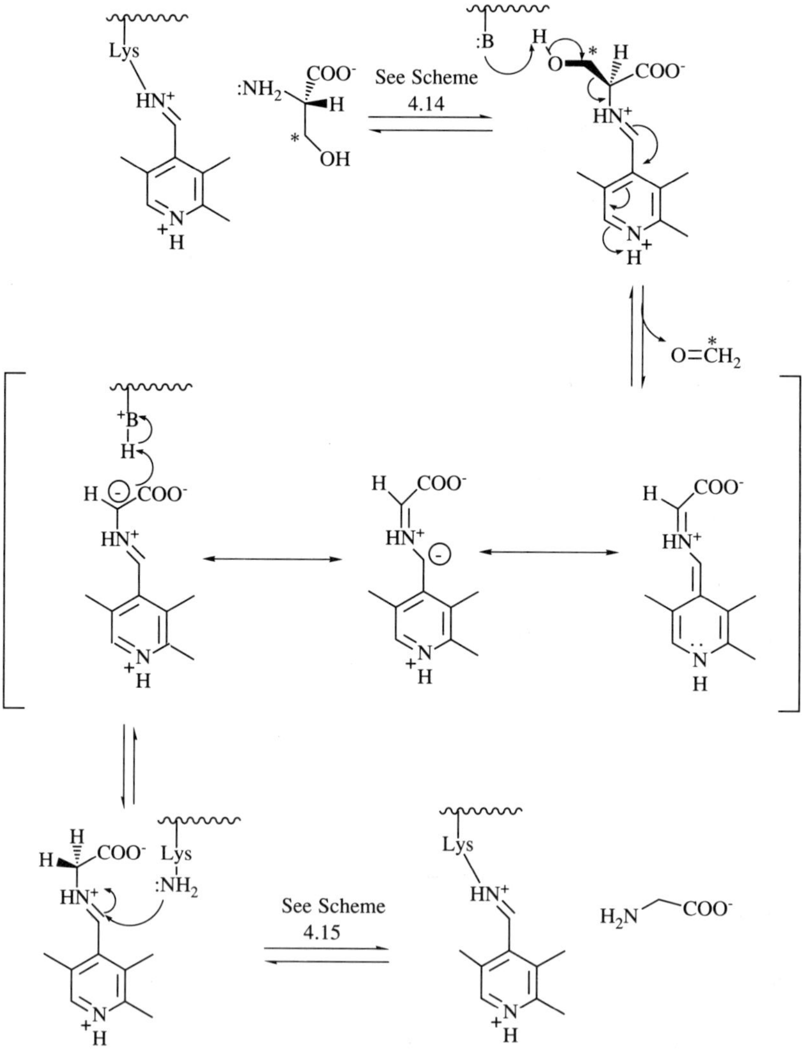

Scheme 4.22. First half reaction for serine hydroxymethylase–catalyzed transfer of formaldehyde from serine to tetrahydrofolate.

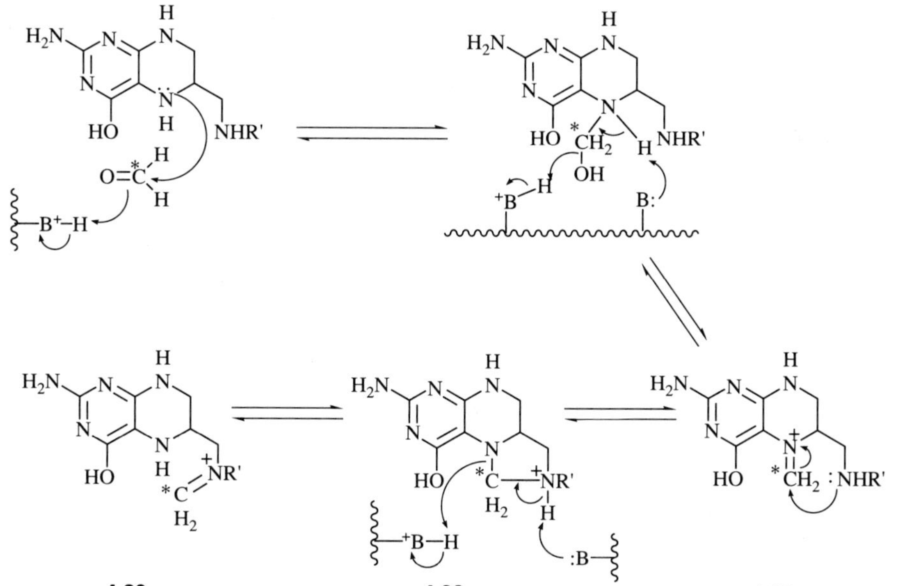

4.29 4.28 4.27

**Scheme 4.23.** Second half reaction for serine hydroxymethylase–catalyzed transfer of formaldehyde from serine to tetrahydrofolate.

tion. In this case the cation formed from C–C bond cleavage is an oxygen-stabilizing one; deprotonation of the hydroxyl gives formaldehyde, so this is an efficient process. In the absence of tetrahydrofolate, serine is not converted to glycine, suggesting that there is a binding site for tetrahydrofolate which, when enzyme bound, triggers the enzyme-catalyzed cleavage of serine[32] (Scheme 4.22). The formaldehyde that is generated from serine degradation reacts with the tetrahydrofolate (Scheme 4.23) prior to escape from the active site. Note that the more basic nitrogen of tetrahydrofolate is the $N^5$ nitrogen[33] which attacks the formaldehyde first[34] (see Scheme 4.23); the $N^{10}$ nitrogen is attached to a phenyl ring which has a carbonyl in the para position, therefore giving it more amide-like character. Dehydration of the carbinolamine intermediate gives $N^5$-methylenetetrahydrofolate (**4.27**), which is in equilibrium with $N^5$,$N^{10}$-methylenetetrahydrofolate (**4.28**) and $N^{10}$-methylenetetrahydrofolate (**4.29**).[34] In solution the cyclic adduct (**4.28**) is favored ($K_{eq} = 3.2 \times 10^4$, pH 7.2).[34]

$N^5$,$N^{10}$-Methylenetetrahydrofolate can be reduced by a NADPH-dependent enzyme ($N^5$,$N^{10}$-methylenetetrahydrofolate reductase, as you might guess) to give $N^5$-methyltetrahydrofolate (**4.30**, Scheme 4.24), or it can be oxidized by a $NADP^+$-dependent enzyme ($N^5$,$N^{10}$-methylenetetrahydrofolate dehydrogenase) to give $N^5$,$N^{10}$-methenyltetrahydrofolate (**4.31**, Scheme 4.25). This can be hydrolyzed (without change in oxidation state) to $N^5$-formyltetrahydro-

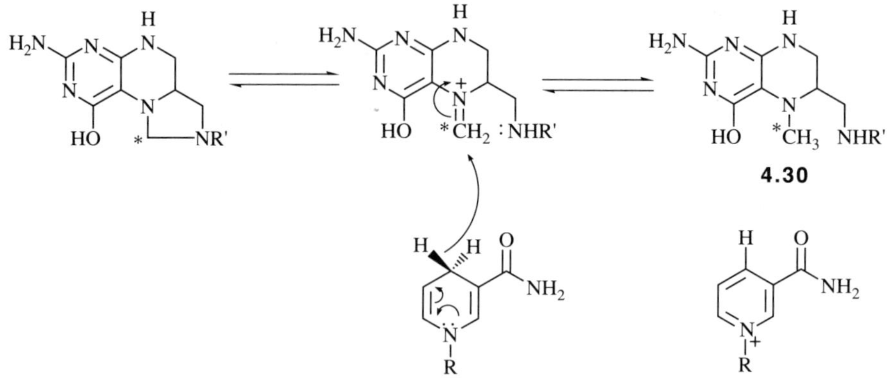

4.30

Scheme 4.24. Reduction of $N^5$,$N^{10}$-methylenetetrahydrofolate to $N^5$-methyltetrahydrofolate.

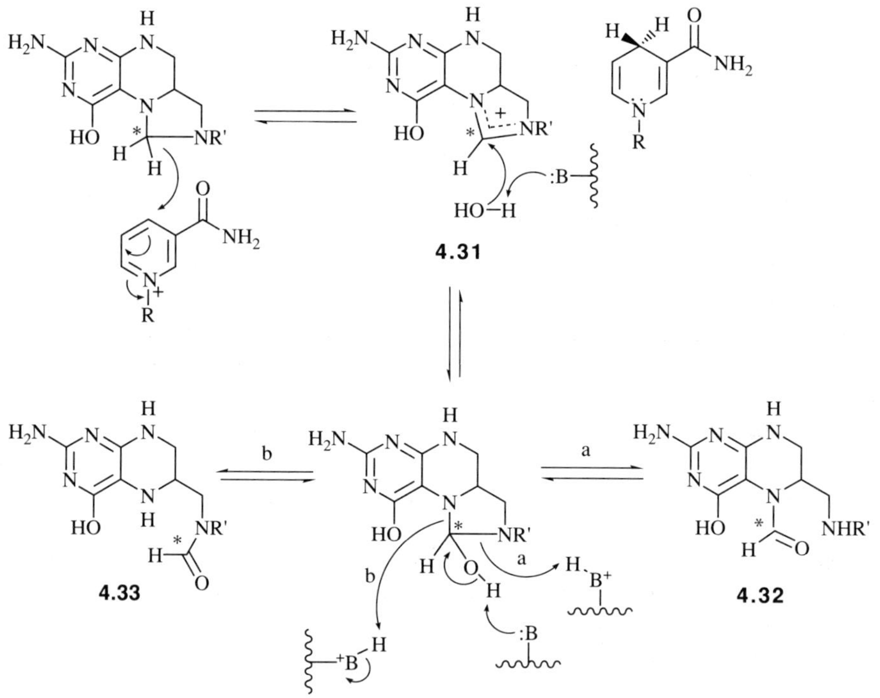

4.31

4.33

4.32

Scheme 4.25. Oxidation of $N^5$,$N^{10}$-methylenetetrahydrofolate to $N^5$,$N^{10}$-methenyltetrahydrofolate and hydrolysis to $N^5$- and $N^{10}$-formyltetrahydrofolate.

folate (**4.32**) or $N^{10}$-formyltetrahydrofolate (**4.33**). All of these forms of the coenzyme are involved in enzyme-catalyzed transfers of one-carbon units at the methanol ($N^5$-methyl-THF, **4.30**) formaldehyde ($N^5$,$N^{10}$-methylene-THF, **4.28**), or formate ($N^5$,$N^{10}$-methenyl-THF, **4.31**) oxidation states. Some enzymes that transfer a one-carbon unit at the formate oxidation state do so with

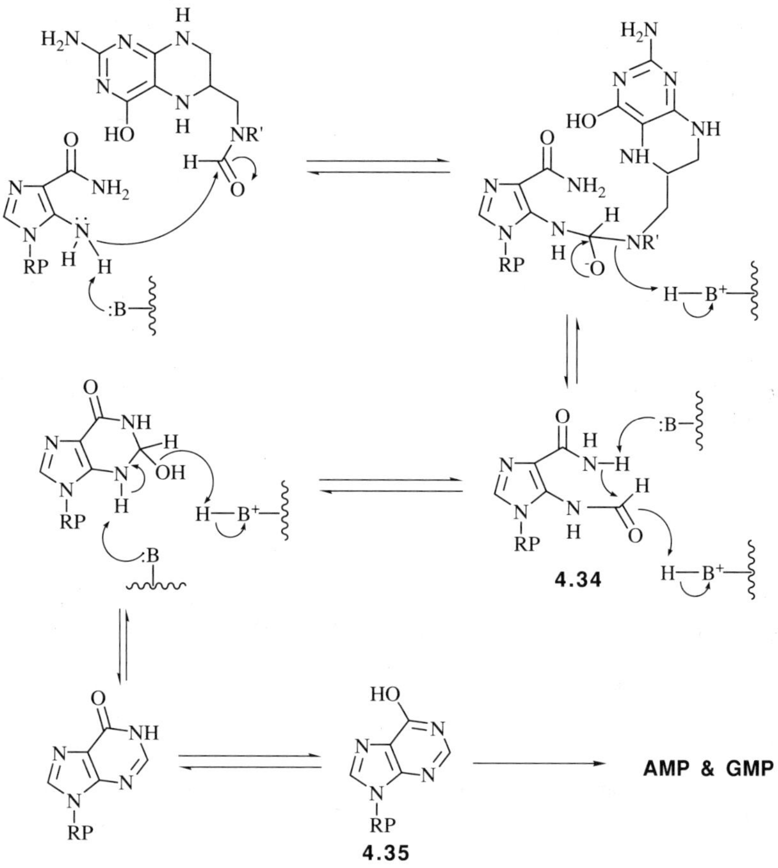

**Scheme 4.26.** $N^{10}$-Formyltetrahydrofolate in the biosynthesis of purines. RP stands for ribose phosphate.

**4.31** as the coenzyme, whereas other enzymes utilize **4.32** or **4.33** as the coenzyme, even though the same unit is transferred.

An example of a one-carbon transfer at the formate oxidation state can be found in the biosynthesis of purines (Scheme 4.26). The actual formyl transfer is completed at **4.34**; the rest of the scheme is given to show how the purine nucleus is formed from **4.34**. The end product, inosine monophosphate (**4.35**), is then converted to adenosine monophosphate and guanosine monophosphate by other enzymes.

Compounds with the general structure **4.36** are purines. These, as well as pyrimidines, are converted to the nucleic acids, which ultimately are important in the biosynthesis of DNA. In Scheme 4.26 it can be seen that the C-2 carbon atom of the purine nucleus is derived from an $N^{10}$-formyltetrahydrofolate-dependent reaction; the C-8 carbon atom of the purine nucleus also is derived from an $N^{10}$-formyltetrahydrofolate-dependent enzyme reaction earlier in the biosynthetic pathway. A one-carbon transfer at the formaldehyde oxidation state also occurs in the biosynthesis of the DNA precursor thymidylate, which is discussed in more detail in Section V,C,3,e of Chapter 5.

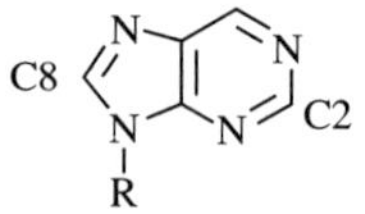

**4.36**

$N^5$-Methyltetrahydrofolate is involved in the enzyme-catalyzed transfer of a one-carbon unit at the methanol oxidation state of homocysteine to give methionine. Methyl transfer in general, however, is not carried out by methyltetrahydrofolate; another methyl transfer agent, *S*-adenosylmethionine, usually is implicated.

## C. Flavin

Unlike pyridoxal 5′-phosphate, which catalyzes reactions of amino acids, the flavin coenzymes, flavin mononucleotide (FMN) and flavin adenine dinucleotide (FAD), catalyze a wide variety of *redox* and *monooxygenation reactions* on diverse classes of compounds (Table 4.5). Both forms of flavin appear to be functionally equivalent, but some enzymes use FMN, others use FAD, and some utilize both.

The highly conjugated isoalloxazine tricyclic ring system of the flavins is an excellent electron acceptor, and this is responsible for the strong redox properties of flavins. Flavins can accept either one electron at a time or two electrons simultaneously (Scheme 4.27); in many overall two-electron oxidations, it is not clear if the reaction proceeds by a single two-electron reaction or by two one-electron transfer steps. The three forms of the coenzyme are the oxidized form ($Fl_{ox}$), the semiquinone form ($Fl^{\dot{-}}$), and the reduced form ($FlH^-$). With some enzymes the flavin coenzyme is covalently bound to an

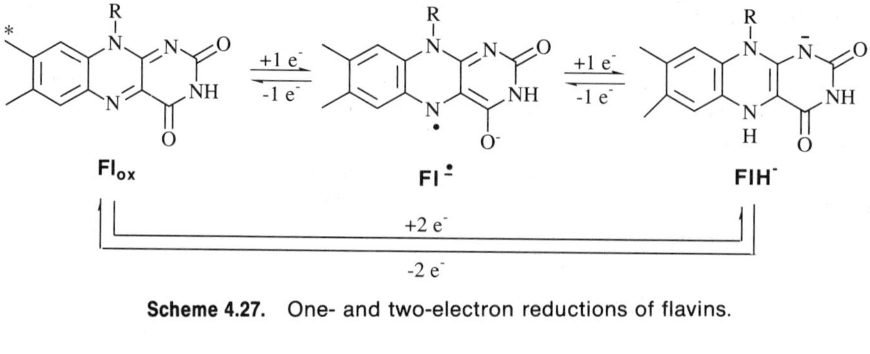

**Scheme 4.27.** One- and two-electron reductions of flavins.

**Table 4.5** Reactions Catalyzed by Flavin-Dependent Enzymes

| Substrate | Product |
|---|---|
| $R-CH(OH)R'$ | $R-C(=O)-R'$ |
| $R-CH(NH_2)R'$ (amines and amino acids) | $[R-C(=NH^+)-R'] \xrightarrow{H_2O} R-C(=O)-R' + NH_4^+$ |
| $RR'CH-CH(R'')-C(=O)-R'''$ | $RR'C=C(R'')-C(=O)R'''$ |
| $RCH(SH)-(CH_2)_n-CH(SH)R'$ | $RCH(CH_2)_nCHR'$ (S—S bridged) |
| $R-C_6H_4-OH$ | $R-C_6H_3(OH)-OH$ (catechol) |
| $R-C(=O)-R'$ | $R-C(=O)-OR'$ |
| $R-CHO$ | $R-C(=O)-OH$ |
| $R-CH(OH)COOH$ | $RCOOH + CO_2$ |

active site cysteine or histidine residue at the 8α position (see asterisk on the 8α-methyl group of the oxidized flavin in Scheme 4.27).

Once the flavin has been reduced, the enzyme requires a second substrate to return the flavin to the oxidized form so that it can accept electrons from another substrate molecule. This is reminiscent of the requirement of the second substrate (an α-keto acid) to return pyridoxamine phosphate to pyridoxal phosphate after PLP-dependent enzyme transamination. In the case of flavoenzymes, however, there are two mechanisms for conversion of reduced flavin back to oxidized flavin. Those enzymes that utilize a one-electron acceptor, such as ubiquinone or cytochrome *b*, proceed by a one-electron mechanism (Scheme 4.28)[35] and are called *dehydrogenases*. Other flavoenzymes utilize $O_2$ as the acceptor, which is converted to $H_2O_2$ with concomitant oxidation of the flavin. These enzymes are called *oxidases*, and are believed to proceed by a two-electron oxidation (Scheme 4.29, pathway a) via the 4a-

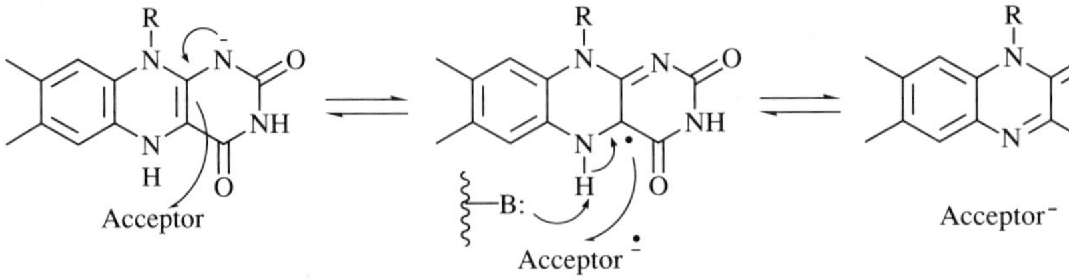

**Scheme 4.28.** Mechanism for dehydrogenase-catalyzed flavin oxidation.

hydroperoxide (**4.37**).[35] Since ground state $O_2$ is in the triplet state, there has to be a spin flip of triplet $O_2$ to singlet $O_2$ in order for pathway a to be viable. A one-electron mechanism (Scheme 4.29, pathway b) skirts that issue. The intermediate semiquinone, after one-electron transfer, can either combine with the superoxide formed (pathway c) to give **4.37** or transfer a second electron to the superoxide to give the oxidized flavin directly (pathway d).

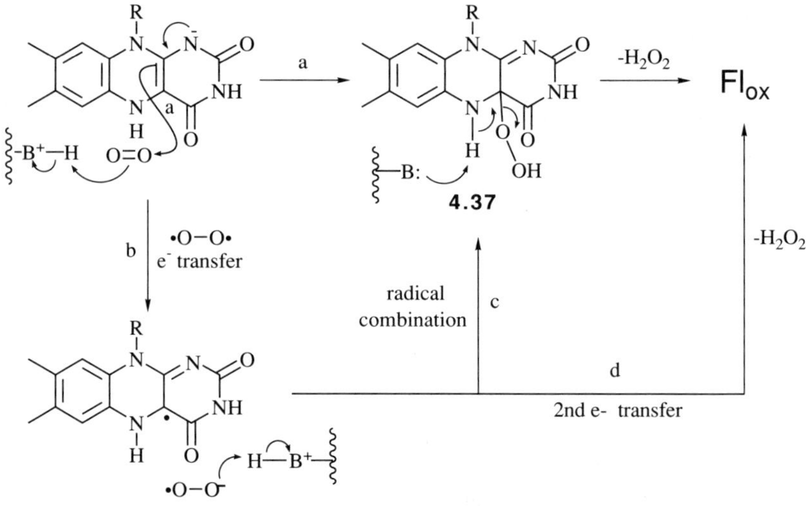

**Scheme 4.29.** Mechanism for oxidase-catalyzed flavin oxidation.

Each of the reactions shown in Table 4.5 is specific for a particular enzyme. Some of the reactions are difficult to carry out nonenzymatically; therefore, although the flavin is essential for the redox reaction, the enzyme is responsible for catalyzing these reactions and for both substrate and reaction specificity.

An example of a two-electron flavin enzyme reaction that has been reported[36] is that of general acyl-CoA dehydrogenase, the FAD-dependent enzyme that catalyzes the second step in the catabolism of fatty acids. Fatty acids in the body are first converted to acyl-coenzyme A derivatives (see Section III,E), which then undergo a dehydrogenation at the $\alpha,\beta$-carbon

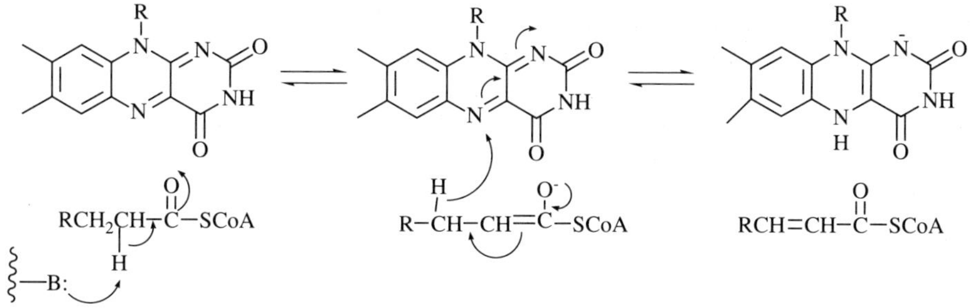

**Scheme 4.30.** Proposed[36] two-electron mechanism for general acyl-CoA dehydrogenase.

atoms. The mechanism proposed[36] (Scheme 4.30) involves enzyme-catalyzed $\alpha$-carbon deprotonation followed by $\beta$-carbon hydride (2 $e^-$) transfer to the enzyme-bound FAD. More recently, however, this two-electron mechanism has been questioned[37]; evidence suggests that after $\alpha$-carbon deprotonation, the resulting enolate may transfer its electrons to the flavin one at a time (Scheme 4.31). The jury is still out on which mechanism is viable.

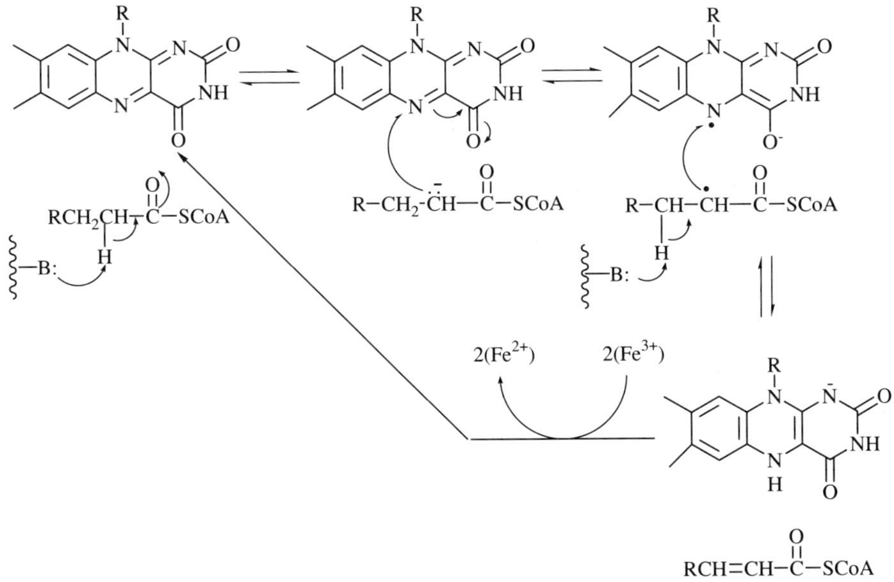

**Scheme 4.31.** Proposed[37] one-electron mechanism for general acyl-CoA dehydrogenase.

A flavoenzyme reaction important in medicinal chemistry that is believed to proceed by two one-electron transfer steps is that catalyzed by monoamine oxidase [amine oxidase (flavin-containing)] (see Section V,C,3,c of Chapter 5 for the importance of this enzyme to psychopharmacology). This is one of the enzymes responsible for the catabolism of various biogenic amine neurotransmitters, such as norepinephrine and dopamine. Evidence based on inactivator studies[38] and spin trapping experiments[39] strongly suggests the involvement of radical intermediates from which the mechanism in Scheme 4.32 has been proposed.[38,39] Note that this exemplifies a flavoenzyme-dependent conversion of an amine to an immonium ion. The immonium salt is hydrolyzed nonenzymatically to the corresponding aldehyde.[40]

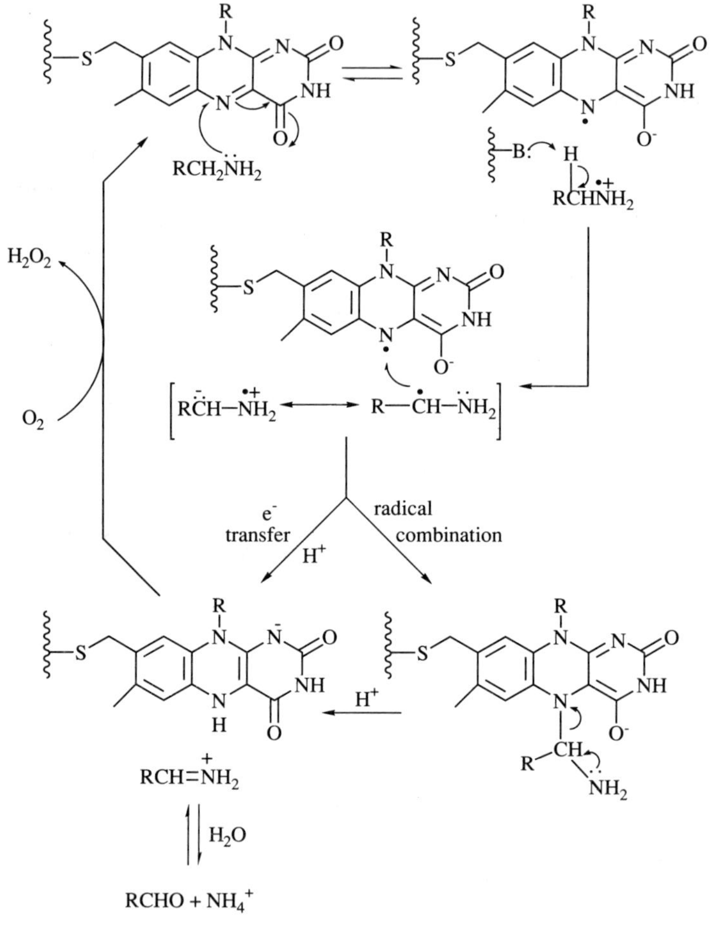

**Scheme 4.32.** Proposed[38,39] one-electron mechanism for monoamine oxidase.

Flavin monooxygenases are members of the class of liver microsomal mixed function oxygenases that are important in the oxygenation of xenobiotics (foreign substances) that enter the body, including drugs (see Chapter 7).[41,42] Unlike flavin oxidases and dehydrogenases, this class of enzymes incorporates an oxygen atom from molecular oxygen into the substrate. It is believed that a flavin 4a-hydroperoxide is an important intermediate in this process,[43] which can be formed by the same mechanisms suggested for the oxidation of reduced flavin (Scheme 4.29, pathways a and b/c). This then requires the flavin monooxygenase to have the flavin in its reduced form. The way this is accomplished is with a stable reducing agent, NADH or NADPH, which converts the oxidized flavin to its reduced form and initiates the oxygenation reaction (Scheme 4.33). The simplest mechanism that can be drawn involves nucleophilic attack by the substrate at the distal oxygen of the flavin hydroperoxide. As several flavin intermediates have been observed spectroscopically for *p*-hydroxybenzoate hydroxylase (4-hydroxybenzoate 3-monooxygenase), a bacterial flavin monooxygenase, the more complicated mechanism shown in Scheme 4.33 has been proposed,[44] and this may be extrapolated to the mammalian enzyme.

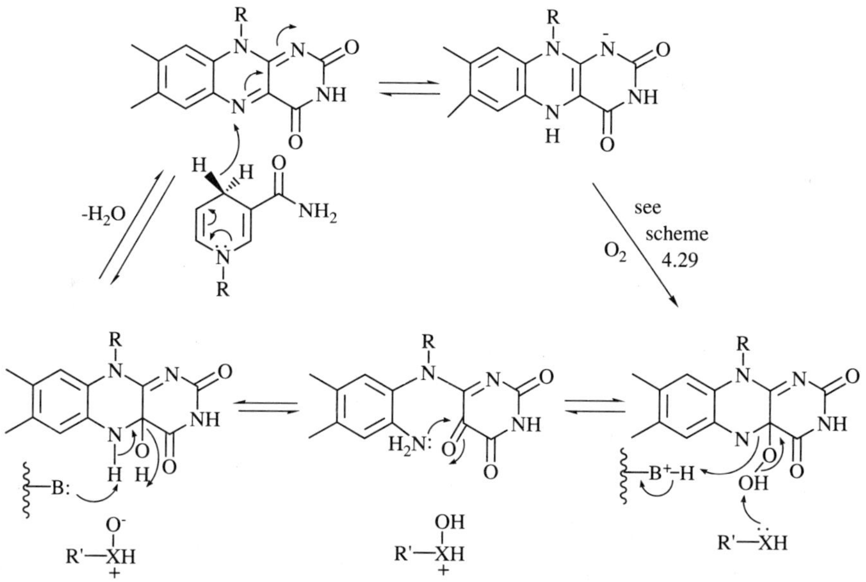

**Scheme 4.33.** Proposed[44] mechanism for a flavin monooxygenase. X is a nucleophile, for example, NH, S, or an aromatic ring.

Hamilton[45] has argued that alkyl hydroperoxides are poor hydroxylating agents in the absence of transition metal catalysts. In order for the flavin 4a-hydroperoxide to become more electrophilic, the oxenoid mechanism shown in Scheme 4.34 has been suggested. In both Scheme 4.33 and Scheme 4.34 R′XH could be any of the substrates for this class of enzymes.

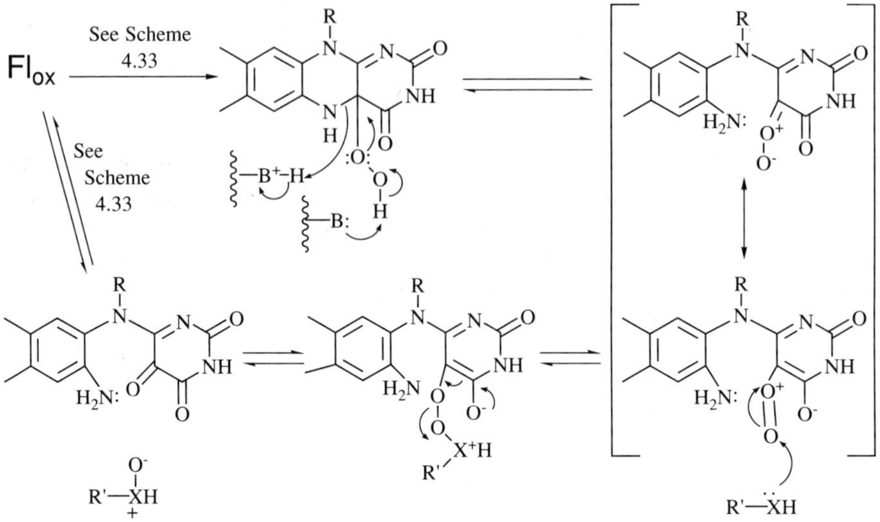

Scheme 4.34. Hamilton[45] mechanism for a flavin monooxygenase.

## D. Heme

Heme, or protoporphyrin IX, is an iron(III)-containing porphyrin cofactor (**4.14**) for a large number of liver microsomal mixed function oxygenases principally in the cytochrome *P*-450 family of enzymes. These enzymes, like flavin monooxygenases, are important in the metabolism of xenobiotics, including drugs (see Chapter 7).[46] As in the case of the flavin monooxygenases, molecular oxygen binds to the heme cofactor (after reduction of the $Fe^{3+}$ to $Fe^{2+}$) and is converted to a reactive form which is used in a variety of oxygenation reactions, especially hydroxylation and epoxidation. The hydroxylation reactions often occur at seemingly unactivated carbon atoms. The mechanism for this class of enzymes is still in debate, but a high-energy iron–oxo species must be involved.[47,48] Formally, this species can be written as **4.39a** (Scheme 4.35), however, because iron(V) is such an exceedingly high oxidation state, it is probably more likely an iron(IV) species (**4.39b**) with a radical cation in the porphyrin ring. The heme is abbreviated as **4.38**, where the peripheral nitrogens represent the four pyrrole nitrogens. The fifth (axial) ligand (not shown) is a cysteine thiolate from the protein in the case of cytochrome *P*-450.[49] As in the case of the flavin monooxygenases, NADPH is required in the heme-dependent enzymes to reduce the flavin coenzymes used to transfer electrons to the heme and heme–oxygen complex. It should be noted that the charges and electrons shown in Scheme 4.35 are only for the purposes of electronic bookkeeping and should not be taken as definitive structures.

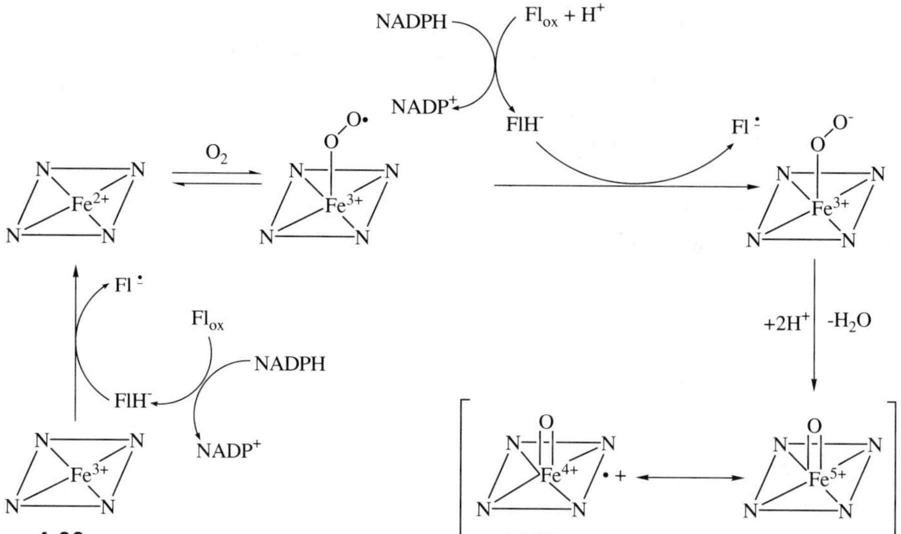

**Scheme 4.35.** Mechanism for formation of the high-energy iron–oxo species in heme-dependent oxygenases.

Scheme 4.36 gives possible mechanisms for the hydroxylation of unactivated hydrocarbons and for epoxidation of alkenes. Chemical model studies,[50] however, suggest that alkene epoxidation may involve an initial charge-transfer complex between the iron–oxo species and the alkene followed by a concerted process; if **4.40** is formed, it is so short-lived that radical reactions are not observed.

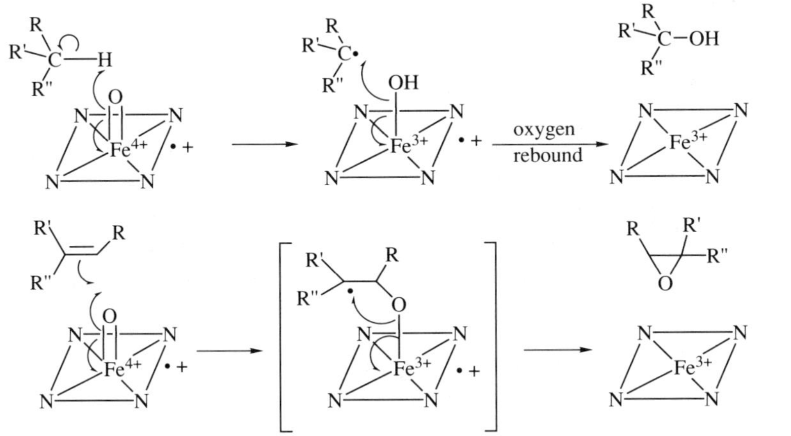

**Scheme 4.36.** Possible mechanisms for heme-dependent hydroxylation and epoxidation reactions.

When more easily oxidizable groups are involved, for example, nitrogen and sulfur atoms, heme-dependent oxygenases can also function as oxidases and proceed by electron transfer mechanisms. Scheme 4.37 shows a hypothetical mechanism for hydroxylation of the antihistamine diphenhydramine (**4.41**), modeled after other work on the cytochrome *P*-450 oxidation of tertiary amines.[48,51,52] Up to the immonium ion (**4.42**), this mechanism is virtually identical to the mechanism shown in Scheme 4.32 for monoamine oxidase. The carbinolamine (**4.43**) readily decomposes to the secondary amine with loss of formaldehyde.

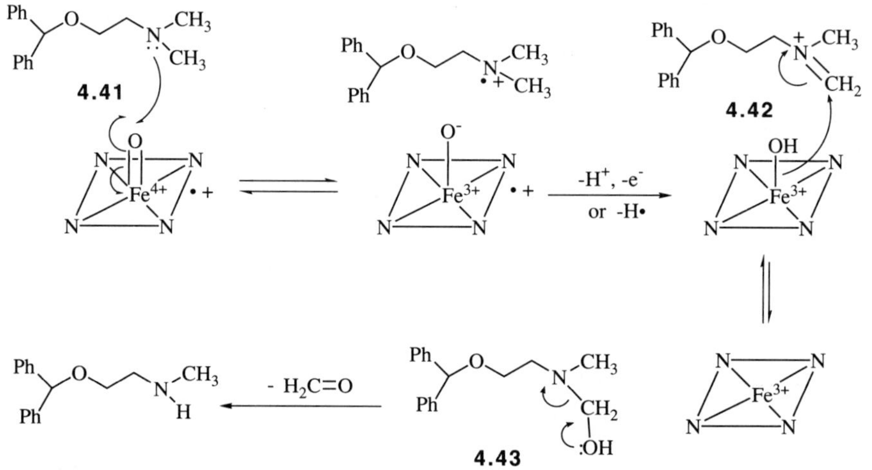

**Scheme 4.37.** Mechanism of $\alpha$-hydroxylation of tertiary amines by heme-dependent cytochrome *P*-450.

## *E. Adenosine Triphosphate and Coenzyme A*

Adenosine triphosphate–dependent enzymes serve several functions involving group transfers to a variety of substrates. A nucleophilic substrate can react with ATP (**4.44**) at the $\gamma$-phosphorus to transfer a phosphoryl group (bond cleavage a), at the $\beta$-phosphorus to transfer a diphosphoryl group (bond cleavage b) or an adenosine diphosphoryl group (bond cleavage c), at the

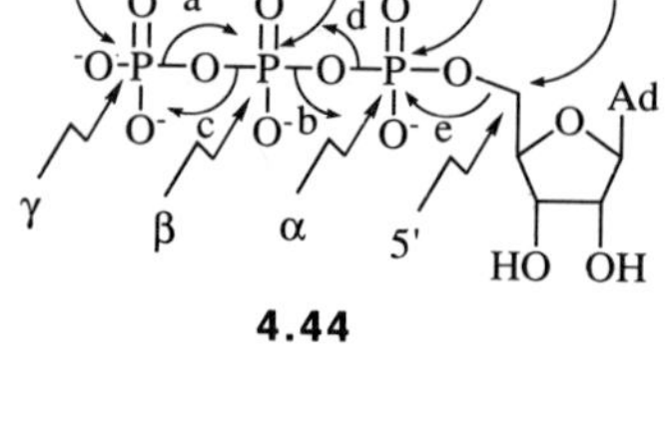

**4.44**

$\alpha$-phosphorus to transfer an adenosine monophosphoryl group (bond cleavage d), or at the 5′ position to transfer an adenosyl group (bond cleavage e).

In the case of the conversion of fatty acids to the corresponding fatty acyl-coenzyme A derivatives, which are substrates for general acyl-CoA dehydrogenase (see Section III,C, Schemes 4.30 and 4.31), ATP is used to activate the carboxylic acid as an adenosine monophosphate ester (Scheme 4.38). This intermediate is a reactive ester which undergoes rapid reaction with the highly nucleophilic thiol group of coenzyme A (**4.13b**). In this case, ATP acts as Mother Nature's form of thionyl chloride, an organic reagent used to convert carboxylic acids to acyl chlorides to make them much more reactive toward nucleophiles.

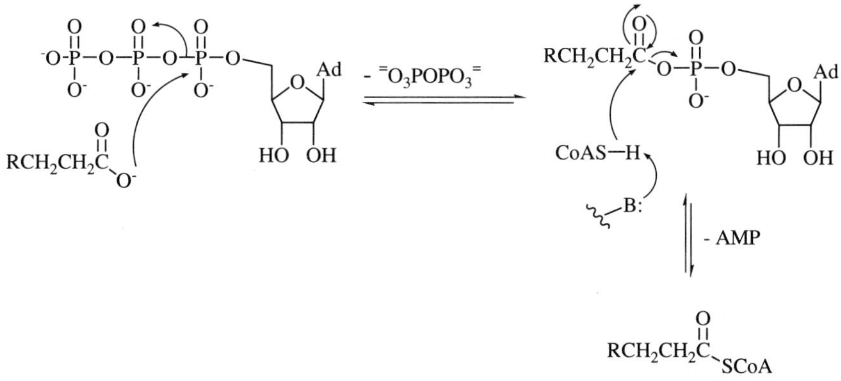

**Scheme 4.38.** Conversion of fatty acids to fatty acyl-CoA esters.

Coenzyme A thioesters serve three important functions for different enzymes. With general acyl-CoA dehydrogenase the thioester group makes the $\alpha$-proton much more acidic than that of a carboxylate, and therefore easier to remove. The $pK_a$ of the $\alpha$-proton of a thioester is lower than that of an ester and considerably lower than that of the $\alpha$-proton of a carboxylate salt. Second, thioesters are much more reactive toward acylation of nucleophilic substrates than are carboxylic acids and oxygen esters, but not so reactive, as in the case of the phosphate esters made from carboxylic acids with ATP, that they cannot be functional in aqueous media. Third, coenzyme A esters are important in the transport of molecules and in their binding to enzymes.

## IV. Enzyme Therapy

Throughout this text small organic and some inorganic molecules are discussed in terms of their design and mechanism of action. In Chapter 5 drugs that inhibit the catalytic action of enzymes are described. Some enzymes, however, are themselves useful as drugs. For the most part, the enzymes that have therapeutic utility catalyze hydrolytic reactions. For example, amylase, ligase, cellulase, trypsin, papain, and pepsin are proteolytic or lipolytic digestive enzymes used for gastrointestinal disorders resulting from poor digestion. Trypsin is also used for degrading necrotic tissue from wounds. Collagenase hydrolyzes collagen in necrotic tissue, but does not attack collagen in healthy tissue. Lactase (β-D-galactosidase) is taken by those having low lactase activity for the hydrolysis of lactose into glucose and galactose. Deoxyribonuclease and fibrinolysin (plasmin) are used to dissolve the DNA and fibrinous material, respectively, in purulent exudates and in blood clots. Because malignant leukemia cells are dependent on an exogenous source of asparagine for survival, whereas normal cells are able to synthesize asparagine, the enzyme asparaginase (L-asparagine amidohydrolase), which hydrolyzes the exogenous asparagine to aspartate, has been successful in the treatment of acute lymphocytic leukemia. Three different enzymes, urokinase, streptokinase, and tissue plasminogen activator (tPA), convert the inactive protein plasminogen to plasmin, a proteolytic enzyme that digests fibrin clots; therefore, these enzymes are effective in the treatment of myocardial infarction, venous and arterial thrombosis, and pulmonary embolism.[53] Excessive bilirubin in the blood is the cause for neonatal jaundice. A blood filter containing immobilized bilirubin oxidase can degrade more than 90% of the bilirubin in the blood in a single pass through the filter.[54] The use of genetic engineering techniques to produce altered active enzymes which have increased stability should lead to increased use of enzymes as drugs. The major drawbacks to the use of enzymes in therapy are enzyme instability (other proteases degrade them) and allergic responses.

In Chapter 5 we examine the design and mechanism of action of enzyme inhibitors as drugs.

## References

1. Sumner, J. B. 1926. *J. Biol. Chem.* **69**, 435.
2. Page, M. I. 1987. *In* "Enzyme Mechanisms" (Page, M. I., and Williams, A., eds.), p. 1. Royal Society of Chemistry, London.
3. Kraut, J. 1988. *Science* **242**, 533.
4. Haldane, J. B. S. 1930. "Enzymes." Longmans, Green, London, 1930 (reprinted in 1965 by MIT Press, Cambridge, Massachusetts).
5. Eyring, H. 1935. *J. Phys. Chem.* **3**, 107.
6. Pauling, L. 1946. *Chem. Eng. News* **24**, 1375; Pauling, L. 1948. *Am. Sci.* **36**, 51.
7. Schowen, R. L. 1978. *In* "Transition States of Biochemical Processes" (Gandour, R. D., and Schowen, R. L., eds.), p. 77. Plenum, New York.
8. Richards, F. M. 1977. *Annu. Rev. Biophys. Bioeng.* **6**, 151.
9. Talalay, P., and Benson, A. 1972. *In* "The Enzymes" (Boyer, P. D., ed.), 3rd Ed., Vol. 6, p. 591. Academic Press, New York.

10. Jencks, W. P. 1969 "Catalysis in Chemistry and Enzymology," Chaps. 1–3, and 5. McGraw-Hill, New York.
11. Jencks, W. P. 1975. *Adv. Enzymol.* **43**, 219.
12. Wolfenden, R., and Frick, L. 1987. *In* "Enzyme Mechanisms" (Page, M. I., and Williams, A., eds.), p. 97. Royal Society of Chemistry, London.
13. Bruice, T. C., and Pandit, U. K. 1960. *J. Am. Chem. Soc.* **82**, 5858.
14. Kirby, A. J. 1980. *Adv. Phys. Org. Chem.* **17**, 183.
15. March, J. 1985. "Advanced Organic Chemistry," 3rd Ed., p. 268. Wiley, New York.
15a. Gerlt, J. A., Kozarich, J. W., Kenyon, G. L., and Gassman, P. G. 1991. *J. Am. Chem. Soc.* **113**, 9667.
16. Kraut, J. 1977. *Annu. Rev. Biochem.* **46**, 331.
17. Bryan, P., Pantoliano, M. W., Quill, S. G., Hsaio, H.-Y., and Poulos, T. 1986. *Proc. Natl. Acad. Sci. U.S.A.* **83**, 3743.
18. Carter, P., and Wells, J. A. 1988. *Nature (London)* **332**, 564.
19. Warshel, A., Naray-Szabo, G., Sussman, F., and Hwang, J.-K. 1989. *Biochemistry* **28**, 3629.
20. Covitz, T., and Westheimer, F. H. 1963. *J. Am. Chem. Soc.* **85**, 1773.
21. Koshland, D. E., Jr., and Neet, K. E. 1968. *Annu. Rev. Biochem.* **37**, 359.
22. Jencks, W. P. 1987. *Cold Spring Harbor Symp. Quant. Biol.* **52**, 65.
23. Waxman, D. J., and Strominger, J. L. 1983. *Annu. Rev. Biochem.* **52**, 825.
24. Marcus, R., and Coulston, A. M. 1985. *In* "Goodman and Gilman's The Pharmacological Basis of Therapeutics" (Gilman, A. G., Goodman, L. S., Rall, T. W., and Murad, F., eds.), p. 1551. Macmillan, New York.
25. Martell, A. E. 1989. *Acc. Chem. Res.* **22**, 115.
26. Leussing, D. L. 1986. *In* "Vitamin $B_6$ Pyridoxal Phosphate" (Dolphin, D., Poulson, R., and Avramović, O., eds.), Part A, p. 69. Wiley, New York.
27. Dunathan, H. C. 1971. *Adv. Enzymol.* **35**, 79.
28. Walsh, C. T. 1989. *J. Biol. Chem.* **264**, 2393.
29. Soda, K., Tanaka, H., and Tanizawa, K. 1986. *In* "Vitamin $B_6$ Pyridoxal Phosphate" (Dolphin, D., Poulson, R., and Avramović, O., eds.), Part B, p. 223. Wiley, New York.
30. Sukhareva, B. S. 1986. *In* "Vitamin $B_6$ Pyridoxal Phosphate" (Dolphin, D., Poulson, R., and Avramović, O., eds.), Part B, p. 325. Wiley, New York.
31. Westheimer, F. H. 1987. *In* "Pyridine Nucleotide Coenzymes" (Dolphin, D., Poulson, R., and Avramović, O., eds.), Part A, p. 253. Wiley, New York.
32. Jordan, P. M., and Akhtar, M. 1970. *Biochem. J.* **116**, 277; Jordan, P. M., El-Obeid, H. A., Corina, D. L., and Akhtar, M. 1966. *J. Chem. Soc., Chem. Commun.*, 73.
33. Kallen, R. G., and Jencks, W. P. 1966. *J. Biol. Chem.* **241**, 5845.
34. Kallen, R. G., and Jencks, W. P. 1966. *J. Biol. Chem.* **241**, 5851.
35. Massey, V., Müller, F., Feldberg, R., Schuman, M., Sullivan, P. A., Howell, L. G., Mayhew, S. G., Matthews, R. G., and Foust, G. P. 1969. *J. Biol. Chem.* **244**, 3999.
36. Ghisla, S., Thorpe, C., and Massey, V. 1984. *Biochemistry* **23**, 3154.
37a. Lenn, N. D., Shih, Y., Stankovich, M. T., and Liu, H.-W. 1989. *J. Am. Chem. Soc.* **111**, 3065.
37b. Lia, M.-t., Liu, L.-d., and Liu, H.-w. 1991. *J. Am. Chem. Soc.* **113**, 7388.
38. Banik, G. M., and Silverman, R. B. 1990. *J. Am. Chem. Soc.* **112**, 4499 and references therein.
39. Yelekci, K., Lu, X., and Silverman, R. B. 1989. *J. Am. Chem. Soc.* **111**, 1138.
40. Hellerman, L., Chuang, H. Y. K., and DeLuce, D. C. 1972. *In* "Monoamine Oxidase—New Vistas" (Costa, E., and Sandler, M., eds.), (*Adv. Biochem. Psychopharmacol.* **5**) p. 327. Raven, New York.
41. Ziegler, D. M. 1980. *In* "Enzymatic Basis of Detoxification" (Jacoby, W. B., ed.), Vol. 1, p. 201. Academic Press, New York.

42. Poulsen, L. L. 1981. *In* "Reviews in Biochemical Toxicology" (Hodogson, E., Bend, J. R., and Philpot, R. M., eds.), Vol. 3, p. 33. Elsevier, New York.
43. Ghisla, S., and Massey, V. 1989. *Eur. J. Biochem.* **181**, 1.
44. Entsch, B., Ballou, D., and Massey, V. 1976. *J. Biol. Chem.* **251**, 2550.
45. Hamilton, G. A. 1974. *In* "Molecular Mechanisms of Oxygen Activation" (Hayaishi, O., ed.), p. 405. Academic Press, New York.
46. Guengerich, F. P., ed. 1987. "Mammalian Cytochromes *P*-450," Vols. 1 and 2. CRC Press, Boca Raton, Florida.
47. Ortiz de Montellano, P. R. 1986. *In* "Cytochrome *P*-450" (Ortiz de Montellano, P. R., ed.), p. 217. Plenum, New York.
48. Guengerich, F. P., and Macdonald, T. L. 1984. *Acc. Chem. Res.* **17**, 9.
49. Dawson, J. H., and Sono, M. 1987. *Chem Rev.* **87**, 1255.
50. Ostovič, D., and Bruice, T. C. 1989. *J. Am. Chem. Soc.* **111**, 6511.
51. Burka, L. T., Guengerich, F. P., Willard, R. J., and Macdonald, T. L. 1985. *J. Am. Chem. Soc.* **107**, 2549.
52. Miwa, G. T., Walsh, J. S., Kedderis, G. L., and Hollenberg, P. F. 1983. *J. Biol. Chem.* **258**, 14445.
53. Haber, E., Quertermous, T., Matsueda, G. R., and Runge, M. S. 1989. *Science* **243**, 51.
54. Lavin, A., Sung, C., Klibanov, A. M., and Langer, R. 1985. *Science* **230**, 543.

## General References

### Enzyme Catalysis

Boyer, P. D., ed. 1970–1987. "The Enzymes," 3rd Ed. Academic Press, New York.
Fersht, A. 1985. "Enzyme Structure and Mechanism," 2nd Ed. Freeman, New York.
Jencks, W. P. 1969. "Catalysis in Chemistry and Enzymology." McGraw-Hill, New York.
Jencks, W. P. 1975. *Adv. Enzymol.* **43**, 219.
Koshland, D. E., Jr., and Neet, K. E. 1968. *Annu. Rev. Biochem.* **37**, 359.
Page, M. I., and Williams, A., eds. 1987. "Enzyme Mechanisms." Royal Society of Chemistry, London.
Segel, I. H. 1975. "Enzyme Kinetics." Wiley, New York.
Walsh, C. 1979. "Enzymatic Reaction Mechanisms," Freeman, San Francisco, California.

### Pyridoxal 5′-Phosphate (PLP)

Dolphin, D., Poulson, R., and Avramović, O., eds. 1986. "Vitamin $B_6$ Pyridoxal Phosphate: Chemical, Biochemical, and Medical Aspects," Parts A and B. Wiley, New York.

### Pyridine Nucleotides (NADH and NADPH)

Dolphin, D., Poulson, R., and Avramović, O., eds. 1987. "Pyridine Nucleotide Coenzymes," Parts A and B. Wiley, New York.

### Flavin

Ghisla, S., and Massey, V. 1989. *Eur. J. Biochem.* **181**, 1, and references therein.
Müller, F. 1983. *Top. Curr. Chem.* **108**, 71.
Walsh, C. 1979. "Enzymatic Reaction Mechanisms." Freeman, San Francisco, California.

## Tetrahydrofolate

Blakley, R. L., and Benkovic, S. J., eds. 1984. "Folates and Pterins," Vol. 1. Wiley, New York.

## Heme

Guengerich, F. P., ed. 1987. "Mammalian Cytochromes *P*-450," Vols. 1 and 2. CRC Press, Boca Raton, Florida.
Guengerich, F. P., and MacDonald, T. L. 1990. *FASEB J.* **4**, 2453.

## Enzyme Therapy

Blohm, D., Bollschweiler, C., and Hillen, H. 1988. *Angew. Chem. Int. Ed. Engl.* **27**, 207.
Holcenberg, J. S., and Roberts, J., eds. 1981. "Enzymes as Drugs." Wiley, New York.
"Physicians' Desk Reference." Medical Economics Company, Oradell, New Jersey.

CHAPTER 5

# Enzyme Inhibition and Inactivation

# I. Why Inhibit an Enzyme?

Many diseases, or at least symptoms of diseases, arise from a deficiency or excess of a specific metabolite in the body, from an infestation of a foreign organism, or from aberrant cell growth. If the metabolite deficiency or excess can be normalized, and if the foreign organisms and aberrant cells can be destroyed, then these disease states will be remedied. All of these situations can be effected by specific enzyme inhibition.

Any compound that slows down or blocks enzyme catalysis is an *enzyme inhibitor*. If the interaction with the *target enzyme* is irreversible (usually covalent), then the compound is referred to as an *enzyme inactivator*. Many drugs function as enzyme inhibitors or inactivators. Consider what happens when an enzyme activity is blocked. The substrates for that enzyme cannot be metabolized, and the metabolic products are not generated (that is, unless there is another enzyme that can metabolize the substrate, and unless there is another metabolic pathway that generates the same product).

Why should these two outcomes be important to drug design? If a cell has a deficiency of the substrate for the target enzyme, and as a result of that deficiency a disease state results, then inhibition of the enzyme would prevent the degradation of the substrate, thereby increasing its concentration. An example of this is the onset of seizures that arises from diminished $\gamma$-aminobutyric acid (GABA) levels in the brain. Inhibition of the enzyme that degrades GABA, namely, GABA aminotransferase (4-aminobutyrate aminotransferase), leads to an anticonvulsant effect. If there is an excess of a particular metabolite that produces a disease state, then inhibition of the enzyme that catalyzes the biosynthesis of that metabolite would diminish its concentration. Excess uric acid can lead to gout. Inhibition of xanthine oxidase, the enzyme that catalyzes the conversion of xanthine to uric acid, decreases the uric acid levels, and results in an antihyperuricemic effect. If the product of an enzyme reaction is required to carry out an important physiological function that the drug is to block, then inhibition of the enzyme decreases the concentration of that product and can interfere with the physiological effect. Prostaglandins are important hormones that are involved in the pathogenesis of inflammation and fever. Inhibition of prostaglandin synthase results in anti-inflammatory, antipyretic, and analgesic effects.

In the case of foreign organisms such as bacteria and parasites, or in the case of tumor cells, inhibition of one of their essential enzymes can prevent important metabolic processes from taking place, resulting in inhibition of growth or replication of the organism or aberrant cell. Inhibition of the bacterial alanine racemase, for example, would prevent the biosynthesis of the peptidoglycan strands and, therefore, the biosynthesis of the bacterial cell wall. Such inhibitory compounds possess antibacterial activity. The use of drugs to combat foreign organisms and aberrant cells is called *chemotherapy*.

Enzyme inhibition is a promising approach for the rational discovery of new leads or drugs. Although there are numerous drugs that exert their therapeutic action by inhibiting specific enzymes, the mechanisms of action of most of these drugs were determined subsequent to the discovery of the therapeutic properties of the drugs. Target enzymes selected for rational drug design are those whose inhibition *in vivo* would lead to the desired therapeutic effect.

There are two general categories of target enzymes. In most cases a potential drug is designed for an enzyme whose inhibition is known to produce a specific pharmacological effect, but for which existing inhibitors have certain undesirable properties such as lack of potency or specificity or exhibition of side effects. A more daring approach is to design inhibitors of enzymes whose inhibition has not yet been established to lead to a desired therapeutic effect. Pursuing this category of enzyme targets requires knowledge of the pathophysiology of disease processes and the ability to identify important metabolites whose function or dysfunction results in a disease state. Not until an inhibitor is obtained will it be possible to determine the real effect of inhibition of that enzyme on the metabolism of the organism. Once an enzyme target is identified, lead compounds must be prepared that can inhibit it completely and specifically.

Of all the protein targets for potential therapeutic use, including hormone and neurotransmitter receptors and carrier proteins, enzymes are the most promising for rational inhibitor design. Enzyme purification is generally a much simpler task than receptor purification; a homogeneous enzyme preparation can be obtained for preliminary screening purposes and, in some cases, may be used to elucidate the active site structure, which is useful for computer-based drug design approaches (see Section II,F of Chapter 2). Furthermore, whereas effective receptor antagonists often bear no structural similarity to agonists, enzyme inhibitors are generally very similar in molecular structure to substrates or products of the target enzyme. Consequently, lead compounds are readily obtainable for enzyme targets. In addition, knowledge of enzyme mechanisms can be used in the design of transition-state analogs and multisubstrate inhibitors (Section IV,D), slow, tight-binding inhibitors (Section IV,C), and mechanism-based enzyme inactivators (Section V,C).

To minimize side effects there are certain properties that ideal enzyme inhibitors and/or enzyme targets should possess. An ideal enzyme inhibitor should be totally specific for the one target enzyme. Since this is rare, if attained at all, highly selective inhibition is a more realistic objective. In some cases, such as infectious diseases, enzyme targets can be identified because of biochemical differences in essential metabolic pathways between foreign organisms and their hosts.[1] In other instances there are substrate specificity differences between enzymes from the two sources that can be utilized in the design of selective enzyme inhibitor drugs. Unfortunately, when dealing with various organisms, and especially with tumor cells, the enzymes that are essential for their growth also are vital to human health. Inhibition of these enzymes can destroy human cells as well. Nonetheless, this approach is taken in various types of chemotherapy. The reason this approach is effective is that foreign organisms and tumor cells replicate at a much faster rate than do most normal human cells (those in the gut, the bone marrow, and the mucosa are exceptions). Consequently, rapidly proliferating cells have an elevated requirement for essential metabolites. *Antimetabolites*, compounds whose structures are similar to those of essential metabolites and which inhibit the metabolizing enzymes, are taken up by the rapidly replicating cells, and, therefore, these cells are selectively inhibited. The *selective toxicity* in this case derives from a kinetic difference rather than a qualitative difference in the metabolism.

An ideal enzyme target in a foreign organism or aberrant cell would be one which is essential for its growth but which is either nonessential for human health or, even better, not even present in humans. This type of selective toxicity would destroy only the foreign organism or aberrant cell, and it would not require the careful administration of drugs that is necessary when the inhibited enzyme is important to human metabolism as well. The penicillins, for example, which inhibit the bacterial peptidoglycan transpeptidase essential for the biosynthesis of the bacterial cell wall, are nontoxic to humans.

Enzyme inhibitors can be grouped into two general categories: reversible and irreversible inhibitors. As the name implies, inhibition of enzyme activity by a *reversible inhibitor* is reversible, suggesting that noncovalent interactions are involved. This is not strictly the case; there also can be reversible covalent interactions. An *irreversible inhibitor*, also called an *inactivator*, is one which prevents the return of enzyme activity for an extended period of time, suggesting the involvement of a covalent bond. This also is not strictly the case; it is possible for noncovalent interactions to be so effective that the enzyme–inhibitor complex is, for all intents and purposes, irreversibly formed.

Prior to a more detailed discussion of each of these types of enzyme inhibitors, we turn our attention to drug resistance and drug synergism, two important concerns in drug design and drug action which are referred to throughout this chapter.

## II. Drug Resistance

### A. What Is Drug Resistance?

*Drug resistance* is when a formerly effective drug dose is no longer effective. This can be a natural resistance or an acquired resistance. Resistance arises mainly by *natural selection*, the replication of a naturally resistant strain after the drug has killed all of the susceptible strains. Since mutagenic drugs generally are not used, resistance by drug-induced mutation seldom occurs. Drug resistance also can develop from gene transfer or gene amplification.

### B. Mechanisms of Drug Resistance

There are seven main mechanisms of drug resistance that arise from natural selection.[2,3]

#### 1. Altered Drug Uptake

One type of resistance involves the ability of the organism to exclude the drug from the site of action by preventing the uptake of the drug. The plasma membrane can adjust its net charge by varying its proportion of anionic (phosphatidylglycerol) to cationic (lysylphosphatidylglycerol) groups.[4] In this way a drug with the same charge can be repelled from the membrane. Aminoglycoside antibiotic resistance can arise from a lack of quinones that mediate the drug transport or from a lack of an electrical potential gradient required to drive the drug across the bacterial membrane.[5]

#### 2. Overproduction of Target Enzyme

Increased target enzyme production is another mechanism for drug resistance. Resistance to inhibitors of the enzyme dihydrofolate reductase by malarial parasites[6] and by malignant white blood cells[7] has been shown to be the result of overproduction of that enzyme in an unaltered form. The drug apparently induces extra copies of the gene encoding the enzyme.[8]

#### 3. Altered Target Enzyme (or Site of Action)

Modification of the target enzyme can result in poorer binding of the drug to the active site. Resistance to the antibiotic trimethoprim (**5.1**) derives from an altered dihydrofolate reductase, the enzyme inhibited by **5.1**.[9] The properties of the singly mutated enzyme differ somewhat from those of the normal enzyme, but it still binds dihydrofolate. Erythromycin resistance results from drug-induced formation of $N^6$,$N^6$-dimethyladenines in the 23S ribosomal RNA, the site of action of that antibiotic.[10] This reduces the affinity of erythromycin for the target RNA.

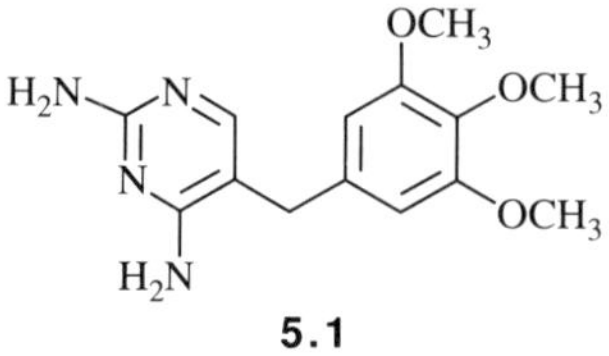

**5.1**

#### 4. Production of a Drug-Destroying Enzyme

Increased production of enzymes that degrade the drug also can occur. In Section V,B,2,a increased β-lactamase production is discussed as a mechanism for penicillin resistance.[3]

#### 5. Deletion of a Prodrug-Activating Enzyme

Another form of resistance derives from the deletion of an enzyme required to convert a prodrug (see Chapter 8) to its active form. Tumor resistance to the antileukemia drug 6-mercaptopurine (**5.2**, R = H) is caused by deletion of hypoxanthine–guanine phosphoribosyltransferase (hypoxanthine phosphoribosyltransferase),[11] the enzyme required to convert **5.2** (R = H) to thioinosine monophosphate (**5.2**, R = ribosyl 5′-monophosphate), the active form of the drug.

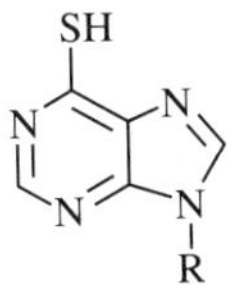

SH

N N N N R

**5.2**

#### 6. Overproduction of Substrate for the Target Enzyme

Overproduction of the substrate for a target enzyme would block the ability of the drug to bind at the active site (see Section IV,A). As mentioned in Section IV,B,1,d this is one mechanism of resistance to sulfa drugs.

#### 7. New Pathway for Formation of Product of the Target Enzyme

If the effect of a drug is to block production of a metabolite by enzyme inhibition, the organism could bypass the effect of the drug by inducing an altered metabolic pathway that produces the same metabolite.

## III. Drug Synergism (Drug Combination)

### *A. What Is Drug Synergism?*

When drugs are given in combination, their effects can be antagonistic, subadditive, additive, or synergistic. *Drug synergism* arises when the therapeutic effect of two or more drugs used in combination is greater than the sum of the effects of the drugs administered individually.

## *B. Mechanisms of Drug Synergism*

### 1. Inhibition of a Drug-Destroying Enzyme

An important mechanism for synergism arises when a compound is used to inhibit an enzyme that is destroying the drug. In this case the compound has no real therapeutic effect; it only protects the drug from being destroyed. This protective action, however, is essential for the activity of the drug.

### 2. Sequential Blocking

A second mechanism for synergism is *sequential blocking*, the inhibition of two or more consecutive steps in a metabolic pathway. The reason this is effective is because it is difficult (particularly with a reversible inhibitor) to inhibit an enzyme completely. If less than 100% of the enzyme activity is blocked, the metabolic pathway has not been shut down. With the combined use of inhibitors of two consecutive enzymes in the pathway, it is possible to block the metabolic pathway virtually completely. Because reversible enzyme inhibition (or receptor antagonism) is hyperbolic in nature, complete inhibition of an enzyme would require a large excess of the drug, and may be toxic. This approach becomes somewhat less important with an irreversible inhibitor which may inhibit an enzyme completely (see Section V,C,3,b for an example of an irreversible inhibitor that does not shut down the target enzyme totally).

### 3. Inhibition of Enzymes in Different Metabolic Pathways

A related approach is the combination of drugs that inhibit enzymes in different metabolic pathways which converge. Blockage of one metabolic pathway may not deplete the cell of the end product of that pathway because there may be an alternate biosynthetic route to the same metabolite. Inhibition of enzymes in both (or several) pathways may lead to synergistic effects.

### 4. Use of Multiple Drugs for the Same Target

A fourth mechanism for synergism is the use of two or more drugs to inhibit the growth of tumor cells or bacterial mutants, since a mutant that is resistant to one drug does not easily undergo further mutation. Typically, only 1 in a culture of $10^7$ bacteria may be resistant to a particular drug.[2] The chance of finding an organism resistant against two different drugs is 1 in $10^{14}$, and against three drugs is 1 in $10^{21}$. Use of multiple antimicrobial agents greatly minimizes the opportunity for a mutant resistant organism to proliferate. For example, the antituberculosis drug, isoniazid (**5.3**), which inhibits the replication of the tubercle bacillus, is used in combination with other antimicrobial agents such as rifampin (**5.4**); these two drugs are sold in combination for the treatment of tuberculosis.

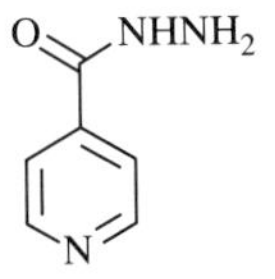

**5.3** 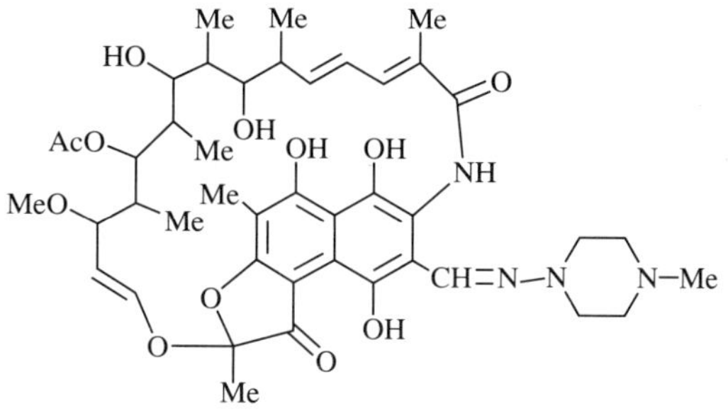 **5.4**

## IV. Reversible Enzyme Inhibitors

### *A. Mechanism of Reversible Inhibition*

The most common enzyme inhibitor drugs are of the reversible type, particularly ones that compete with the substrate for active site binding. These are known as *competitive reversible inhibitors*, compounds that have structures similar to those of the substrates or products of the target enzymes, and which bind at the substrate binding sites, thereby blocking substrate binding.

As in the case of the interaction of a substrate with an enzyme, an inhibitor (I) also can form a complex with an enzyme (E) (Scheme 5.1). The equilibrium constant $K_i$ ($k_{off}/k_{on}$) is a *dissociation* constant for breakdown of the E·I complex; therefore, the smaller the $K_i$ value for I, the better the inhibitor. Generally, this equilibrium is established very rapidly, but in some cases (see Section IV,C) it can be relatively slow.

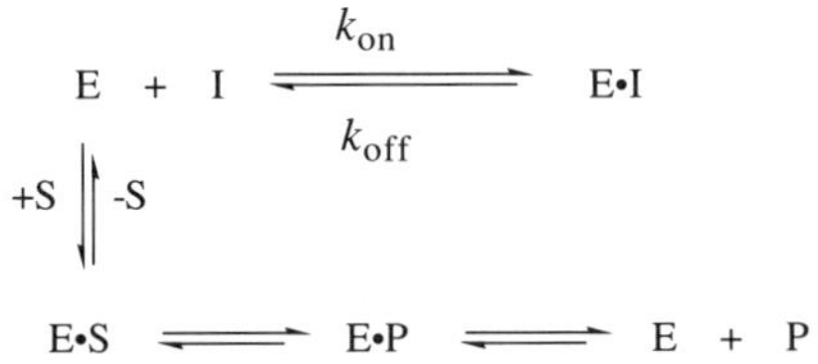

$$\mathrm{E + I} \underset{k_{off}}{\overset{k_{on}}{\rightleftharpoons}} \mathrm{E{\bullet}I}$$

$$\mathrm{E + S} \rightleftharpoons \mathrm{E{\bullet}S} \rightleftharpoons \mathrm{E{\bullet}P} \rightleftharpoons \mathrm{E + P}$$

**Scheme 5.1.** Kinetic scheme for competitive enzyme inhibition.

When the inhibitor binds at the active site (the substrate binding site), then the inhibitor is a competitive inhibitor. Formation of the E·I complex prevents the binding of substrate and, therefore, shuts down the catalytic conversion of the substrate to product. The inhibitor, however, also may act as a substrate and may be converted to a metabolically useless product. In the context of drug design, this is not a favorable process because the product formed may be toxic or may lead to other toxic metabolites.

Interaction of the inhibitor with the enzyme can occur at a site other than the substrate binding site and still result in inhibition of substrate turnover. When this occurs, usually as a result of an inhibitor-induced conformational change in the enzyme to give a form of the enzyme that does not bind the substrate properly, then the inhibitor is a *noncompetitive reversible inhibitor*. Unless something is known about this *allosteric binding site* in the enzyme, it is not possible to design noncompetitive enzyme inhibitors. Consequently, the discussion in this chapter is limited to the design and mechanism of action of competitive enzyme inhibitors.

The equilibrium shown in Scheme 5.1 will depend on the concentrations of the enzyme, the inhibitor, and the substrate. Since the enzyme concentration is usually low and fixed, the equilibrium constant, and, therefore, the E·I concentration, will principally depend on the inhibitor and substrate concentrations. As the concentration of the inhibitor is increased, it drives the equilibrium toward the E·I complex. Because the substrate and the inhibitor bind to the enzyme at the same site, they both cannot interact with the enzyme simultaneously. Conversely, when the inhibitor concentration diminishes, the E·I complex concentration diminishes, and the effect of the inhibitor can be overcome by the substrate (increasing the substrate concentration also would displace the equilibrium from the E·I complex toward increased E·S complex formation).

If the enzyme inhibitor is a drug, the maximal pharmacological effect will occur when the drug concentration is maintained at a saturating level at the target enzyme active site. As the drug is metabolized (see Chapter 7), and the concentration of I diminishes, repeated administration of the drug is required to maintain the integrity of the E·I complex. This accounts for why drugs often need to be taken several times a day. In order to increase the potency of reversible inhibitors and, thereby, reduce the dosage of the drug, the binding interactions with the target enzyme should be optimized (i.e., the inhibitor should have a low $K_i$ value).

When a drug is designed to be an enzyme inhibitor, it generally will be a competitive inhibitor since the lead compound often will be the substrate for the target enzyme. If the three-dimensional structure of the target enzyme is known, then the enzyme can be the lead, and molecular graphics–based techniques (see Section II,F in Chapter 2) can be used for drug design. Because an enzyme is just a specific type of receptor, an analogy can be made between agonists, partial agonists, and antagonists with good substrates, poor substrates, and competitive inhibitors, respectively.

## B. Selected Examples of Competitive Reversible Inhibitor Drugs

### 1. Sulfonamide Antibacterial Agents (Sulfa Drugs)

***a. Lead Discovery.*** At the beginning of the twentieth century Paul Ehrlich showed that various azo dyes were effective agents against trypanosomiasis in mice; however, none was effective in man. Then, in the early 1930s Gerhard Domagk, head of bacteriological and pathological research at the Bayer Company in Germany, who was trying to find agents against streptococci, tested a variety of azo dyes. One of the dyes, prontosil (**5.5**), showed dramatically positive results and successfully protected mice against streptococcal infections.[12] Bayer was unwilling to move rapidly on getting prontosil into the drug market. As Albert[13] tells it, when, in late 1935, Domagk's daughter cut her hand and was about to die of a streptococcal infection, her father gave her prontosil. Her recovery was rapid,[14] and the effectiveness of the drug became quite credible.

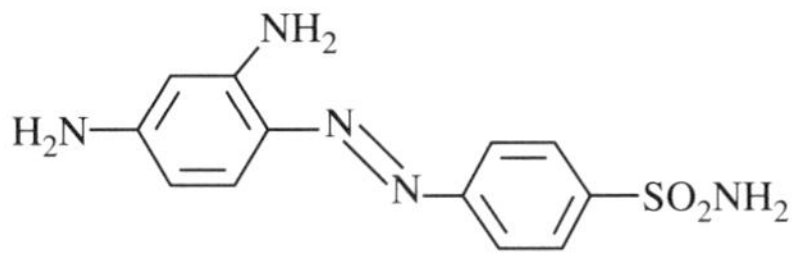

**5.5**

An unexpected property of prontosil, however, was that it had no activity against bacteria *in vitro*. Tréfouël and co-workers[15] found that if a reducing agent was added to prontosil, then it was effective *in vitro*. They suggested that the reason for the lack of *in vitro* activity, but high *in vivo* activity, was that prontosil was metabolized by reduction to the active antibacterial agent, namely, *p*-aminobenzenesulfonamide (also called sulfanilamide) (**5.6**). Furthermore, they demonstrated that sulfanilamide was as effective as prontosil in protecting mice against streptococcal infections, and that it exerted a *bacteriostatic effect in vitro*. Unlike a *bactericidal* agent, which kills bacteria, a bacteriostatic drug inhibits further growth of the bacteria, thus allowing the host defenses to catch up in their fight against the bacteria. Prontosil, then, is an early example of a *prodrug* (see Chapter 8), a compound that requires metabolic activation to be effective.

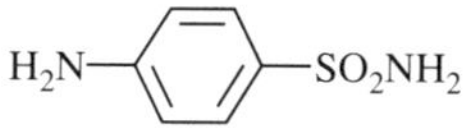

**5.6**

***b. Lead Modification.*** The discovery of prontosil marks the beginning of modern chemotherapy. During the next decade thousands of sulfonamides were synthesized and tested as antibacterial agents. These were the first structure–activity studies, and they demonstrated the importance of molecular modification in drug design. Also, this was one of the first examples where new lead compounds for other diseases were revealed from side effects observed during pharmacological and clinical studies (see Section II,C of Chapter 2). These early studies led to the development of new antidiabetic and diuretic agents. Another important scientific advance that was derived from work with sulfonamides was a simple method for the assay of these compounds in body fluids and tissues.[16] Furthermore, it was shown that the antibacterial effect of sulfanilamide was proportional to its concentration in the blood, and that at a given dose this varied from patient to patient. This was the beginning of the monitoring of blood drug levels during chemotherapy treatment, and it led to the initiation of the routine use of *pharmacokinetics*, the study of the absorption, distribution, and excretion of drugs, in drug development programs. Proper drug dosage requirements could now be calculated.

***c. Mechanism of Action.*** On the basis of the work by Stamp,[17] who showed that bacteria and other organisms contained a heat-stable substance that inhibited the antibacterial action of sulfonamides, Woods[18] in 1940 reported a breakthrough in the determination of the mechanism of action of this class of drugs. He hypothesized that because enzymes are inhibited by compounds whose structures resemble those of their substrates, the inhibitory substance is a substrate for an essential enzyme and it has a structure similar to that of sulfanilamide. After various chemical tests, and having a vague notion of the possible structure of this inhibitory substance, he deduced that it must be *p*-aminobenzoic acid (**5.7**), and he proceeded to show that **5.7** was a potent inhibitor of sulfanilamide-induced bacteriostasis. The results of his experiments showed that sulfanilamide was competitive with *p*-aminobenzoic acid for microbial growth. To maintain growth with increasing concentrations of sulfanilamide, it is necessary also to increase the concentration of *p*-aminobenzoic acid. Selbie[19] found that coadministration of *p*-aminobenzoic acid and sulfanilamide into streptococcal-infected mice prevented the antibacterial action of the drug.

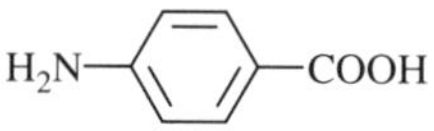

$H_2N-C_6H_4-COOH$

**5.7**

The observation of competitive inhibition by sulfanilamide was the basis for Fildes[20] to propose his theory of *antimetabolites*, compounds that block enzymes in metabolic pathways. He proposed a rational approach to chemotherapy, namely, enzyme inhibitor design, and suggested that the molecular basis for enzyme inhibition was that either the inhibitor combines with the enzyme and displaces its substrate or coenzyme or it combines directly with the substrate or coenzyme.

In the mid-1940s Miller and co-workers[21] demonstrated that sulfanilamide inhibited folic acid biosynthesis, and in 1948 Nimmo-Smith *et al.*[22] showed that the inhibition of folic acid biosynthesis by sulfonamides was competitively reversed by *p*-aminobenzoic acid. Two enzymes from *Escherichia coli* were purified by Richey and Brown,[23] one that catalyzed the diphosphorylation of 2-amino-4-hydroxy-6-hydroxymethyl-7,8-dihydropteridine (**5.8**) and another (dihydropteroate synthase) that catalyzed the synthesis of dihydrofolate (**5.9**) from the diphosphate of **5.8** and *p*-aminobenzoic acid (Scheme 5.2). The name of the enzyme stems from the fact that folic acid is a derivative of the pterin ring system (**5.10**). Because of the structural similarity of sulfanilamide to *p*-aminobenzoic acid, it is a potent competitive inhibitor of the second enzyme. The reversibility of the inhibition was demonstrated by Weisman and Brown,[24] who suggested that sulfonamides were incorporated into the dihydrofolate. This was verified by Bock *et al.*[25] who incubated dihydropteroate synthase with the diphosphate of **5.8** and [$^{35}$S]sulfamethoxazole (**5.11**) and identified the product as **5.12** (Scheme 5.3). Therefore, this is an example of competitive reversible inhibition in which the inhibitor also is a substrate. However, the product (**5.12**) cannot produce dihydrofolate, and, therefore, the organism cannot get the tetrahydrofolate needed as a coenzyme to make purines (see Section II,B of Chapter 4) which are needed for DNA biosynthesis. This is why the sulfonamides are bacteriostatic, not bactericidal. Inhibition of tetrahydrofolate biosynthesis only inhibits replication; it does not kill existing bacteria.

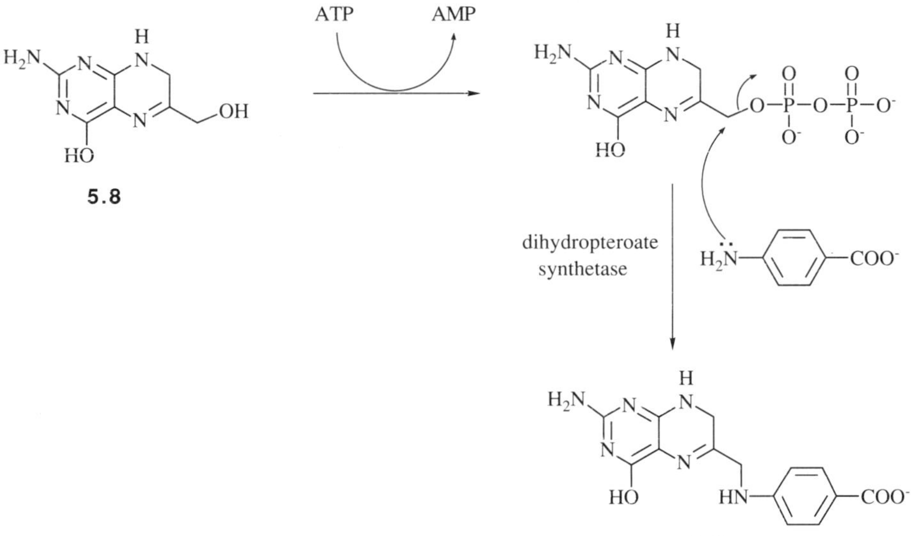

ATP AMP

5.8

dihydropteroate synthetase

5.9

**Scheme 5.2.** Biosynthesis of bacterial dihydrofolic acid.

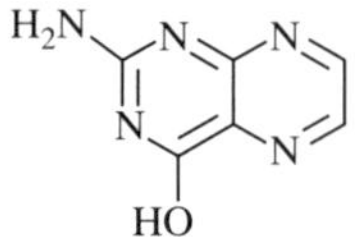

**5.10**

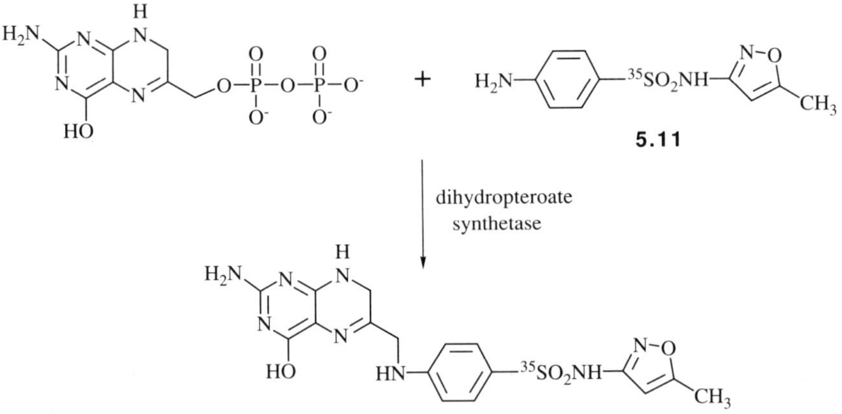

dihydropteroate synthetase

**5.11**

**5.12**

**Scheme 5.3.** Dihydropteroate synthase-catalyzed conversion of **5.11** to **5.12**.

Inhibitors of dihydropteroate synthase, however, have no effect on humans, as we are incapable of biosynthesizing folic acid and, therefore, do not have that enzyme. Folic acid is a vitamin and must be eaten by humans. Furthermore, because bacteria biosynthesize their folate, they do not have a transport system for it.[26] Consequently, we can eat all of the folate we want, and the bacteria cannot utilize it. This is another example of selective toxicity, inhibition of the growth of a foreign organism without affecting the host, and it falls into the category of an ideal enzyme inhibitor (Section I). It is interesting to note that sulfonamides are not effective with pus-forming infections because pus contains many compounds that are the end products of tetrahydrofolate-dependent reactions, such as purines, methionine, and thymidine. In this case, inhibition of folate biosynthesis is unimportant, and pus can contribute to bacterial sustenance.

***d. Drug Resistance.*** One major limitation to the use of sulfonamide antibacterial drugs is the development of drug resistance (see Section II). There are principally three mechanisms of sulfonamide drug resistance. One mechanism is that organisms can overproduce *p*-aminobenzoic acid.[27] A second mechanism is the result of a plasmid-mediated synthesis of a less sensitive dihydropteroate synthase, one which binds *p*-aminobenzoic acid normally but binds sulfonamides several thousand times less tightly than the normal enzyme can.[28] The third mechanism involves altered permeability to the sulfonamides.[29]

***e. Drug Synergism.*** Combination therapy of sulfadoxine (**5.13**) with pyrimethamine (**5.14**) or sulfamethoxazole (**5.11**) with trimethoprim (**5.1**)[30] has been shown to be quite effective for the treatment of malaria and bacterial infections, respectively. Both of these combinations are examples of a sequential blocking mechanism[31] (Section III,B,2). The sulfa drugs (**5.13** and **5.11**) inhibit dihydropteroate synthase, which catalyzes the synthesis of dihydrofolate, and **5.14** and **5.1** inhibit dihydrofolate reductase, which catalyzes the synthesis of tetrahydrofolate (see Section III,B of Chapter 4). Trimethoprim has the desirable property of being a very tight-binding inhibitor of bacterial dihydrofolate reductase but a poor inhibitor of mammalian dihydrofolate reductase; the $IC_{50}$ is 5 n$M$ for the former and $2.6 \times 10^5$ n$M$ for the latter.[32]

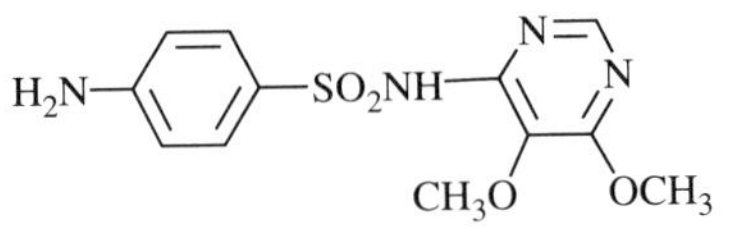

$H_2N$—C$_6$H$_4$—$SO_2NH$— (pyrimidine: N, N; $CH_3O$, $OCH_3$)

**5.13**

$H_2N$, N, N, $H_2N$, Cl

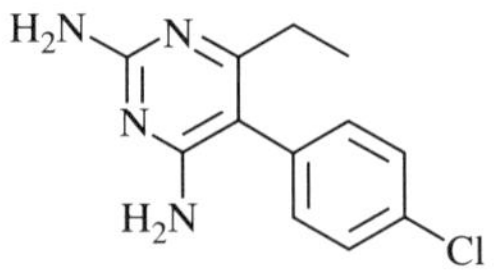

**5.14**

## 2. Lovastatin (Mevinolin) and Simvastatin, Antihypercholesterolemic Drugs

***a. Cholesterol and Its Effects.*** Coronary heart disease is the leading cause of death in the United States and other Western countries; about one-half of all deaths in the United States can be attributed to atherosclerosis,[33] which results from the buildup of fatty deposits called plaque on the inner walls of arteries. The major component of atherosclerotic plaque is cholesterol. In humans, more than one-half of the total body cholesterol is derived from its *de novo* biosynthesis in the liver.[34] Cholesterol biosynthesis requires more than 20 enzymatic steps starting from acetyl-CoA. The rate-determining step is the conversion of 3-hydroxy-3-methylglutaryl coenzyme A (HMG-CoA; **5.15**) to mevalonic acid (**5.16**), catalyzed by HMG-CoA reductase (Scheme 5.4). Because hypercholesterolemia is a primary risk factor for coronary heart disease,[35] and because the overall rate of cholesterol biosynthesis is a function of this enzyme, efforts were initiated to inhibit HMG-CoA reductase as a means of lowering plasma cholesterol levels.

***b. Lead Discovery.*** Endo and co-workers[36] at the Sankyo Company in Tokyo tested 8000 strains of microorganisms for metabolites that inhibited sterol biosynthesis *in vitro*, and they discovered three active compounds in culture broths of the fungus *Penicillium citrinum*. The most active compound,

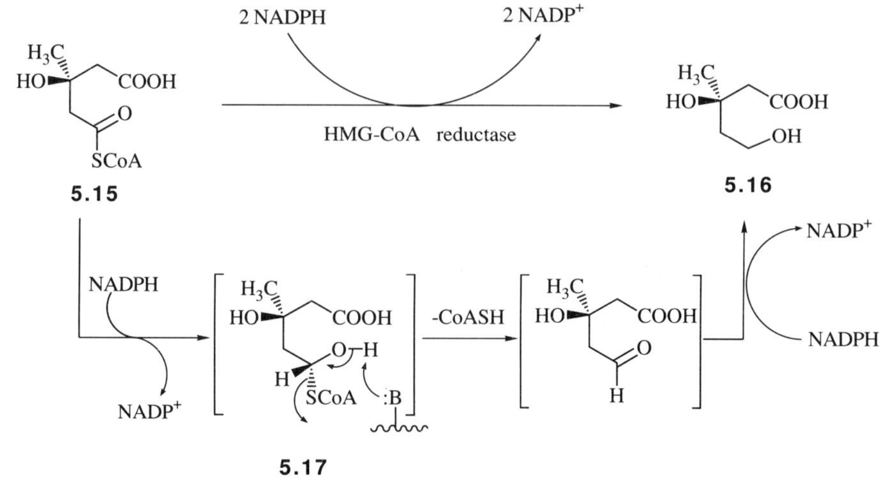

**Scheme 5.4.** HMG-CoA reductase, the rate-determining enzyme in *de novo* cholesterol biosynthesis.

called mevastatin,[36] also was isolated from broths of *Penicillium brevicompactum* by Brown and co-workers[37] at Beecham Pharmaceuticals in England, who named it compactin (**5.18**, R = H). A second, more active, compound was isolated by Endo[38] from the fungus *Monascus ruber*, which he named monacolin K; the same compound was isolated by a group at Merck from *Aspergillus terreus*,[39] which they named mevinolin (**5.18**, R = $CH_3$). Mevinolin is now referred to as lovastatin. Several related metabolites also were isolated from cultures of these fungi,[40] including dihydrocompactin from *P. citrinum* [**5.19**, R = H, R′ = (*S*)-$CH_3CH_2CH(CH_3)CO_2^-$], dihydromevinolin from *A. terreus* [**5.19**, R = $CH_3$, R′ = (*S*)-$CH_3CH_2CH(CH_3)CO_2^-$], and dihydromonacolin L from a mutant strain of *M. ruber* (**5.19**, R = $CH_3$, R′ = H).

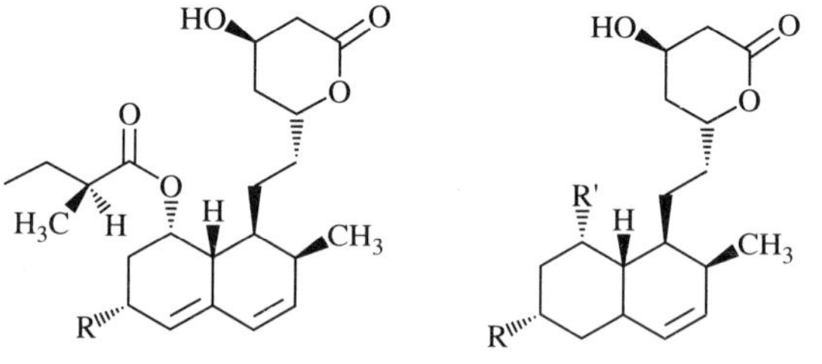

**5.18** **5.19**

***c. Mechanism of Action.*** Compactin[41] and mevinolin[39] are potent competitive reversible inhibitors of HMG-CoA reductase. The $K_i$ for compactin is $1.4 \times 10^{-9}$ *M* and that for mevinolin is $6.4 \times 10^{-10}$ *M* (rat liver enzyme); for comparison, the $K_m$ for HMG-CoA is about $10^{-5}$ *M*. Therefore, the affinity of

HMG-CoA reductase for compactin and mevinolin is 7140 and 16,700 times, respectively, greater than for its substrate. Compactin and mevinolin do not affect any other enzyme in cholesterol synthesis except HMG-CoA reductase.[36]

It may not be immediately obvious why mevinolin mimics the structure of HMG-CoA. The reason is that the active form is not that shown in **5.18**, but rather, the hydrolysis product, that is, the open chain 3,5-dihydroxyvaleric acid form (**5.20**). This form mimics the structure of the proposed intermediate **5.17** (Scheme 5.4) in the reduction of HMG-CoA by HMG-CoA reductase.

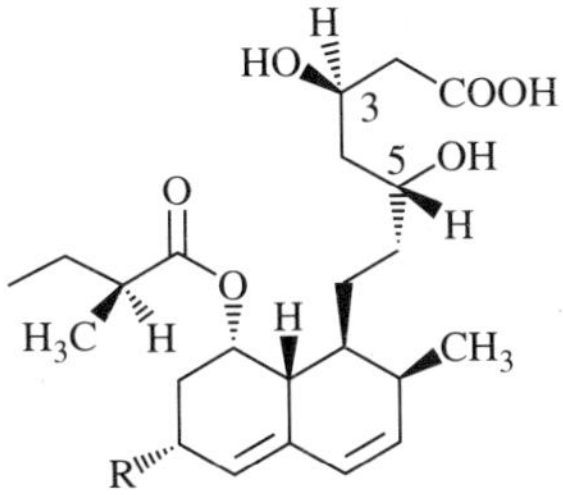

**5.20**

Enzyme studies with compactin and analogs indicated that there are two important binding domains at the active site, the hydroxymethylglutaryl binding domain, to which the upper part of **5.20** binds, and a hydrophobic pocket located adjacent to the active site to which the decalin (lower) part of **5.20** binds.[42] The high affinity of compactin and its analogs to HMG-CoA reductase derives from the simultaneous interactions of the two parts of these inhibitors with the two binding domains on the enzyme. As a result of these interactions, dissociation of the E · I complex is very slow. A kinetic analysis of the on and off rate constants ($k_{on}$ and $k_{off}$, respectively; see Scheme 5.1) for HMG-CoA and compactin (i.e., the rate constants for the binding and dissociation of compactin to HMG-CoA reductase) with yeast HMG-CoA reductase gave values of $1.9 \times 10^5\ M^{-1}\ \text{sec}^{-1}$ and $0.11\ \text{sec}^{-1}$ for HMG-CoA and $2.7 \times 10^7\ M^{-1}\ \text{sec}^{-1}$ and $6.5 \times 10^{-3}$ for compactin, respectively. Therefore, compactin binds faster and dissociates slower than does HMG-CoA, which accounts for the difference in the $K_m$ and $K_i$ ($k_{off}/k_{on}$) values. It, also, may be possible to classify these inhibitors as transition state analogs (see Section IV,D).

***d. Lead Modification.*** Numerous structural modifications were made on compactin and mevinolin in order to determine the importance of the lactone moiety and its stereochemistry, the ability of the lactone moiety to be opened to the dihydroxy acid, the optimal length and structure of the moiety bridging the lactone and the lipophilic groups, and the size and shape of the lipophilic group. It was found that potency was greatly reduced[43] unless a carboxylate anion could be formed and the hydroxyl groups were left unsubstituted in an erythro relationship. Insertion of a bridging unit other than ethyl or (*E*)-ethenyl between the 5-carbinol moiety and the lipophilic moiety also diminishes the potency.[43] Modifications of the lower (lipophilic) part of compactin in most cases led to compounds with considerably lower potencies,[44,45] except for certain substituted biphenyl analogs.[45] If the substituted biphenyl rings were constrained as fluorenylidine moieties, the corresponding potencies decreased.[46] When the hydroxyl group in the lactone ring was replaced by amino and thio groups, diminished potencies were observed.[47]

Modification of the 2(*S*)-methylbutyryl ester side chain [$CH_3CH_2CH(CH_3)CO_2^-$] of mevinolin indicated that introduction of an additional aliphatic group on the carbon $\alpha$ to the carbonyl group increased the potency of mevinolin.[48] When the side chain was $CH_3CH_2C(CH_3)_2CO_2^-$ (now marketed as simvastatin), the potency was about 2.5 times greater than mevinolin. The lactone epimer of mevinolin (the epimer of the carbon adjacent to the lactone oxygen in **5.18**) has less than 1/10,000th the potency of mevinolin.[49,50] Modifications in the 3,5-dihydroxyvaleric acid moiety of analogs of **5.20** resulted in lower potencies,[40,50] except when the 5-hydroxyl group was replaced by a 5-keto group, in which case potencies comparable to the parent compounds were observed.[50] Presumably, the 5-keto group becomes reduced, but it is not known if that occurs by HMG-CoA reductase and, if it does, whether it occurs prior to inhibition or whether it is the cause for inhibition.

### 3. Captopril, Enalapril, and Lisinopril, Antihypertensive Drugs

***a. Humoral Mechanism for Hypertension.*** The elucidation of the molecular details of the *renin–angiotensin* system, one of the humoral mechanisms for blood pressure control, began over 50 years ago.[51] Angiotensinogen, an $\alpha$-globulin produced by the liver,[52] is hydrolyzed by the proteolytic enzyme renin to a decapeptide, angiotensin I (Scheme 5.5), which has little, if any, biological activity. The C-terminal histidylleucine dipeptide is cleaved from angiotensin I by angiotensin-converting enzyme (or ACE, dipeptidyl carboxypeptidase I), mainly in the lungs and blood vessels to give the octapeptide angiotensin II. This peptide is responsible for the increase in blood pressure by several mechanisms. It acts as a hormone and is a very potent vasoconstrictor. Furthermore, it is the main physiological stimulus for the release of another hormone, aldosterone (**5.21**), which regulates the electrolyte balance of body fluids by promoting excretion of potassium ions and retention of sodium ions and water.[53] Both vasoconstriction and sodium ion/water retention lead to an increase in blood pressure. To make matters even worse, in addition to cleaving angiotensin I to angiotensin II, ACE also catalyzes the hydrolysis of the two C-terminal dipeptides from the potent vasodilator nonapeptide bradykinin (Arg-Pro-Pro-Gly-Phe-Ser-Pro-Phe-Arg). Consequently,

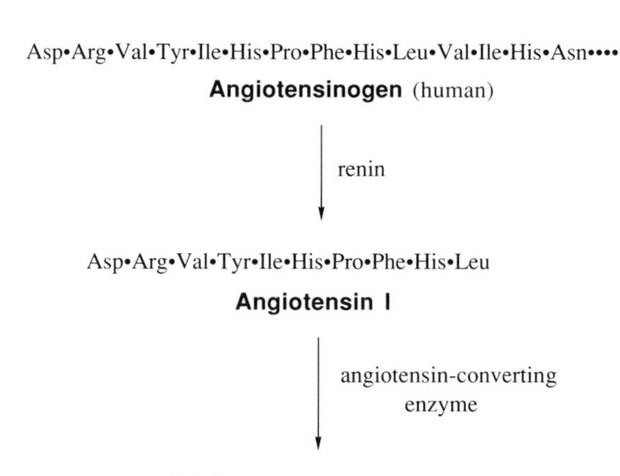


**Angiotensin II**

**Scheme 5.5.** The renin–angiotensin system.

the action of ACE results in the generation of a potent hypertensive agent (angiotensin II), which also stimulates the release of another hypertensive agent (aldosterone), and destroys a potent antihypertensive agent (bradykinin). All of these outcomes of ACE action result in *hypertension*, an increase in blood pressure. Angiotensin-converting enzyme, therefore, is an important target for the design of antihypertensive agents; inhibition of ACE would shut down its three hypertensive actions.

**5.21**

***b. Lead Discovery.*** In 1965 Ferreira[54] reported that a mixture of peptides in the venom of the South American pit viper *Bothrops jararaca* potentiated the action of bradykinin by inhibition of some bradykininase activity. Bakhle and co-workers[55] subsequently showed that these peptides also inhibited the conversion of angiotensin I to angiotensin II. Nine active peptides were isolated from the venom; the structure of a pentapeptide (Pyr-Lys-Trp-Ala-Pro, where Pyr is L-pyroglutamate) was identified.[56] This peptide was shown to inhibit the conversion of angiotensin I to II and bradykinin degradation *in vitro*[57] and *in vivo*.[58] The structures of six more of the peptides were determined by Ondetti and co-workers.[59] The peptide with the greatest *in vitro* activity was the pentapeptide,[60] but a nonapeptide (Pyr-Trp-Pro-Arg-Pro-Gln-Ile-Pro-Pro) called teprotide had the greatest *in vivo* potency[61] and was effective in lowering blood pressure.[62] Five other active peptides were isolated from the venom of the Japanese pit viper, *Agkistrodon halys blomhoffii*.[63] Because these compounds were peptides, they were not effective when administered orally, but they laid the foundation for the design of orally active angiotensin-converting enzyme inhibitors.

***c. Lead Modification and Mechanism of Action.*** The fact that N-acylated tripeptides are substrates of ACE indicated that it may be possible to prepare a small, orally active ACE inhibitor. After testing numerous peptides as competitive inhibitors of ACE, it was concluded that proline was best in the C-terminal position and alanine was best in the penultimate position.[62] An aromatic amino acid is preferred in the antepenultimate position.

When the search for a potent inhibitor of ACE was initiated at Squibb and Merck pharmaceutical companies, the enzyme had not yet been purified. Because the enzyme was inhibited by EDTA and other chelating agents, particularly bidentate ligands, it was believed to be a metalloenzyme. In fact, ACE purified to homogeneity from rabbit lung[64] was shown to contain 1 gram-atom of zinc ion per mole of protein. The zinc ion is believed to be a cofactor that assists in the catalytic hydrolysis of the peptide bond by coordination to the carbonyl oxygen (Fig. 5.1). This makes the carbonyl more electrophilic and activates it for nucleophilic attack and peptide bond cleavage.

Because the structure of the enzyme was not known, it was not obvious what peptidelike structures would be the best inhibitors. It was hypothesized that the mechanism and active site of ACE may resemble those of carboxypeptidase A, another zinc-containing peptidase whose X-ray structure was known.[65] Three important binding interactions between carboxypeptidase A and peptides are a carboxylate-binding group, a group that binds the C-terminal amino acid side chain, and the zinc ion that coordinates to the carbonyl of the penultimate (the scissile) peptide bond (Fig. 5.2).[66] (*R*)-2-Benzylsuccinic acid, which can bind at all three of these sites, is a potent inhibitor of carboxypeptidase A[67] (Fig. 5.2). The extreme potency of inhibition of carboxypeptidase A by (*R*)-2-benzylsuccinic acid was suggested[67] to be derived from the resemblance of this inhibitor to the *collected products* (Fig. 5.3) of hydrolysis of the substrate, and, therefore, it combines all of their individual binding characteristics into a single molecule (see Sections IV,C,2; IV,D,1; and IV,D,2,a).

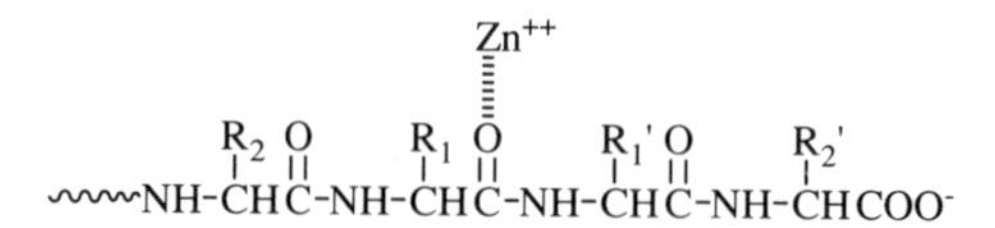


**Figure 5.1.** Importance of Zn(II) in angiotensin-converting enzyme catalysis.

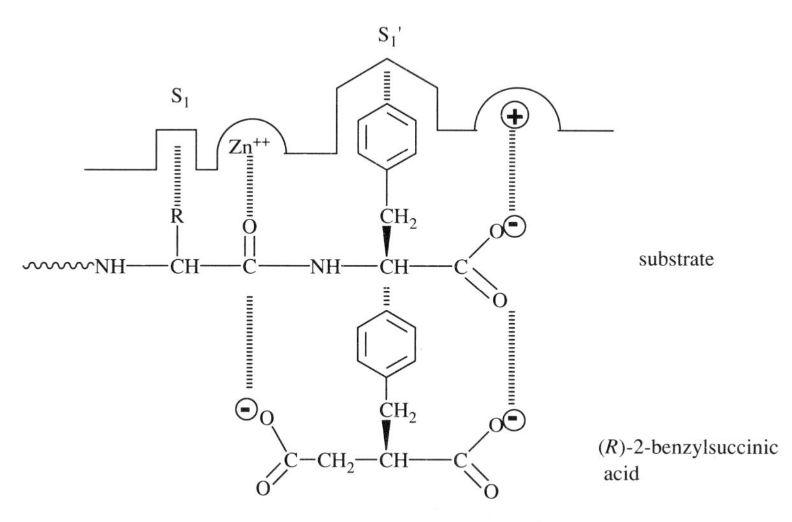


**Figure 5.2.** Hypothetical active site of carboxypeptidase A. [Adapted with permission from Cushman, D. W., Cheung, H. S., Sabo, E. F., Ondetti, M. A. (1977). *Biochemistry* **16,** 5484. Copyright © 1977 American Chemical Society.]

With (*R*)-2-benzylsuccinic acid as a model, and the known effectiveness of a C-terminal proline for ACE inhibition, a series of carboxyalkanoylproline derivatives (**5.22**) were tested as inhibitors of ACE.[66] Note that in order to avoid having an orally unstable dipeptide, the N-terminal amino group was substituted by an isosteric $CH_2$ group to which the Zn(II)-coordinating car-

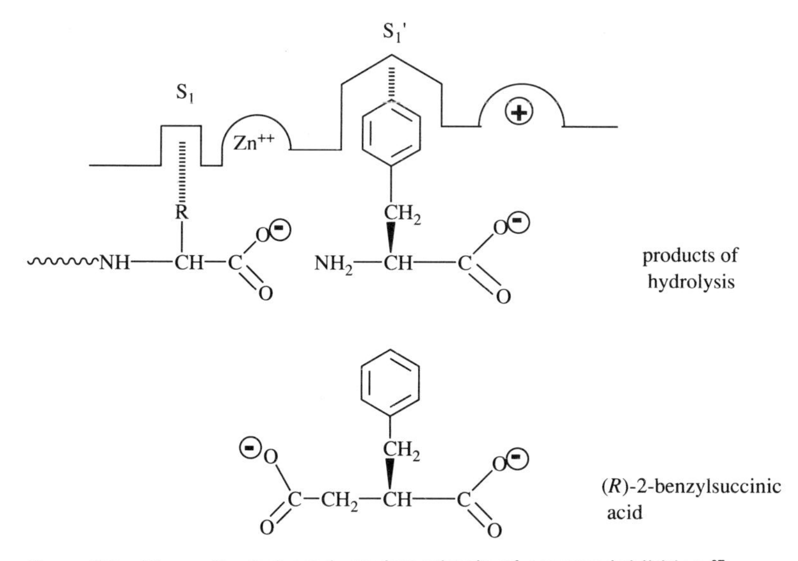


**Figure 5.3.** The collected products hypothesis of enzyme inhibition.[67]

boxylate was attached. Although the results were encouraging, all of these compounds were only weak inhibitors of ACE. To increase the potency of the compounds, a better Zn(II)-coordinating ligand, a thiol group, was substituted for the carboxylate (**5.23**).[66] These compounds were very potent inhibitors of ACE.

$^{-}OOC$—alkyl—C(=O)—N (proline, $CO_2H$)

**5.22**

HS—alkyl—C(=O)—N (proline, $CO_2H$)

**5.23**

Figure 5.4 shows a hypothesized depiction of the interaction of **5.22** and **5.23** with ACE. Note that carboxypeptidase A is a C-terminal *exopeptidase* (it cleaves the C-terminal amino acid), whereas ACE is a C-terminal *endopeptidase* or, more precisely, a *dipeptidyl carboxypeptidase* (it cleaves a C-terminal dipeptide). Therefore, the active site of ACE (Fig. 5.4) has two additional binding sites than carboxypeptidase A has between the Zn(II) and the group that interacts with the C-terminal carboxylate group (Fig. 5.2). The compound that had the best binding properties was **5.24** (captopril), a competitive inhibitor of ACE with a $K_i$ of $1.7 \times 10^{-9}$ $M$. Furthermore, captopril is highly specific for angiotensin-converting enzyme; the $K_i$ values for captopril with carboxypeptidase A and carboxypeptidase B, two other Zn(II)-containing peptidases, are $6.2 \times 10^{-4}$ and $2.5 \times 10^{-4}$ $M$, respectively.[68]

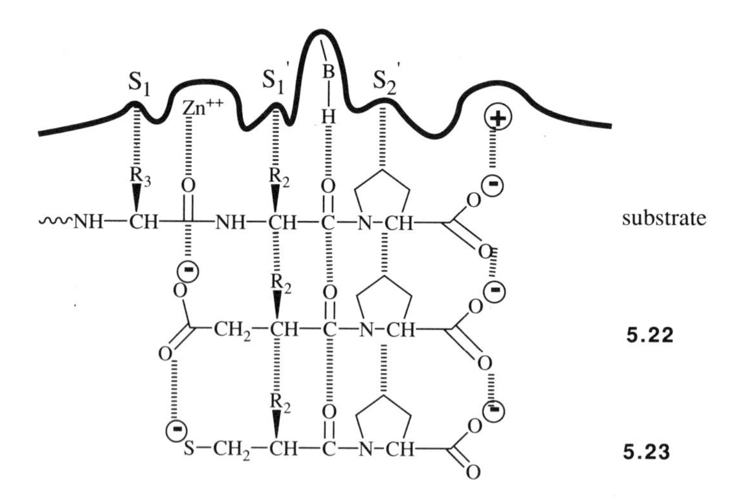


**Figure 5.4.** Hypothetical binding of carboxyalkylproline and mercaptoalkylproline derivatives to angiotensin-converting enzyme. [Adapted with permission from Cushman, D. W., Cheung, H. S., Sabo, E. F., and Ondetti, M. A. (1977). *Biochemistry* **16,** 5484. Copyright © 1977 American Chemical Society.]

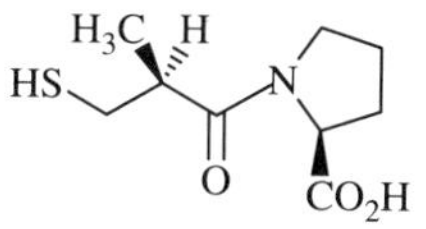

**5.24**

Presumably, the reason for the specificity is that in one small molecule it has many functional groups not present in other peptidases that can regio- and stereospecifically interact with appropriate groups at the active site of ACE (compare Figs. 5.2 and 5.4). The carboxylate group of the inhibitor can be stabilized by an electrostatic interaction with a cationic group on the enzyme, the amide carbonyl can be hydrogen bonded to a hydrogen donor group, the sulfhydryl can be liganded to the zinc ion, and the proline and (*S*)-methyl group can be involved in stereospecific hydrophobic and van der Waals interactions. All of these interactions must be important because deletion or alteration of any of these groups raises the $K_i$ considerably (Table 5.1). A myriad of analogs of this basic structure, including compounds with Zn(II)-coordinating ligands other than carboxylate and thiol groups, has been synthesized and tested as ACE inhibitors (see General References).

Captopril was the first ACE inhibitor on the drug market, and it was shown to be effective for the treatment of both hypertension and congestive heart failure. Given alone, captopril can normalize the blood pressure of about 50% of the hypertensive population. When given in combination with a diuretic, such as hydrochlorothiazide (**5.25**) (remember, angiotensin II releases aldosterone which causes water retention), this effectiveness can be extended to 90% of the hypertensive population. In more severe cases, a $\beta$-blocker, an antagonist for the $\beta$-adrenergic receptor, which triggers vasodilation, may be used in a triple therapy with captopril and a diuretic.

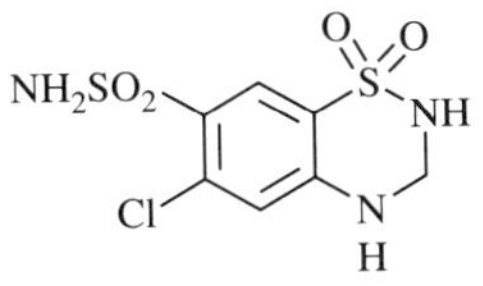

**5.25**

Two principal side effects may arise from the use of captopril, namely, rashes and loss of taste. Both of these side effects are reversible upon drug withdrawal or reduction of the dose.[69] Considering the potential lethality of hypertension, a minor rash or loss of taste would seem insignificant to the benefits of the therapy. However, hypertension is a disease without a symptom (that is, until it is too late); generally, a patient discovers he has this disease when his physician determines it after taking his blood pressure. Because of this lack of immediate discomfort, there may be difficulties getting

**Table 5.1** Effect on $K_i$ of Structural Modification of Captopril

| Analog | Relative $K_i$ |
|---|---|
| 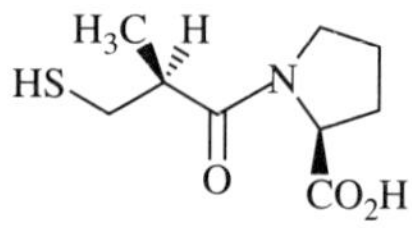 $H_3C$ H HS N O $CO_2H$ (captopril) | 1.0 |
| 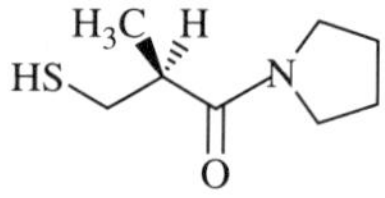 $H_3C$ H HS N O | 12,500 |
| 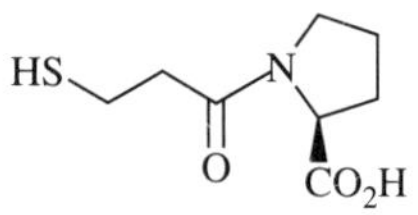 HS N O $CO_2H$ | 10 |
| 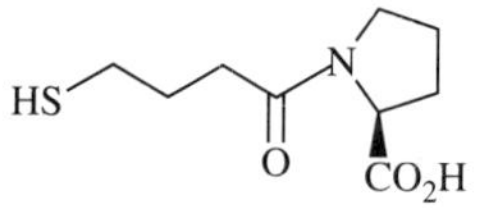 HS N O $CO_2H$ | 12,000 |
| 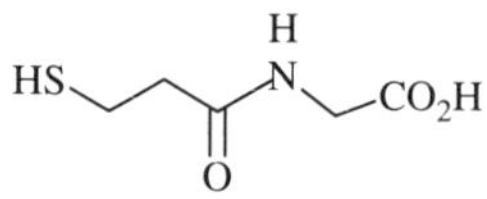 H HS N $CO_2H$ O | 120 |
| 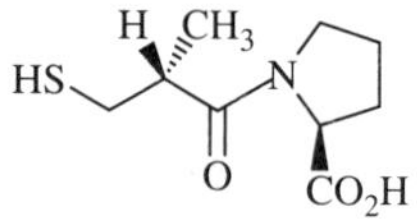 H $CH_3$ HS N O $CO_2H$ | 120 |
| 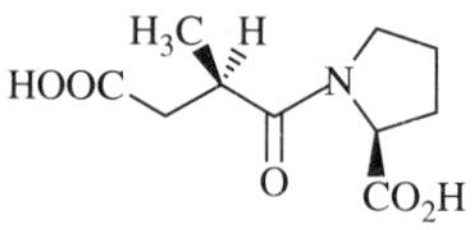 $H_3C$ H HOOC N O $CO_2H$ | 1,100 |

the patient to comply with the therapy, especially if unpleasant side effects arise when the drug is taken.

Consequently, the Merck group investigated the cause for the side effects. Because similar side effects arise when penicillamine is administered, it was hypothesized that the thiol group may be responsible.[70] Furthermore, deletion of this functional group should give inhibitors with greater metabolic stability as thiols undergo facile *in vivo* oxidation to disulfides. The approach taken[70] was to attempt to increase the previously found weak potency of the carboxyalkanoylproline analogs by adding groups that can interact with additional sites on the enzyme.

If the carboxyalkanoylproline derivatives are collected product inhibitors (see Fig. 5.3), then there are two features that can be built into these analogs to make them look more productlike. One is to make them structurally more similar to dipeptides by substituting an NH for a $CH_2$ such as **5.26** (R = R′ = H). Disappointingly, however, this compound had less than twice the potency

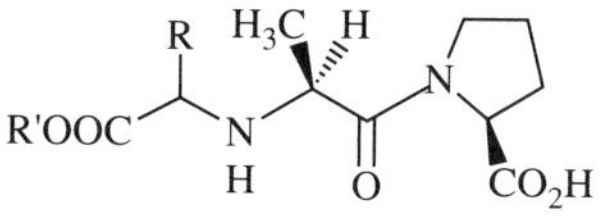

**5.26**

of the isostere with a $CH_2$ in place of NH. The reason for this could be compensatory factors. An NH (or its protonated form) is much more hydrophilic than a $CH_2$ group. Therefore, an additional hydrophobic group should be added to counterbalance this hydrophilic effect. When a methyl group was appended (**5.26**, R = $CH_3$, R′ = H), the potency increased about 55-fold. As the other feature which could make these compounds structurally more similar to the products would be to append a group that might interact with the substrate $S_1$ subsite, the R group of **5.26** was modified further,[70] and **5.26** [R = (*S*)-$PhCH_2CH_2$, R′ = H], called enalaprilat, emerged as the viable drug candidate. The $IC_{50}$ (the concentration that gives 50% enzyme inhibition) for enalaprilat is 19 times lower than that for captopril, suggesting that there are increased interactions of enalaprilat with ACE. These may be hydrophobic interactions of the phenylethyl group with the $S_1$ subsite (see Fig. 5.5; also, see Section IV,D,2,a for an alternative explanation of the potency of enalaprilat). Furthermore, the undesirable side effects of captopril are rare with enalaprilat.

Enalaprilat, however, is poorly absorbed orally and, therefore, must be given by intravenous injection. This problem was remedied simply by conversion of the carboxyl group to an ethyl ester [**5.26**, R = (*S*)-$PhCH_2CH_2$, R′ = $CH_3CH_2$], giving enalapril, which has excellent oral activity. As the *in vitro* $IC_{50}$ for enalapril is $10^3$ times higher than that for enalaprilat,[70] the ethyl ester group must be hydrolyzed by esterases in the body to liberate the active form of the drug, namely enalaprilat. Enalapril, then, is an example of a *prodrug*, a compound that requires metabolic activation for activity (see Chapter 8). The effect of esterification is to lower the $pK_a$ of the NH group (the $pK_a$ is 5.5 in enalapril but 7.6 in enalaprilat)[71] and to remove the charge of the carboxylate, both of which would increase membrane transport.

Another compound that was prepared by the Merck chemists as an alternative to enalaprilat was the lysylproline analog called lisinopril (**5.27**). Note the stereochemistry shown (*S*,*S*,*S*) is that found in the most potent isomer of lisinopril, and it is also the stereochemistry of enalaprilat. Lisinopril is more slowly and less completely absorbed than enalapril, but its longer oral dura-

substrate

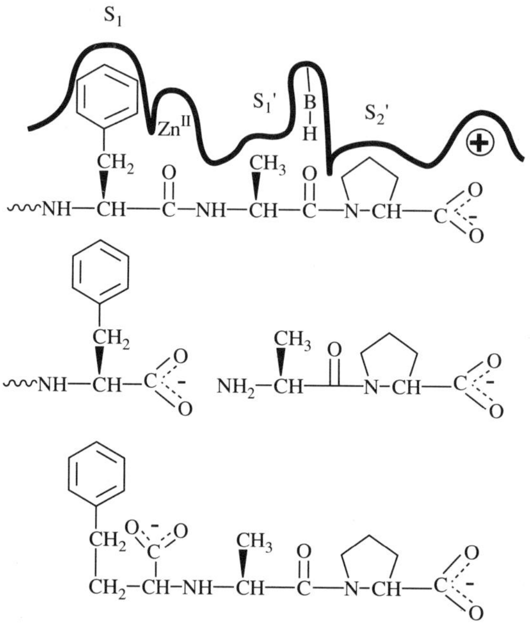

product

**5.26**
(R = $PhCH_2CH_2$
R' = —)

**Figure 5.5.** Hypothetical interactions of enalaprilat with angiotensin-converting enzyme.

tion of action, and the fact that it does not require metabolic activation, made it an attractive alternative.

All three of these ACE inhibitors, captopril, enalapril, and lisinopril, are effective antihypertensive drugs. Whereas captopril is taken in doses of 25–50 mg twice or thrice a day, enalapril and lisinopril are taken in a single daily dose of 20–40 mg. All have a relatively low incidence of side effects.

Squibb more recently has reported a new potent class of ACE inhibitors, the (hydroxyphosphinyloxy)acyl amino acids. Compound **5.28** had oral activity superior to that of captopril.[72] It is apparent by comparison of **5.27** and **5.28** that similar binding interactions with ACE are possible.

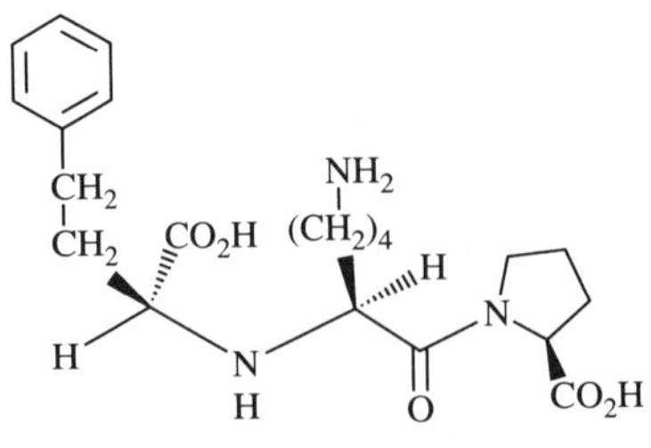

**5.27** 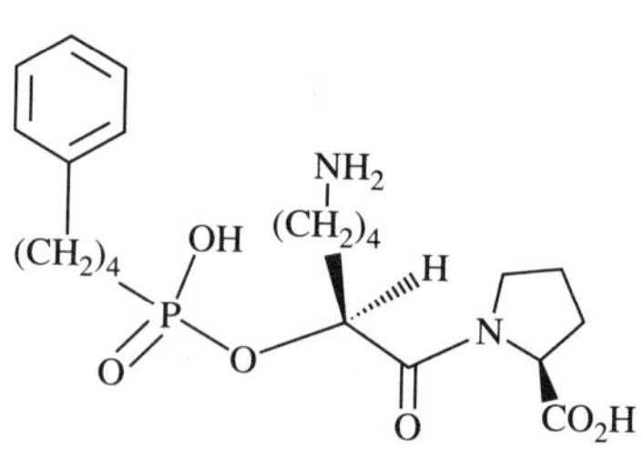 **5.28**

## C. Slow, Tight-Binding Inhibitors

### 1. Theoretical Basis

*Slow binding inhibitors* are inhibitors for which the equilibrium between enzyme and inhibitor is reached slowly; *tight-binding inhibitors* are inhibitors for which substantial inhibition occurs when the concentrations of inhibitor and enzyme are comparable.[73,74] *Slow, tight-binding inhibitors* have both properties. These inhibitors can bind noncovalently[73,74] or covalently[75,76]; when a covalent bond is formed, a slowly reversible adduct may be involved. Because binding is slow and tight, time-dependent loss of enzyme activity is observed, which is reminiscent of the kinetics for irreversible inhibitors (see Section V,B,1).

The reason for slow binding inhibition is not known. One possibility is that these inhibitors are such good analogs of the substrate that they induce a conformational change in the enzyme which resembles that associated with the transition state; typically these compounds are transition state analogs (see Section IV,D). If this is the case, then the inhibitor binding would be slow because it does not have all of the essential structural features of the substrate transition state geometry. The dissociation would be even slower because the dissociation rate is not enhanced by product formation. The conformational change may result from a change in the protonation state of the enzyme[77] or from the displacement of an essential water molecule by the inhibitor.[74]

### 2. Enalaprilat

The interaction of enalaprilat with purified rabbit lung angiotensin-converting enzyme was studied, and the kinetics indicate that it is a slow, tight-binding inhibitor.[78] The rate constant for the formation of the E·I complex ($k_{on}$) at pH 7.5 was determined to be $2 \times 10^6\ M^{-1}\ sec^{-1}$, which is at least two orders of magnitude smaller than expected for a diffusion-controlled reaction. Steady-state kinetics gave the value of the $K_i$ as $1.8 \times 10^{-10}\ M$. As $K_i$ equals $k_{off}/k_{on}$, the value of $k_{off}$ should be $3.6 \times 10^{-4}\ sec^{-1}$, which is in satisfactory agreement with the measured $k_{off}$ of $1.6 \times 10^{-4}\ sec^{-1}$. The small $k_{off}$ value for this noncovalent E·I complex emphasizes the strong affinity of enalaprilat for angiotensin-converting enzyme.

### 3. Peptidyl Trifluoromethyl Ketone Inhibitors of Human Leukocyte Elastase

Human leukocyte elastase and cathepsin G are serine proteases that are released normally by the immune system neutrophils in the lungs to digest dead lung tissue and destroy invading bacteria. Natural inhibitors of these enzymes ($\alpha_1$-protease inhibitor[79] and bronchial mucous inhibitor[80]) also are released to prevent these enzymes from destroying the key structural protein component of the lung, elastin, and lung connective tissue. It is hypothesized that an imbalance in the protease and protease inhibitor concentrations (the protease/antiprotease hypothesis)[81] may be the cause for emphysema. An imbalance could arise because of a genetic deficiency in $\alpha_1$-protease inhibitor or from inhalation of cigarette smoke which oxidizes Met-358 at the active site of the inhibitor.[82]

A drug design approach which would return the imbalance to normal would be to discover an inhibitor of leukocyte elastase and cathepsin G in order to mimic the action of their natural antiproteases.[83] Chemists at Stuart Pharmaceuticals have developed a class of peptidyl trifluoromethyl ketones with the structures X-Val-$CF_3$, X-Pro-Val-$CF_3$, X-Val-Pro-Val-$CF_3$, and X-Lys(Z)-Val-Pro-Val-$CF_3$, where X is *N*-(methoxysuccinyl) and Z is *N*-(carbobenzoxy).[76] The most potent analog, *Z*-Lys(Z)-Val-Pro-Val-$CF_3$, had a $K_i$ below $10^{-10}$ *M*. All of these compounds were shown to be competitive slow, tight-binding inhibitors of human leukocyte elastase. The kinetic constants for the most potent analog were $k_{on} = 8 \times 10^4\ M^{-1}\ sec^{-1}$ and $k_{off} = <10^{-5}\ sec^{-1}$ (obtained from the equation $k_{off} = k_{on}K_i$).

The design of peptidyl trifluoromethyl ketones as inhibitors of serine proteases stems from the work of Abeles and co-workers[75,84] who suggested that because trifluoromethyl ketones exist almost exclusively as the hydrate in water, and because they enhance nucleophilic addition to the carbonyl, then fluoroketones may form a reasonably stable, but reversible, hemiketal with the active site serine of a serine protease (Scheme 5.6).[76] This inhibition mechanism allows for various possible conformational changes in the enzyme that may occur when the enzyme is in different protonation forms. The (E · I)′ complex may involve a reorientation of the inhibitor in the active site to allow for more efficient attack of the serine at the trifluoromethyl carbonyl. According to this hypothetical mechanism, the covalently bound inhibitor only dissociates when the active site imidazole is protonated. This could account for the small $k_{off}$. These inhibitors are an example of covalent reversible inhibition.

## D. Transition State Analogs and Multisubstrate Analogs

### 1. Theoretical Basis

As discussed in Chapter 4 (Section I,B,2), an enzyme accelerates the rate of a reaction by stabilizing the transition state, which lowers the free energy of activation. The enzyme achieves this rate enhancement by changing its conformation so that the strongest interactions occur between the substrate and enzyme active site at the transition state of the reaction. Some enzymes act by straining or distorting the substrate toward the transition state. The catalysis-by-strain hypothesis led to early observations that some enzyme inhibitors

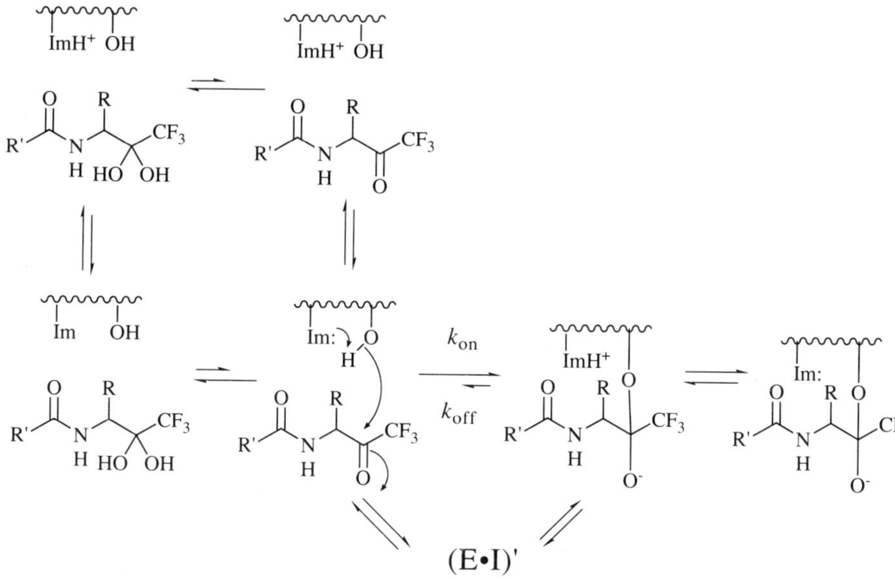

Scheme 5.6. Hypothetical mechanism for slow, tight-binding inhibition of peptidyl trifluoromethyl ketones with serine proteases. Im is the imidazole of histidine.

owe their effectiveness to a resemblance to the strained species. Bernhard and Orgel[85] theorized that inhibitor molecules resembling the transition state species would be much more tightly bound to the enzyme than would be the substrate. Therefore, a potent enzyme inhibitor would be a stable compound whose structure resembles that of the substrate at a postulated transition state (or transient intermediate) of the reaction rather than that at the ground state. A compound of this type would bind much more tightly to the enzyme and is called a *transition state analog inhibitor*. Jencks[86] was the first to suggest the existence of transition state analog inhibitors, and he cited several possible literature examples; Wolfenden[87a] and Lienhard[87b] developed the concept further. Values for dissociation constants ($K_i$) of $10^{-15}$ $M$ for enzyme–transition state complexes may not be unreasonable given the normal range of $10^{-3}$–$10^{-5}$ $M$ for dissociation constants of enzyme–substrate complexes ($K_m$).

In order to design such an inhibitor, the mechanism of the enzyme reaction must be understood, so that a theoretical structure for the substrate at the transition state can be hypothesized. Because many enzyme-catalyzed reactions have similar transition states (e.g., the different serine proteases), the basic structure of a transition state analog for one enzyme can be modified to meet the specificity requirements of another enzyme in the same mechanistic class and, thereby, generate a transition state analog for the other enzyme. This modification may be as simple as changing an amino acid in a peptidyl transition state analog inhibitor for one protease so that it conforms to the peptide specificity requirement of another protease. This is the approach that was taken to obtain serine protease specificity for the peptidyl trifluoromethyl ketone inhibitors of Abeles and co-workers[75,85] (see Section IV,C,3). Christianson and Lipscomb[88] have renamed reversible inhibitors that undergo a bond-forming reaction with the enzyme prior to the observation of the enzyme–inhibitor complex as *reaction coordinate analogs*. The peptidyl trifluoromethyl ketones (see Section IV,C,3) would be an example of this type of inhibitor.

When more than one substrate is involved in the enzyme reaction, a single stable compound can be designed that has a structure similar to that of the two or more substrates at the transition state of the reaction. This special case of a transition state analog is termed a *multisubstrate analog inhibitor*. Because of the magnitude of the binding interactions of transition state analogs with enzymes, they typically exhibit kinetics for slow, tight-binding inhibition.

### 2. Transition State Analogs

***a. Enalaprilat.*** In Section IV,B,3,c enalaprilat was shown to be a very potent competitive reversible inhibitor of angiotensin-converting enzyme. The rationalization for its binding effectiveness was that it had multiple binding interactions with the substrate and product binding sites. The resemblance of enalaprilat to the substrate and products of the enzyme reaction (Fig. 5.5) supported this notion. Both of these rationalizations are ground state arguments, but transition state theory suggests that the most effective interactions occur at the transition state of the reaction. With that in mind, let us consider a potential mechanism for angiotensin-converting enzyme–catalyzed substrate hydrolysis (Scheme 5.7) and see if the transition state structure is relevant. It is not known if a general base mechanism, as shown in Scheme 5.7, or a covalent catalytic mechanism is involved. Enalaprilat has been drawn beneath transition states 1 and 2 ($\ddagger_1$ and $\ddagger_2$) to show how the structures are related. An enzyme conformational change at the transition state (shown as a blocked enzyme instead of a rounded enzyme in Scheme 5.7) could increase the binding interactions. The resemblance of enalaprilat to the transition state structure could account for the observation that it is a slow, tight-binding inhibitor of the enzyme (see Section IV,C,2).

***b. Pentostatin.*** Pentostatin (2′-deoxycoformycin; **5.29**), an antineoplastic agent isolated from fermentation broths of *Streptomyces antibioticus*, is a potent inhibitor of the enzyme adenosine deaminase (adenosine aminohydrolase).[89] The $K_i$ is $2.5 \times 10^{-12}$ *M*, which is $10^7$ times lower than the $K_m$ for adenosine! As in the case of enalaprilat (Sections IV,B,3,c and IV,C,2), pentostatin is a slow, tight-binding inhibitor (Section IV,C).[90] The $k_{on}$ with human erythrocyte adenosine deaminase is $2.6 \times 10^6\ M^{-1}\ \text{sec}^{-1}$, and the $k_{off}$ is $6.6 \times 10^{-6}\ \text{sec}^{-1}$. The very small $k_{off}$ value is reflected in the very low $K_i$ value.

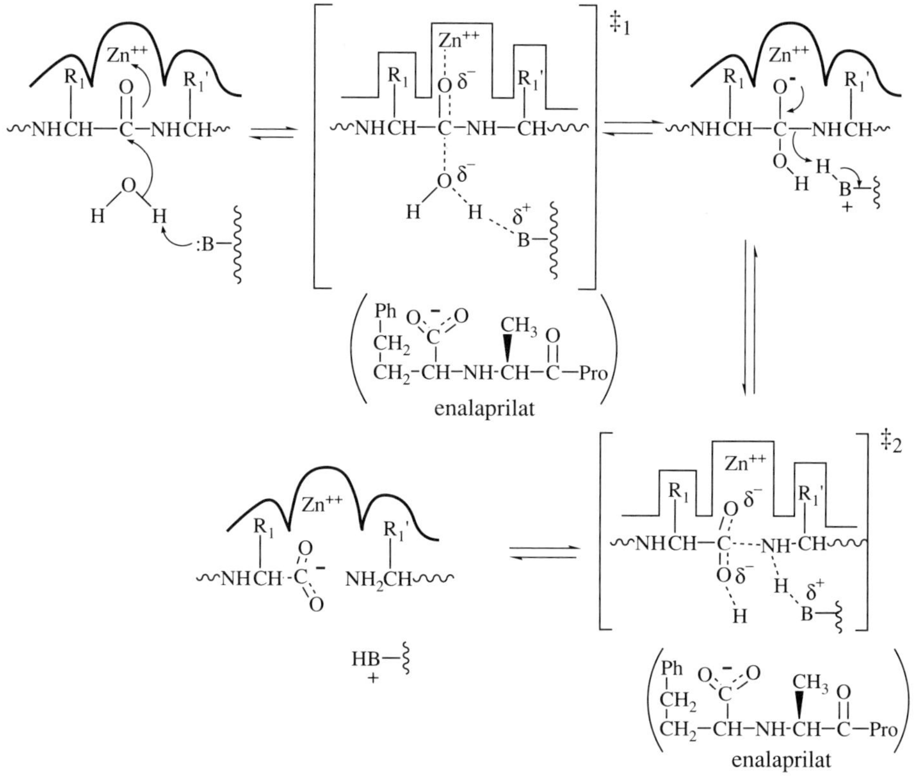

Scheme 5.7. Hypothetical mechanism for angiotensin-converting enzyme–catalyzed peptide hydrolysis.

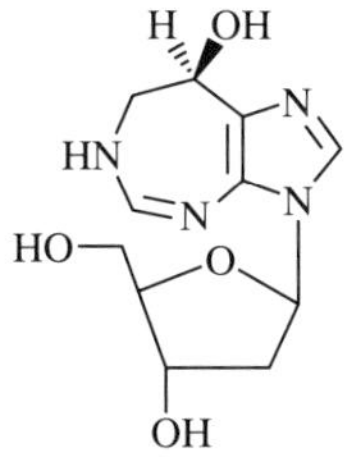

5.29

Pentostatin is an analog of the natural nucleoside 2′-deoxyinosine in which the purine is modified to contain a seven-membered ring with two $sp^3$ carbon atoms. It is believed that this compound mimics the transition state structure of the substrates adenosine and 2′-deoxyadenosine during their hydrolysis to inosine and 2′-deoxyinosine, respectively, by adenosine deaminase (see Scheme 5.8 for a hypothetical mechanism). In this case the resemblance of **5.29** to the intermediate **5.30** (Scheme 5.8) is clearer than its similarity to the

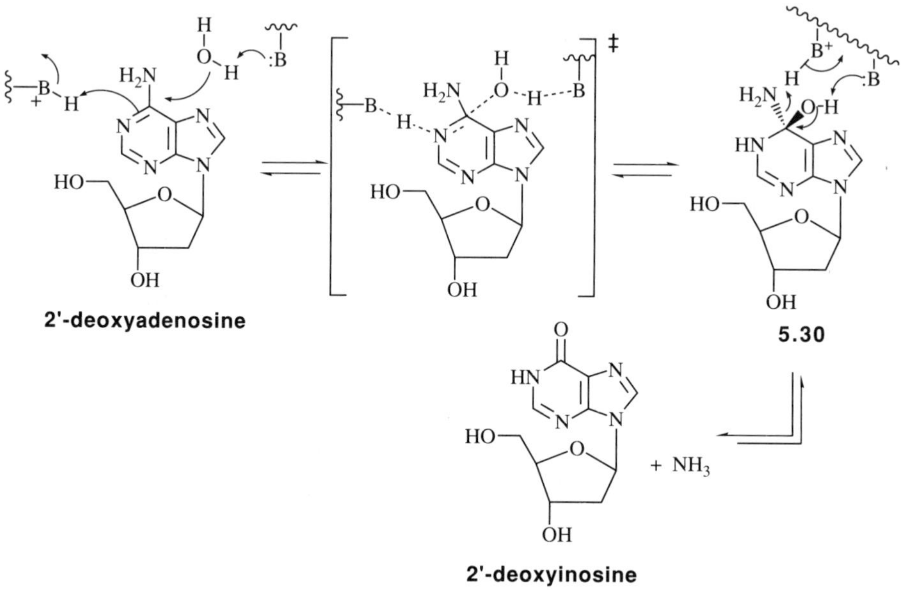

2'-deoxyadenosine

5.30

2'-deoxyinosine

+ $NH_3$

**Scheme 5.8.** Hypothetical mechanism for adenosine deaminase–catalyzed hydrolysis of 2′-deoxyadenosine.

transition state shown; however, a late transition state would look more like **5.30**.

It is not clear why inhibition of this enzyme should result in selective lymphotoxicity. One hypothesis is 2′-deoxyadenosine accumulates which, in turn, is an inhibitor of ribonucleotide reductase and *S*-adenosylhomocysteine hydrolase (adenosylhomocysteinase).[91] Ribonucleotide diphosphate reductase, which catalyzes the conversion of ribonucleotides to the corresponding 2′-deoxyribonucleotides, is essential for DNA biosynthesis. Inhibition of this enzyme leads to inhibition of DNA biosynthesis. *S*-Adenosylhomocysteine competitively inhibits most of the methyltransferases that utilize *S*-adenosylmethionine as the methyl-donating agent. This, apparently, is a mechanism for the regulation of these methyltransferases. Inhibition of *S*-adenosylhomocysteine hydrolase, the enzyme that degrades *S*-adenosylhomocysteine, results in an accumulation of *S*-adenosylhomocysteine, which inhibits the growth and replication of various tumors (and viruses), particularly those requiring a methylated 5′-cap structure on their messenger ribonucleic acids (mRNAs). In addition, various lymphocytic functions are suppressed by the accumulation of extracellular adenosine.

A major obstacle to the success of antipurines, enzyme inhibitors that mimic purines and block their metabolism, is acquired resistance. Unlike adenosine and various other 2′-deoxyribonucleosides, which are converted directly to the corresponding 5′-monophosphates by nucleoside kinases, a similar reaction does not occur with inosine or 2′-deoxyinosine. Instead, they are converted by purine-nucleoside phosphorylase to hypoxanthine (inosine without the sugar moiety), which is transformed into the corresponding nucleotide by hypoxanthine–guanine phosphoribosyltransferase (HGPRT, hypoxanthine phosphoribosyltransferase). The most common mechanism for antipurine drug resistance is a lack of the enzyme HGPRT or an altered HGPRT that binds the substrates poorly.

Animals treated with pentostatin show marked immunosuppression. Synergistic effects are observed when pentostatin is used in combination with other antipurines, especially the antiviral drug vidarabine (ara-A; **5.31**) which is degraded by adenosine deaminase. Although pentostatin has induced remissions of nodular lymphomas and lymphocytic leukemia, some unexplained deaths occurred in clinical trials; consequently, further study has been restricted.[92]

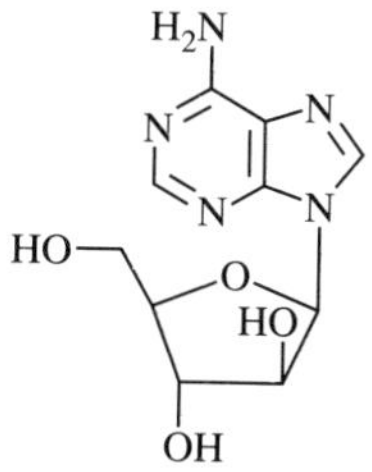

$H_2N$ N N N N HO O HO OH

**5.31**

### 3. Multisubstrate Analogs: *N*-Phosphonoacetyl-L-Aspartate

One of the first steps in the *de novo* biosynthesis of pyrimidines is the condensation of carbamyl phosphate (**5.32**) and L-aspartic acid, catalyzed by aspartate transcarbamylase (aspartate carbamoyltransferase), which produces *N*-carbamyl-L-aspartate (**5.33**, Scheme 5.9). Below the transition state structure is drawn *N*-phosphonoacetyl-L-aspartate (**5.34**, PALA), which is a stable compound (the isosteric exchange of a $CH_2$ for the O prohibits loss of the $PO_3^{2-}$ moiety) that resembles the transition state for condensation of the two substrates.[93]

PALA was ineffective in clinical trials as a result of tumor resistance.[94] When tumor cells acquired the ability to utilize preformed circulating pyrimidine nucleosides, they no longer needed to have a *de novo* metabolic pathway for pyrimidines. Other mechanisms of resistance are increased carbamyl phosphate or aspartate transcarbamylase production. To overcome resistance to PALA, nitrobenzylthioinosine, which inhibits the diffusion of nucleoside transport, and should block the uptake of the preformed pyrimidines, was

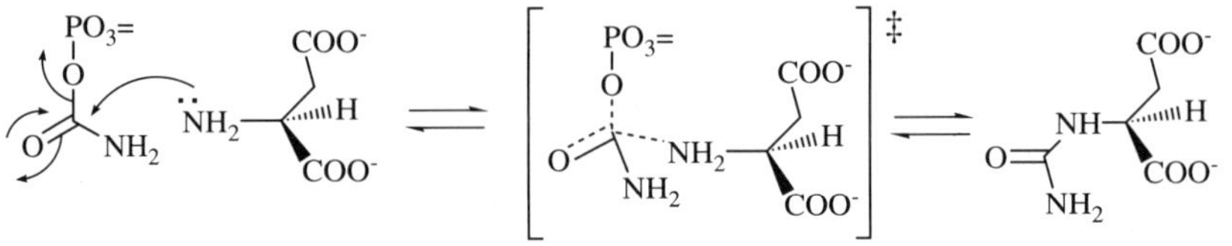

**5.33**

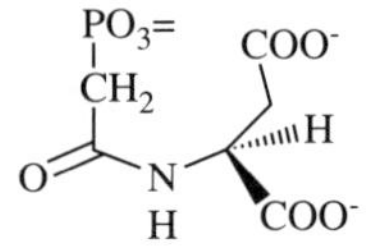

**5.34**

**Scheme 5.9.** Hypothetical mechanism for the reaction catalyzed by aspartate transcarbamylase.

tested for its synergistic effect with PALA. These two compounds were synergistic *in vitro*, but they were too toxic at effective dosages to be used *in vivo*.

## V. Irreversible Enzyme Inhibitors

### A. Potential of Irreversible Inhibition

A reversible enzyme inhibitor is effective as long as a suitable concentration of the inhibitor is present to drive the equilibrium $E + I \rightleftharpoons E \cdot I$ to the right (see Section IV,A). Therefore, a reversible inhibitor drug is effective only while the drug concentration is maintained at a high enough level to sustain the enzyme–drug complex. Because of drug metabolism and excretion (see Chapter 7), repetitive administration of the drug is required.

A competitive *irreversible enzyme inhibitor*, also known as an *active site–directed irreversible inhibitor* or an *enzyme inactivator*, is a compound whose structure is similar to that of the substrate or product of the target enzyme and which generally forms a covalent bond to an active site residue (a slow, tight-binding inhibitor, however, often is a noncovalent inhibitor that can be functionally irreversibly bound). In the case of irreversible inhibition it is not necessary to sustain the inhibitor concentration in order to retain the enzyme–inhibitor interaction. Because this is an irreversible reaction, once the target enzyme has reacted with the irreversible inhibitor, the complex cannot dissociate (again, there are exceptions), and, therefore, the enzyme remains inactive, even in the absence of additional inhibitor. This effect could translate into the requirement for smaller and fewer doses of the drug. Even though the target enzyme is destroyed by the irreversible inhibitor, it does not mean that only one dose of the drug would be sufficient to inhibit the enzyme permanently. Our genes are constantly encoding more proteins; as the enzyme loses activity, additional copies of the enzyme are synthesized, but this process can take hours or even days.[95] In some cases, however, particularly where genetic translation of the target enzyme is slow, it may be safer to design reversible inhibitors whose effects can be controlled more effectively by termination of their administration.

One may wonder what the effect on metabolism would be if a particular enzyme activity were completely inhibited for an extended period of time. Consider the case of aspirin, an irreversible inhibitor of prostaglandin synthase (see Section V,B,2,b). If the quantity of aspirin consumed in the United States (20 thousand tons annually, at least, prior to the announcement that it may cause Reye's syndrome in children[96]) were averaged over the entire population, then every man, woman, and child would be taking about 200 mg of aspirin every day, enough to shut down human prostaglandin biosynthesis for the entire country permanently! Suffice it to say that there are many irreversible enzyme inhibitors on the drug market.

The term irreversible is a loose one; either a very stable covalent bond or a labile bond may be formed between the drug and the enzyme active site. As pointed out earlier, some tight-binding reversible inhibitors also are functionally irreversible. As long as the enzyme remains nonfunctional long enough to produce the desired pharmacological effect, it is considered irreversibly inhibited. The two principal types of enzyme inactivators are reactive compounds called affinity labeling agents and unreactive compounds that are activated by the target enzyme known as mechanism-based enzyme inactivators.

## B. Affinity Labeling Agents

### 1. Mechanism of Action

An *affinity labeling agent* is a reactive compound that has a structure similar to that of the substrate for a target enzyme. Subsequent to reversible E·I complex formation, it reacts with active site nucleophiles (amino acid side chains), generally by acylation or alkylation ($S_N2$) mechanisms, thereby forming a stable covalent bond to the enzyme (Scheme 5.10). Note that this reaction scheme is similar to that for the conversion of a substrate to a product (see Section I,B,1 of Chapter 4); instead of a $k_{cat}$, a catalytic rate constant for product formation, there is a $k_{inact}$, an inactivation rate constant for enzyme inactivation. On the assumption that the equilibrium for reversible E·I complex formation ($K_I$) is rapid and the rate of dissociation of the E·I complex

$$E + I \underset{}{\overset{K_I}{\rightleftharpoons}} E\cdot I \xrightarrow{k_{inact}} E\text{-}I$$

**Scheme 5.10.** Basic kinetic scheme for an affinity labeling agent.

($k_{off}$) is fast relative to that of the covalent bond-forming reaction, then $k_{inact}$ will be the rate-determining step. In this case, unlike simple reversible inhibition, there will be a time-dependent loss of enzyme activity (as is the case with slow, tight-binding reversible inhibitors because of the relatively small $k_{off}$).

The rate of inactivation is proportional to low concentrations of inhibitor but becomes independent at high concentrations. As is the case with substrates, the inhibitor also can reach *enzyme saturation* when $k_{inact}$ is slow relative to $k_{off}$. Once all of the enzyme molecules are tied up in an E·I complex, the addition of more inhibitor will have no effect on the rate of inactivation.

Because an affinity labeling agent contains a reactive functional group, not only can it react with the active site of the target enzyme, but it can also react with thousands of nucleophiles associated with many other enzymes and biomolecules in the body. Consequently, these inactivators are potentially quite toxic. In fact, many cancer chemotherapy drugs (see Chapter 6) are affinity labeling agents, and they are quite toxic. Therefore, they are not as useful in drug design as other types of enzyme inhibitors.

There are several principal reasons why these reactive molecules, nonetheless, can be effective drugs. First, once the inactivator forms an E·I complex, a unimolecular reaction ensues (the E·I complex is now a single molecule), which can be many orders of magnitude ($10^8$ times; see Section II,A of Chapter 4) more rapid than nonspecific bimolecular reactions with nucleophiles on other proteins. Furthermore, the inactivator may form an E·I complex with other enzymes, but if there is no nucleophile near the reactive functional group, no reaction will take place. Third, in the case of antitumor agents, mimics of DNA precursors are rapidly transported to the appropriate site and, therefore, they are preferentially concentrated at the desired target.

The key to the effective design of affinity labeling agents as drugs is specificity of binding. If the molecule has a very low $K_i$ for the target enzyme, then E·I complex formation will be favored, and the selective reactivity will be enhanced. Another approach to increase the selectivity of this class of inactivators is to modulate the reactivity of the active functional group. The effectiveness of this approach can be seen by comparing the relatively moderate reactivity of the functional groups in the nontoxic affinity labeling agents described in Section V,B,2 with the highly reactive functional groups in some of the cancer chemotherapeutic drugs in Chapter 6 (Section III,B). A third feature that would increase the potential effectiveness of an affinity labeling agent can be built into the molecule, if something is known about the location of the active site nucleophiles. In cases where a nucleophile is known to be at a particular position relative to the bound substrate, the reactive functional group can be incorporated into the affinity labeling agent so that it is near that site when the inactivator is bound to the target enzyme. This increases the probability for reaction with the target enzyme by approximation.

Because many enzymes involved in DNA biosynthesis utilize substrates with similar structures, high concentrations of reversible inhibitors may block multiple enzymes. A lower concentration of an irreversible inhibitor may be more selective, if it only reacts with those enzymes having appropriately juxtaposed active site nucleophiles. Even when all of the design factors are taken into consideration, nonspecific reactions still can take place which may result in side effects. Nonetheless, because there are affinity labeling agents being used effectively as drugs, this class of inactivators needs to be considered seriously as potential drug candidates.

### 2. Selected Affinity Labeling Agents

***a. Penicillins and Cephalosporins/Cephamycins.*** Penicillins have the general structure **5.35**; for example, **5.35a** is penicillin G, **5.35b** is penicillin V, **5.35c** (R′ = H) is oxacillin, **5.35c** (R′ = Cl) is cloxacillin, **5.35d** (R′ = H) is ampicillin, and **5.35d** (R′ = OH) is amoxicillin. The differences in these derivatives (other than structure) are related to absorption properties, resistance to penicillinases, and specificity for organisms for which they are most effective.

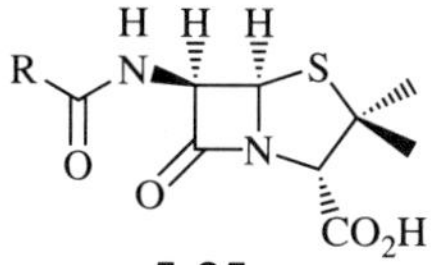

**5.35**

**a**, R = $PhCH_2$ -

**b**, R = $PhOCH_2$ -

**c**, R =

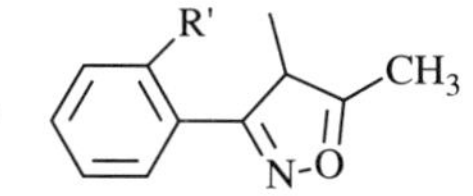

**d**, R =

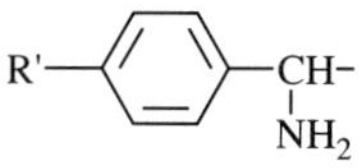

The structures of some cephalosporins and cephamycins are shown in **5.36**; for example, **5.36a** is cefazolin (injectable), **5.36b** is cefoxitin (injectable), **5.36c** is cefaclor (oral), and **5.36d** is ceftizoxime (injectable). The analogs where X is H are cephalosporins, and those where X is $OCH_3$, such as cefoxitin (**5.36b**), are called cephamycins. Cephalosporins and cephamycins are classified by *generations* which are based on general features of antimicrobial activity.[97] Cefazolin is a first generation cephalosporin, cefoxitin and cefaclor are second generation cephalosporins, and ceftizoxime is a third generation cephalosporin. Modifications in the structure of these antibiotics have been extensive,[98] so much so that essentially every atom excluding the lactam nitrogen has been replaced or modified in the search for improved antibiotics.

The discovery of penicillin was described in Chapter 2 (Section I,A,1). Penicillins and cephalosporins are bactericidal (they kill existing bacteria), unlike the sulfonamides (Section IV,B,1), which are bacteriostatic. They are ideal drugs in that they inactivate an enzyme that is essential for bacterial growth but that does not exist in animals, namely, the peptidoglycan transpeptidase. This enzyme catalyzes the cross-linking of the peptidoglycan to

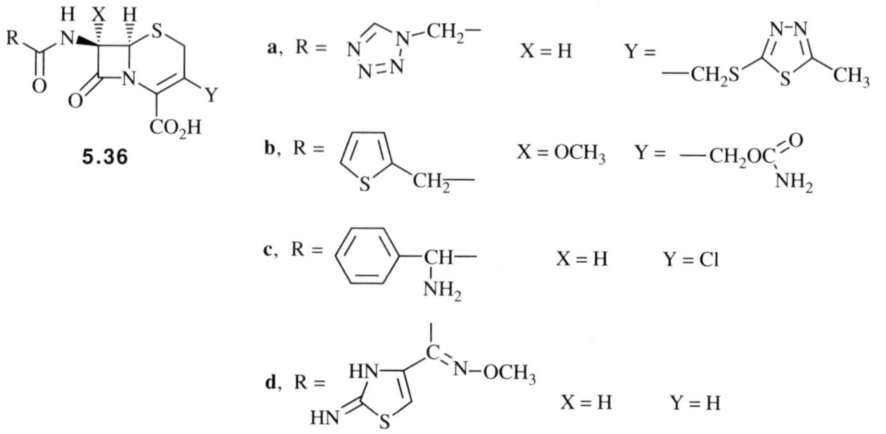

**5.36**

a, R = (tetrazol-1-yl)$CH_2$—, X = H, Y = —$CH_2S$-(5-methyl-1,3,4-thiadiazol-2-yl)

b, R = (2-thienyl)$CH_2$—, X = $OCH_3$, Y = —$CH_2OC(=O)NH_2$

c, R = $C_6H_5CH(NH_2)$—, X = H, Y = Cl

d, R = (2-imino-thiazolinyl)C(=N–$OCH_3$)—, X = H, Y = H

form the bacterial cell wall. As animal cells do not have cell walls, there is no need for this enzyme in animals. When bacterial cell wall biosynthesis is interrupted by penicillins, the high internal pressure can no longer be sustained, and the bacteria burst.

The peptidoglycan is a branched polymer of alternating $\beta$-D-*N*-acetylglucosamine (NAG) and $\beta$-D-*N*-acetylmuramic acid (NAM) residues. Attached to what was the carboxylic acid group of NAM is a polypeptide chain that varies in structure according to the strain of bacteria. One example is shown in Scheme 5.11 with the mechanism for cross-linkage of the peptidoglycan. The terminal D-alanyl-D-alanine residues of the NAM side chain of the peptidoglycan bind to the transpeptidase, which initially acts as a serine protease, clipping the terminal peptide bond and making a serine ester. Cross-linkage of this ester with another peptidoglycan strand builds the cell wall.

The transpeptidase has not been purified, but D-alanine carboxypeptidase, an enzyme believed to be the actual *in vivo* transpeptidase that became uncoupled during purification, has been purified.[99] This enzyme acts freely as a serine protease and carries out the first half of the transpeptidase reaction, namely, the formation of the serine ester. The acceptor molecule, then, is water instead of the amino group of another peptidoglycan chain. This leads to the hydrolyzed product rather than the cross-linked one. An active site serine has been identified as the residue involved in catalysis.[100]

By comparison of a molecular model of penicillin with that of D-alanyl-D-alanine (the terminal dipeptide of the peptidoglycan side chain), Tipper and Strominger[101] suggested that penicillin could mimic the structures of the terminus of the peptidoglycan and bind at the active site of transpeptidase (Fig. 5.6). The $N^a$ to $N^b$ distances (3.3 Å) and the $N^b$ to carboxylate carbon distances (2.5 Å) in both molecules are identical. The $N^a$ to carboxylate carbon distance is 5.4 Å in the penicillins and 5.7 Å in D-alanyl-D-alanine. The $\beta$-lactam carbonyl may be further activated by torsional effects of the thiazoli-

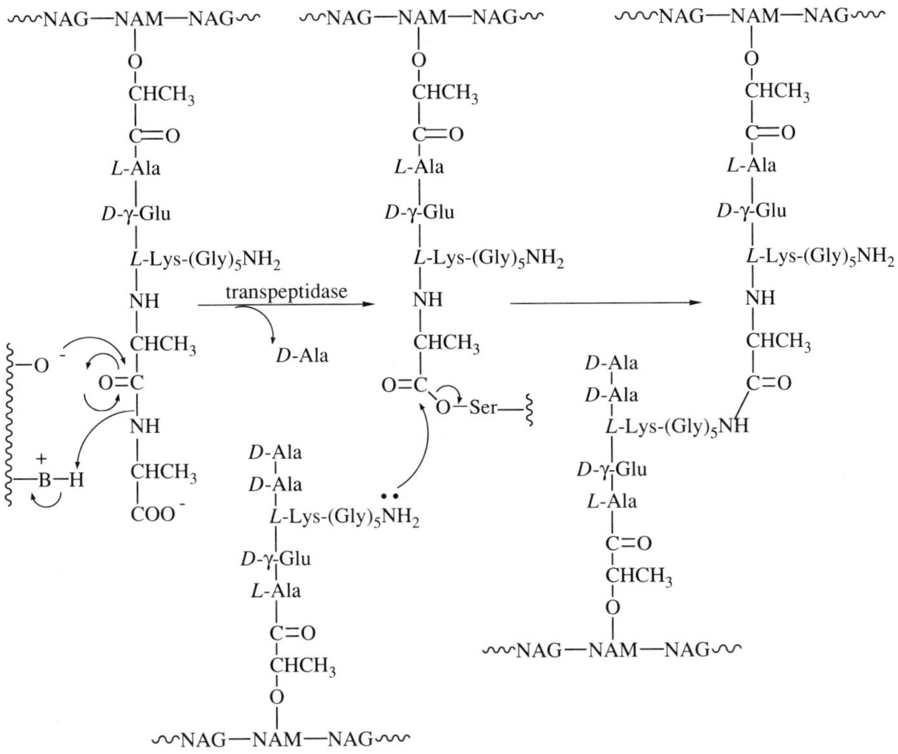

Scheme 5.11. Cross-linkage of bacterial cell wall peptidoglycan.

dine ring in the penicillins. This carbonyl corresponds to the carbonyl in the acyl D-alanyl-D-alanine that acylates the active site serine, and, therefore, penicillins also could acylate the transpeptidase serine residue[102] (Scheme 5.12). The bulk of the penicillin molecule when it is attached to the active site precludes hydrolysis or transamidation either for steric reasons or because it

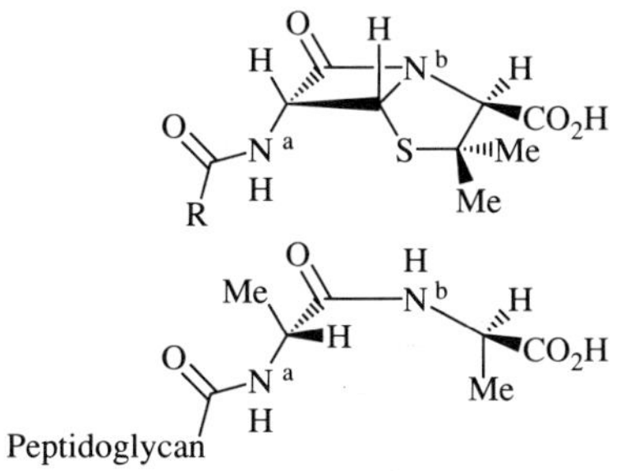

Figure 5.6. Comparison of the structure of penicillins with acyl D-alanyl-D-alanine.

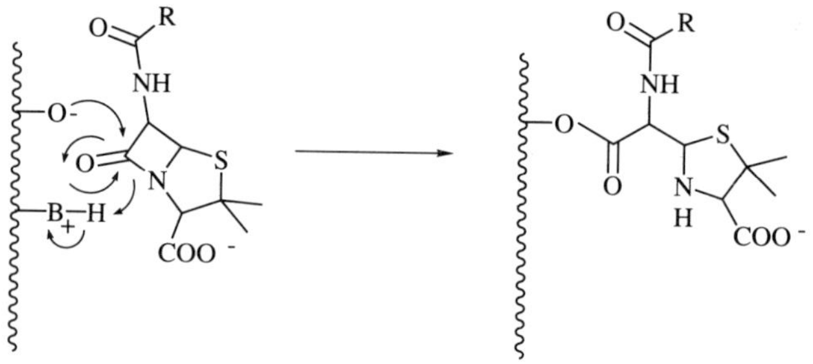

Scheme 5.12. Acylation of transpeptidase by penicillins.

induces a conformational change in the enzyme so that these processes cannot occur. Covalent binding at the active site prevents the substrate from binding. Similar arguments could be made for cephalosporins. The double bond in the dihydrothiazine ring also may activate the β-lactam carbonyl (Scheme 5.13).[103]

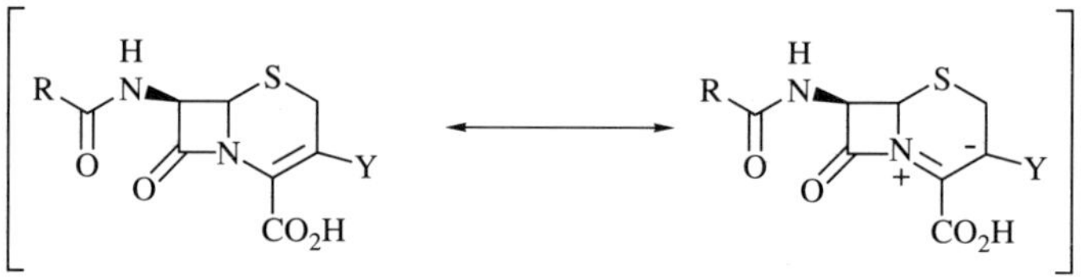

Scheme 5.13. Activation of the β-lactam carbonyl of cephalosporins.

On the basis of the structural similarity of penicillins to acyl D-alanyl-D-alanine (see Fig. 5.6), it was predicted[101] that 6-methylpenicillin (a methyl on the $sp^3$ carbon adjacent to $N^a$) would be a more potent inhibitor than the parent molecule. However, both 6-methylpenicillin and 7-methylcephalosporin (the numbering is different for cephalosporins because it has one more carbon in the ring, but the methyl in 7-methylcephalosporin corresponds to that in 6-methylpenicillin) were synthesized and were shown to be inactive.[104] Since the corresponding 7-methoxycephalosporins (i.e., cephamycins) are better inhibitors than the parent cephalosporins, it is not clear why the methyl analogs are poor inhibitors.

The beauty of the penicillins (and cephalosporins) is that they are not exceedingly reactive; consequently, few nonspecific acylation reactions occur. Their modulated reactivity and nontoxicity make them ideal drugs. If it were not for allergic responses and problems associated with drug resistance, penicillins might be considered nutritious foods, comprised of various carboxylic acid derivatives (the RCO side chains), cysteine, and the essential amino acid valine.

Penicillins are "wonder drugs" in their activity against a variety of bacteria; however, many strains of bacteria have become resistant to penicillins.

The principal cause for resistance, as a result of gene transfer and recombination, is the excretion of the enzyme *β-lactamase*. This enzyme catalyzes the hydrolysis of β-lactams, presumably by having the ability to hydrolyze the acylated serine residue, a process that the susceptible strains cannot carry out.[105] Other important causes for resistance are that the transpeptidase becomes less susceptible to acylation by penicillins, and there is a change in the outer membrane permeability of the penicillins into the periplasm.

Because the major cause for resistance is the excretion of β-lactamases, an obvious approach to drug synergism would be the combination of a penicillin with a β-lactamase inhibitor. In the 1970s certain naturally occurring β-lactams that did not have the general penicillin or cephalosporin structure were isolated from various organisms and were found to be potent inactivators of β-lactamases (see Section V,C,3,g). These compounds are used in combination with penicillins to destroy penicillin-resistant strains of bacteria. The β-lactamase inhibitor has no antibiotic activity, but it protects the penicillin from destruction before it can interfere with cell wall biosynthesis.

***b. Aspirin.*** Hippocrates recommended the use of willow bark for pain during childbirth.[106] It is stated in the *Papyrus Ebens* (circa 1550 B.C.) that dried leaves of myrtle provide a remedy for rheumatic pain.[106] A boiled vinegar extract of willow leaves was suggested by Aulus Cornelius Celsus (A.D. 30) for relief of pain.[107] In 1763 Reverend Edmund Stone of England announced his findings that the bark of the willow (*Salix alba vulgaris*) provided an excellent substitute for Peruvian bark (*Cinchona* bark, a source of quinine) in the treatment of fevers.[108] The connection between the two barks was discovered by his tasting them and making the observation that they both had a similar bitter taste. The bitter active ingredient with antipyretic activity in the willow bark was called salicin, which was first isolated in 1829 by Leroux. Upon hydrolysis, salicin produced glucose and salicylic alcohol, which was metabolized to salicylic acid. Sodium salicylate was first used for the treatment of rheumatic fever and as an antipyretic agent in 1875. The father of a chemist employed by Bayer Company suffered from severe rheumatoid arthritis and pleaded with his son to search for a less irritating drug than the sodium salicylate he was using. Many salicylate derivatives were synthesized, and acetylsalicylate (aspirin, **5.37**) was the best. He gave it to his father, who responded well, and in 1899 Bayer introduced aspirin as an antipyretic (fevers), anti-inflammatory (arthritis), and analgesic (pain) agent.[109] The trade

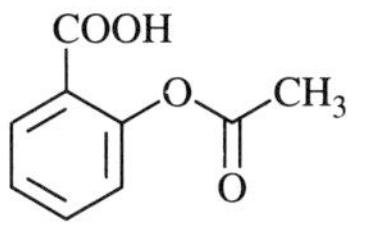

COOH

O

$CH_3$

O

**5.37**

name aspirin was coined by adding an "a" for acetyl to spirin for *Spiraea*, the plant species from which salicylic acid was once prepared.

The mechanism of action was initially reported by Vane[110] and by Smith and Willis[111] to be the result of inhibition of prostaglandin biosynthesis. Prostaglandins (PGs) are derived from arachidonic acid (**5.38**) by the action of prostaglandin synthase (also known as cyclooxygenase) (Scheme 5.14). At the time of the initial report there was evidence that various prostaglandins were involved in the pathogenesis of inflammation and fever. It is now known that all mammalian cells (except erythrocytes) have microsomal enzymes which catalyze the biosynthesis of prostaglandins, and that prostaglandins are always released when cells are damaged, being found in increased concentrations in inflammatory exudates. When prostaglandins are injected into animals, the effects are reminiscent of those observed during inflammatory responses, namely, redness of the skin (erythema) and increased local blood flow. A long lasting vasodilatory action also is prevalent. Prostaglandins can cause headache and vascular pain when infused in man. Elevation of body temperature during infection also is mediated by the release of prostaglandins.

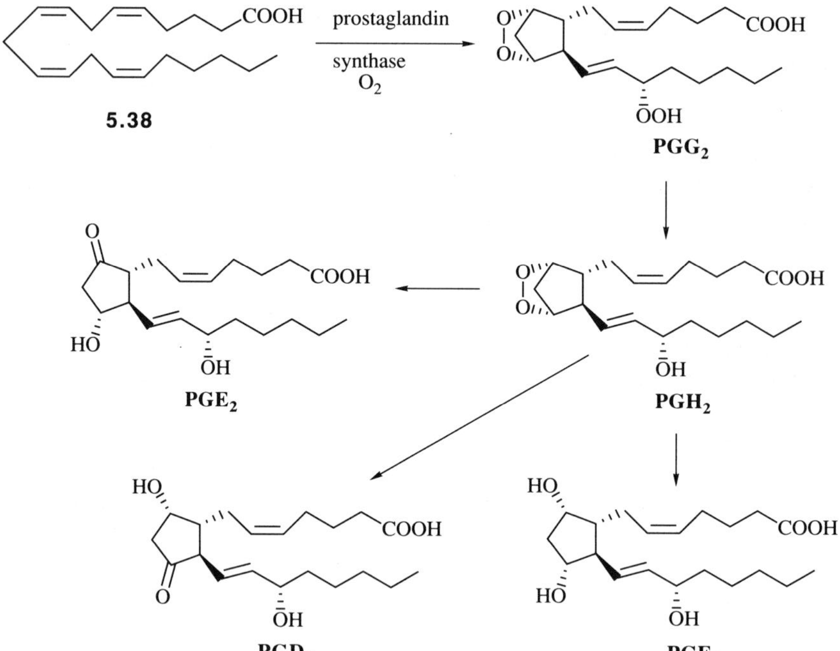

**Scheme 5.14.** Biosynthesis of prostaglandins (PGs).

From the above discussion it is apparent that inhibition of prostaglandin synthase, the enzyme responsible for the biosynthesis of all of the prostaglandins and related compounds [$PGH_2$ also is converted by prostacyclin synthase to prostacyclin (**5.39**) and by thromboxane synthase to thromboxane $A_2$ (**5.40**)], would be a desirable approach for the design of anti-inflammatory, analgesic, and antipyretic drugs. Inhibition of platelet cyclooxygenase is particularly effective at blocking prostaglandin biosynthesis because, unlike most other cells, platelets cannot regenerate the enzyme, as they have little or no capacity for protein biosynthesis. Therefore, a single dose of 40 mg per day of aspirin is sufficient to destroy the cyclooxygenase for the life of the platelet!

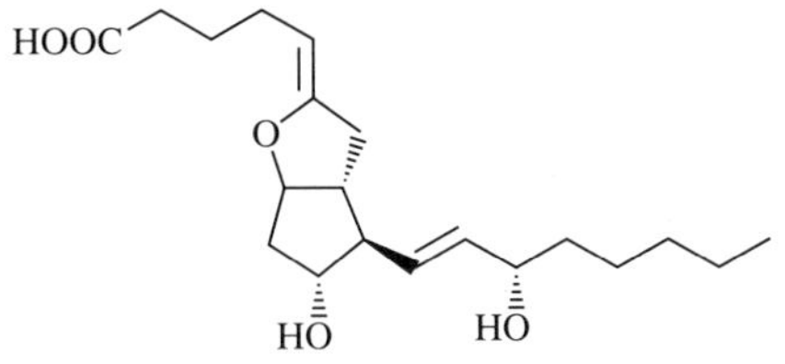

HOOC O HO HO

**5.39**

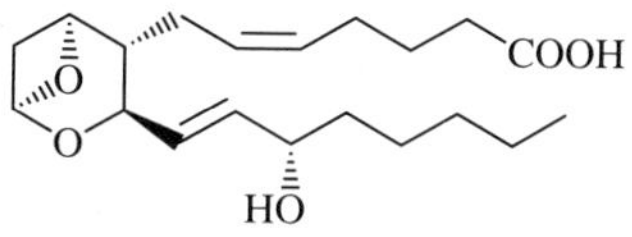

O O COOH HO

**5.40**

Sheep vesicular gland prostaglandin synthase was shown to be irreversibly inactivated by aspirin.[112] When microsomes of sheep seminal vesicles were treated with aspirin tritiated in the methyl of the acetyl group, acetylation of a single protein was observed. The same experiment carried out with aspirin tritiated in the benzene ring resulted in no tritium incorporation, suggesting that acetylation was occurring. Incubation of purified prostaglandin synthase with [*acetyl*-$^3$H]aspirin led to irreversible inactivation with incorporation of one acetyl group per enzyme molecule.[113,114] Thermolysin digestion of the tritiated enzyme gave the labeled dipeptide Phe-Ser, where the serine hydroxyl group had become acetylated.[114] Pepsin digestion of the tritiated enzyme gave a 22-amino acid labeled peptide; tryptic digestion of this peptide gave a tritiated decapeptide in which the serine residue was acetylated.[115] The most straightforward mechanism for acetylation would be a transesterification mechanism by aspirin acting as an affinity labeling agent (Scheme 5.15).

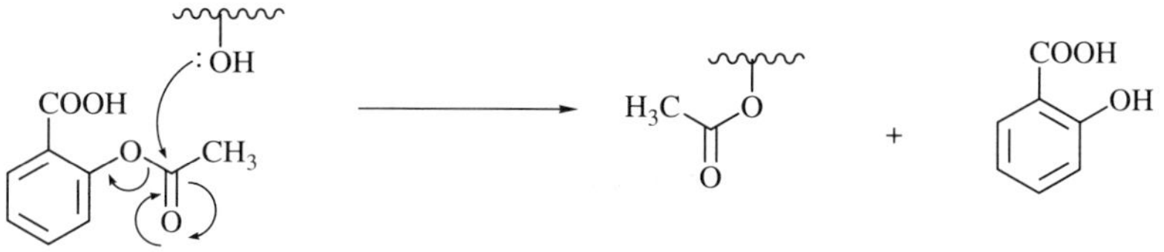

: OH COOH O $CH_3$ O → $H_3C$ O O + COOH OH

**Scheme 5.15.** Hypothetical mechanism for inactivation of prostaglandin synthase by aspirin.

Both of the examples given for affinity labeling agents involved acylation mechanisms. The affinity labeling agents that proceed by alkylation reactions are discussed in Chapter 6.

### C. Mechanism-Based Enzyme Inactivators

#### 1. Theoretical Aspects

A *mechanism-based enzyme inactivator* is an unreactive compound that bears a structural similarity to the substrate or product for a specific enzyme. Once this compound binds to the active site, then the target enzyme, via its normal catalytic mechanism, converts it to a product that is generally very reactive. Prior to escape from the active site, this product, in almost all cases, forms a covalent bond to the target enzyme. It is not essential for the inactivator to form a covalent bond to the enzyme (a tight-binding inhibitor may form), but the target enzyme must transform the inactivator into the actual inactivating species, and inactivation must occur prior to the release of this species from the active site. Because there is an additional step in the inactivation process relative to that for an affinity labeling agent (see Section V,B,1), the kinetic scheme for mechanism-based inactivation differs from that for affinity labeling (Scheme 5.16). Provided $k_4$ is a fast step and the equilibrium $k_1/k_{-1}$ is set up rapidly, then $k_2$ is the inactivation rate constant ($k_{\text{inact}}$) that determines the rate of the inactivation process. The two key features of this type of inactivator that differentiate it from an affinity labeling agent are its initial unreactivity and the requirement for the enzyme to catalyze a reaction on it, thereby converting it to a product, namely, the actual inactivator species. Often this inactivator species is quite reactive and therefore acts as an affinity labeling agent which already is at the active site of the target enzyme. Inactivation ($k_4$ in Scheme 5.16) does not necessarily occur every time the inactivator is transformed into the inactivating species; sometimes it escapes from the active site ($k_3$ in Scheme 5.16). The ratio of the number of turnovers that gives a released product per inactivation event, $k_3/k_4$, is called the *partition ratio*.

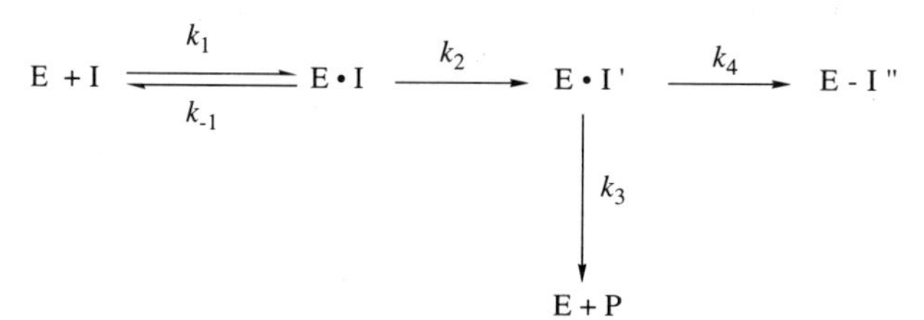


**Scheme 5.16.** Kinetic scheme for simple mechanism-based enzyme inactivation.

#### 2. Potential Advantages in Drug Design

Because of the generally high reactivity of affinity labeling agents, they can react with enzymes and biomolecules other than the target enzyme. When this occurs, toxicity and side effects can arise. However, mechanism-based enzyme inactivators are unreactive compounds, and this is the key feature that makes them so amenable to drug design. Consequently, nonspecific alkylations and acylations of other proteins should not be a problem. In the ideal case only the target enzyme will be capable of catalyzing the appropriate conversion of the inactivator to the activated species, and inactivation will result with every turnover, that is, the partition ratio (see Section V,C,1) will be 0 (no metabolites formed per inactivation event). This may be quite important for potential drug use. If the partition ratio is greater than 0, the released activated species may react with other proteins, possibly resulting in a toxic effect. In this case the inactivator is called a *metabolically activated inactivator.*[116] Alternatively, the released species may be hydrolyzed by the aqueous medium prior to reaction with other biomolecules, but the product formed may be toxic or may be metabolized further to other toxic substances. Under the ideal conditions mentioned above, the inactivator would be a strong drug candidate because it would be highly enzyme specific and low in toxicity. In fact, $\alpha$-difluoromethylornithine (eflornithine), a specific mechanism-based inactivator of ornithine decarboxylase (see Section V,C,3,b) used in the treatment of protozoal infections, has been administered to patients in amounts of 30 g per day for several weeks with only minor side effects.[117]

Although there are currently no drugs on the American drug market that were rationally designed as mechanism-based enzyme inactivators, there are quite a few drugs in current medical use that were determined ex post facto to be mechanism-based inactivators. Some known mechanism-based inactivators include the antidepressant drugs tranylcypromine (see Section V,C,3,c) and phenelzine, the antihypertensive drugs hydralazine and pargyline, and the antiparkinsonian drug L-deprenyl or selegiline (all inactivate monoamine oxidase); clavulanic acid (see Section V,C,3,g), a compound used to protect penicillins and cephalosporins against bacterial degradation (inactivates $\beta$-lactamases); the antitumor drug 5-fluoro-2′-deoxyuridylate (see Section V,C,3,e) and the antiviral agent trifluridine (see Section V,C,3,f) (both inactivate thymidylate synthase); the antihyperuricemic agent allopurinol (see Section V,C,3,h) (inactivates xanthine oxidase); the antithyroid drugs methimazole, methylthiouracil, and propylthiouracil (inactivate thyroid peroxidase); and the antibiotic chloramphenicol, the antifertility drug norethindrone, the anesthetics halothane and fluoroxene, the sedative ethclorvynol, the diuretic and antihypertensive agent spironolactone, the pituitary suppressant danazol, the pigmentation agent methoxsalen, and the hypnotic novonal (all inactivate cytochrome *P*-450). Those drugs that inactivate cytochrome *P*-450, however, do not derive their medicinal effects as a result of that inactivation. Examples of some enzymes that already have been targeted for mechanism-based enzyme inactivation and the therapeutic goals of inactivation are listed in Table 5.2.

Because the activation of mechanism-based inactivators depends on the catalytic mechanism of their target enzyme, these inactivators can be designed by a rational organic mechanistic approach. In the next section several

**Table 5.2** Enzymes with Potential Medicinal Importance Already Targeted for Mechanism-Based Enzyme Inactivation

| Enzyme | Therapeutic goal |
|---|---|
| *S*-Adenosylhomocysteine hydrolase | Antiviral agent |
| *S*-Adenosylmethionine decarboxylase | Antitumor agent |
| Alanine racemase | Antibacterial agent |
| D-Amino-acid aminotransferase | Antibacterial agent |
| γ-Aminobutyric acid aminotransferase | Anticonvulsant agent |
| Arginine decarboxylase | Antibacterial agent |
| Aromatase | Anticancer agent |
| Aromatic-L-amino-acid decarboxylase | Synergistic with antiparkinsonian drug |
| Cysteine proteases | Multiple sclerosis, muscular dystrophy, antibacterial, antiviral agent, myocardial infarction, bone resorption, antitumor agent |
| Dihydrofolate reductase | Anticancer agent, antibacterial agent, antiprotozoal agent |
| Dihydroorotate dehydrogenase | Antiparasitic agent, anticancer agent |
| DNA polymerase I | Antiviral agent |
| Dopamine β-hydroxylase | Antihypertensive agent, pheochromocytoma agent |
| α-Glucosidase | Anti-AIDS agent |
| β-Glucosidase | Diabetes/antiobesity agent |
| Histidine decarboxylase | Antihistamine, antiulcer agent |
| β-Lactamase | Synergistic with antibiotics |
| Lipoxygenase | Anti-inflammatory agent |
| Monoamine oxidase | Antidepressant agent, antihypertensive agent, antiparkinsonian agent |
| Ornithine decarboxylase | Anticancer agent, antiprotozoal agent |
| Ribonucleotide reductase | Antiviral, antitumor agent |
| Serine hydroxymethylase | Antitumor agent |
| Serine proteases | Treatment of emphysema, inflammation, arthritis, adult respiratory distress syndrome, anticoagulant agent, pancreatitis, certain degenerative skin disorders, antiviral agent, digestive disorders |
| Testosterone 5α-reductase | Anticancer agent |
| Thymidylate synthase | Anticancer agent |
| Thyroid peroxidase | Antithyroid agent |
| Xanthine oxidase | Antihyperuricemic agent |

mechanism-based inactivators are discussed, first in terms of the medicinal relevance of their target enzyme, then from a mechanistic point of view.

## 3. Selected Examples of Mechanism-Based Enzyme Inactivators

***a. Vigabatrin, an Anticonvulsant Agent.*** *Epilepsy* is a disease that was described over 4000 years ago in early Babylonian and Hebrew writings. Full clinical descriptions were written in Hippocrates' monograph *On the Sacred Disease* in about 400 B.C.[118] If it is broadly defined as any central nervous system disease characterized by recurring convulsive seizures, then 0.5–1% of the world population has epilepsy. It is categorized as *primary* or *idiopathic* when no cause for the seizure is known, and *secondary* or *symptomatic* when the etiology has been identified. Symptomatic epilepsy can result from specific physiological phenomena such as brain tumors, syphilis, cerebral arteriosclerosis, multiple sclerosis, Buerger's disease, Pick's disease, Alzheimer's disease, sunstroke or heatstroke, acute intoxication, lead poisoning, head trauma, vitamin $B_6$ deficiency, hypoglycemia, and labor.

The biochemical mechanism leading to central nervous system electrical discharges and epilepsy are unknown, but there may be multiple mechanisms involved. However, it has been shown that convulsions arise when there is an imbalance in two principal neurotransmitters in the brain, L-glutamic acid, an excitatory neurotransmitter, and $\gamma$-aminobutyric acid (GABA), an inhibitory neurotransmitter. The concentration of GABA is regulated by two pyridoxal 5′-phosphate (PLP)-dependent enzymes, L-glutamic acid decarboxylase (GAD, glutamate decarboxylase), which converts glutamate to GABA, and GABA aminotransferase (GABA-AT, 4-aminobutyrate aminotransferase), which degrades GABA to succinic semialdehyde (Scheme 5.17). Although succinic semialdehyde is toxic to cells, there is no buildup of this metabolite because it is efficiently oxidized to succinic acid by the enzyme succinic semialdehyde dehydrogenase {SSADH, succinate-semialdehyde dehydrogenase [NAD(P)$^+$]}.

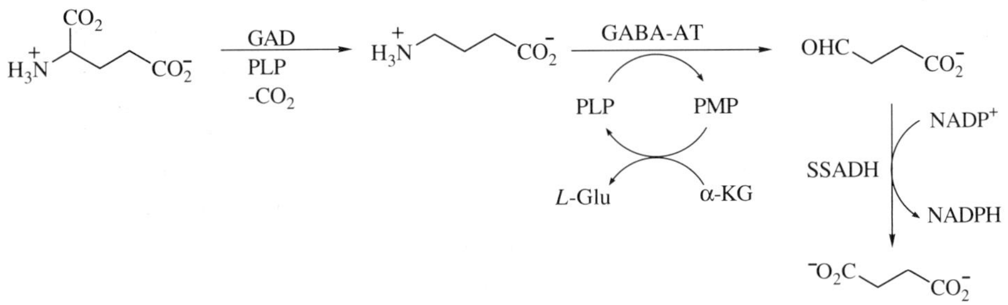

**Scheme 5.17.** Metabolism of L-glutamic acid.

GABA system dysfunction has been implicated in the symptoms associated with epilepsy, Huntington's disease, Parkinson's disease, and tardive dyskinesia. When the concentration of GABA diminishes below a threshold level in the brain, convulsions begin. If a convulsion is induced in an animal, and GABA is injected directly into the brain, the convulsions cease. It would seem, then, that an ideal anticonvulsant agent would be GABA; however, peripheral administration of GABA produces no anticonvulsant effect. This was shown to be the result of the failure of GABA, under normal circumstances, to cross the *blood–brain barrier*, a membrane that surrounds the capillaries of the circulatory system in the brain and protects it from passive diffusion of undesirable chemicals from the bloodstream. Another approach for increasing the brain GABA concentration, however, would be to design a compound capable of permeating the blood–brain barrier that subsequently inactivates GABA aminotransferase, the enzyme that catalyzes the degradation of GABA. Provided that glutamate decarboxylase also is not inhibited, GABA concentrations should rise. This, in fact, has been shown to be an effective approach to the design of anticonvulsant agents.[119a] Compounds that both cross the blood–brain barrier and inhibit GABA aminotransferase *in vitro* have been reported to increase whole brain GABA levels *in vivo* and possess anticonvulsant activity. The anticonvulsant effect does not correlate with whole brain GABA levels, but it does correlate with an increase in the GABA concentration at the nerve terminals of the substantia nigra.[119b]

The mechanism for the PLP-dependent aminotransferase reactions was discussed earlier (see Section III,A,3 of Chapter 4). On the basis of this mechanism researchers at Merrell Dow Pharmaceuticals[120a] designed 4-amino-5-hexenoic acid (vigabatrin, **5.41**; Scheme 5.18). This is the first rationally designed mechanism-based inactivator drug (currently, it is in medical use in Europe). The inactivation mechanism[120b] is shown in Scheme 5.18. By comparison of Scheme 5.18 with Scheme 4.18 in Chapter 4 (in Scheme 4.18 an $\alpha$-amino acid is used, but here a $\gamma$-amino acid is the example), it is apparent that identical mechanisms are proposed up to compound **5.43** (Scheme 5.18) and compound **4.25** (Scheme 4.18). In the case of normal substrate turnover, hydrolysis of **4.25** gives pyridoxamine 5′-phosphate (PMP) and the keto acid (Scheme 4.18). The same hydrolysis could occur with **5.43** to give the corresponding products, PMP and **5.45**. However, **5.43** is a potent electrophile, a Michael acceptor, which can undergo conjugate addition by an active site nucleophile ($X^-$) and produce inactivated enzyme (**5.44a** or **5.44b**). The mechanism in Scheme 5.18 appears to be relevant for about 70% of the inactivation pathways. The other 30% of the inactivation is accounted for by an allylic isomerization and enamine rearrangement leading to **5.41a** (pathway b).[120b] Note that vigabatrin is an unreactive compound that is converted by the normal catalytic mechanism of the target enzyme to a reactive compound (**5.43**) which attaches to the enzyme. This is the typical course of events for a mechanism-based inactivator.

It may seem strange that GABA does not cross the blood–brain barrier, but vigabatrin, which also is a small charged molecule, can diffuse through that lipophilic membrane. The attachment of a vinyl substituent to GABA, apparently, has two effects that permit this compound to cross the blood–brain barrier. First, the vinyl substituent increases the lipophilicity of the molecule. Second, it is an electron-withdrawing substituent which would have the effect of lowering the $pK_a$ of the amino group. This would increase the concentration of the nonzwitterionic form (**5.46b**, Scheme 5.19), which is more lipophilic than the zwitterionic form (**5.46a**) because it is uncharged.

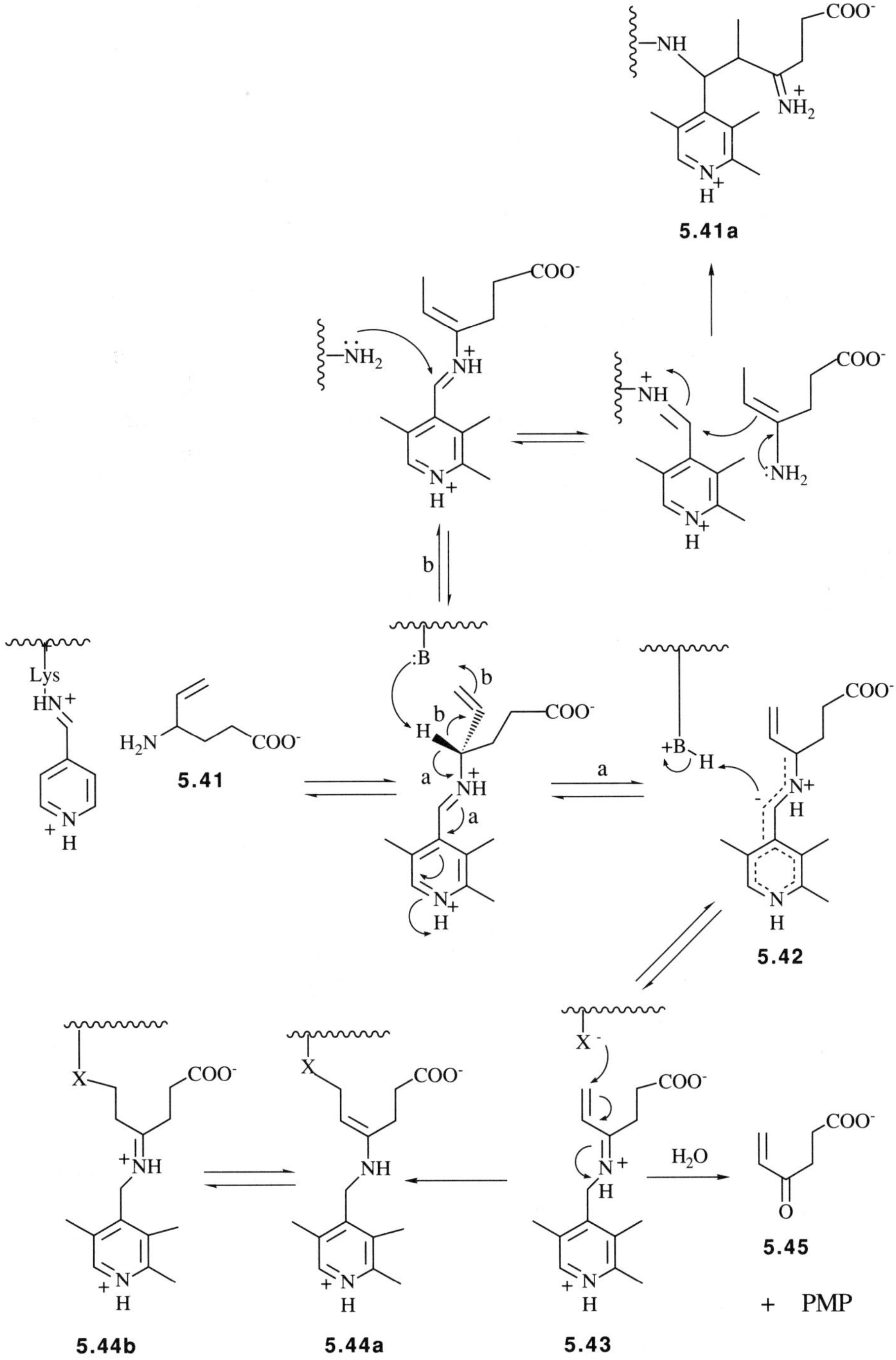

Scheme 5.18. Proposed mechanism for the inactivation of GABA aminotransferase by vigabatrin.

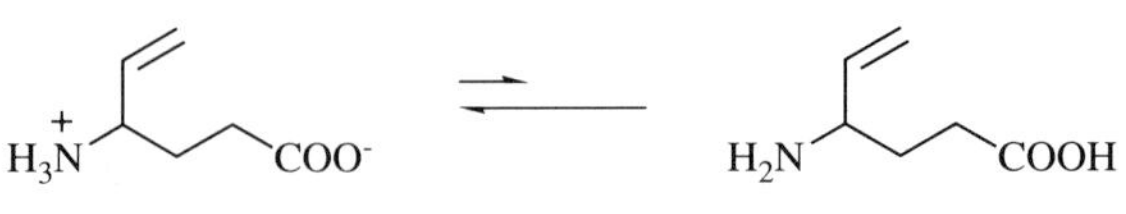

$H_3\overset{+}{N}$ ... $COO^-$ ⇌ $H_2N$ ... $COOH$

**5.46a** **5.46b**

**Scheme 5.19.** Zwitterionic and nonzwitterionic forms of vigabatrin.

***b. Eflornithine, an Antiprotozoal Agent.*** The polyamines spermidine (**5.47**) and spermine (**5.48**) and their precursor putrescine (**5.49**) are important regulators of cell growth, division, and differentiation. The mechanisms by which they do this are unclear, but they appear to be required for DNA synthesis. Rapidly growing cells have much higher levels of polyamines (and ornithine decarboxylase; see below) than do slowly growing or quiescent cells. When quiescent cells are stimulated, the polyamine and ornithine decarboxylase levels increase prior to an increase in the levels of DNA, RNA, and protein. The polycationic nature of the polyamines may be responsible for their interaction with cellular structures that have negatively charged groups such as DNA.

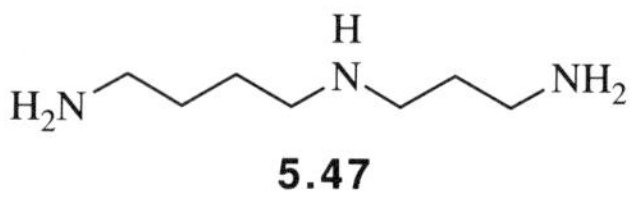

$H_2N$ ... $\overset{H}{N}$ ... $NH_2$ $H_2N$ ... $\underset{H}{N}$ ... $\overset{H}{N}$ ... $NH_2$

**5.47** 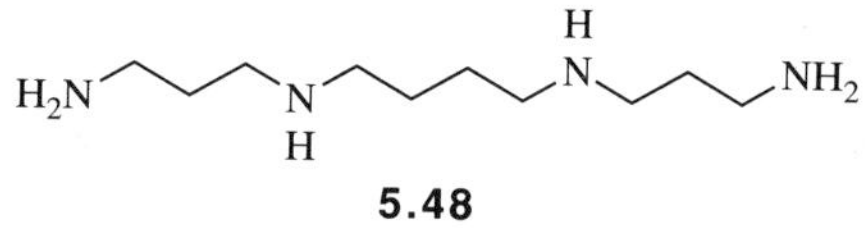 **5.48**

$H_2N$ ... $NH_2$

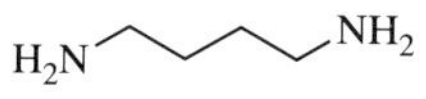

**5.49**

Ornithine decarboxylase, the enzyme that catalyzes the conversion of ornithine to putrescine (**5.49**) is the rate-limiting step in polyamine biosynthesis (Scheme 5.20). Spermidine (**5.47**) is produced by the spermidine synthase–catalyzed reaction of putrescine with *S*-adenosylhomocysteamine (**5.51**), which is derived from the decarboxylation of *S*-adenosylmethionine (**5.50**; SAM) in a reaction catalyzed by *S*-adenosylmethionine decarboxylase. Another aminopropyltransferase, namely, spermine synthase, catalyzes the reaction of spermidine with **5.51** to produce spermine (**5.48**).

Because polyamines are important for rapid cell growth, inhibition of polyamine biosynthesis should be an effective approach for the design of antitumor and antimicrobial agents. In fact, inactivation of ornithine decarboxylase by the potent mechanism-based inactivator eflornithine ($\alpha$-difluoromethylornithine or DFMO, **5.52**), results in virtually complete reduction in the putrescine and spermidine content; however, the spermine concentration is only slightly affected. There are at least two reasons for the latter observation. Inactivation of ornithine decarboxylase activity by

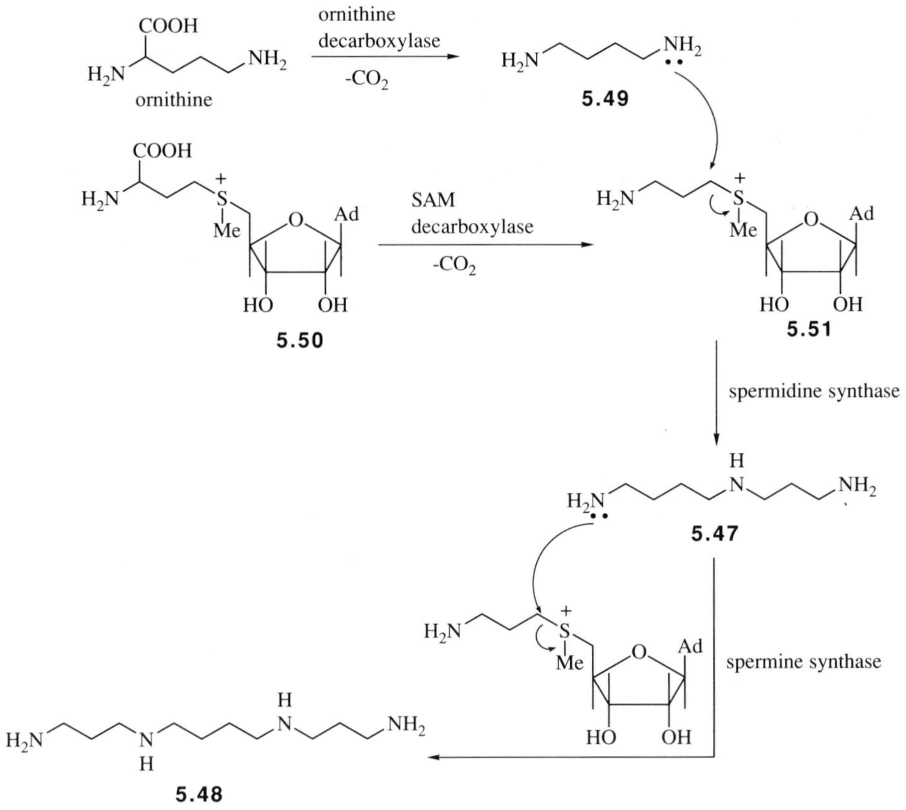

**Scheme 5.20.** Polyamine biosynthesis.

eflornithine *in vivo* may not be complete because gene synthesis of ornithine decarboxylase has a half-life of only 30 min, and, also, the endogenous ornithine that is present competitively protects the enzyme from the inactivator. Furthermore, inactivation of ornithine decarboxylase induces an increase in the levels of *S*-adenosylmethionine decarboxylase, which leads to higher *S*-adenosylhomocysteamine levels. This can be efficiently used to convert the existing putrescine and spermidine to spermine.[121] The inability of eflornithine to shut down all polyamine biosynthesis may be responsible for its discouragingly poor antitumor effects observed in clinical trials.[117] However, eflornithine has been found to have great value in the treatment of certain protozoal infestations such as *Trypanosoma brucei rhodesiense*, which

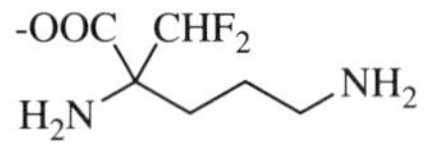

**5.52**

causes African sleeping sickness, and *Pneumocystis carinii*, the microorganism that produces pneumonia in acquired immunodeficiency syndrome (AIDS) patients.[117]

The mechanism for the PLP-dependent decarboxylases was discussed in Chapter 4 (see Section III,A,2). Ornithine decarboxylase catalyzes the decarboxylation of eflornithine (**5.52**), producing a reactive product that inactivates the enzyme; a possible inactivation mechanism is shown in Scheme 5.21. According to the mechanism for PLP-dependent decarboxylases (see Scheme

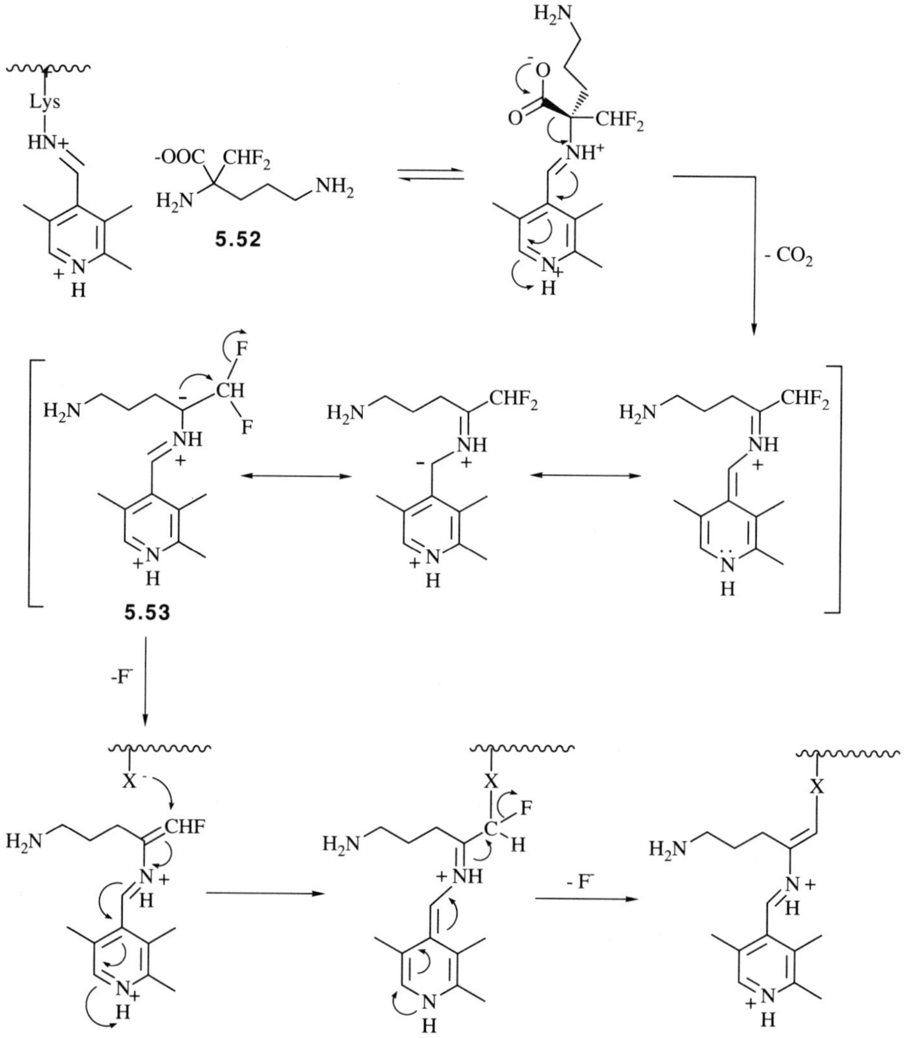

**Scheme 5.21.** Hypothetical mechanism for inactivation of ornithine decarboxylase by eflornithine.

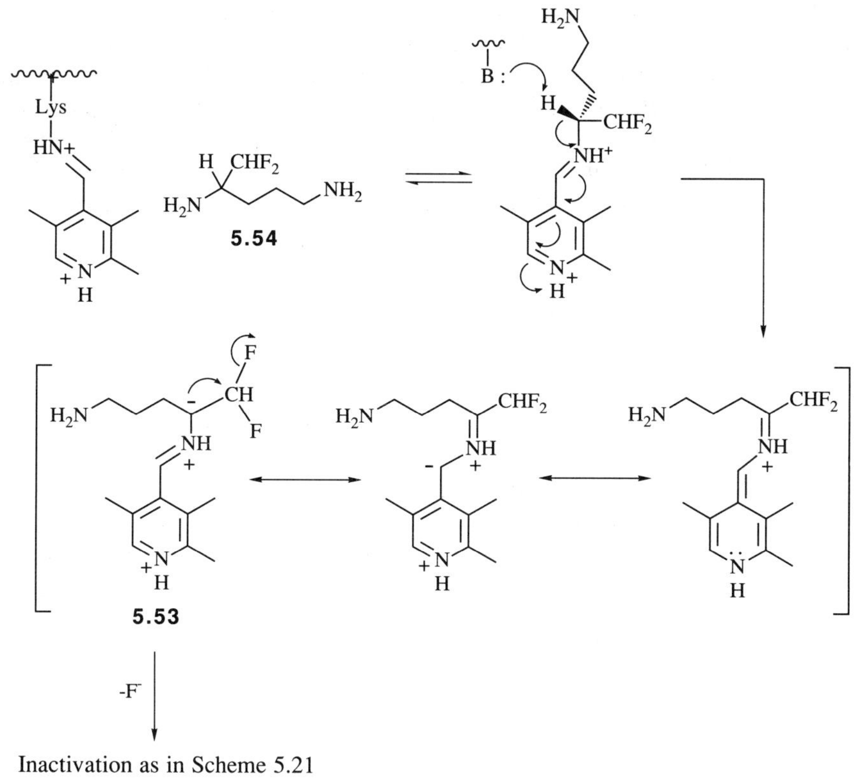

Inactivation as in Scheme 5.21

**Scheme 5.22.** Hypothetical mechanism for inactivation of ornithine decarboxylase by $\alpha$-difluoromethylputrescine.

4.16), the only irreversible step is the one where $CO_2$ is released. Therefore, if you incubate the decarboxylase with the product amine, it catalyzes the reverse reaction up to the loss of $CO_2$, that is, the removal of the $\alpha$-proton. This is the microscopic reverse of the reaction that occurs once the $CO_2$ is released. The *principle of microscopic reversibility* states that for a reversible reaction the same mechanistic pathway will be followed in the forward and reverse reactions. Therefore, a mechanism-based inactivator that has a structure similar to that of the product of the ornithine decarboxylase reaction (the amine) should undergo catalytic $\alpha$-deprotonation. $\alpha$-Difluoromethylputrescine (**5.54**) was shown to inactivate ornithine decarboxylase,[122] presumably because deprotonation gives the same intermediate that is obtained by decarboxylation of eflornithine (compare **5.53** in Scheme 5.21 with **5.53** in Scheme 5.22). For any target enzyme that is reversible, mechanism-based inactivators should be designed for both the forward (substratelike) and back (productlike) reactions. Depending on the metabolic pathway involved, enzyme selectivity may be more favorable for one over the other.

***c. Tranylcypromine, an Antidepressant Agent.*** The modern era of therapeutics for the treatment of depression began in the late 1950s with the introduction of both the monoamine oxidase [MAO, amine oxidase (flavin-containing)] inhibitors and the tricyclic antidepressants. The first MAO inhibitor was iproniazid (**5.55**), which initially was used as an antituberculosis drug until it was observed that patients taking it exhibited excitement and euphoria.[123a] In 1952 Zeller *et al.*[123b] showed that iproniazid was a potent inhibitor of MAO, and clinical studies were underway in the late 1950s.[124]

$$\text{(4-pyridyl)}-\overset{O}{\overset{\|}{C}}NHNHCH(CH_3)_2$$

**5.55**

The brain concentrations of various biogenic (pressor) amines such as norepinephrine, serotonin, and dopamine were found to be depleted in chronically depressed individuals. A correlation was observed between an increase in the concentrations of these brain biogenic amines and the onset of an antidepressant effect.[125] This was believed to be the result of MAO inhibition, since MAO is one of the enzymes responsible for the catabolism of these biogenic amines. By the early 1960s several MAO inactivators were being used clinically for the treatment of depression.

Unfortunately, it was found that in some cases there was a cardiovascular side effect which led to the deaths of several patients. Consequently, these drugs were removed from the market until the cause of death could be ascertained. Within a few months the problem was understood. It was determined that all of those who died while taking an MAO inhibitor had two things in common: they had all died from a hypertensive crisis, and, prior to their deaths, they had eaten foods containing a high tyramine content (e.g., cheese, wine, beer, and yeast products). The connection between these observations is that the ingested tyramine triggers the release of norepinephrine, a potent vasoconstrictor, which raises the blood pressure. Under normal conditions the excess norepinephrine is degraded by MAO and catecholamine *O*-methyltransferase (catechol methyltransferase). If the MAO is inactivated, then the norepinephrine does not get degraded fast enough, the blood pressure keeps rising, and this can lead to a hypertensive crisis. This series of events has been termed the *cheese effect* because of the high tyramine content found in certain cheeses. Since the MAO inhibitors were not toxic, except when taken with certain foods, these drugs were allowed to be returned to the drug market, but they were prescribed with strict dietary regulations. Because of this inconvenience when using these drugs, and the discovery of the tricyclic antidepressants (which block the reuptake of biogenic amines at the nerve terminals), MAO inhibitors are not the drugs of choice, except for those types of depression that do not respond to tricyclic antidepressants or when treating phobic–anxiety disorders, which respond well to MAO inhibitors.

More recently there has been a resurgence of interest by the pharmaceutical industry in MAO inhibitors,[126] because it is now known that MAO exists in two isozymic forms, termed MAO A and MAO B. The main difference in the two isozymes is their selectivity for the oxidation of the various biogenic amines. Since the antidepressant effect is related to increased concentrations of brain serotonin and norepinephrine, both of which are MAO A substrates, compounds that selectively inhibit MAO A possess antidepressant activity; selective inhibitors of MAO B show potent antiparkinsonian properties (see Section V,C,3,d).[127] In order to have an antidepressant drug without the cheese effect, however, it is necessary to inhibit brain MAO A selectively without inhibition of peripheral MAO A, particularly MAO A in the gastrointestinal tract and sympathetic nerve terminals. Inhibition of brain MAO A increases the brain serotonin and norepinephrine concentrations which leads to the antidepressant effect; the peripheral MAO A must remain active to degrade the peripheral tyramine and norepinephrine that cause the undesirable cardiovascular effects.

Tranylcypromine (*trans*-2-phenylcyclopropylamine, **5.56**), a nonselective MAO A/B inactivator that exhibits a cheese effect, was one of the first MAO inactivators approved for clinical use; many other cyclopropylamine analogs show antidepressant activity.[128] The one-electron mechanism for MAO was discussed in Chapter 4 (Section III,C, Scheme 4.32). If tranylcypromine acts as a substrate, and one-electron transfer from the amino group to the flavin occurs, the resulting cyclopropylaminyl radical will undergo rapid cleavage,[129] and the benzylic radical produced should combine with an active site radical (Scheme 5.23).[130] It was suggested[131] that an adduct with an active site cysteine may have formed. In Scheme 5.23 a mechanism for formation of a cysteine adduct is shown.

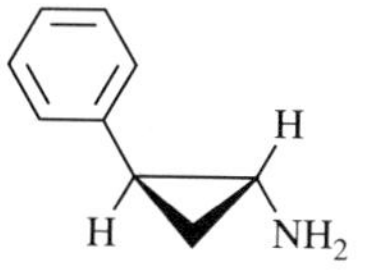

**5.56**

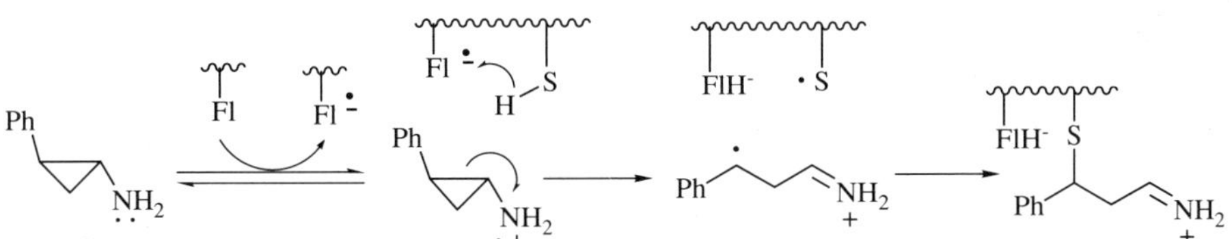

**5.56**

**Scheme 5.23.** Proposed[130] mechanism for the inactivation of monoamine oxidase by tranylcypromine.

***d. Selegiline (L-Deprenyl), an Antiparkinsonian Drug.*** Parkinson's disease, a degenerative neurological disease afflicting more than a half-million people in the United States, is characterized by chronic, progressive motor dysfunction resulting in severe tremors, rigidity, and akinesia. The symptoms of Parkinson's disease arise from the degeneration of dopaminergic neurons in the substantia nigra and a marked reduction in the concentration of the pyridoxal 5′-phosphate–dependent aromatic L-amino acid decarboxylase, the enzyme that catalyzes the conversion of L-dopa to the inhibitory neurotransmitter dopamine. Because dopamine is metabolized primarily by monoamine oxidase B in man (see Section V,C,3,c), and Parkinson's disease is characterized by a reduction in the brain dopamine concentration (see Section II,B,6 of Chapter 8), selective inactivation of MAO B has been shown to be an effective approach to increase the dopamine concentration and, thereby, treat this disease. Actually, as described in Section II,B,6 of Chapter 8, a MAO B–selective inactivator is used in combination with the antiparkinsonian drug L-dopa. Selective inhibition of MAO B does not interfere with the MAO A–catalyzed degradation of tyramine and norepinephrine; therefore, no cardiovascular side effects (the cheese effect; see Section V,C,3,c) are observed with the use of a MAO B–selective inactivator. The earliest MAO B–selective inactivator, selegiline [L-(−)deprenyl] (**5.57**),[132] was approved in 1989 by the U.S. Food and Drug Administration for the treatment of Parkinson's disease in the United States.

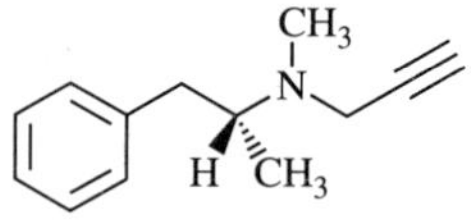

$CH_3$ N H $CH_3$

**5.57**

Little was known about the cause for Parkinson's disease until 1977 when a previously healthy 23-year-old man, who was a chronic street drug user, was referred to the National Institute for Mental Health for investigation of symptoms of what appeared to be Parkinson's disease.[133] Although the patient responded favorably to the usual treatment for Parkinson's disease (L-dopa/carbidopa; see Chapter 8), the speed with which and the age at which the symptoms developed were inconsistent with the paradigm of Parkinson's disease as a regressive geriatric disease. When this man later died of a drug overdose, an autopsy showed the same extensive destruction of the substantia nigra that is found in idiopathic parkinsonism. In 1982 four young Californians who had tried some "synthetic heroin" also developed symptoms of an advanced case of Parkinson's disease, including near total immobility.[134] It was found that the drug they had been taking was a *designer drug*, a synthetic narcotic that has a structure designed to be a variation of an existing illegal narcotic; because the new structure is not listed as a controlled substance, it is not an illegal drug.

In this case the "designers" of the street drug were using the illegal analgesic meperidine (**5.58**) as the basis for their structure modification. The compound synthesized, 1-methyl-4-phenyl-4-propionoxypiperidine (MPPP; **5.59**), was referred to as a "reverse ester" of meperidine (note that the ester oxygen in **5.59** is on the opposite side of the carbonyl from that in **5.58**). However, by "reversing" the ester, the drug designers converted a stable ethoxycarbonyl group of meperidine to a propionoxy group, a good leaving group, in MPPP. In fact, MPPP decomposes upon heating or in the presence of acids with elimination of propionate to give 1-methyl-4-phenyl-1,2,5,6-tetrahydropyridine (MPTP; **5.60**). Analysis of several samples of the designer drug (called "new heroin") revealed the MPTP contamination.

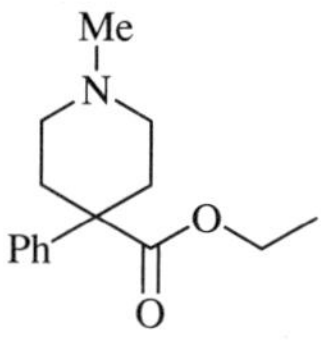

Me N Ph O O **5.58** | Me N Ph O O **5.59** | Me N Ph **5.60**

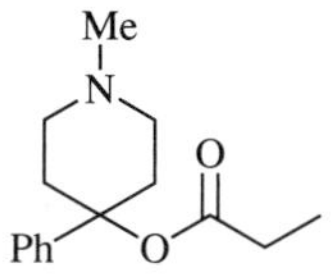

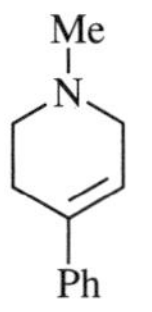

The same MPTP contamination was identified in the samples of drugs taken by the 23-year-old in 1977, but the drug samples at that time were found to exhibit no neurotoxicity in rats, so it was thought that MPTP was not responsible for the neurological effects. However, when the young California drug addicts were observed to have similar symptoms, tests in primates[135a,b] and mice[135c] showed that MPTP produced the same neurological symptoms and histological changes in the substantia nigra as those observed with idiopathic parkinsonism. Rats, it is now known, are remarkably resistant to the neurotoxic effects of MPTP, which explains why the earlier tests in rats were negative.

The observation that an industrial chemist developed Parkinson's disease after synthesizing large amounts of MPTP as a starting material led to the suggestion that cutaneous absorption or vapor inhalation may be significant pathways for introduction of the neurotoxin. Because of this it can be hypothesized that Parkinson's disease is actually an environmental disease arising from long-term slow degeneration of dopaminergic neurons by ingested or inhaled neurotoxins similar to MPTP. Because the symptoms of Parkinson's disease do not appear until at least 80% of the dopaminergic neurons are destroyed, it is reasonable that this disease is associated with the elderly. Opponents of this hypothesis note that the interregional and subregional patterns of striated dopamine loss by MPTP differ from those of idiopathic Parkinson's disease.[136] However, results of surveys taken throughout the world suggest that environmental toxins may have an important role in the etiology of Parkinson's disease.[137] Furthermore, it is interesting that many frequently used medicines have structures related to MPTP; some, particularly neuroleptics, produce parkinsonian side effects.[138] Also, genetic factors do not appear to be important in most cases of Parkinson's disease.

Once a connection was made between MPTP and Parkinson's disease, a vast amount of research with MPTP was initiated, and it soon was realized that the neurotoxic agent was actually a metabolite of MPTP. Pretreatment of animals with the MAO B–selective inactivator selegiline was shown to protect the animals from the neurotoxic effects (both the disease symptoms and the damage to the substantia nigra) of MPTP, whereas the MAO A–selective agent clorgyline (**5.61**) did not.[139,140] This indicated that the neurotoxic metabolite was generated by MAO B oxidation of MPTP. The two metabolites produced by MAO B are 1-methyl-4-phenyl-2,3-dihydropyridinium ion (**5.62**, $MPDP^+$) and 1-methyl-4-phenylpyridinium ion (**5.63**, $MPP^+$).[140] The latter compound (**5.63**) accumulates in selected areas in the brain and, therefore, is believed to be the actual neurotoxic agent.[141] Because MPTP is a neutral molecule, it can cross the blood–brain barrier and enter the brain; once it is oxidized by MAO B, the pyridinium ion (**5.63**) cannot diffuse out of the brain. Selective toxicity of MPTP appears to be the result of the transport of $MPP^+$, but not MPTP, into dopamine neurons via an amine uptake system.[142]

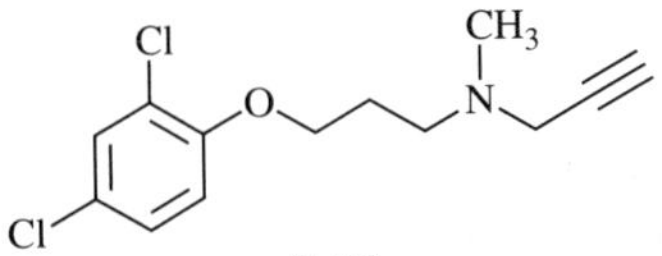

**5.61** 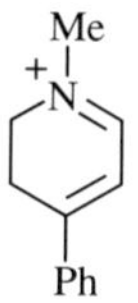 **5.62** 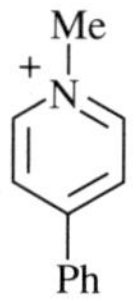 **5.63**

It is apparent, then, that, by inactivation of MAO B, selegiline is important both to the prevention of the oxidation of neurotoxin precursors such as MPTP and to the degradation of dopamine. Although the mechanism of inactivation of MAO by selegiline has not been studied, that of an analog, 3-dimethylamino-1-propyne (**5.64**), has; this compound becomes attached to the $N^5$ of the flavin.[143] Because of the evidence for a one-electron mechanism for MAO-catalyzed oxidations (see Section III,C in Chapter 4), the inactivation mechanisms by selegiline shown in Scheme 5.24 seem most reasonable.

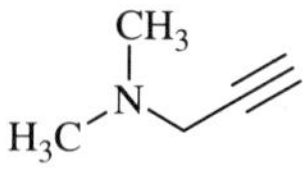

**5.64**

***e. 5-Fluoro-2′-deoxyuridylate, Floxuridine, and 5-Fluorouracil, Antitumor Agents.*** *Cancer* is a group of diseases characterized by abnormal and uncontrolled cell division. Neither the etiology nor the way in which it causes death is understood in most cases. One important approach to *antineoplastic* (antitumor) agents is the design of compounds with structures related to those of pyrimidines and purines that are involved in the biosynthesis of DNA.

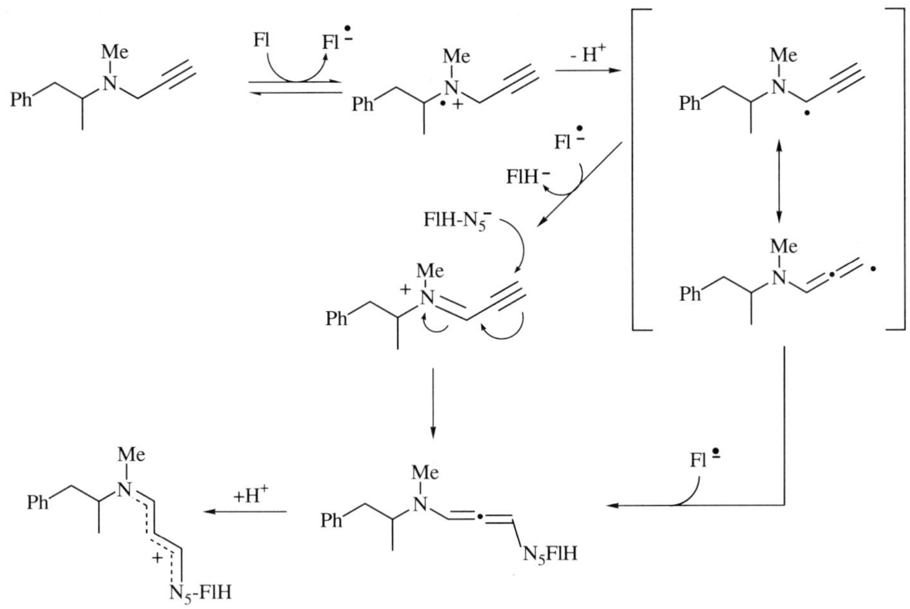

**Scheme 5.24.** Possible mechanisms of inactivation of monoamine oxidase B by selegiline.

These compounds are known as *antimetabolites* because they interfere with the formation or utilization of a normal cellular metabolite. This interference generally results from the inhibition of an enzyme in the biosynthetic pathway of the metabolite or from incorporation, as a false building block, into vital macromolecules such as proteins or nucleic acids. Antimetabolites usually are obtained by making a small structural change in the metabolite, for example, a bioisosteric interchange (see Section II,D,4 in Chapter 2).

5-Fluorouracil (**5.65**), its 2′-deoxyribonucleoside floxuridine (**5.66**), and its 2′-deoxyribonucleotide 5-fluoro-2′-deoxyuridylate (**5.67**), are potent antimetabolites of uracil and its congeners and are potent antineoplastic agents. 5-Fluorouracil itself is not active, but it is converted *in vivo* to the 2′-deoxynucleotide (**5.67**), which is the active form. Because the van der Waals radius of

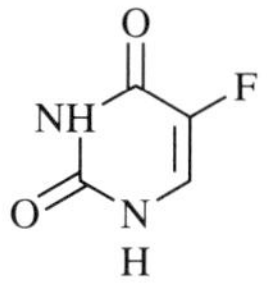

**5.65** 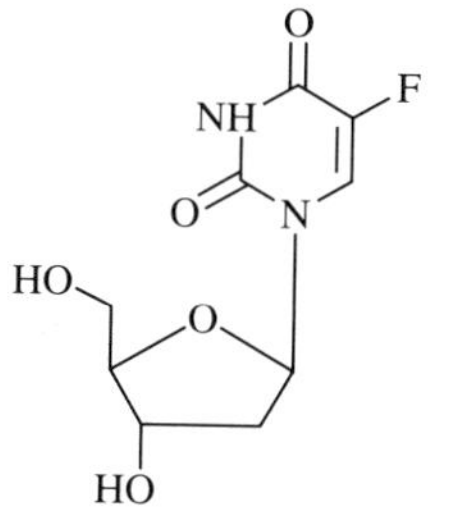 **5.66** 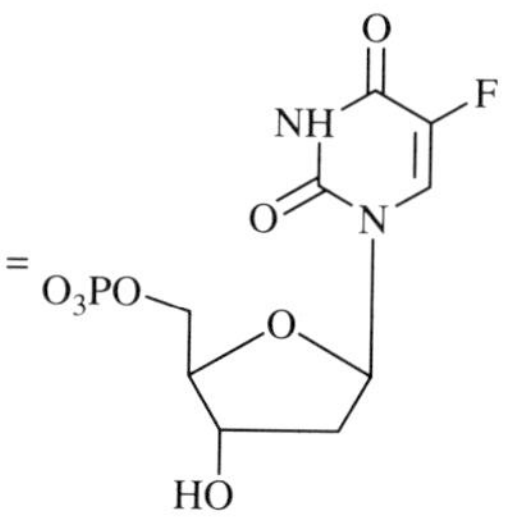 **5.67**

fluorine (1.35 Å) is similar to that of hydrogen (1.20 Å), 5-fluorouracil and its metabolites are recognized by enzymes that act on uracil and its metabolites. There are several pathways for this *in vivo* activation. A minor pathway to the intermediate 5-fluorouridylate (**5.69**) begins with the conversion of 5-fluorouracil (**5.65**) to 5-fluorouridine (**5.68**), catalyzed by uridine phosphorylase, followed by further conversion of **5.68** to **5.69**, catalyzed by uridine kinase (Scheme 5.25). The major pathway to **5.69** is the direct conversion of 5-fluorouracil, which is catalyzed by orotate phosphoribosyltransferase. 5-

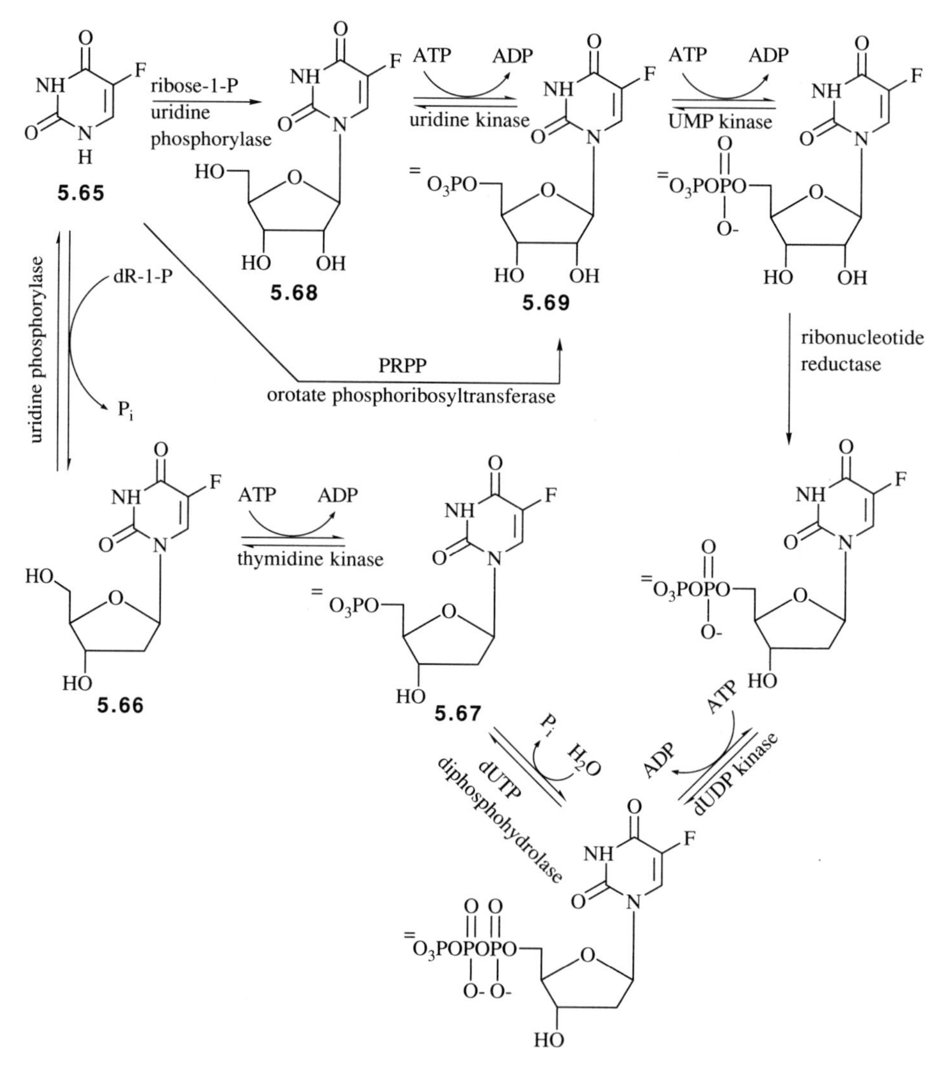


**Scheme 5.25.** Metabolism of 5-fluorouracil.

Fluoro-2′-deoxyuridylate (**5.67**) is produced from **5.69** by the circuitous route shown in Scheme 5.25 or by direct conversion of **5.65** to its 2′-deoxyribonucleoside floxuridine (**5.66**), catalyzed by uridine phosphorylase, followed by 5′-phosphorylation, which is catalyzed by thymidine kinase (Scheme 5.25). However, when **5.66** is administered rapidly, it is converted back to **5.65** faster than it is phosphorylated to **5.67**. Under these circumstances attempts to use floxuridine to bypass the long metabolic route for conversion of **5.65** to **5.67** are unsuccessful. Continuous intra-arterial infusion of floxuridine, however, enhances the direct conversion of **5.66** to **5.67**.

The principal site of action of **5.67** is thymidylate synthase, the enzyme that catalyzes the last step in *de novo* biosynthesis of thymidylate, namely, the conversion of 2′-deoxyuridylate to 2′-deoxythymidylate (referred to as just thymidylate). The reaction catalyzed by thymidylate synthase is the only *de novo* source of thymidylate, which is an essential constituent of the DNA. Therefore, inhibition of thymidylate synthase in tumor cells inhibits DNA biosynthesis and produces what is known as *thymineless death* of the cell.[144] Unfortunately, normal cells also require thymidylate synthase for *de novo* synthesis of their thymidylate. Nonetheless, inhibitors of thymidylate synthase are effective antineoplastic agents. There are several reasons for the selective toxicity against tumor cells; all are related to the difference in the rates of cell division for normal and abnormal cells. Because aberrant cells replicate much more rapidly than do most normal cells, the rapidly proliferating tumor cells have a higher requirement for DNA synthesis than do the slower proliferating normal cells. This means that the activity of thymidylate synthase is elevated in tumor cells relative to normal cells. As uracil is one of the precursors of thymidylate, it and 5-fluorouracil are taken up into tumor cells much more efficiently than into normal cells. Finally, and possibly most importantly, enzymes that degrade uracil in normal cells also degrade 5-fluorouracil, and these degradation processes do not take place in cancer cells.[145] The adverse side effects accompanying the use of 5-fluorouracil in humans generally arise from the inhibition of thymidylate synthase and destruction of the rapidly proliferating normal cells of the intestines, the bone marrow, and the mucosa. The effects of anticancer drugs are discussed in more detail in Chapter 6.

Unlike other tetrahydrofolate-dependent enzymes (see Section III,B in Chapter 4), thymidylate synthase utilizes methylenetetrahydrofolate both as a one-carbon donor and as a reducing agent (Scheme 5.26).[146] An active site cysteine residue undergoes Michael addition to the 6 position of 2′-deoxyuridylate (**5.70**, dRP is deoxyribose phosphate) to give an enolate (**5.71**) that attacks $N^5$,$N^{10}$-methylenetetrahydrofolate and forms a ternary complex (**5.72**) of the enzyme, the substrate, and the coenzyme. Enzyme-catalyzed removal of the C-5 proton leads to $\beta$-elimination of tetrahydrofolate (**5.73**). Oxidation of the tetrahydrofolate (a hydride mechanism is shown, but a one-electron mech-

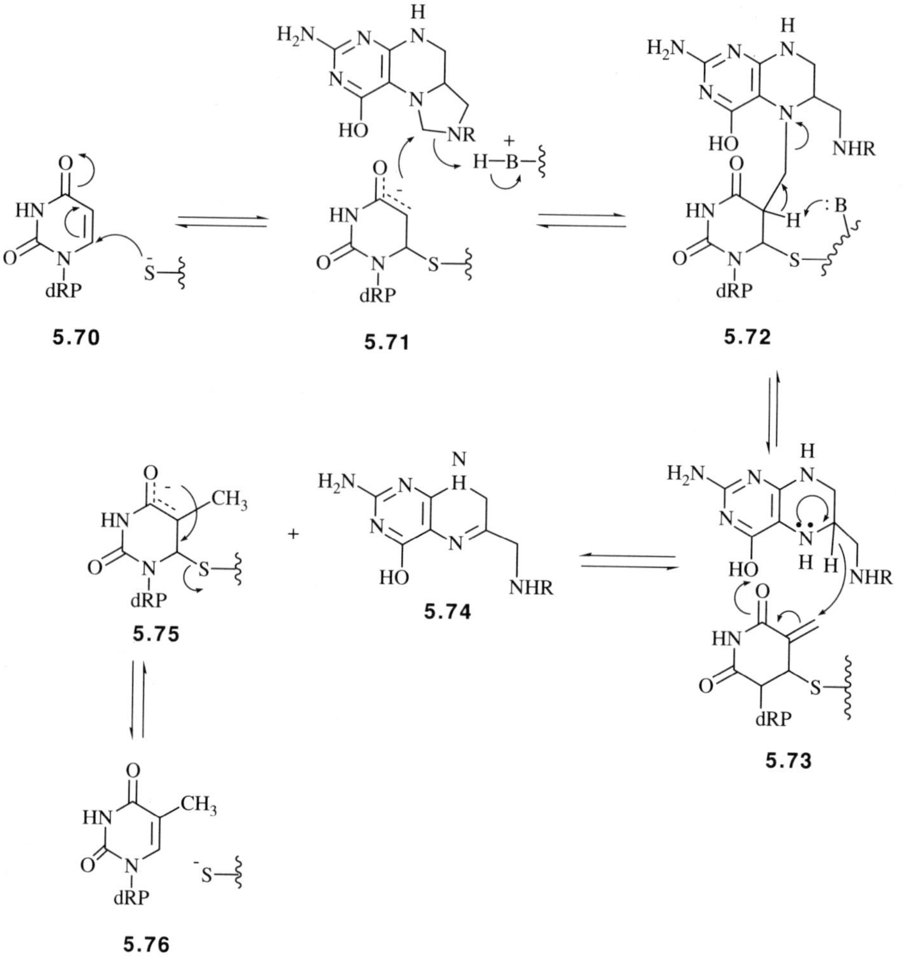

Scheme 5.26. Mechanism proposed for thymidylate synthase (dRP is deoxyribose phosphate).

anism, namely, first transfer of an N-5 nitrogen nonbonded electron to the alkene followed by hydrogen atom transfer, is possible) gives dihydrofolate (**5.74**) and the enzyme-bound thymidylate enolate (**5.75**). Reversal of the first step releases the active site cysteine residue and produces thymidylate (**5.76**). This reaction changes the oxidation state of the coenzyme. Because of this, another enzyme, dihydrofolate reductase, is required to reduce the dihydrofolate back to tetrahydrofolate (see Scheme 4.21 in Chapter 4).

5-Fluoro-2′-deoxyuridylate inactivates thymidylate synthase because once the ternary complex (**5.77**) forms, there is no C-5 proton that the enzyme can remove to eliminate the tetrahydrofolate (Scheme 5.27).[147] Consequently, the enzyme remains as the ternary complex. Note that in this mechanism the inactivator is not converted to a reactive compound that attaches to the enzyme. Instead, it first attaches to the enzyme, then requires condensation with 5,10-methylenetetrahydrofolate to generate a stable complex. A mechanism-based inactivator, therefore, does not necessarily require activation. It only requires the enzyme to catalyze a reaction on it that leads to inactivation. If the enzyme were inactivated without the requirement of $N^5,N^{10}$-methylenetetrahydrofolate, that is, by simple Michael addition of the active site cysteine to the C-6 position of the inactivator, then **5.67** would be an affinity labeling agent.

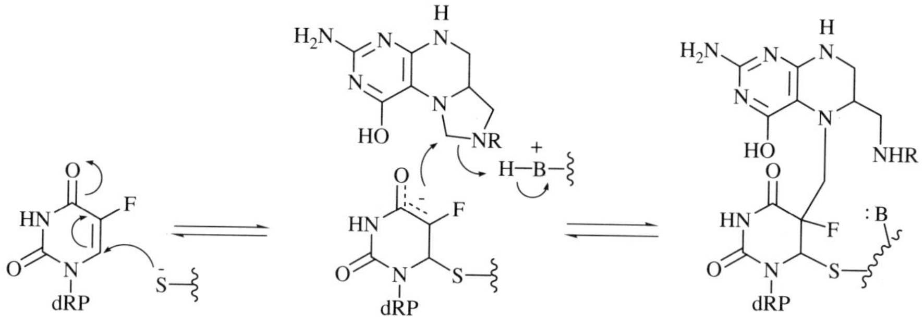

5.67 5.77

Scheme 5.27. Mechanism proposed for the inactivation of thymidylate synthase by 5-fluoro-2′-deoxyuridylate.

Because dihydrofolate reductase is essential for the regeneration of tetrahydrofolate from the dihydrofolate produced in the thymidylate synthase reaction, drug synergism occurs when inhibitors of dihydrofolate reductase and thymidylate synthase are used in combination.

***f. Trifluridine, an Antiviral Agent.*** Another inactivator of thymidylate synthase is trifluridine (5-trifluoromethyl-2′-deoxyuridine; **5.78**), which is used principally for topical therapy of herpes simplex virus infection of the eyes. On the basis of enzyme inactivation and chemical model studies, an inactivation mechanism was proposed[148] (Scheme 5.28).

***g. Clavulanate and Sulbactam, Synergistic Agents for Penicillins.*** As discussed in Section V,B,2,a, many strains of bacteria have become resistant to penicillins and cephalosporins because they produce $\beta$-lactamases which hydrolyze the drugs before they can destroy the bacteria. A compound that inactivates the $\beta$-lactamase, therefore, would protect $\beta$-lactam antibacterials from premature destruction. In 1976 clavulanic acid (**5.79**) was isolated from *Streptomyces clavuligerus*,[149] and it was found to be a potent inactivator of $\beta$-

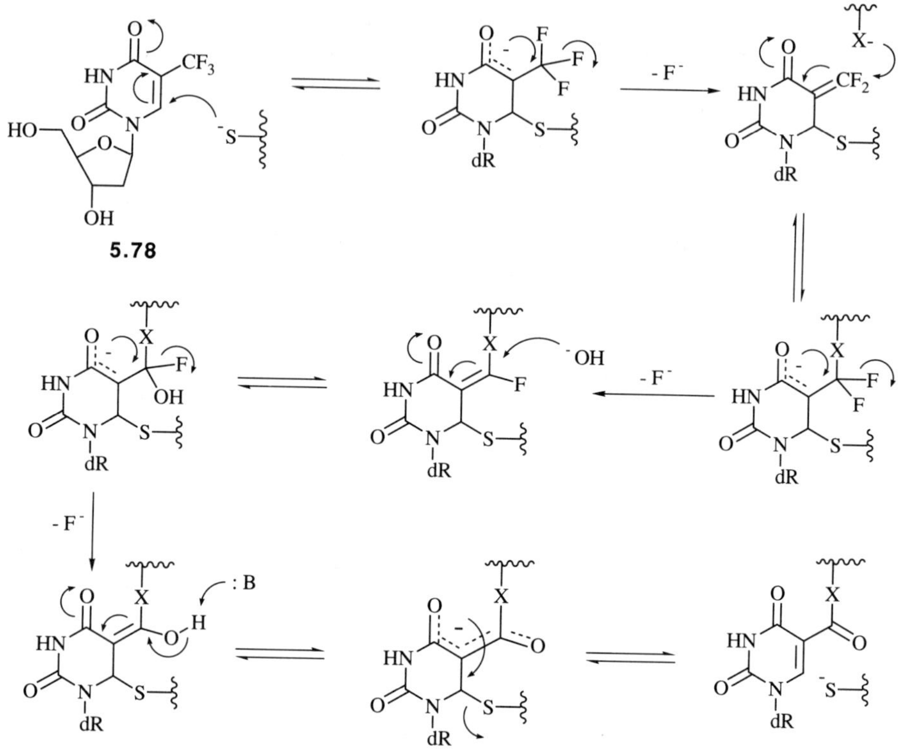

5.78

Scheme 5.28. Mechanism proposed for the inactivation of thymidylate synthase by trifluridine (dR is 2′-deoxyribose).

lactamases from a variety of microorganisms.[150,151] Shortly thereafter, another $\beta$-lactamase inhibitor, penicillanic acid sulfone, now called sulbactam (**5.80**), was reported.[152]

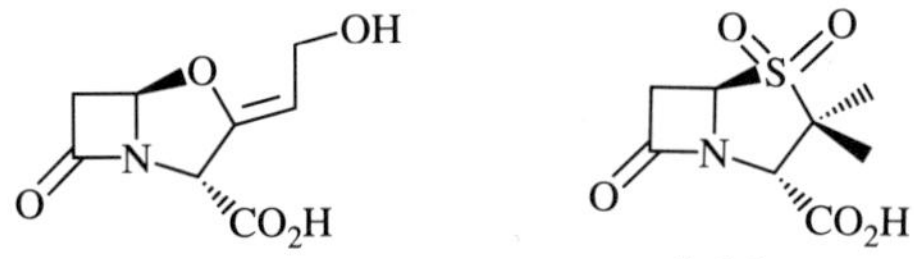

5.79 5.80

Both clavulanate[153] and sulbactam[154] are potent mechanism-based inactivators of $\beta$-lactamases.[105] Inactivation mechanisms proposed for each are similar (see Scheme 5.29 for clavulanate inactivation and Scheme 5.30 for sulbactam inactivation). Both of these $\beta$-lactamase inactivators are used in combination with penicillins for the treatment of penicillin-resistant bacterial infections; the combination of amoxicillin (**5.35d**, R′ = OH) with clavulanate is sold as Augmentin® and ampicillin (**5.35d**, R′ = H) plus sulbactam are in Unasyn® (in unison, get it?).

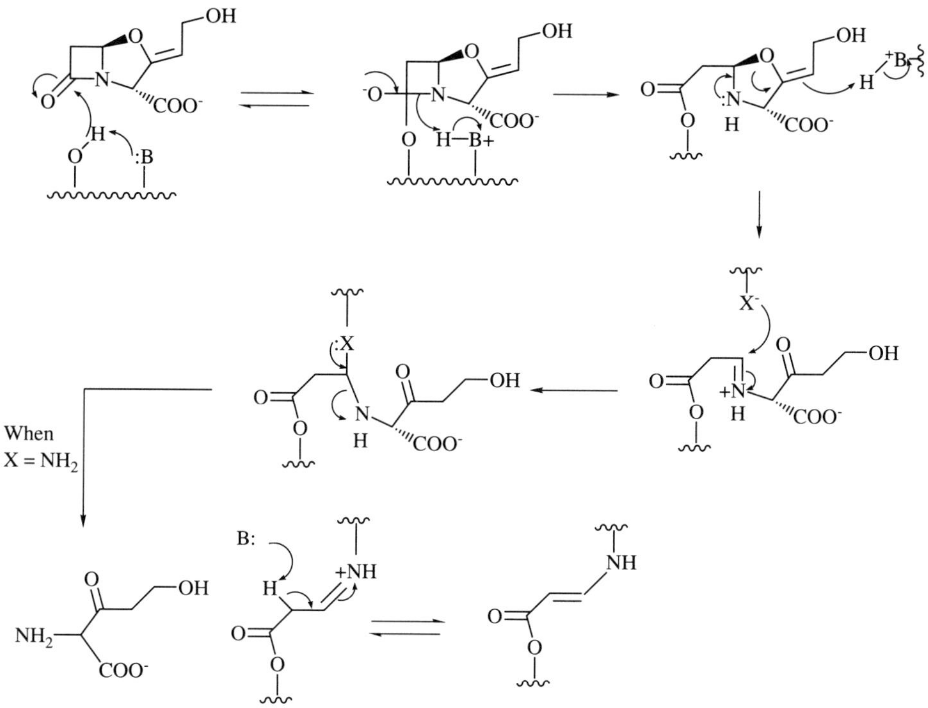

Scheme 5.29. Proposed[153] mechanism for inactivation of $\beta$-lactamases by clavulanate.

***h. Allopurinol, an Antihyperuricemia Agent.*** Uric acid (**5.85**) in man is formed primarily by the xanthine oxidase–catalyzed oxidation of hypoxanthine (**5.83**) and xanthine (**5.84**); **5.83** and **5.84** are derived from the hydrolysis of adenine (**5.81**) and guanine (**5.82**), respectively (Scheme 5.31). The uric acid produced is excreted through the kidneys into the urine; the urinary content of purines is almost exclusively uric acid. In acidic solutions uric acid is quite insoluble; at physiological pH its solubility is only slightly better, but it tends to form supersaturated solutions. When this point is exceeded, crystallization of uric acid can occur, particularly in the joints, the connective tissues, and the kidneys. This hyperuricemia is generally the cause for the disease called gout. Often gout sufferers will complain of pain in their big toe before other parts of their body. Since the big toe is one of the coldest parts of the body, uric acid crystallization occurs there first.

One effective approach to alleviate this problem is to administer a compound that inhibits the biosynthesis of uric acid so that the crystallized uric acid gradually can redissolve and be excreted. An inhibitor of xanthine oxidase, namely, allopurinol (**5.86**, Scheme 5.32), has been shown to be very effective. Its mechanism of inactivation of xanthine oxidase is shown in

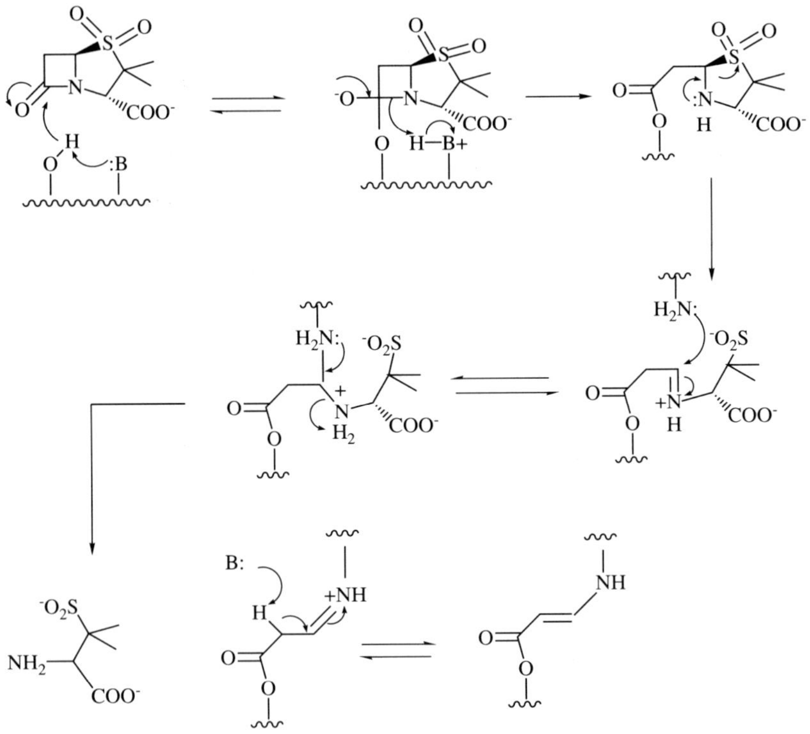

**Scheme 5.30.** Proposed[154] mechanism for inactivation of $\beta$-lactamases by sulbactam.

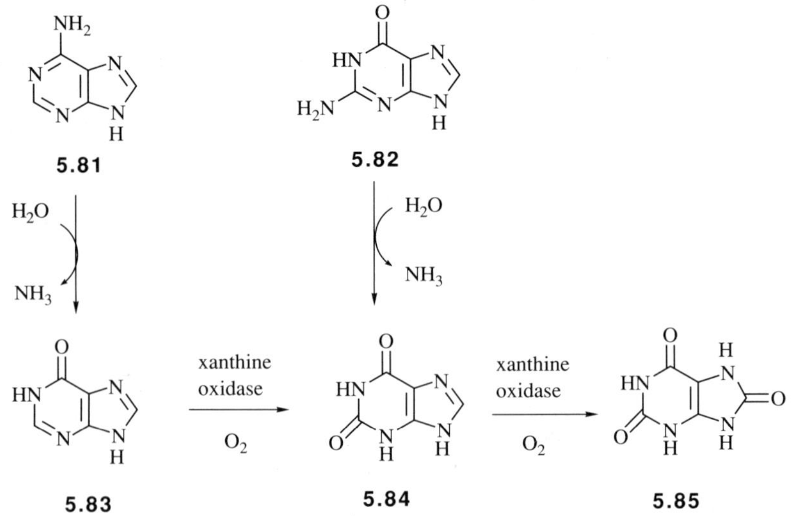
5.81
5.82
$H_2O$
$NH_3$
$H_2O$
$NH_3$
xanthine oxidase
$O_2$
xanthine oxidase
$O_2$
5.83
5.84
5.85

**Scheme 5.31.** Purine metabolism.

Scheme 5.32. Allopurinol is oxidized by xanthine oxidase to alloxanthine (oxypurinol; **5.87**) with concomitant reduction of the Mo(VI) cofactor to Mo(IV). The product (**5.87**) is a potent, tight-binding inhibitor of the enzyme, but only when it is in the reduced [Mo(IV)] state.[155]

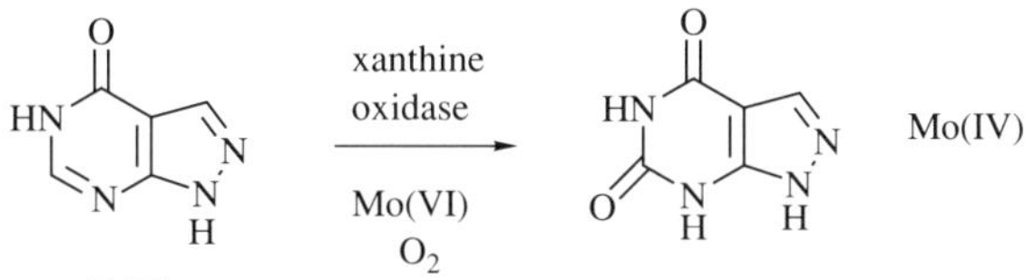

xanthine oxidase, Mo(VI), $O_2$

**5.86** → **5.87** Mo(IV)

**Scheme 5.32.** Mechanism of inactivation of xanthine oxidase by allopurinol.

Note that, unlike other mechanism-based enzyme inactivators, allopurinol undergoes enzyme-catalyzed oxidation to a noncovalent tight-binding inhibitor. It is not essential for a compound to form a covalent bond to the target enzyme to be classified as a mechanism-based inactivator. The compound only needs to be converted by the target enzyme to a product that, without prior release from the active site, inactivates the enzyme.

***i. Chloramphenicol, an Antibiotic.*** Chloramphenicol (**5.88**) is produced by *Streptomyces venezuelae* and is an important drug in the treatment of typhoid fever and other types of salmonella infections; it also is used to combat bacterial meningitis, anaerobic bacterial infections, and rickettsial diseases such as Rocky Mountain spotted fever. However, because it is quite toxic, it is prescribed only for infections that cannot be treated effectively by other antimicrobial agents.

Chloramphenicol is an example of a mechanism-based enzyme inactivator in which the enzyme inactivation is unrelated to its pharmacological activity. However, the enzyme that it inactivates, cytochrome *P*-450, is essential for drug metabolism (see Chapter 7). Inactivation of cytochrome *P*-450 has the effect of prolonging the half-life of drugs that are metabolized by it. Although this, at first sight, may appear to be beneficial, it can lead to severe toxicity and even death as a result of the inability of the body to metabolize and excrete certain other drugs.

The mechanism of inactivation of cytochrome *P*-450, a heme-dependent enzyme (see Section III,D of Chapter 4), by chloramphenicol (**5.88**)[156] is shown in Scheme 5.33. Hydroxylation of the dichloromethyl group gives an unstable species that eliminates HCl to give the corresponding oxamoyl chloride (**5.89**), which readily acylates an active site lysine residue. A major form of bacterial resistance to chloramphenicol is by production of chloramphenicol acetyltransferase, an enzyme that catalyzes the acetyl-CoA–dependent acetylation of one or both of its hydroxyl groups and deactivates the drug.[157] An inhibitor of chloramphenicol acetyltransferase would be a useful approach

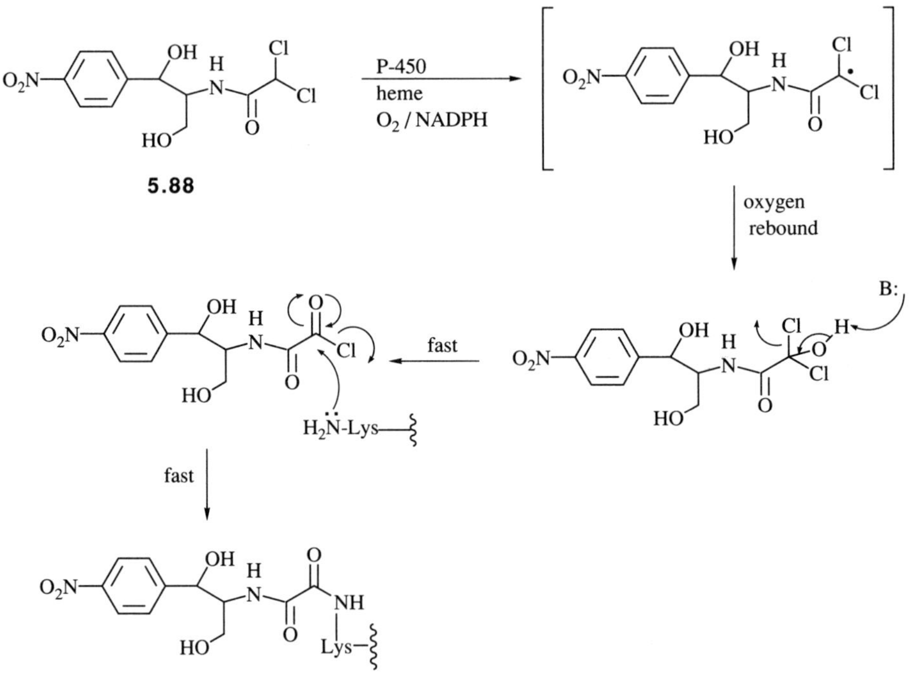

Scheme 5.33. Mechanism of inactivation of cytochrome *P*-450 by chloramphenicol.[156]

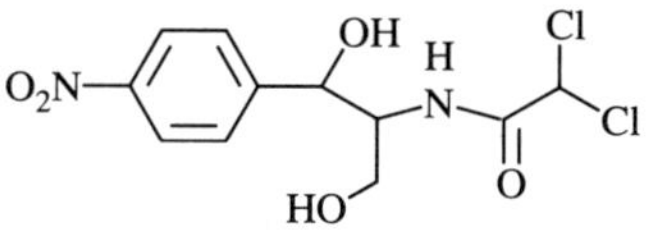

5.88

for drug synergism; however, resistance to chloramphenicol is not a significant problem.

Next, in Chapter 6, we will consider drug interactions with another type of receptor, DNA.

## References


1. Cohen, S. S. 1979. *Science* **205,** 964.
2. Albert, A. 1985. "Selective Toxicity," 7th Ed., p. 256. Chapman & Hall, London.
3. Lowe III, J. A. 1982. *Annu. Rep. Med. Chem.* **17,** 119.
4. Haest, C. W. M., de Gier, J., op den Kamp, J. A. F., Bartels, P., and Van Deenen, L. L. M. 1972. *Biochim. Biophys. Acta* **255,** 720.
5. Bryan, C. E., and Kwan, Š. 1981. *J. Antimicrob. Chemother.* **8** (Suppl. D.), 1.

6. Kan, S., and Siddiqui, W. 1979. *J. Protozool.* **26,** 660.
7. Bertino, J., Cashmore, A., Fink, N., Calabresi, P., and Lefkowitz, E. 1965. *Clin. Pharmacol. Ther. (St Louis)* **6,** 763.
8. Alt, F. W., Kellems, R. E., Bertino, J. R., and Schimke, R. T. 1978. *J. Biol. Chem.* **253,** 1357.
9. Then, R. L., and Hermann, F. 1981. *Chemotherapy* **27,** 192.
10. Weisblum, B. 1975. *In* "Microbiology—1974" (Schlessinger, D., ed.), p. 199. American Society for Microbiology, Washington, D.C.
11. Harrap, K. 1976. *In* "Scientific Foundations of Oncology" (Symington, T., and Carter, R., eds.), p. 641. Heinemann, London.
12. Domagk, G. 1935. *Dtsch. Med. Wochenschr.* **61,** 250.
13. Albert, A. 1985. "Selective Toxicity," 7th Ed., p. 220. Chapman & Hall, London.
14. Domagk, G. 1936. *Klin. Wochenschr.* **15,** 1585.
15. Tréfouël, J., Tréfouël, M. J., Nitti, F., and Bovet, D. 1935. *C. R. Seances Soc. Biol. Ses Fil.* **120,** 756.
16. Marshall, E. K., Jr. 1937. *J. Biol. Chem.* **122,** 263; Bratton, A. C., and Marshall, E. K., Jr. 1939. *J. Biol. Chem.* **128,** 537.
17. Stamp, T. C. 1939. *Lancet* **2,** 10.
18. Woods, D. D. 1940. *Br. J. Exp. Pathol.* **21,** 74.
19. Selbie, F. R. 1940. *Br. J. Exp. Pathol.* **21,** 90.
20. Fildes, P. 1940. *Lancet* **1,** 955.
21. Miller, A. K. 1944. *Proc. Soc. Exp. Pathol. Med.* **57,** 151; Miller, A. K., Bruno, P., and Berglund, R. M. 1947. *J. Bacteriol.* **54,** 9 (G20).
22. Nimmo-Smith, R. H., Lascelles, J., and Woods, D. D. 1948. *Br. J. Exp. Pathol.* **29,** 264.
23. Richey, D. P., and Brown, G. M. 1969. *J. Biol. Chem.* **244,** 1582.
24. Weisman, R. A., and Brown, G. M. 1964. *J. Biol. Chem.* **239,** 326.
25. Bock, L., Miller, G. H., Schaper, K.-J., and Seydel, J. K. 1974. *J. Med. Chem.* **17,** 23.
26. Wood, R. C., Ferone, R., and Hitchings, G. H. 1961. *Biochem. Pharmacol.* **6,** 113.
27. Landy, M., and Gerstung, R. B. 1944. *J. Bacteriol.* **47,** 448.
28. Wise, E. M., Jr., and Abou-Donia, M. M. 1975. *Proc. Natl. Acad. Sci. U.S.A.* **72,** 2621; Sköld, O. 1976. *Antimicrob. Agents Chemother.* **9,** 49.
29. Swedberg, G., and Sköld, O. 1980. *J. Bacteriol.* **142,** 1; Nagate, T., Inoue, M., Inoue, K., and Mitsuhashi, S. 1978. *Microbiol. Immun.* **22,** 367.
30. Wormser, G. P., Keusch, G. T., and Rennie, C. H. 1982. *Drugs* **24,** 459.
31. Hitchings, G. H., and Burchall, J. J. 1965. *Adv. Enzymol.* **27,** 417; Anand, N. 1983. *In* "Inhibition of Folate Metabolism in Chemotherapy" (Hitchings, G. H., ed.), p. 25. Springer-Verlag, Berlin.
32. Ferone, R., Burchall, J. J., and Hitchings, G. H. 1969. *Mol. Pharmacol.* **5,** 49.
33. Brown, M. S., and Goldstein, J. L. 1985. *In* "Goodman and Gilman's The Pharmacological Basis of Therapeutics" (Gilman, A. G., Goodman, L. S., Rall, T. W., and Murad, F., ed.), 7th Ed., p. 827. Macmillan, New York.
34. Grundy, S. M. 1978. *West. J. Med.* **128,** 13.
35. Stamler, J. 1978. *Arch. Surg. (Chicago)* **113,** 21; Havel, R. J., Goldstein, J. L., and Brown, M. S. 1980. *In* "Metabolic Control and Disease" (Bundy, P. K., and Rosenberg, L. E., eds.), p. 393. Saunders, Philadelphia, Pennsylvania.
36. Endo, A., Kuroda, M., and Tsujita, Y. 1976. *J. Antibiot.* **29,** 1346; Endo, A., Tsujita, Y., Kuroda, M., and Tanzawa, K. 1977. *Eur. J. Biochem.* **77,** 31.
37. Brown, A. G., Smale, T. C., King, T. J., Hasenkamp, R., and Thompson, R. H. 1976. *J. Chem. Soc., Perkin Trans. 1,* 1165.
38. Endo, A. 1979. *J. Antibiot.* **32,** 852; Endo, A. 1980. *J. Antibiot.* **33,** 334.
39. Alberts, A. W., Chen, J., Kuron, G., Hunt, V., Huff, J., Hoffman, C., Rothrock, J., Lopez, M., Joshua, H., Harris, E., Patchett, A., Monaghan, R., Currie, S., Stapley, E., Albers-

Schonberg, G., Hensens, O., Hirschfield, J., Hoogsteen, K., Liesch, J., and Springer, J. 1980. *Proc. Natl. Acad. Sci. U.S.A.* **77,** 3957.

40. Endo, A. 1985. *J. Med. Chem.* **28,** 401.
41. Tanzawa, K., and Endo, A. 1979. *Eur. J. Biochem.* **98,** 195.
42. Nakamura, C. E., and Abeles, R. H. 1985. *Biochemistry* **24,** 1364.
43. Stokker, G. E., Hoffman, W. F., Alberts, A. W., Cragoe, E. J., Jr., Deana, A. A., Gilfillan, J. L., Huff, J. W., Novello, F. C., Prugh, J. D., Smith, R. L., and Willard, A. K. 1985. *J. Med. Chem.* **28,** 347.
44. Hoffman, W. F., Alberts, A. W., Cragoe, E. J., Jr., Deana, A. A., Evans, B. E., Gilfillan, J. L., Gould, N. P., Huff, J. W., Novello, F. C., Prugh, J. D., Rittle, K. E., Smith, R. L., Stokker, G. E., and Willard, A. K. 1986. *J. Med. Chem.* **29,** 159.
45. Stokker, G. E., Alberts, A. W., Anderson, P. S., Cragoe, E. J., Jr., Deana, A. A., Gilfillan, J. L., Hirschfield, J., Holtz, W. J., Hoffman, W. F., Huff, J. W., Lee, T. J., Novello, F. C., Prugh, J. D., Rooney, C. S., Smith, R. L., and Willard, A. K. 1986. *J. Med. Chem.* **29,** 170.
46. Stokker, G. E., Alberts, A. W., Gilfillan, J. L., Huff, J. W., and Smith, R. L. 1986. *J. Med. Chem.* **29,** 852.
47. Bartmann, W., Beck, G., Granzer, E., Jendralla, H., Kerekjarto, B. V., and Wess, G. 1986. *Tetrahedron Lett.* **27,** 4709.
48. Hoffman, W. F., Alberts, A. W., Anderson, P. S., Chen, J. S., Smith, R. L., and Willard, A. K. 1986. *J. Med. Chem.* **29,** 849.
49. Stokker, G. E., Rooney, C. S., Wiggins, J. M., and Hirschfield, J. 1986. *J. Org. Chem.* **51,** 4931.
50. Heathcock, C. H., Hadley, C. R., Rosen, T., Theisen, P. D., and Hecker, S. J. 1987. *J. Med. Chem.* **30,** 1858.
51. Fasciolo, J. C. 1977. *In* "Hypertension" (Genest, J., Koiw, E., and Kuchel, O., eds.), p. 134. McGraw-Hill, New York; Skeggs, L. T., Dorer, F. E., Kahn, J. R., Lentz, K. E., and Levine, M. 1981. *In* "Biochemical Regulation of Blood Pressure" (Soffer, R. L., ed.), p. 3. Wiley, New York.
52. Tewksbury, D. A., Dart, R. A., and Travis, J. 1981. *Biochem. Biophys. Res. Commun.* **99,** 1311; Tewksbury, D. 1981. *In* "Biochemical Regulation of Blood Pressure" (Soffer, R. L., ed.), p. 95. Wiley, New York.
53. Peach, M. J. 1977. *Physiol. Rev.* **57,** 313; Peach, M. J. 1981. *Biochem. Pharmacol.* **30,** 2745.
54. Ferreira, S. H. 1965. *Br. J. Pharmacol. Chemother.* **24,** 163.
55. Bakhle, Y. S. 1968. *Nature (London)* **220,** 919; Bakhle, Y. S., Reynard, A. M., and Vane, J. R. 1969. *Nature (London)* **222,** 956.
56. Ferreira, S. H., Bartelt, D. C., and Greene, L. J. 1970. *Biochemistry* **9,** 2583.
57. Ferreira, S. H., Greene, L. J., Alabaster, V. A., Bakhle, Y. S., and Vane, J. R. 1970. *Nature (London)* **225,** 379.
58. Stewart, J. M., Ferreira, S. H., and Greene, L. J. 1971. *Biochem. Pharmacol.* **20,** 1557.
59. Ondetti, M. A., Williams, N. J., Sabo, E. F., Pluščec, J., Weaver, E. R., and Kocy, O. 1971. *Biochemistry* **10,** 4033.
60. Cheung, H. S., and Cushman, D. W. 1973. *Biochim. Biophys. Acta* **293,** 451.
61. Cushman, D. W., and Cheung, H. S. 1972. *In* "Hypertension" (Genest, J., and Koiw, E., ed.), p. 532. Springer, Berlin.
62. Ondetti, M. A., and Cushman, D. W. 1981. *In* "Biochemical Regulation of Blood Pressure" (Soffer, R. L., ed.), p. 165. Wiley, New York.
63. Kato, H., and Suzuki, T. 1971. *Biochemistry* **10,** 972.
64. Das, M., and Soffer, R. L. 1975. *J. Biol. Chem.* **250,** 6762.
65. Quiocho, F. A., and Lipscomb, W. N. 1971. *Adv. Protein Chem.* **25,** 1.
66. Cushman, D. W., Cheung, H. S., Sabo, E. F., and Ondetti, M. A. 1977. *Biochemistry* **16,** 5484.

67. Byers, L. D., and Wolfenden, R. 1972. *J. Biol. Chem.* **247,** 606; Byers, L. D., and Wolfenden, R. 1973. *Biochemistry* **12,** 2070.
68. Ondetti, M. A., Cushman, D. W., Sabo, E. F., and Cheung, H. S. 1979. *In* "Drug Action and Design: Mechanism-Based Enzyme Inhibitors" (Kalman, T. I., ed.), p. 271. Elsevier/North-Holland, New York.
69. Atkinson, A. B., and Robertson, J. I. S. 1979. *Lancet* **2,** 836.
70. Patchett, A. A., Harris, E., Tristram, E. W., Wyvratt, M. J., Wu, M. T., Taub, D., Peterson, E. R., Ikeler, T. J., ten Broeke, J., Payne, L. G., Ondeyka, D. L., Thorsett, E. D., Greenlee, W. J., Lohr, N. S., Hoffsommer, R. D., Joshua, H., Ruyle, W. V., Rothrock, J. W., Aster, S. D., Maycock, A. L., Robinson, F. M., Hirschmann, R., Sweet, C. S., Ulm, E. H., Gross, D. M., Vassil, T. C., and Stone, C. A. 1980. *Nature* (*London*) **288,** 280.
71. Wyvratt, M. J., and Patchett, A. A. 1985. *Med. Res. Rev.* **4,** 483.
72. Karanewsky, D. S., Badia, M. C., Cushman, D. W., DeForrest, J. M., Dejneka, T., Loots, M. J., Perri, M. G., Petrillo, E. W., Jr., and Powell, J. R. 1988. *J. Med. Chem.* **31,** 204.
73. Sculley, M. J., and Morrison, J. F. 1986. *Biochim. Biophys. Acta* **874,** 44; Schloss, J. V. 1988. *Acc. Chem. Res.* **21,** 348; Morrison, J. F., and Walsh, C. T. 1988. *Adv. Enzymol.* **61,** 201.
74. Rich, D. H. 1985. *J. Med. Chem.* **28,** 263.
75. Imperiali, B., and Abeles, R. H. 1986. *Biochemistry* **25,** 3760.
76. Stein, R. L., Strimpler, A. M., Edwards, P. D., Lewis, J. J., Mauger, R. C., Schwartz, J. A., Stein, M. M., Trainor, D. A., Wildonger, R. A., and Zottola, M. A. 1987. *Biochemistry* **26,** 2682.
77. Bartlett, P. A., and Marlowe, C. K. 1987. *Science* **235,** 569.
78. Bull, H. G., Thornberry, N. A., Cordes, M. H. J., Patchett, A. A., and Cordes, E. H. 1985. *J. Biol. Chem.* **260,** 2952.
79. Travis, J., and Salvesen, G. S. 1983. *Annu. Rev. Biochem.* **52,** 655.
80. Stockley, R. A., Morrison, H. M., Smith, S., and Tetley, T. 1984. *Hoppe-Seyler's Z. Physiol. Chem.* **365,** 587.
81. Bignon, J., and Scarpa, G. L., eds. 1981. "Biochemistry, Pathology and Genetics of Pulmonary Emphysema." Pergamon, New York; Mittman, C., and Taylor, J. C. eds. 1988. "Pulmonary Emphysema and Proteolysis." Academic Press, New York.
82. Johnson, D., and Travis, J. 1978. *J. Biol. Chem.* **253,** 7142; Beatty, K., Matheson, N., and Travis, J. 1984. *Hoppe Seyler's Z. Physiol. Chem.* **365,** 731.
83. Groutas, W. C. 1987. *Med. Res. Rev.* **7,** 227.
84. Gelb, M. H., Svaren, J. P., and Abeles, R. H. 1985. *Biochemistry* **24,** 1813.
85. Bernhard, S. A., and Orgel, L. E. 1959. *Science* **130,** 625.
86. Jencks, W. P. 1966. *In* "Current Aspects of Biochemical Energetics" (Kennedy, E. P., ed.), p. 273. Academic Press, New York.
87a. Wolfenden, R. 1976. *Annu. Rev. Biophys. Bioeng.* **5,** 271; Wolfenden, R. 1969. *Nature* (*London*) **223,** 704; Wolfenden, R. 1977. *In* "Methods in Enzymology" (Jakoby, W. B., and Wilchek, M., eds.), Vol. 46, p. 15. Academic Press, New York.
87b. Lienhard, G. E. 1973. *Science* **180,** 149; Lienhard, G. E. 1972. *Annu. Rep. Med. Chem.* **7,** 249.
88. Christianson, D. W., and Lipscomb, W. N. 1989. *Acc. Chem. Res.* **22,** 62.
89. Loo, T. L., and Nelson, J. A. 1982. *In* "Cancer Medicine" (Holland, J. F., and Frei III, E., eds.), 2nd Ed., p. 790. Lea & Febiger, Philadelphia, Pennsylvania; McCormack, J. J., and Johns, D. G. 1982. "Pharmacologic Principles of Cancer Treatment" (Chabner, B. A., ed.), p. 213. Saunders, Philadelphia, Pennsylvania.
90. Agarwal, R. P., Spector, T., and Parks, R. E., Jr. 1977. *Biochem. Pharmacol.* **26,** 359.
91. Berne, R. M., Rall, T. W., and Rubio, R., eds. 1983. "Regulatory Functions of Adenosine." Nijhoff, Boston, Massachusetts.

92. Tritsch, G. L., ed. 1985. "Adenosine Deaminase in Disorders of Purine Metabolism and in Immune Deficiency," Vol. 451. New York Academy of Sciences, New York.
93. Stark, G. R., and Bartlett, P. A. 1983. *Pharmacol. Ther.* **23,** 45; Collins, K. D., and Stark, G. R. 1971. *J. Biol. Chem.* **246,** 6599.
94. Erlichman, C., and Vidgen, D. 1984. *Biochem. Pharmacol.* **33,** 3177.
95. Lippert, B., Jung, M. J., and Metcalf, B. W. 1980. *Brain Res. Bull.* **5** (Suppl. 2), 375.
96. Committee on Infectious Diseases. 1982. *Pediatrics* **69,** 810.
97. Mandell, G. L. 1985. *In* "Principles and Practice of Infectious Diseases" (Mandell, G. L., Douglas, R. G., Jr., and Bennett, J. E., eds.), 2nd Ed., p. 180. Wiley, New York.
98. Sammes, P. G., ed. 1980. "Topics in Antibiotic Chemistry," Vol. 4. Ellis Horwood, Chichester; Brown, A. G., and Roberts, S. M., eds. 1985. "Recent Advances in the Chemistry of β-Lactam Antibiotics." Royal Society of Chemistry, London.
99. Waxman, D. J., and Strominger, J. L. 1983. *Annu. Rev. Biochem.* **52,** 825.
100. Yocum, R. R., Rasmussen, J. R., and Strominger, J. L. 1980. *J. Biol. Chem.* **255,** 3977.
101. Tipper, D. J., and Strominger, J. L. 1965. *Proc. Natl. Acad. Sci. U.S.A.* **54,** 1133.
102. Izaki, K., Matsuhashi, M., and Strominger, J. L. 1968. *J. Biol. Chem.* **243,** 3180.
103. Sweet, R. M., and Dahl, K. F. 1970. *J. Am. Chem. Soc.* **92,** 5489.
104. Böhme, E. H. W., Applegate, H. E., Toeplitz, B., Dolfini, J. E., and Gougoutas, J. Z. 1971. *J. Am. Chem. Soc.* **93,** 4324.
105. Knowles, J. R. 1985. *Acc. Chem. Res.* **18,** 97; Kelly, J. A., Dideberg, O., Charlier, P., Wery, J. P., Libert, M., Moews, P. C., Knox, J. R., Duez, C., Fraipoint, C., Joris, B., Dusart, J., Frère, J. M., and Ghuysen, J. M. 1986. *Science* **231,** 1429.
106. Gross, M., and Greenberg, L. A. 1948. "The Salicylates. A Critical Bibliographic Review." Hillhouse, New Haven, Connecticut.
107. Margotta, R. 1968. "An Illustrated History of Medicine." Paul Hamlyn, Middlesex, England.
108. Stone, E. 1763. *Philos. Trans. R. Soc., London* **53,** 195.
109. Martin, B. K. 1963. *In* "Salicylates, An International Symposium" (Dixon, A. St. J., Martin, B. K., Smith, M. J. H., and Wood, P. H. N., eds.) P. G. Little, Brown and Co., Boston.
110. Vane, J. R. 1971. *Nature (London) New Biol.* **231,** 232.
111. Smith, J. B., and Willis, A. L. 1971. *Nature (London) New Biol.* **231,** 235.
112. Roth, G. J., Stanford, N., and Majerus, P. W. 1975. *Proc. Natl. Acad. Sci. U.S.A.* **72,** 3073.
113. Hemler, M., Lands, W. E. M., and Smith, W. L. 1976. *J. Biol. Chem.* **251,** 5575.
114. Van der Ouderaa, F. J., Buytenhek, M., Nugteren, D. H., and Van Dorp, D. A. 1980. *Eur. J. Biochem.* **109,** 1.
115. Roth, G. J., Machuga, E. T., and Ozols, J. 1983. *Biochemistry* **22,** 4672.
116. Nelson, S. D. 1982. *J. Med. Chem.* **25,** 753.
117. Schechter, P. J., Barlow, J. L. R., and Sjoerdsma, A. 1987. *In* "Inhibition of Polyamine Metabolism: Biological Significance and Basis for New Therapies" (McCann, P. P., Pegg, A. E., and Sjoerdsma, A., eds.), p. 345. Academic Press, Orlando, Florida.
118. Isaacson, E. I., and Delgado, J. N. 1981. *In* "Burger's Medicinal Chemistry" (Wolff, M. E., ed.), 4th Ed., Part 3, p. 829. Wiley, New York.
119a. Nanavati, S. M., and Silverman, R. B. 1989. *J. Med. Chem.* **32,** 2413.
119b. Iadarola, M. J., and Gale, K. 1982. *Science* **218,** 1237.
120a. Lippert, B., Metcalf, B. W., Jung, M. J., and Casara, P. 1977. *Eur. J. Biochem.* **74,** 441.
120b. Nanavati, S. M., and Silverman, R. B. 1991. *J. Am. Chem. Soc.* **113,** 9341.
121. Pegg, A. E. 1988. *Cancer Res.* **48,** 759.
122. Danzin, C., Bey, P., Schirlin, D., and Claverie, N. 1982. *Biochem. Pharmacol.* **31,** 3871.
123a. Selikoff, I. J., Robitzek, E. H., and Ornstein, G. G. 1952. *Q. Bull. Sea View Hosp.* **13,** 17 and 27.

123b. Zeller, E. A., Barsky, J., Fouts, J. P., Kirchheimer, W. F., and Van Orden, L. S. 1952. *Experientia* **8,** 349.

124. Zeller, E. A., ed. 1959. *Ann. N.Y. Acad. Sci.* **80,** 551.

125. Ganrot, P. O., Rosengren, E., and Gottfries, C. G. 1962. *Experientia* **18,** 260.

126. Dostert, P. L., Strolin Benedetti, M., and Tipton, K. F. 1989. *Med. Res. Rev.* **9,** 45.

127. Palfreyman, M. G., McDonald, I. A., Bey, P., Schechter, P. J., and Sjoerdsma, A. 1988. *Prog. Neuro-Psychopharmacol. Biol. Psychiatry* **12,** 967; McDonald, I. A., Bey, P., and Palfreyman, M. G. 1989. *In* "Design of Enzyme Inhibitors as Drugs" (Sandler, M., and Smith, H. J., eds.), p. 227. Oxford Univ. Press, Oxford.

128. Kaiser, C., and Setler, P. E. 1981. *In* "Burger's Medicinal Chemistry" (Wolff, M. E., ed.), 4th Ed., Part 3, p. 997. Wiley, New York.

129. Maeda, Y., and Ingold, K. U. 1980. *J. Am. Chem. Soc.* **102,** 328.

130. Silverman, R. B. 1983. *J. Biol. Chem.* **258,** 14766.

131. Paech, C., Salach, J. I., and Singer, T. P. 1980. *J. Biol. Chem.* **255,** 2700.

132. Riederer, P., and Przuntek, H., eds. 1987. "MAO-B Inhibitor Selegiline (*R*-(−)-Deprenyl)." Springer-Verlag, Wien, Austria.

133. Davis, G. C., Williams, A. C., Markey, S. P., Ebert, M. H., Caine, E. D., Reichert, C. M., and Kopin, I. J. 1979. *Psychiatry Res.* **1,** 249.

134. Langston, J. W., Ballard, P., Tetrud, J. W., and Irwin, I. 1983. *Science* **219,** 979.

135a. Burns, R. S., Chiueh, C. C., Markey, S. P., Ebert, M. H., Jacobowitz, D. M., and Kopin, I. J. 1983. *Proc. Natl. Acad. Sci. U.S.A.* **80,** 4546.

135b. Langston, J. W., Forno, L. S., Robert, C. J., and Irwin, I. 1984. *Brain Res.* **292,** 390.

135c. Heikkila, R. E., Hess, A., and Duvoisin, R. C. 1984. *Science* **224,** 1451.

136. Hornykiewicz, O. 1989. *Prog. Neuro-Psychopharmacol. Biol. Psychiat.* **13,** 319.

137. Tanner, C. M. 1989. *Trends Neurosci.* **12,** 49.

138. Markey, S. P., and Schmuff, N. R. 1986. *Med. Res. Rev.* **6,** 389.

139. Langston, W. B., Irwin, I., Langston, E. B., and Forno, L. S. 1984. *Science* **225,** 1480; Heikkila, R. E., Manzino, L., Cabbat, F. S., and Duvoisin, R. C. 1984. *Nature (London)* **311,** 467.

140. Chiba, K., Trevor, A., and Castagnoli, N., Jr. 1984. *Biochem. Biophys. Res. Commun.* **120,** 574.

141. Markey, S. P., Johannessen, J. N., Chiueh, C. C., Burns, R. S., and Herkenham, M. A. 1984. *Nature (London)* **311,** 464.

142. Javitch, J. A., D'Amato, R. J., Strittmater, S. M., and Snyder, S. H. 1985. *Proc. Natl. Acad. Sci. U.S.A.* **82,** 2173.

143. Maycock, A. L., Abeles, R. H., Salach, J. I., and Singer, T. P. 1976. *Biochemistry* **15,** 114.

144. Cohen, S. S. 1971. *Ann. N.Y. Acad. Sci.* **186,** 292.

145. Mukherjee, K. L., and Heidelberger, C. 1960. *J. Biol. Chem.* **235,** 433.

146. Douglas, K. T. 1987. *Med. Res. Rev.* **4,** 441; Benkovic, S. J. 1980. *Annu. Rev. Biochem.* **49,** 227.

147. Santi, D. V., McHenry, C. S., Raines, R. T., and Ivanetich, K. M. 1987. *Biochemistry* **26,** 8606; Silverman, R. B. 1988. "Mechanism-Based Enzyme Inactivation: Chemistry and Enzymology," Vol. 1, p. 59. CRC Press, Boca Raton, Florida.

148. Wataya, Y., Sonobe, Y., Maeda, M., Yamaizumi, Z., Aida, M., and Santi, D. V. 1987. *J. Chem. Soc., Perkin Trans. 1,* 2141.

149. Brown, A. G., Butterworth, D., Cole, M., Hanscomb, G., Hood, J. D., Reading, C., and Rolinson, G. N. 1976. *J. Antibiot.* **29,** 668.

150. Silverman, R. B. 1988. "Mechanism-Based Enzyme Inactivation: Chemistry and Enzymology," Vol. 1, p. 135. CRC Press, Boca Raton, Florida.

151. Cartwright, S. J., and Waley, S. G. 1983. *Med. Res. Rev.* **3,** 341.

152. English, A. R., Retsema, J. A., Girard, J. A., Lynch, J. E., and Barth, W. E. 1978. *Antimicrob. Agents Chemother.* **14,** 414.

153. Charnas, R. L., and Knowles, J. R. 1981. *Biochemistry* **20,** 3214.
154. Brenner, D. G., and Knowles, J. R. 1984. *Biochemistry* **23,** 5833.
155. Massey, V., Komai, H., Palmer, G., and Elion, G. B. 1970. *J. Biol. Chem.* **245,** 2837; Cha, S., Agarwal, R. P., and Parks, R. E., Jr. 1975. *Biochem. Pharmacol.* **24,** 2187.
156. Pohl, L. R., and Krishna, G. 1978. *Biochem. Pharmacol.* **27,** 335; Halpert, J. 1982. *Mol. Pharmacol.* **21,** 166.
157. Shaw, W. V. 1975. *In* "Methods in Enzymology" (Hash, J. H., ed.), Vol. 43, p. 737. Academic Press, New York; Gaffney, D. F., and Foster, T. J. 1978. *J. Gen. Microbiol.* **109,** 351.

## General References

### Drug Resistance and Synergism; Chemotherapy

Albert, A. 1985. "Selective Toxicity," 7th Ed. Chapman & Hall, London.

### Sulfonamides

Anand, A. 1979. *In* "Burger's Medicinal Chemistry" (Wolff, M. E., ed.), 4th Ed., Part 2, p. 1. Wiley, New York.

### HMG-CoA Reductase Inhibitors

Endo, A. 1988. *Klin. Wochenschr.* **66,** 421.
Grundy, S. M. 1988. *N. Engl. J. Med.* **319,** 24.

### Angiotensin-Converting Enzyme

Ondetti, M. A., and Cushman, D. W. 1981. *J. Med. Chem.* **24,** 355.
Ondetti, M. A., and Cushman, D. W. 1984. *Crit. Rev. Biochem.* **16,** 381.
Patchett, A. A., and Cordes, E. H. 1985. *Adv. Enzymol.* **57,** 1.
Petrillo, E. W., Jr., and Ondetti, M. A. 1982. *Med. Res. Rev.* **2,** 1.
Wyvratt, M. J., and Patchett, A. A. 1985. *Med. Res. Rev.* **5,** 483.

### Slow, Tight-Binding Inhibitors

Morrison, J. F., and Walsh, C. T. 1988. *Adv. Enzymol.* **61,** 201.
Schloss, J. V. 1988. *Acc. Chem. Res.* **21,** 348.
Sculley, M. J., and Morrison, J. F. 1986. *Biochim. Biophys. Acta* **874,** 44.

### Transition State Analogs

Andrews, P. R., and Winkler, D. A. 1984. *In* "Drug Design: Fact or Fantasy?" (Jolles, G., and Wooldridge, K. R. H., eds.), p. 145. Academic Press, London.
Wolfenden, R. 1976. *Annu. Rev. Biophys. Bioeng.* **5,** 271; Wolfenden, R. 1977. *In* "Methods in Enzymology (Jakoby, W. B., and Wilchek, M., eds.), Vol. 46, p. 15. Academic Press, New York.

## Multisubstrate Analogs

Broom, A. D. 1989. *J. Med. Chem.* **32,** 2.

## Penicillins and Cephalosporins

Mandell, G. L., and Sande, M. A. 1985. *In* "Goodman and Gilman's The Pharmacological Basis of Therapeutics" (Gilman, A. G., Goodman, L. S., Rall, T. W., and Murad, F., eds.), 7th Ed., p. 1115. Macmillan, New York.
Morin, R. B., and Gorman, M., eds. 1982. "Chemistry and Biology of β-Lactam Antibiotics." Academic Press, New York.

## Aspirin

Flower, R. J., Moncada, S., and Vane, J. R. 1985. *In* "Goodman and Gilman's The Pharmacological Basis of Therapeutics" (Gilman, A. G., Goodman, L. S., Rall, T. W., and Murad, F., eds.), 7th Ed., p. 674. Macmillan, New York.
Rainsford, K. D. 1984. "Aspirin and the Salicylates." Butterworth, London.

## Mechanism-Based Enzyme Inactivators

Silverman, R. B. 1988. "Mechanism-Based Enzyme Inactivation: Chemistry and Enzymology," Vols. 1 and 2. CRC Press. Boca Raton, Florida.

## Polyamines

McCann, P. P., Pegg, A. E., and Sjoerdsma, A., eds. 1987. "Inhibition of Polyamine Metabolism. Biological Significance and Basis for New Therapies." Academic Press, Orlando, Florida.
Pegg, A. E. 1988. *Cancer Res.* **48,** 759.
Tabor, C. W., and Tabor, H. 1984. *Annu. Rev. Biochem.* **53,** 749.

## Monoamine Oxidase Inhibitors

Dostert, P. L., Strolin Benedetti, M., and Tipton, K. F. 1989. *Med. Res. Rev.* **9,** 45.
Fowler, C. J., and Ross, S. B. 1984. *Med. Res. Rev.* **4,** 323.

## β-Lactamase Inhibitors

Cartwright, S. J., and Waley, S. G. 1983. *Med. Res. Rev.* **3,** 341.
Knowles, J. R. 1985. *Acc. Chem. Res.* **18,** 97.

CHAPTER 6

# DNA



## I. Introduction

### A. Basis for DNA-Interactive Drugs

Another receptor (broadly defined) with which drugs can interact is deoxyribonucleic acid or DNA, the polynucleotide that carries the genetic information in cells. Because this receptor is so vital to human functioning, and because there are so few differences between normal DNA and DNA from other cells, drugs that interact with this receptor (*DNA-interactive drugs*) are generally very toxic to normal cells. Therefore, these drugs are reserved only for life-threatening diseases such as cancers and viral infections. Unlike the case of the design of drugs that act on enzymes in a foreign organism, there is little that is useful to direct the design of selective agents against abnormal DNA. The principal feature of cancer cells which differs from that of most normal cells is that cancer cells undergo a rapid, abnormal, and uncontrolled cell division. Because the cells are continually undergoing mitosis, there is a constant need for DNA (and its precursors). The difference, then, is mostly quantitative rather than qualitative, although genes coding for differentiation in cancer cells appear to be shut off or inadequately expressed, while genes coding for cell proliferation are expressed when they should not be. Also, because of the similarity of normal and abnormal cells, a compound that reacts with a cancer cell, most likely, will react with a normal cell as well. However, because of rapid cell division, cancer cell mitosis can be halted preferentially to that found in normal cells where there is sufficient time for the triggering of repair mechanisms.[1,2] In general, anticancer drugs are most effective against malignant tumors with a large proportion of rapidly dividing cells, such as leukemias and lymphomas. Unfortunately, the most common tumors are the solid tumors, which have a small proportion of rapidly dividing cells.

This is not a chapter on antitumor agents, but rather on the organic chemistry of DNA-interactive drugs. Therefore, only a few drugs have been selected as representative examples to demonstrate the organic chemistry involved. Some of the principles of antitumor drug design are discussed elsewhere in this book (Section V,C,3,e of Chapter 5; Sections II,B,2,a–c, II,B,3,d, and II,B,4 of Chapter 8).

### B. Toxicity of DNA-Interactive Drugs

The *toxicity* associated with cancer drugs usually is observed in those parts of the body where rapid cell division normally occurs, such as in the bone marrow, the gastrointestinal (GI) tract, the mucosa, and the hair. The clinical effectiveness of a cancer drug requires that it generally be administered at doses in the toxic range so that it kills tumor cells but allows enough normal cells in the critical tissues, such as the bone marrow and GI tract, to survive, thereby allowing recovery to be possible. There is some evidence, however, that the nausea and vomiting which often occur from these toxic agents are triggered by the central nervous system rather than as a result of destruction of cells in the GI tract.[3]

Even though cancer drugs are very cytotoxic, they must be administered repeatedly over a relatively long period of time to assure that all of the malignant cells have been eradicated. According to the *fractional cell kill hypothesis*, a given drug concentration that is applied for a defined time period will kill a constant fraction of the cell population, independent of the absolute number of cells. Therefore, each cycle of treatment will kill a specific fraction of the remaining cells, and the effectiveness of the treatment is a direct function of the dose of the drug administered and the frequency of repetition. Furthermore, it is now known that single drug treatments are only partially effective and produce responses of short duration. When complete remission is obtained with these drugs, it is only short-lived, and relapse is associated with resistance to the original drug. Because of this, combination chemotherapy was adopted.

### C. Combination Chemotherapy

The introduction of cyclic *combination chemotherapy* for acute childhood lymphatic leukemia in the late 1950s marked the turning point in effective treatment of neoplastic disease. The improved effectiveness of combination chemotherapy compared to single agent treatment is derived from a variety of reasons: initial resistance to any single agent is frequent; initially responsive tumors rapidly acquire resistance after drug administration, probably because of selection for the preexisting resistant tumor cells in the cell population; anticancer drugs themselves increase the rate of mutation of cells into resistant forms; multiple drugs having different mechanisms of action allow independent cell killing by each agent; cells resistant to one drug may be sensitive to another; if drugs have nonoverlapping toxicities, each can be used at full dosage and the effectiveness of each drug will be maintained in combination. Also, unlike enzymes, which require gene-encoded resynthesis in order to achieve restored activity after inactivation, covalent modification of DNA can be reversed by repair enzymes.[1,2] In repair-proficient tumor cells it is possible to potentiate the cytotoxic effects of DNA-reactive drugs with a combination of alkylating drugs and inhibitors of DNA repair.[4]

### D. Drug Interactions

The most significant problem associated with the use of combination chemotherapy is *drug interactions*; overlapping toxicities are of primary concern. For example, drugs that cause renal toxicity must be used cautiously or not at all with other drugs that depend on renal elimination as their primary mechanism of excretion. The order of administration also is important. An example (unrelated to DNA-interactive drugs) is the synergistic effects that are obtained when methotrexate (an inhibitor of dihydrofolate reductase) precedes 5-fluorouracil (an inhibitor of thymidylate synthase; see Section V,C,3,e of Chapter 5),[5] probably because of increased activation of 5-fluorouracil to its nucleotide form. The opposite order of administration leads to initial inactivation of thymidylate synthase so that the intracellular stores of tetrahydrofolate are not consumed (remember, thymidylate synthase consumes tetrahydrofolate in the form of methylenetetrahydrofolate when it converts deoxyuridylate to thymidylate). That being the case, inhibition of dihydrofolate reductase, the enzyme that catalyzes the resynthesis of tetrahydrofolate, then becomes inconsequential.

### E. Drug Resistance

As indicated earlier, the prime reason for the utilization of combination chemotherapy is to avoid *drug resistance*, which generally arises because of one or more of the following reasons: selection of cells that have increased expression of membrane glycoproteins; increases in levels of cytoplasmic thiols; increases in deactivating enzymes or decreases in activating enzymes (see Chapter 8) by changes in specific gene sequences; and increases in DNA repair. All of these mechanisms of resistance involve gene alterations. *Membrane glycoproteins* (or *P-glycoproteins*) are responsible for the efflux of natural product drugs from cells and represent a type of *multidrug resistance* (MDR).[6,7] These P-glycoproteins bind and extrude drugs from tumor cells. Also, by increasing pools of cytoplasmic thiols, such as glutathione, the cell increases its ability to destroy reactive electrophilic anticancer drugs (see Chapter 7, Section IV,C,5).[8] More specifically, the gene encoding the family of glutathione *S*-transferases, which catalyze the reaction of glutathione with electrophilic compounds, may be altered so that these enzymes are overproduced (*gene amplification*). As described in more detail in Chapter 8 (Section II,B,2), many drugs that covalently bind to DNA require enzymatic activation (*prodrugs*). The gene encoding these enzymes may be altered so that certain tumor cells no longer produce sufficient quantities of the activating enzymes to allow the drugs to be effective. Finally, once the DNA has been modified, a resistant cell could produce DNA repair enzymes[1,2] able to excise the mutation in the DNA and repair the polynucleotide strands.

Specific mechanisms of resistance are discussed in the appropriate sections for the different classes of drugs that interact with DNA. Before we discuss these DNA-interactive drugs, however, we need to consider the structure and properties of DNA.

## II. DNA Structure and Properties

### A. Basis for the Structure of DNA

The elucidation of the structure of DNA by Watson and Crick[9] was the culmination and synthesis of experimental results reported by a large number of scientists over several years.[10] Todd and co-workers[11] established that the four deoxyribonucleotides, containing the two *purine bases* adenine (A) and guanine (G) and the two *pyrimidine bases* cytosine (C) and thymine (T), are linked by bonds joining the 5′-phosphate group of one nucleotide to a 3′-hydroxyl group on the sugar of the adjacent nucleotide to form 3′,5′-phosphodiester linkages (**6.1**). Chargaff and co-workers[12] showed that the ratios of A/T and G/C are always equal to 1 regardless of the base composition of the DNA.

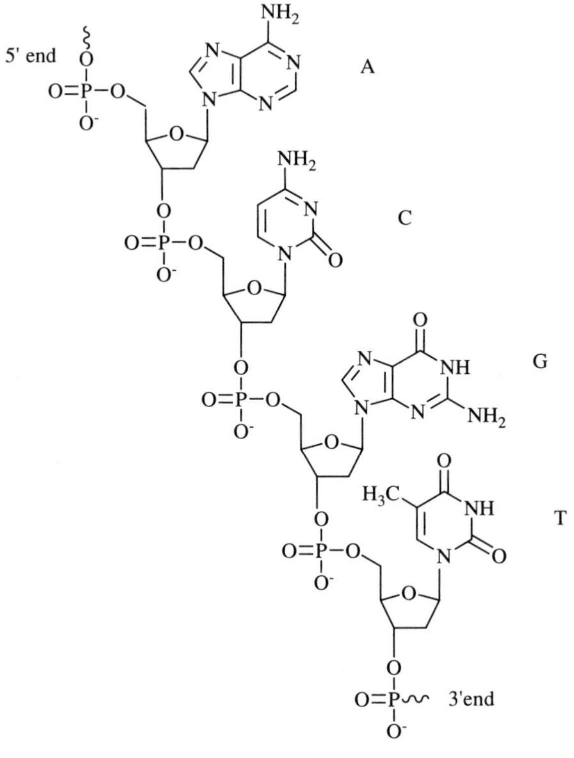

5' end
$NH_2$
A
$NH_2$
C
G
$NH_2$
$H_3C$
T
3'end

**6.1**

The first X-ray photographs of fibrous DNA exhibited a very strong meridional reflection at 3.4 Å distance, suggesting that the bases are stacked on each other.[13] On the basis of electrotitrimetric studies, Gulland[14] concluded that the nucleotide bases are linked by hydrogen bonding. X-Ray data by Wilkins[15] indicated that DNA is a helical molecule able to adopt a variety of conformations. All of these data were digested by Watson and Crick,[9] who then proposed that there are specific hydrogen-bonded base pairs of adenine with thymine (**6.2**) and guanine with cytosine (**6.3**), which explained the results of Gulland[14] and Chargaff and co-workers,[12] and that these base pairs are stacked at 3.4 Å distance according to the work of Astbury.[13] Furthermore,

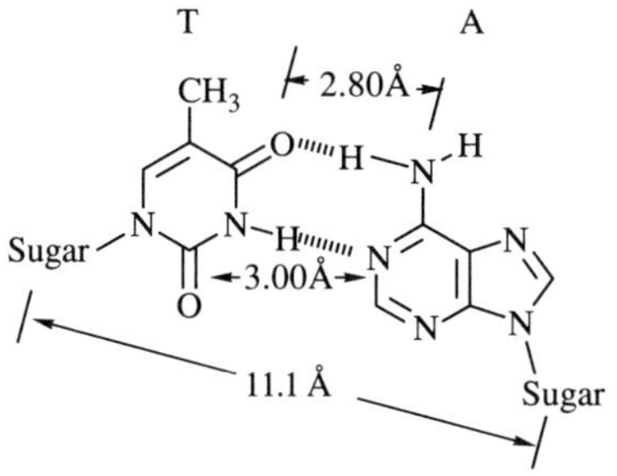

 

T A C G

2.80Å 3.00Å 11.1 Å Sugar

2.90Å 3.00Å 2.90Å 10.8 Å Sugar

**6.2** 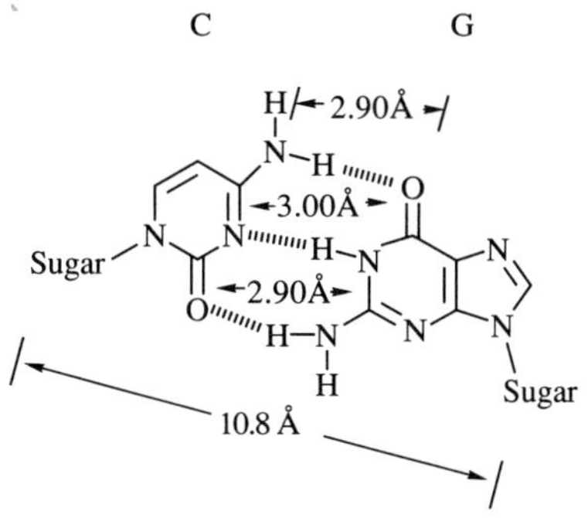 **6.3**

right-handed rotation between adjacent base pairs by about 36° produces a double helix with 10 base pairs per turn.

A model of the helix was constructed using dimensions and conformations of the individual nucleotides based on the structure of cytidine.[16] The bases are located along the axis of the helix, with sugar–phosphate backbones winding in an antiparallel orientation along the periphery (Fig. 6.1). Because the sugar and phosphate groups are always linked together by 3′,5′-phosphodiester bonds, this part of DNA is very regular; however, the order of the nucleotides along the chain varies from one DNA molecule to another. The purine and pyrimidine bases are flat and tend to stack above each other approximately perpendicular to the helical axis; this base stacking is stabilized mainly by London dispersion forces[17] and by hydrophobic effects.[18] The two chains of the double helix are held together by hydrogen bonds between the bases.

All of the bases of the DNA are on the inside of the double helix, and the sugar–phosphates are on the outside; therefore, the bases on one strand are close to those on the other. Because of this fit, specific base pairings between a large purine base (either A or G) on one chain and a smaller pyrimidine base (T or C) on the other chain are essential. Base pairing between two purines would occupy too much space to allow a regular helix, and base pairing between two pyrimidines would occupy too little space. In fact, hydrogen bonds between guanine and cytosine or adenine and thymine are more effective than any other combination. Therefore, *complementary base pairs* (also called *Watson–Crick base pairs*) form between guanines and cytosines or adenines and thymines only, resulting in a complementary relation between sequences of bases on the two polynucleotide strands of the double helix. For example, if one strand has the sequence 5′-TGCATG-3′, then the complementary strand must have the sequence 3′-ACGTAC-5′ (note that the chains are antiparallel). As one might predict, because there are three hydrogen bonds between G and C base pairs and only two hydrogen bonds between A and T base pairs, the former are more stable.

The two glycosidic bonds that connect the base pair to its sugar rings are

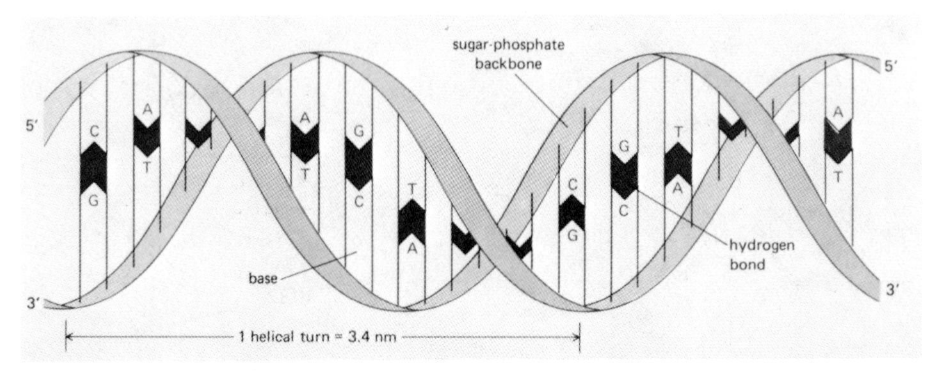


**Figure 6.1.** DNA structure. [Reproduced with permission from Alberts B., Bray, D., Lewis, J., Raff, M., Roberts, K., and Watson, J. D. (1989). Molecular Biology of the Cell, 2nd Ed., p. 99. Garland Publishing, New York. Copyright © 1989 Garland Publishing.]

not directly opposite each other, and, therefore, the two sugar–phosphate backbones of the double helix are not equally spaced along the helical axis. As a result, the grooves that are formed between the backbones are not of equal size; the larger groove is called the *major groove* and the smaller one is called the *minor groove* (Fig. 6.2). The floor of the major groove is filled with base pair nitrogen and oxygen atoms which project inward from their sugar–phosphate backbones toward the center of the DNA. The floor of the minor groove is filled with nitrogen and oxygen atoms of base pairs that project outward from their sugar–phosphate backbones toward the outer edge of the DNA.

### B. Base Tautomerization

Because of the importance of hydrogen bonding to the structure of DNA, we need to consider the tautomerism of the different heterocyclic bases (Fig. 6.3), which depends largely on the dielectric constant of the medium and on the p$K$ of the respective heteroatoms.[19] As can be seen in Fig. 6.3, a change in the tautomeric form would have disastrous consequences with regard to hydrogen bonding, as groups that are electron donors in one tautomeric form become electron acceptors in another form and protons are moved to different positions on the heterocyclic ring. The more stable tautomeric form for the bases having an amino substituent (A, C, and G) is the amino form, not the imino form. The oxygen atoms of guanine and thymine strongly prefer to be in the keto form rather than the enol form.[20] Apparently, these four bases are ideal for maximizing the population of the appropriate tautomeric forms for complementary base recognition. For example, naturally occurring isoguanosine could form a complementary base pair with synthetic isocytidine (**6.4**);

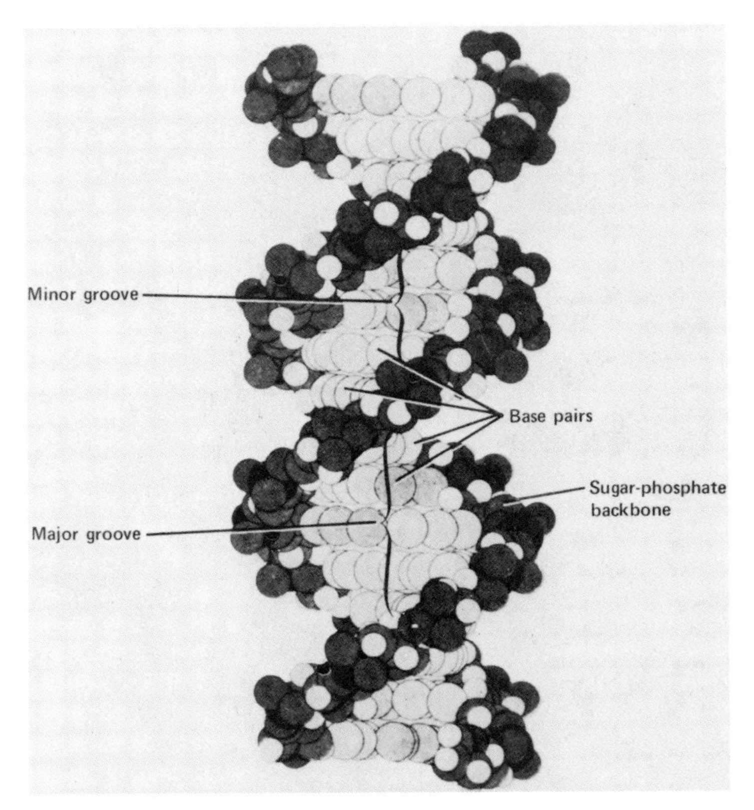


**Figure 6.2.** Major and minor grooves of DNA. (From DNA REPLICATION, by A. Kornberg. Copyright © 1980 by W. H. Freeman and Company. Reprinted with permission.)

however, isoguanosine has an unusual keto–enol tautomerism that is largely dependent on solvent polarity and on temperature.[21] Simple modifications of the bases, such as replacement of the carbonyl group in purines by a thiocarbonyl group increases the enol population to about 7%[22]; this would have a significant effect on base pairing.

**6.4**

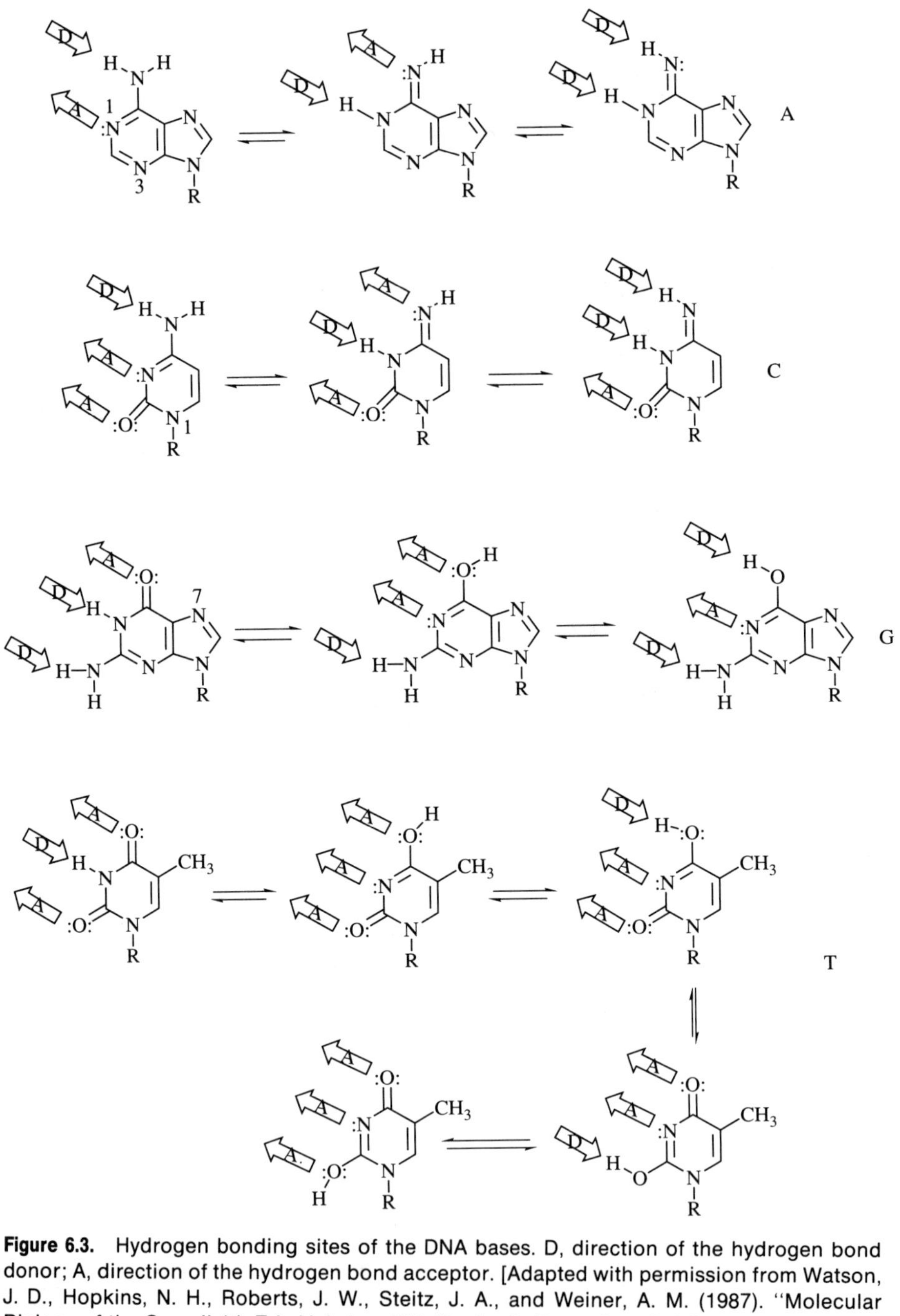

Figure 6.3. Hydrogen bonding sites of the DNA bases. D, direction of the hydrogen bond donor; A, direction of the hydrogen bond acceptor. [Adapted with permission from Watson, J. D., Hopkins, N. H., Roberts, J. W., Steitz, J. A., and Weiner, A. M. (1987). "Molecular Biology of the Gene," 4th Ed., Vol. 1, p. 243. Benjamin/Cummings Publishing, Menlo Park, California. Copyright © 1987 Benjamin/Cummings Publishing Company.]

### C. DNA Shapes

DNA exists in a variety of sizes and shapes. The length of the DNA that an organism contains varies from micrometers to several centimeters in size. In human somatic cells each of the 46 chromosomes consists of a single DNA duplex molecule about 4 cm long. If the chromosomes in each somatic cell were placed end to end, the DNA would stretch almost 2 m long! The 2-m-long DNA is packed into a nucleus that is only 0.5 μm in diameter with the help of small, richly basic proteins called *histones*, which, by electrostatic interactions between the positively charged lysine and arginine residues of the histone and the negatively charged phosphates of DNA, fold the polyanionic DNA into an ordered compact form known as *chromatin*. Some DNA exists in single-stranded or triple-stranded (*triplex*) as well as the much more common double-stranded (*duplex*) form.

Some DNA molecules are linear and others are circular. Linear DNA can freely rotate until the ends become covalently linked to form circular DNA; then, the absolute number of times the DNA chains twist about each other (called the *linkage number*) cannot change. To accommodate further changes in the number of base pairs per turn of the duplex DNA, the circular DNA must twist, as when a rubber band is twisted, into *supercoiled DNA* (Fig. 6.4A shows how double-stranded DNA is cut, twisted, and rejoined to produce supercoiling). Untwisting of the double helix prior to rejoining the ends in circular DNA usually leads to *negative supercoiling* (left-handed direction); overtwisting results in *positive supercoiling* (right-handed direction).

Virtually all duplex DNA within cells exists as chromatin in the negative supercoiled state, which is the direction opposite that of the twist of the double helix (see Section II,D). Because supercoiled DNA is a higher energy state than uncoiled DNA, the cutting (called *nicking*) of one of the DNA strands of supercoiled DNA converts it to *relaxed DNA*. The enzymes that interconvert supercoiled DNA forms and relaxed DNA forms are called *DNA topoisomerases*. These nuclear enzymes catalyze the conversion of one topological isomer of DNA to another and function to resolve topological problems in DNA such as overwinding and underwinding, catenation and decantenation (Fig. 6.4B) and knotting and unknotting (Figure 6.4B and 6.4C), which normally arise during replication, transcription, recombination, and other DNA processes.[23]

There are two principal topoisomerases that regulate the state of supercoiling of intracellular DNA. One, known as *DNA topoisomerase I*, catalyzes a transient break of one strand of duplex DNA and allows the unbroken, complementary strand to pass through the enzyme-linked strand, thereby resulting in DNA relaxation by one positive turn. The other is called *DNA topoisomerase II* (or, in the case of the bacterial enzyme, *DNA gyrase*), and it catalyzes the cleavage of both strands of the duplex DNA, with a four base

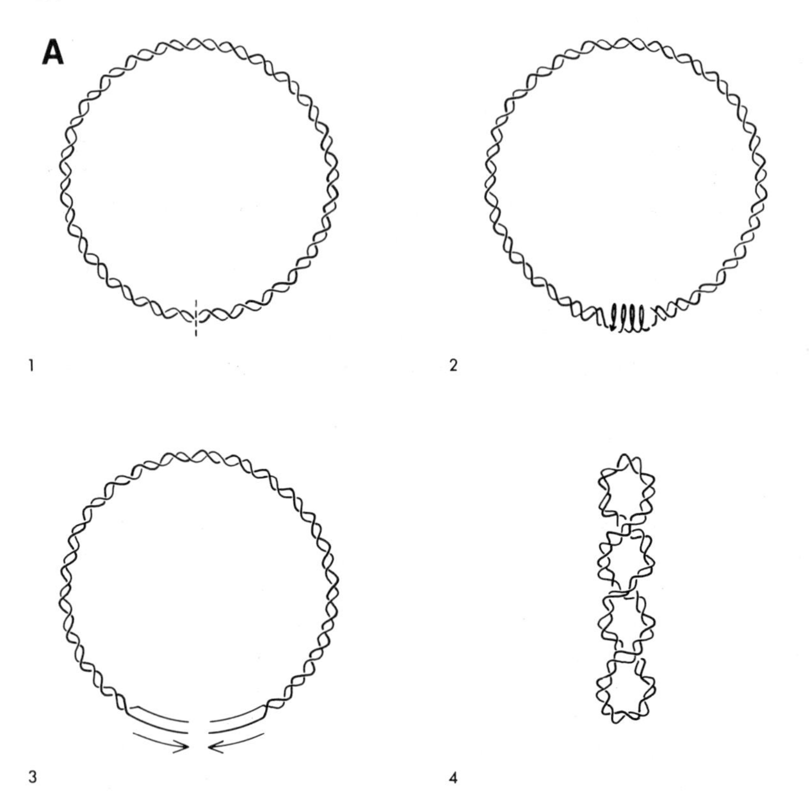


**Figure 6.4A.** Conversion of Double-Stranded DNA to Supercoiled DNA [Reproduced with permission from Watson, J. D., Hopkins, N. H., Roberts, J. W., Steitz, J. A., and Weiner, A. M. (1987). "Molecular Biology of the Gene," 4th ed., Vol. I, p. 258. Benjamin/Cummings Publishing, Menlo Park, CA. Copyright © 1987 Benjamin/Cummings Publishing Company.]

pair stagger between the nicks, then allows another DNA duplex to pass through the break prior to resealing the strands. This results in the supercoiling of the DNA in the negative direction or relaxation of positively supercoiled DNA, and it changes the linkage number by −2. DNA topoisomerase I has no requirement for an energy cofactor to complete the rejoining process, but DNA topoisomerase II requires ATP and Mg(II) for the "strand passing" activity.[24]

The mechanisms for DNA strand cleavage by DNA topoisomerase I and II are quite similar (Scheme 6.1). A tyrosyl group on the enzyme attacks the phosphodiester bond. In the case of DNA topoisomerase I, a covalent bond to the 3′-phosphoryl end (pathway a) results[25]; with DNA topoisomerase II a 5′-phosphotyrosine bond (pathway b) forms.[26]

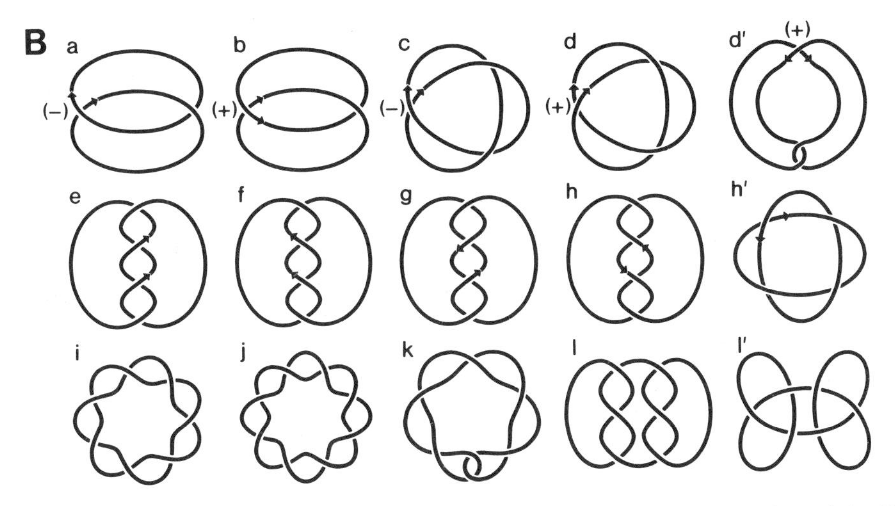


**Figure 6.4B.** Catenane and Knot Catalogue. Arrows indicate the orientation of the DNA primary sequence. a and b, singly-linked catenanes; c and d, simplest knot, the trefoil; e–h, multiply interwound torus catenanes; i, right-handed torus knot with 7 nodes; j, right-handed torus catenane with 8 nodes; k, right-handed twist knot with 7 nodes; l, 6-noded knot composed of 2 trefoils. [Reproduced with permission from Wasserman, S. A., and Cozzarelli, N. R., (1986). *Science* **232,** 951. Copyright © 1986 by the AAAS.]

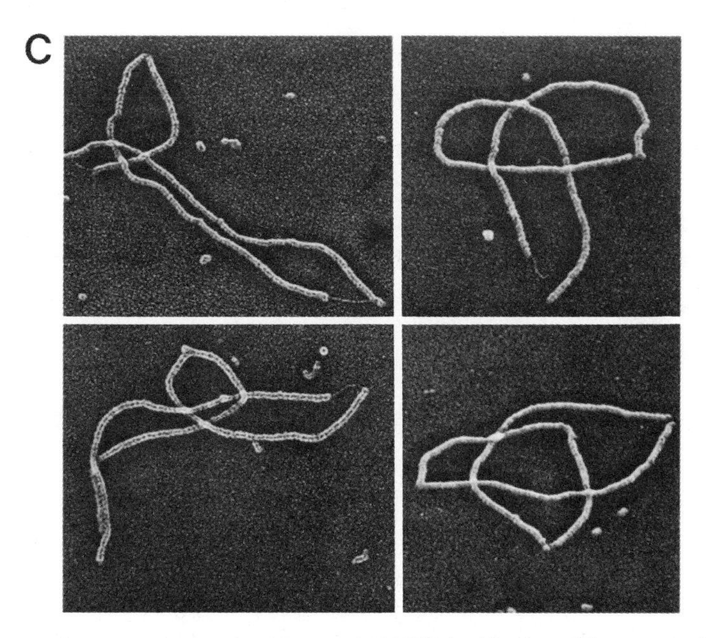


**Figure 6.4C.** Visualization of Trefoil DNA by Electron Microscopy. [Reproduced with permission from Griffith, J. D., and Nash, H. A. (1985). *Proc. Natl. Acad. Sci. U.S.A.* **82,** 3124.]

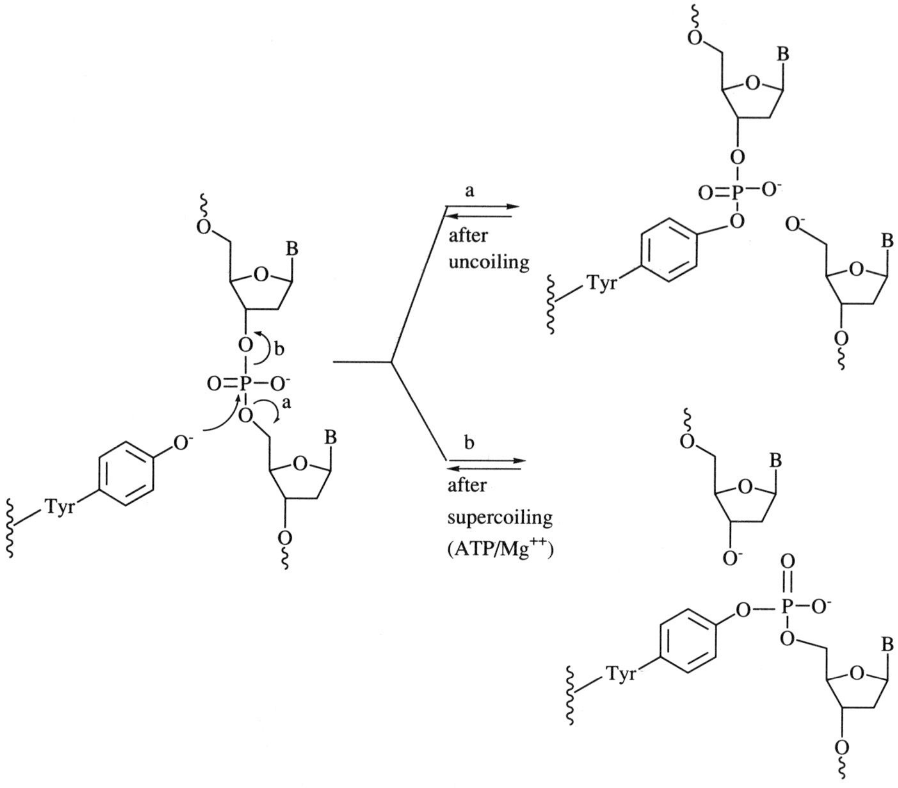

**Scheme 6.1.** DNA topoisomerase–catalyzed strand cleavage.

Along the DNA strands there can be bends, kinks, loops, or crosslike projections called *cruciforms*. There are a wide range of structural variations, including DNA that has a left-handed turn instead of the usual right-handed turn (Section II,D).

### D. DNA Conformations

There are three general helical conformations of DNA; two are right-handed DNA (*A-DNA* and *B-DNA*), and one is a left-handed DNA (*Z-DNA*). Each conformation involves a helix made up of two antiparallel polynucleotide strands with the bases paired through Watson–Crick hydrogen bonding, but the overall shapes of the helices are quite different (Figs. 6.5–6.7).

The right-handed forms differ in the distance required to make a complete

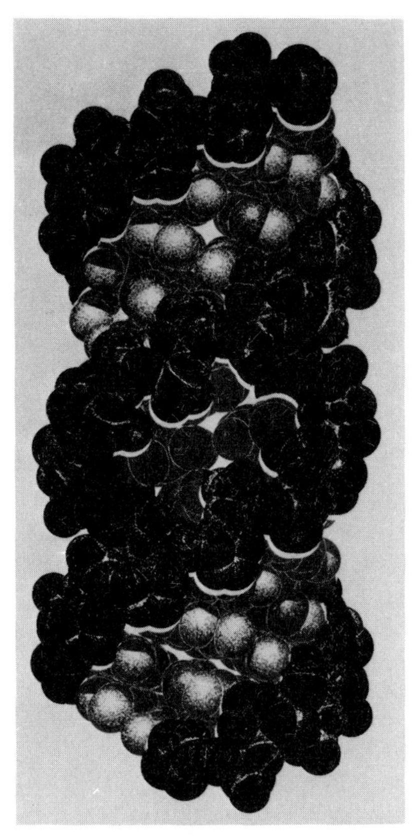

**Figure 6.5.** A-DNA. (Reprinted with permission from Saenger, W. 1984. "Principles of Nucleic Acid Structure," p. 257. Springer-Verlag, New York. Copyright © Springer-Verlag, Inc.)

helical turn (called the *pitch*), differ in the way the sugar groups are bent or puckered, differ in the angle of tilt that the base pairs make with the helical axis, and differ in the dimensions of the grooves. The predominant form is B-DNA, but in environments with low hydration A-DNA occurs. In contrast to B-DNA, individual residues in A-DNA display uniform structural features; nucleotides in A-DNA have more narrowly confined conformations. Whereas there are 11 nucleotides in one helical turn in A-DNA, there are only 10 base pairs per pitch in B-DNA. Therefore, A-DNA is shorter and squatter than B-DNA.

Z-DNA, a minor component of the DNA of a cell, is a left-handed double helix having 12 base pairs per helical turn. It has only a minor groove because the major groove is filled with cytosine C-5 and guanine N-7 and C-8 atoms. In A- and B-DNA the glycosyl bond is always oriented anti (**6.5**). In Z-DNA the

**Figure 6.6.** B-DNA. (Reprinted with permission from Saenger, W. 1984. "Principles of Nucleic Acid Structure," p. 262. Springer-Verlag, New York. Copyright © 1984 Springer-Verlag, Inc.)

glycosyl bond connecting the base to the deoxyribose group is oriented anti at the pyrimidine residues but syn at the purine residues (**6.6**). This alternating anti–syn configuration gives the backbone (a line connecting phosphorus atoms) an overall zig-zag appearance, hence, Z-DNA.

**6.5** **6.6**

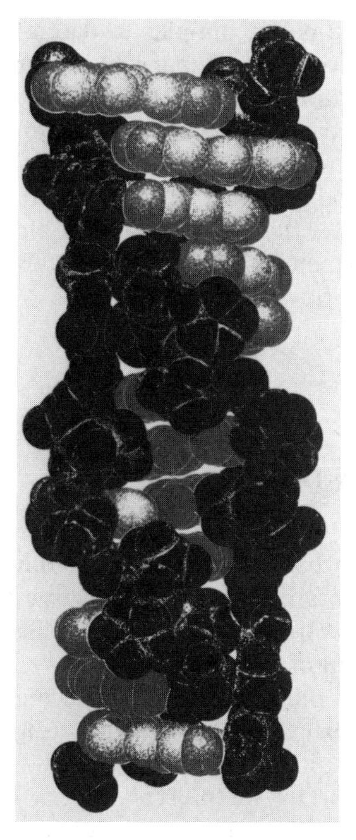

**Figure 6.7.** Z-DNA. (Reprinted with permission from Saenger, W. 1984. "Principles of Nucleic Acid Structure," p. 286. Springer-Verlag, New York. Copyright © 1984 Springer-Verlag, Inc.)

With this brief introduction to the structure of DNA we can now explore the different mechanisms by which drugs interact with DNA.

## III. Classes of Drugs That Interact with DNA

In general there are three major classes of clinically important DNA-interactive drugs: the *intercalators*, which insert between the base pairs of the double helix, thereby unwinding it; the *alkylators*, which react covalently with DNA bases; and the DNA *strand breakers*, which generate reactive radicals that produce cleavage of the polynucleotide strands. The ideal DNA-interactive drug would be a nonpeptide molecule (so that it can diffuse through membranes and not be degraded by peptidases) that is targeted for a specific sequence and site size[27]; however, proteins are the only examples to date of drugs with unambiguous DNA sequence recognition. The primary sequence recognition by proteins results from complementary hydrogen bonding between amino acid residues on the protein and nucleic acid bases in the major and minor grooves of DNA.[28] Proteins generally use major groove interactions with B-DNA because there are more donor and acceptor sites for hydrogen bonding than in the minor groove. Consequently, the major groove would be a preferred target for the design of primary sequence-specific DNA-interactive drugs; however, it is not yet understood how proteins are able to recognize specific sequences.

## A. DNA Intercalators

### 1. Intercalation and Topoisomerase-Induced DNA Damage

Flat, generally aromatic or heteroaromatic molecules bind to DNA by inserting (i.e., intercalating) and stacking between the base pairs of the double helix. The principal driving forces for intercalation are stacking and charge-transfer interactions, but hydrogen bonding and electrostatic forces also play a role in stabilization.[29] *Intercalation*, first described in 1961 by Lerman,[30] is a noncovalent interaction in which the drug is held rigidly perpendicular to the helix axis. This causes the base pairs to separate vertically, thereby distorting the sugar–phosphate backbone and decreasing the pitch of the helix (Fig. 6.8 shows the intercalation of ethidium bromide into B-DNA). Intercalation, apparently, is an energetically favorable process, because it occurs so readily. Presumably, the van der Waals forces that hold the intercalated molecules to the base pairs are stronger than those found between the stacked base pairs. Much of the binding energy is the result of the removal of the drug molecule from the aqueous medium and a hydrophobic effect. Intercalation occurs preferentially (by 7–13 kcal/mol) into pyrimidine-3′,5′-purine sequences rather than into purine-3′,5′-pyrimidine sequences.[31]

In general, intercalation does not disrupt the Watson–Crick hydrogen bonding. It does, however, destroy the regular helical structure, unwinds the DNA at the site of binding, and, as a result of this, interferes with the action of DNA-binding enzymes such as DNA topoisomerases, which alter the degree of supercoiling of DNA, and DNA polymerases, which catalyze the elongation of the DNA chain in the 5′ to 3′ direction and also correct mistakes in the DNA by clipping out (via hydrolysis of the phosphodiester bond) mismatched residues at the terminus.

Although the drugs in this section are categorized as being intercalators, it is now believed that intercalation of a drug into DNA is only the first step in the events that eventually lead to DNA damage by other mechanisms. For

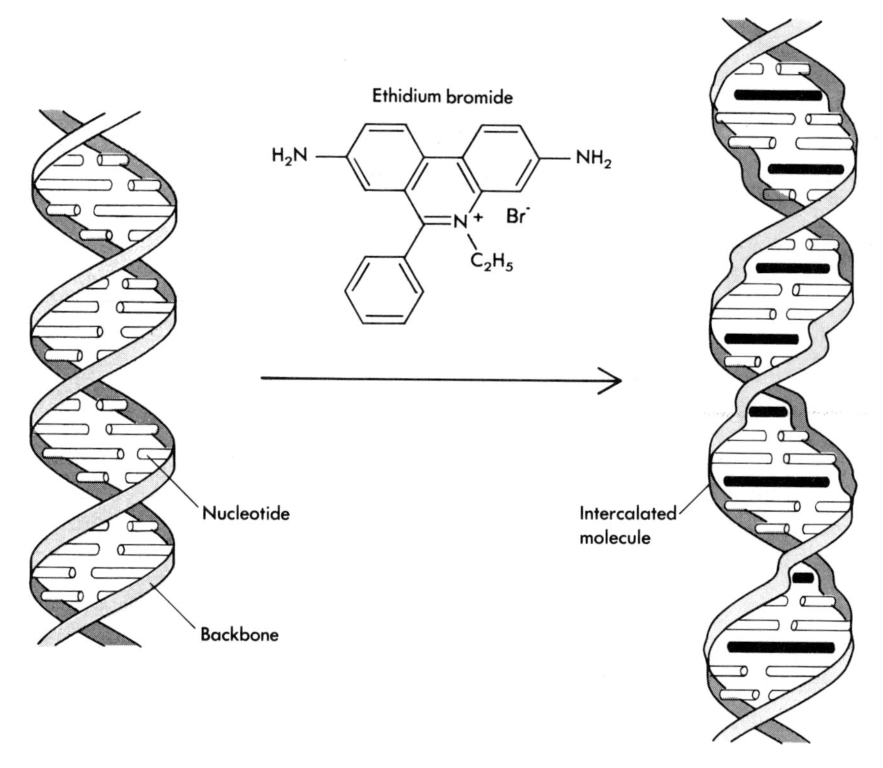


**Figure 6.8.** Intercalation of ethidium bromide into B-DNA. (Adapted with permission from Watson, J. D., Hopkins, N. H., Roberts, J. W., Steitz, J. A., and Weiner, A. M. 1987. "Molecular Biology of the Gene," 4th Ed., Vol. 1, p. 255. Benjamin/Cummings Publ., Menlo Park, California. Copyright © 1987 Benjamin/Cummings Publishing Company.)

many classes of antitumor agents, there is an involvement of the DNA topoisomerases (see Section II,C) subsequent to intercalation. Mammalian DNA topoisomerase I is the target for the antitumor agent camptothecin (**6.6a**),[32] whereas DNA topoisomerase II is the target for a variety of classes of

**6.6a** **6.6b**

antitumor drugs, such as anthracyclines, anthracenediones, acridines, actinomycins, and ellipticines.[23] The quinolone antibacterial drugs, such as nalidixic acid (**6.6b**), act on bacterial DNA topoisomerase II. The evidence to date suggests that drugs which cause topoisomerase-induced DNA damage interfere with the breakage–rejoining reaction (see Scheme 6.1) by trapping a key covalent reaction intermediate (possibly the tyrosine adducts shown in Scheme 6.1), termed the *cleavable complex*.[33] The cleavable complex may be stabilized by the formation of a reversible nonproductive (noncleavable) drug–DNA–topoisomerase ternary complex. It has been hypothesized that this ternary complex may collide with transcription and replication complexes; on collision, the ternary complex may lose its reversibility and generate lethal double-strand DNA breaks.[34] It is not clear if the drug binds to DNA first, then topoisomerase II forms the ternary complex, or if the drug binds to a topoisomerase II–DNA complex.

A study of a series of anthracycline analogs showed that DNA intercalation of these compounds is required, but not sufficient, for topoisomerase II-targeted activity.[35] There was a strong correlation between the potency of intercalation and cleavable complex formation. However, there does not appear to be a correlation between DNA intercalation and antitumor activity. Some strong intercalators do not induce cleavable complexes, possibly because of certain structural requirements for the binding of the intercalated drug to the topoisomerase. Also, epipodophyllotoxins, such as the anticancer drug etoposide (**6.6c**), are nonintercalating DNA topoisomerase II poisons.

Me O O O HO OH O O O O O O MeO OMe OH

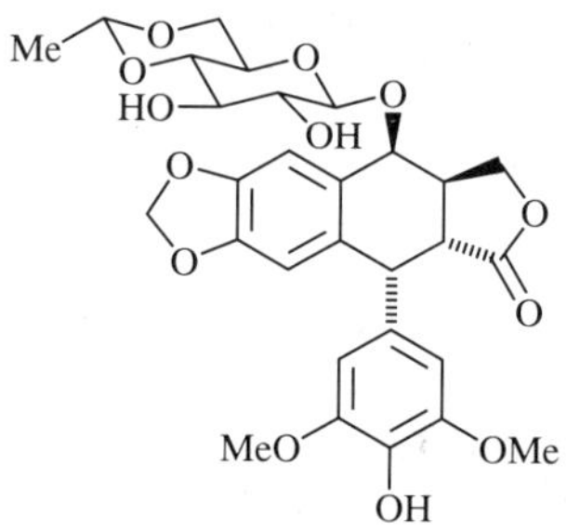

**6.6c**

Although the mechanism of topoisomerase II-induced DNA damage is not clear, the ternary complex appears to be lethal to proliferating cells. Selective sensitivity of proliferating tumor cells to the cytotoxic effects of DNA topoisomerase II poisons may be the result of the high levels of DNA topoisomerase II found in proliferating cells and the very low levels found in quiescent cells.

Intercalation may not be the direct cause for DNA damage, but it does produce a conformational change (unwinding) in the double helix. This, then, can result in the positioning of the drug in the DNA appropriately for binding with the topoisomerase in the ternary complex,[34] or it can position the drug for subsequent reactions, as discussed in Sections III,B and III,C.

### 2. Selected Examples of DNA Intercalators

Three classes of drug molecules that have been well characterized as intercalators of DNA are the acridines (**6.7**), the actinomycins (**6.8**), and the anthracyclines (**6.9**).

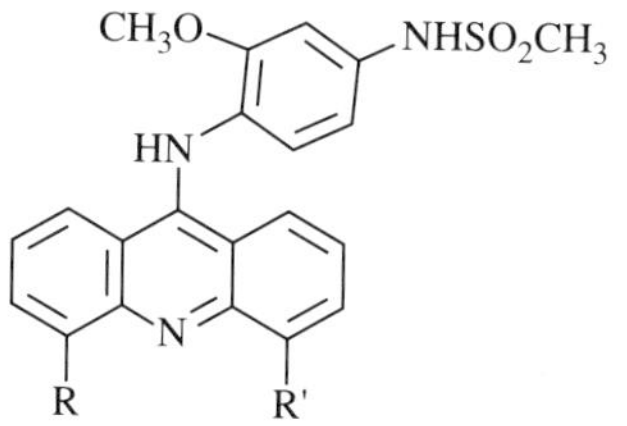

**6.7** 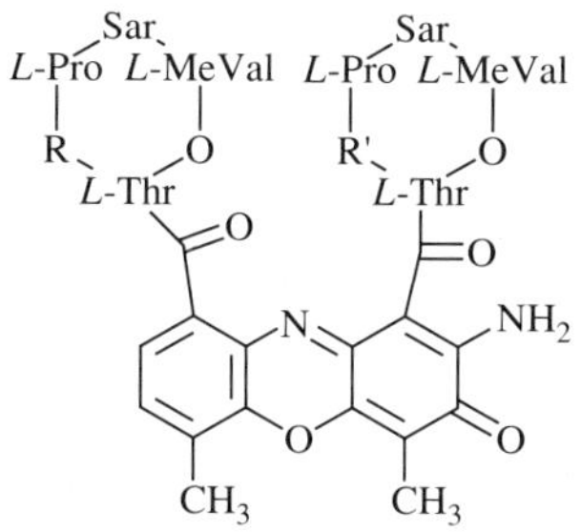 **6.8**

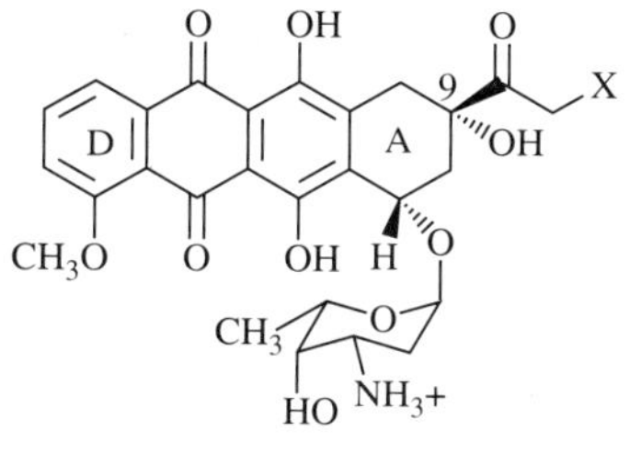

**6.9**

***a. Amsacrine, an Acridine Analog.*** Acridine compounds, which were byproducts of aniline dye manufacture, were first used in clinical medicine in the late nineteenth century against malaria.[36] By the First World War acridine derivatives such as proflavine (**6.10**) were in widespread use as local antibacterial agents. After the Second World War another acridine derivative, aminacrine (**6.11**), was the principal acridine antibacterial agent used.[37] In the 1960s and early 1970s a variety of anilino-substituted analogs of 9-anilinoacridine (**6.12**) were prepared and tested for antitumor activity[38] on the basis of the reported antitumor activity of **6.12** (R = H, R′ = $Me_2N$).[39] Although the 3,4-diamino analog had good antitumor activity, it was unstable to air oxidation. On the basis of structure–activity relationships it was reasoned that an electron donor group was needed on the anilino ring; consequently, a sulfonamide group, which would be partially anionic at physiological pH, was selected. In fact, **6.12** (R = H, R′ = $NHSO_2Me$) was equally as active as the 3,4-diamino analog.[40] This was used as a lead compound, and it was found that the most active analogs had other electron-donating substituents in addition to the sulfonamide group. Amsacrine (**6.12**, R = OMe, R′ = $NHSO_2Me$) was the most potent of those tested[41]; its main use is now in the treatment of leukemia.[42]

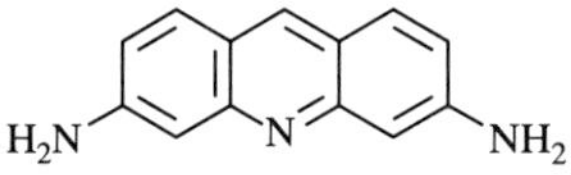

$H_2N$ N $NH_2$ **6.10**

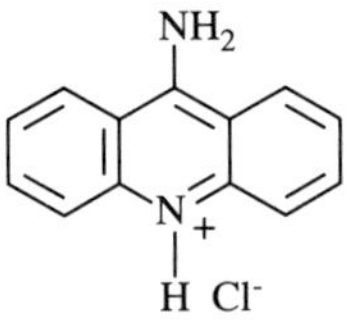

$NH_2$ $N^+$ H $Cl^-$ **6.11**

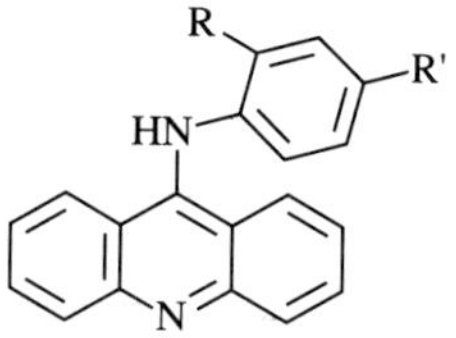

R R′ HN N **6.12**

Paradoxically, although a very wide variation of structures of 9-anilinoacridines can be tolerated with retention of antitumor activity, among the active derivatives large differences in potency are observed with small changes in structure.[43] The antitumor activity was parabolically related to drug lipophilicity as measured by log $P$ values (see Section II,E,2,b of Chapter 2); compounds with log $P$ values close to that of amsacrine were most active. There also is a close correlation between the electronic properties ($\sigma$ constant; See Section II,E,2,a of Chapter 2) of groups at the para position of the anilino ring and acridine p$K_a$ values.[41] Furthermore, when lipophilic and electronic effects of a series of bulky substituents at various positions on the 9-anilinoacridine framework are taken into account, the steric effects of the group play a dominant role.[41]

These results are consistent with the mode of action of 9-anilinoacridines as intercalators of double-stranded (duplex) DNA.[44] Earlier studies showed that these compounds unwound closed circular duplex DNA.[45] By analogy with the crystal structure of 9-aminoacridine bound to a dinucleotide,[46] Denny *et al.*[47] hypothesized that the anilino ring, which lies almost at right angles to the plane of the acridine chromophore, is lodged in the minor groove with the 1′-substituent (the sulfonamide group) pointing tangentially away from the helix. The sulfonamide may interact with a second macromolecule, such as a regulatory protein, and this ternary complex could mediate the biological effects of the 9-anilinoacridines. More recent studies of the rates of dissociation of amsacrine from DNA suggest that the anilino group may bind in the major groove.[48]

Amsacrine lacks broad-spectrum clinical activity and is difficult to formulate because of its low aqueous solubility. It was thought that the relatively high $pK_a$ of the compound (8.02 for the acridine nitrogen) was important in limiting *in vivo* distribution; consequently, analogs with improved solubility and high DNA binding, but with a lower $pK_a$ value were sought. A compound was found (**6.7**, R = Me, R′ = CONHMe) that had all of the desirable physicochemical properties, showed superior antileukemic activity compared with amsacrine, and was broader in its spectrum of action.[49]

***b. Dactinomycin, the Parent Actinomycin Analog.*** Actinomycin D (now called dactinomycin; **6.8**, R = R′ = D-Val) was the first of a family of chromopeptide antibiotics isolated from a culture of a *Streptomyces* strain in 1940.[50] In 1952 these compounds were found to have antitumor activity and were used clinically.[51] Dactinomycin binds to double-stranded DNA and, depending on its concentration, inhibits DNA-directed RNA synthesis or DNA synthesis. RNA chain initiation is not prevented, but chain elongation is blocked.[52] The phenoxazone chromophore intercalates between bases in the DNA.[53] Binding depends on the presence of guanine; the 2-amino group of guanine is important for the formation of a stable drug–DNA complex.[54]

X-Ray crystal structures of a 1:2 complex of dactinomycin with deoxyguanosine,[55] deoxyguanylyl-3′,5′-deoxycytidine (Fig. 6.9),[55,56] and a complex

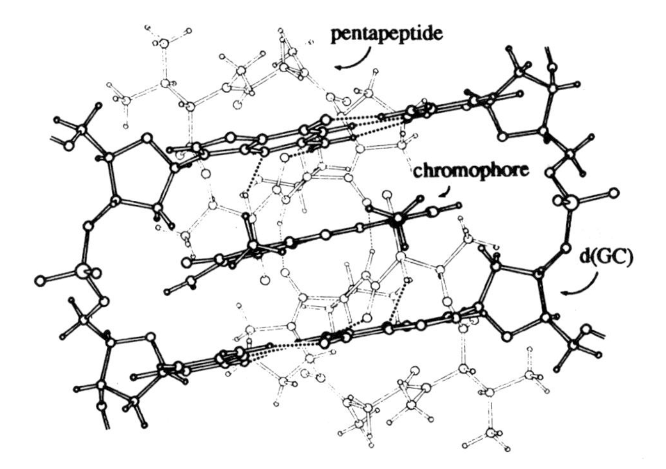


**Figure 6.9.** X-Ray structure of a 1 : 2 complex of dactinomycin with d(GC). (Reprinted with permission of Academic Press from Sobell, H. M., and Jain, S. C. 1972. *J. Mol. Biol.* **68**, 21. Copyright © 1972 Academic Press.)

of dactinomycin with d(ATGCAT)[55,57] are models for the intercalation of dactinomycin into DNA. These structures suggest that the phenoxazone ring can intercalate between deoxyguanosines and that the cyclic peptide substituents can be involved in strong hydrogen bonding and hydrophobic interactions with DNA.[58] In particular, there are two crucial hydrogen bonds that stabilize the DNA-binding complex.[55] One strong hydrogen bond exists between neighboring cyclic pentapeptide chains connecting the N–H of one D-valine residue with the C=O of the other D-valine residue. Another strong hydrogen bond connects the guanine 2-amino group with the carbonyl oxygen of the L-threonine residue. A weaker hydrogen bond connects the guanine N-3 ring nitrogen with the NH group on this same L-threonine residue. Stacking forces are primarily responsible for the recognition and preferential binding of a guanine base to dactinomycin.[59]

The biological activity appears to depend on the very slow rate of DNA–dactinomycin dissociation, which reflects the intermolecular hydrogen bonds, the planar interactions between the purine rings and the chromophore, and the numerous van der Waals interactions between the polypeptide side chains and the DNA. The peptide substituents, which lie in the minor groove, may block the progression of the RNA polymerase along the DNA. Resistance to dactinomycin is associated with an overly active efflux pump, mediated by overexpression of the P170 membrane glycoprotein, and impaired drug uptake.[60] The cell membrane composition may affect the rate of drug diffusion.[61] There also appears to be a correlation between the ability of the cell to retain dactinomycin and the effectiveness of the drug.[62]

***c. Doxorubicin (Adriamycin) and Daunorubicin (Daunomycin), Anthracycline Antitumor Antibiotics.*** The anthracycline class of antitumor antibiotics exemplified by doxorubicin (**6.9**, X = OH; previously called adriamycin) and daunorubicin (**6.9**, X = H; also called daunomycin) are isolated from different species of *Streptomyces*. Although these two compounds differ by only one hydroxyl group, there is a major difference in their antitumor activity. Whereas daunorubicin is active only against leukemia, doxorubicin is active against leukemia as well as a broad spectrum of solid tumors.

There is some controversy as to whether the mechanism of action of these compounds is related to their ability to intercalate into the DNA or to cause DNA strand breakage. The vast majority of the intracellular drug is found in the nucleus, where it intercalates into the DNA double helix (and forms the ternary complex with DNA topoisomerase II; see Section III,A,1), with consequent inhibition of replication and transcription. X-Ray[63] and NMR[64] studies of model daunorubicin–oligonucleotide complexes show that the oligonucleotides form a six-base pair right-handed double helix with two daunorubicin molecules intercalated in the d(CpG) sequences. The tetracyclic

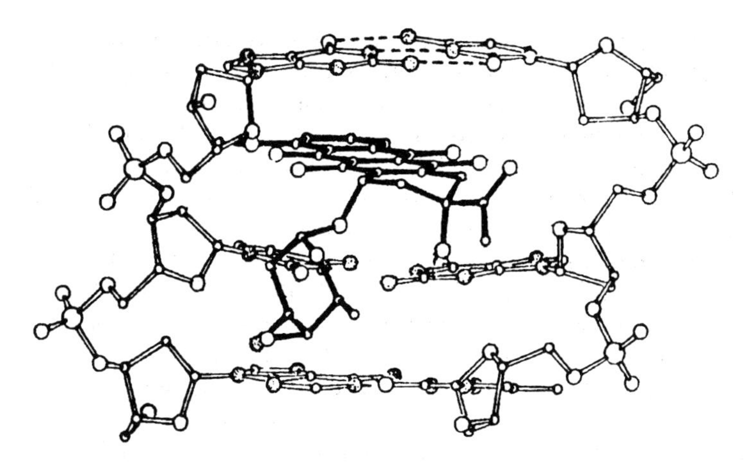

**Figure 6.10.** X-Ray structure of daunorubicin intercalated into an oligonucleotide. [Reprinted with permission from Myers, C. E., Jr., and Chabner, B. A. 1990. *In* "Cancer Chemotherapy: Principles and Practice" (Chabner, B. A., and Collins, J. M., eds.), p. 356. Lippincott, Philadelphia, Pennsylvania. Copyright © 1990 Lippincott.]

chromophore is oriented orthogonally to the long dimension of the DNA base pairs, and the ring that has the amino sugar substituent (A ring) protrudes into the minor groove (Fig. 6.10). Substituents on the A ring hydrogen bond to base pairs above and below the intercalation site. The amino sugar nearly covers the minor groove, but without bonding to the DNA. Ring D protrudes into the major groove. The complex is stabilized by stacking energies and by hydrogen bonding of the hydroxyl and carbonyl groups at C-9 of the A ring.

Earlier X-ray studies of the DNA–daunorubicin complex[65] indicated that the ionized amino group was close to the deoxyribose phosphate chain, suggesting that a strong electrostatic interaction could take place between the drug and the negatively charged DNA phosphate away from the intercalation site. However, later X-ray diffraction analysis[63] indicated that there is no interaction at all between the amino group of the drug and any part of the double helix; it sits in the center of the minor groove. This is consistent with structure–activity studies[66] which indicate that modification of the amino group does not necessarily affect biological activity.

The same mechanisms of drug resistance that were found for dactinomycin (Section III,A,2,b) apply to these drugs as well. In addition to intercalation, and topoisomerase II-induced DNA damage, another mechanism of action of the anthracycline antitumor antibiotics involves radical-induced DNA strand breakage; this mechanism is discussed in Section III,C,1.

***d. Bisintercalating Agents.*** Once success with intercalating agents was realized, it was thought that *bifunctional intercalating agents* (also called *bisintercalating agents*), in which there are two potential intercalating molecules tethered together so that each could intercalate into different DNA strands, would have enhanced affinity for DNA and slower dissociation rates. In general the DNA affinity of bisintercalating agents is greater than that of their monointercalating counterparts, in some cases approaching values typical of those observed with natural repressor proteins. However, this affinity is often the result of the polycationic nature of the high-affinity bisintercalators and not of the intercalating group. For example the quinoxaline antibiotics, which have two potential intercalator molecules tethered to a cyclic peptide, fail to bind to DNA without the cyclic peptide. The rigidity and length of the linker chain between the two intercalator molecules are important to constrain the bisintercalator in the ideal configuration to form a sandwich with two base pairs.[67] The synthetic diacridines with flexible linker chains are ineffective. For the most part, however, much of the expected enhancement in free energy of bisintercalation is not observed.[68] This is probably the result of the unfavorable entropy associated with loss of rotational freedom. In general these agents do not show a remarkable improvement in specificity compared to that found for the monofunctional intercalators.

## B. DNA Alkylators

The difference between the DNA alkylators and the DNA intercalators is akin to the difference between irreversible and reversible enzyme inhibitors (see Chapter 5). The intercalators (reversible enzyme inhibitors) bind to the DNA (enzyme) with noncovalent interactions. The DNA alkylators (irreversible inhibitors) react with the DNA (enzyme) to form covalent bonds. Five of the more important classes of alkylating agents utilized in cancer chemotherapy are nitrogen mustards, ethylenimines, methanesulfonic acid esters, nitrosoureas, and triazenes. The triazenes are mentioned in Section II,B,2,a of Chapter 8 on prodrugs, and the others are discussed here.

### 1. Nitrogen Mustards

***a. Lead Discovery.*** Sulfur mustard (**6.13**) is a highly toxic nerve gas that was used in World Wars I and II. Autopsies of soldiers killed in World War I by sulfur mustard revealed leukopenia (low white blood cell count), bone marrow aplasia (defective development), dissolution of lymphoid tissues, and ulceration of the gastrointestinal tract.[69] All of these lesions indicate that sulfur mustard has a profound effect on rapidly dividing cells and suggest that related compounds may be effective as antitumor agents. In fact, in 1931 sulfur mustard was injected directly into tumors in humans,[70] but this procedure turned out to be too toxic for systemic use. Because of the potential antitumor effects of sulfur mustards, a less toxic form was sought. Gilman and others examined the antitumor effects of nitrogen mustards (**6.14**), less reactive alkylating agents, and in 1942 the first clinical trials of a nitrogen mustard were initiated. However, this research was classified during World War II, so the usefulness of nitrogen mustard in the treatment of cancer was not known until 1946, when these studies became declassified and Gilman published a review of his findings.[71] Soon thereafter several other summaries of clinical research carried out during the war appeared.[72] This work marks the beginning of modern cancer chemotherapy.

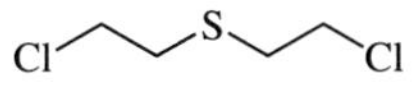

Cl–CH2CH2–S–CH2CH2–Cl        Cl–CH2CH2–N(R)–CH2CH2–Cl

**6.13** 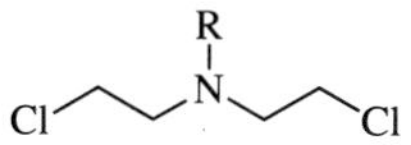        **6.14**

Prior to a discussion of lead modification and other classes of alkylating agents, let us take a brief excursion into the chemistry of alkylating agents in general.

***b. Chemistry of Alkylating Agents.*** According to Ross[73] a biological alkylating agent is a compound that can replace a hydrogen atom with an alkyl group under physiological conditions (pH 7.4, 37° C, aqueous solution). These alkylation reactions are generally described in terms of substitution reactions by N, O, and S heteroatomic nucleophiles with the electrophilic alkylating agent, although Michael addition reactions also may be important. The two most common types of nucleophilic substitution reactions are $S_N1$ (Scheme 6.2A), a stepwise reaction via an intermediate carbenium ion, and $S_N2$ (Scheme 6.2B), a concerted reaction. In general, the relative rates of nucleophilic substitution at physiological pH is in the order thiolate > amino > phosphate > carboxylate.[74] For DNA the most reactive nucleophilic sites are N-7 of guanine > N-3 of adenine > N-1 of adenine > N-1 of cytosine[75] [see

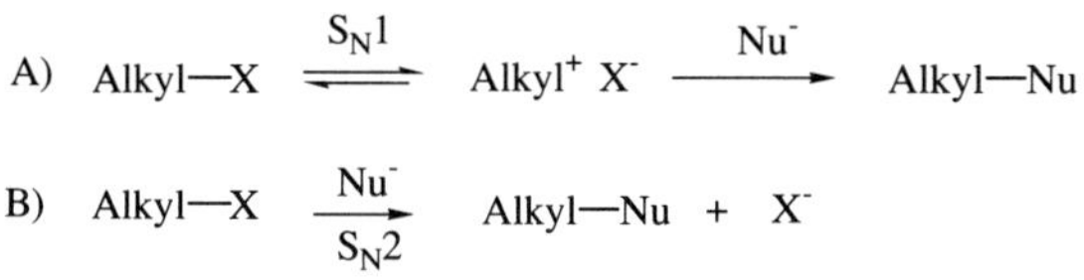

$$\text{A)}\quad \text{Alkyl—X} \underset{}{\overset{S_N1}{\rightleftharpoons}} \text{Alkyl}^+\ X^- \xrightarrow{Nu^-} \text{Alkyl—Nu}$$

$$\text{B)}\quad \text{Alkyl—X} \xrightarrow[S_N2]{Nu^-} \text{Alkyl—Nu} + X^-$$

**Scheme 6.2.** Nucleophilic substitution.

**6.15** and **6.16** for the numbering system of purines (A and G) and pyrimidines (C and T), respectively]. The N-3 of cytosine, the O-6 of guanine, and the phosphate groups also can be alkylated. Quantum mechanical calculations[76] confirm that the N-7 position of guanine is the most nucleophilic site. The reactivity of various nucleophilic sites on DNA is strongly controlled by steric, electronic, and hydrogen-bonding effects. For example, some of the nucleophilic sites are in the interior of the DNA double helix and are sterically blocked. Also, only nucleophilic centers in the major and minor grooves or in the walls of the double helix are accessible to alkylating agents. In addition to these steric effects, the nucleophilicity of various sites on the purine and pyrimidine bases of the DNA is diminished because of their involvement in Watson–Crick hydrogen bonding.

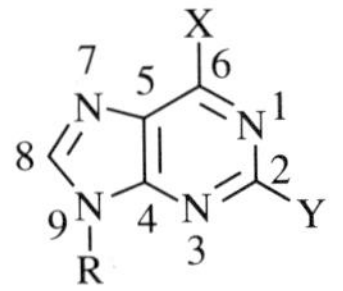

**6.15** 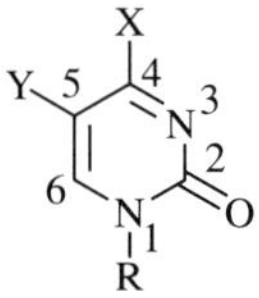 **6.16**

The reaction order for nucleophilic substitution depends on the chemical structure of the alkylating agent. Simple alkylating agents such as ethylenimines (see Section III,B,2) and methanesulfonates (see Section III,B,3) undergo $S_N2$ reactions. Alkylating agents such as nitrogen mustards, which have a nucleophile capable of *anchimeric assistance* (neighboring group participation), can undergo $S_N1$- or $S_N2$-type reactions, depending on the relative rates of the aziridinium ion formation and the nucleophilic attack on the aziridinium ion (Scheme 6.3).[77] When aziridinium ion formation is fast, the overall reac-

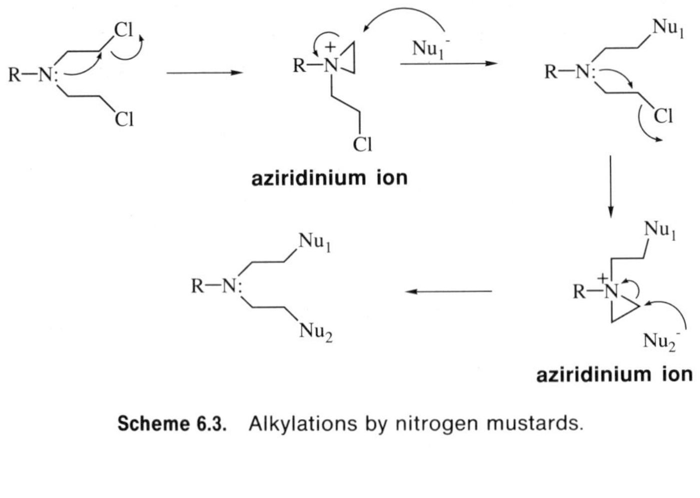

aziridinium ion

aziridinium ion

**Scheme 6.3.** Alkylations by nitrogen mustards.

tion rate is second order ($S_N2$), but when aziridinium ion formation is slower than nucleophilic attack, the overall reaction is first order ($S_N1$).

In the case of the nitrogen mustards, which are bifunctional alkylating agents (i.e., they have two electrophilic sites), the DNA undergoes intrastrand and interstrand *cross-linking*.[78] Although there does not appear to be a direct correlation between the chemical reactivity of the alkylating agent and the therapeutic or toxic effects,[79,80] the compounds that are able to cross-link DNA are much more effective than singly alkylating agents.[81,82] There is a rough correlation between antitumor efficacy and ability to induce mutations and inhibit DNA synthesis.[81] Bardos *et al.*[80] also found a relationship between the rate of solvolysis of the alkylating agent and its cytotoxic effect on tumor cells *in vitro*. The differences in the effectiveness of the various alkylating agents probably result from differences in pharmacokinetic factors, lipid solubility, ability to penetrate the central nervous system, membrane transport properties, detoxification reactions (see Chapter 7), and specific enzymatic reactions capable of repairing alkylated sites on the DNA.[83]

***c. Lead Modification.*** The prototype of the nitrogen mustards is mechlorethamine (**6.14**, R = $CH_3$),[84] which is still used in the treatment of advanced Hodgkin's disease. Mechlorethamine is a bifunctional alkylating agent that reacts with the N-7 of two different guanines in DNA,[85] producing an interstrand cross-link (**6.17**) by the mechanism shown in Scheme 6.3. The formation of the N-7 ammonium ion makes the guanine more acidic and, therefore, shifts the equilibrium in favor of the enol tautomer. Because guanine in this tautomeric form can make base pairs with thymine residues instead of cytosine (**6.18**), this leads to miscoding during replication. Furthermore, the N-7 alkylated guanine is susceptible to hydrolysis, which results in the destruction of the purine nucleus (Scheme 6.4). In addition to forming interstrand cross-links, it also is possible for the second chloroethyl group of

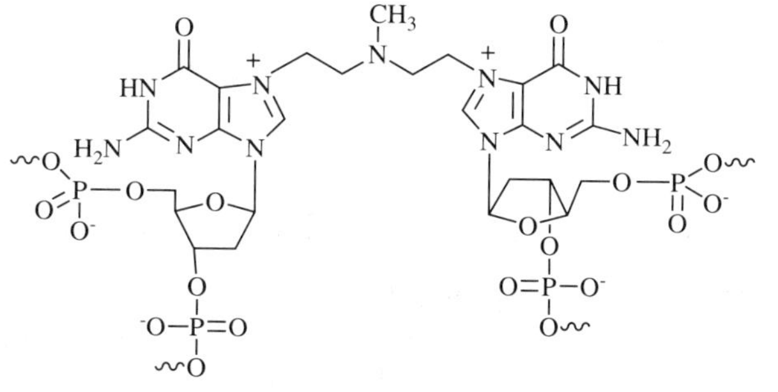

**6.17**

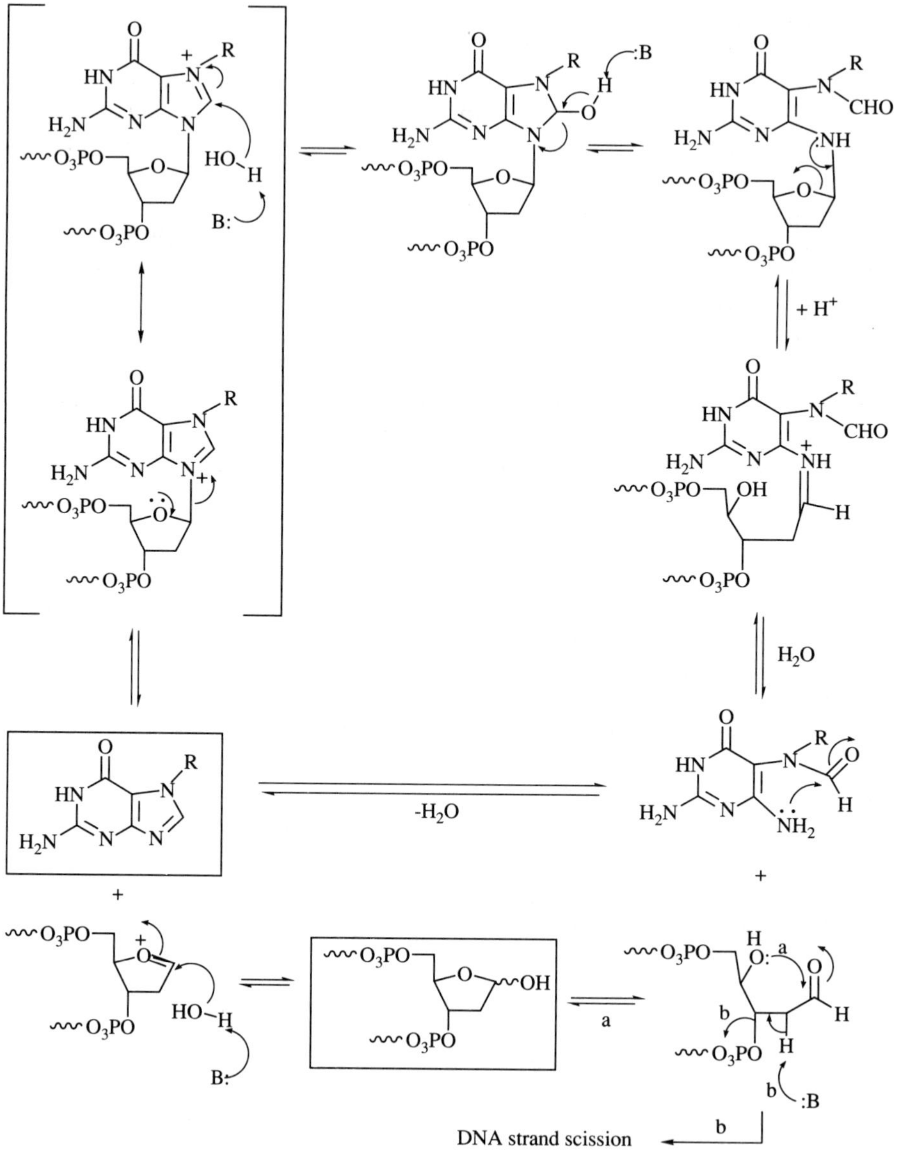

Scheme 6.4. Depurination of N-7 alkylated guanines in DNA.

the nitrogen mustard to react with a thiol or amino group of a protein, resulting in a DNA–protein cross-link. Any of these reactions could explain both the mutagenic and cytotoxic effects of the nitrogen mustards.

Mechlorethamine is quite unstable to hydrolysis. In fact, it is so reactive with water that it is marketed as a dry solid (HCl salt), and aqueous solutions

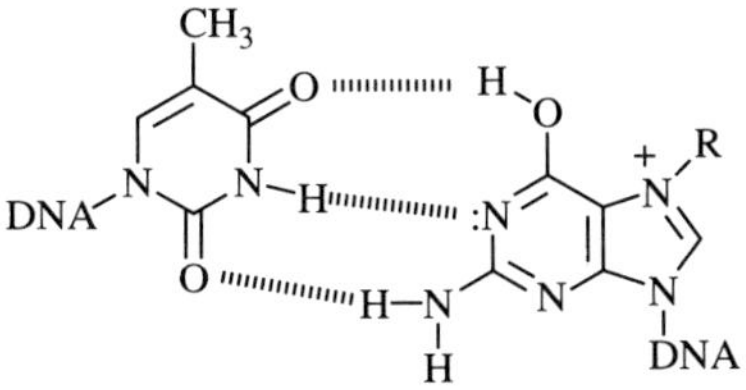

CH3 O H O R N N DNA N N H N N N H N N DNA O H

**6.18**

are prepared immediately prior to injection; within minutes after administration, mechlorethamine reacts completely in the body. Because of this reactivity, a more stable analog was sought. Substitution of the methyl group of mechlorethamine with an electron-withdrawing group, such as an aryl substituent (**6.19**), makes the nitrogen less nucleophilic and, therefore, slows down the rate of aziridinium ion formation (see Scheme 6.3); this decreases the reactivity of the nitrogen mustard.[86] As a result of this stabilization, some of these compounds could be administered orally, and they would be able to undergo absorption and distribution before extensive alkylation occurred. Simple aryl-substituted nitrogen mustards were not sufficiently water soluble for intravenous administration, but the solubility problem was solved with the use of carboxylate-containing aryl substituents. Direct substitution of the phenyl with a carboxylate (**6.19**, R = $CO_2H$), however, gave a compound that was too stable, and, therefore, not very active. To increase the reactivity, the electron-withdrawing effect of the carboxylate was attenuated by separating it from the phenyl group with methylenes. The optimal number of methylenes was found to be three, giving the antitumor drug chlorambucil [**6.19**, R = $(CH_2)_3CO_2H$].[87]

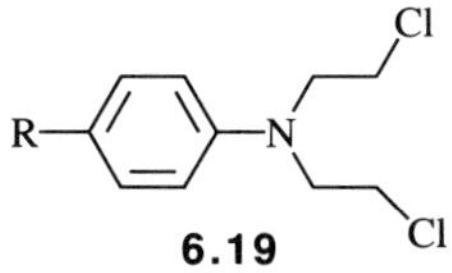

R N Cl Cl

**6.19**

Other nitrogen mustard analogs were prepared in an attempt to obtain an anticancer drug that would be targeted for a particular tissue. Because L-phenylalanine is a precursor to melanin, it was thought that L-phenylalanine nitrogen mustard [**6.19**, R = $CH_2CH(NH_2)CO_2H$; melphalan] might accumulate in melanomas. Although this analog is an effective, orally active anticancer drug (for multiple myelomas), it is not active against melanomas. The L-isomer (melphalan), the D-isomer (medphalan), and the racemic mixture (merphalan) have approximately equal potencies.[88] This lack of enantiospecificity suggests that there is no active transport of these compounds into cancer cells; however, a leucine carrier system appears to be involved in melphalan transport.[89]

***d. Drug Resistance.*** A major problem that limits the effectiveness of alkylating agents in general is resistance in tumor cells. With the use of cells selected for resistance in culture and with repeatedly transplanted animal tumors, mechanisms of resistance have been found to include decreased drug entry into the cell, increased repair of the drug defect, increased content of nonprotein sulfhydryls such as glutathione (to react with the alkylating agent), and increased levels of metabolic enzymes such as glutathione *S*-transferase. It is not clear, however, which of these mechanisms are clinically significant.

### 2. Ethylenimines

Because the reactive intermediate involved in DNA alkylation by nitrogen mustards (Scheme 6.3, Section III,B,1,b) is an aziridinium ion, an obvious extension of the nitrogen mustards is aziridines (ethylenimines). Protonated ethylenimines are highly reactive (they are aziridinium ions), and they would not be effective drugs. When electron-withdrawing groups are substituted on the aziridine nitrogen, however, the $pK_a$ of the nitrogen is lowered to a point where the aziridine is not protonated at physiological pH. These aziridines are much less reactive. In general, two ethylenimine groups per molecule are required for antitumor activity; compounds with three or four aziridines are not significantly more active.[90] Examples of antitumor ethylenimines include triethylenemelamine (**6.20**), carboquone (**6.21**), and diaziquone (**6.22**). By appropriate addition of lipophilic substituents to the benzoquinone ethylenimines, antitumor activity in the central nervous system can be achieved.[91]

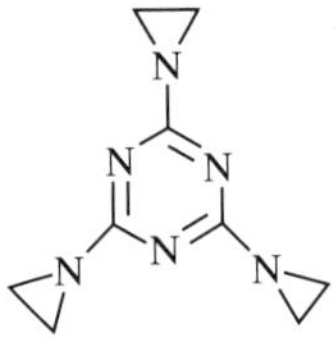

N N N N N N N NH₂ O OMe O O O CH₃ N N NHCO₂Et EtO₂CNH O O

**6.20** 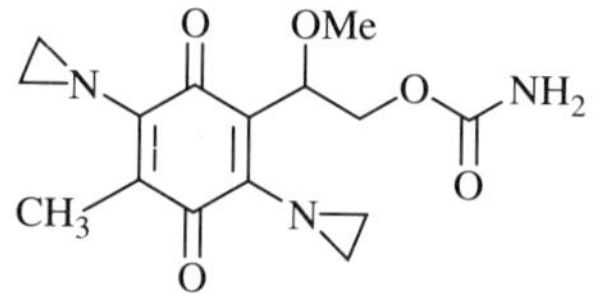 **6.21** 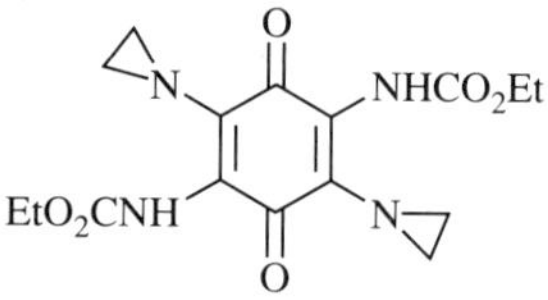 **6.22**

### 3. Methanesulfonates

The most prominent example of the methanesulfonate class of alkylating agents is the bifunctional anticancer drug busulfan[92] (**6.23**, $n = 4$). Compounds with one to eight methylene groups (**6.23**, $n = 1–8$) have antitumor activity, but maximum activity is obtained with four methylenes.[93] Alkylation of the N-7 position of guanine was demonstrated.[85,94] Unlike the nitrogen mustards (Section III,B,1), however, intrastrand, not interstrand, cross-links form.[78,95]

$$CH_3O_2SO{-}(CH_2)_n{-}OSO_2CH_3$$

**6.23**

## 4. Nitrosoureas

The nitrosoureas (**6.24**) were developed from the lead compound *N*-methyl-*N*-nitrosourea (**6.24**, R = $CH_3$, R′ = H), which exhibited modest antitumor activity in animal tumor models.[96] Analogs with 2-chloroethyl substituents, such as carmustine (**6.24**, R = R′ = $CH_2CH_2Cl$; BCNU) and lomustine (**6.24**, R = $CH_2CH_2Cl$, R′ = cyclohexyl; CCNU), were found to possess much greater antitumor activity.[97] Because of their lipophilicity, the 2-chloroethyl analogs were able to cross the blood–brain barrier, and, consequently, they have been used in the treatment of brain tumors. Despite the potency of these antitumor drugs, they are less desirable than others because of a severe problem of delayed and cumulative bone marrow toxicity.

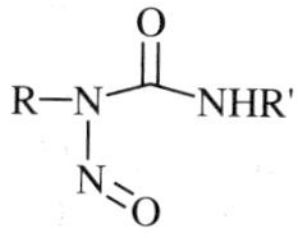

**6.24**

Extensive mechanistic studies have been carried out on nitrosoureas. Decomposition of the first active anticancer nitrosourea, **6.24** (R = $CH_3$, R′ = H), produces methyldiazonium ion (**6.25**, Scheme 6.5), a potent methylating agent, and isocyanic acid (**6.26**), a carbamoylating agent.[98] Evidence that diazomethane is not the alkylating agent was provided by a model study[99] showing that under physiological conditions 1-trideuteriomethyl-3-nitro-1-nitrosoguanidine (**6.27**, Scheme 6.6) alkylates nucleophiles with the trideuteriomethyl group intact; if diazomethane were the alkylating agent, dideute-

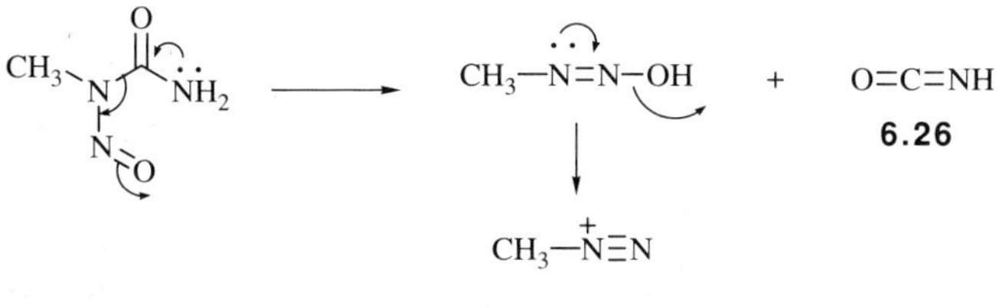

**6.26**

**6.25**

**Scheme 6.5.** Decomposition of *N*-methyl-*N*-nitrosourea.

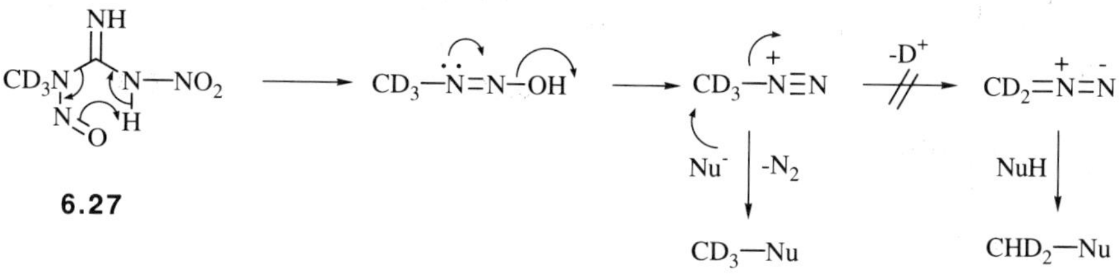

**6.27**

**Scheme 6.6.** Model for the activation of nitrosoureas.

riomethyl groups would have resulted. It is now known that *N*-nitrosoamides (**6.28**) and *N*-nitrosourethanes (**6.29**), which produce alkylating but not carbamoylating species, do have anticancer activity.[100] Furthermore, certain nitrosoureas that have little carbamoylating activity also are quite active, but nitrosoureas with no detectable alkylating activity are either very weakly active or inactive.[100] Therefore, the alkylating, not carbamoylating, product appears to be the principal species responsible for the anticancer activity.

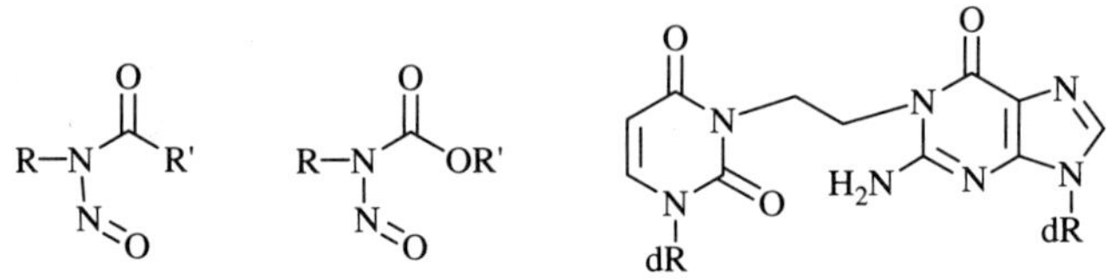

**6.28** **6.29** **6.30**

The 2-chloroethyl-substituted analogs (**6.24**, R = $CH_2CH_2Cl$) react with DNA and produce an interstrand cross-link between a guanine on one strand and a cytosine residue on another.[101] 1-($N^3$-Deoxycytidyl)-2-($N'$-deoxyguanosinyl)ethane (**6.30**) was isolated from the reaction of *N*,*N'*-bis(2-chloroethyl)-*N*-nitrosourea (carmustine; **6.24**, R = R' = $CH_2CH_2Cl$). As the same cross-link occurs with the mono-2-chloroethyl-substituted analog **6.24** (R = $CH_2CH_2Cl$, R' = cyclohexyl), the mechanism shown in Scheme 6.7 was proposed.[101] To rationalize the regioselectivity of these alkylating agents, a

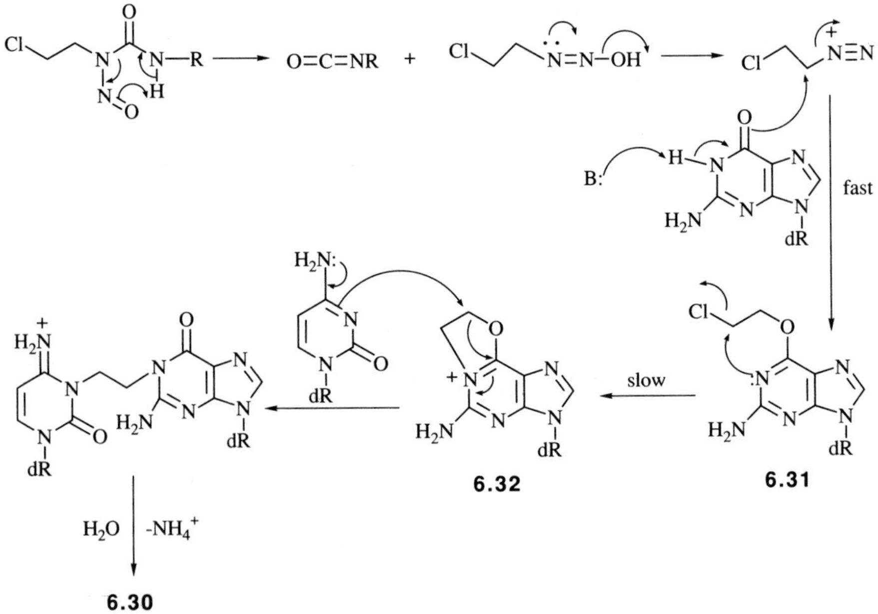

**Scheme 6.7.** Mechanism proposed for cross-linking of DNA by (2-chloroethyl)nitrosoureas.

kinetic analysis of the reaction was carried out, and an alternative reaction mechanism was suggested[102] (Scheme 6.8). The principal difference in these mechanisms is that in Scheme 6.8 a nucleoside on the DNA reacts with the intact drug to form a tetrahedral intermediate (**6.33**), which after cyclization to **6.34**, undergoes reaction with the O-6 of a guanine to give **6.35**, the precursor to the diazonium ion that leads to interstrand cross-linking (see Scheme 6.7). Evidence for the cyclization **6.31** to **6.32** (Scheme 6.7) and nucleophilic attack to give interstrand cross-linking comes from model chemistry for this reaction.[103] Evidence for the intermediacy of an O-6 guanine adduct such as **6.31** (Scheme 6.7) or **6.35** (Scheme 6.8) is based on the observation that cell lines capable of excising O-6 guanine adducts were resistant to cross-link formation[104] and that the addition of rat liver $O^6$-alkylguanine–DNA alkyltransferase (the enzyme that catalyzes the excision of O-6 guanine adducts) prevents formation of the cross-links.[105]

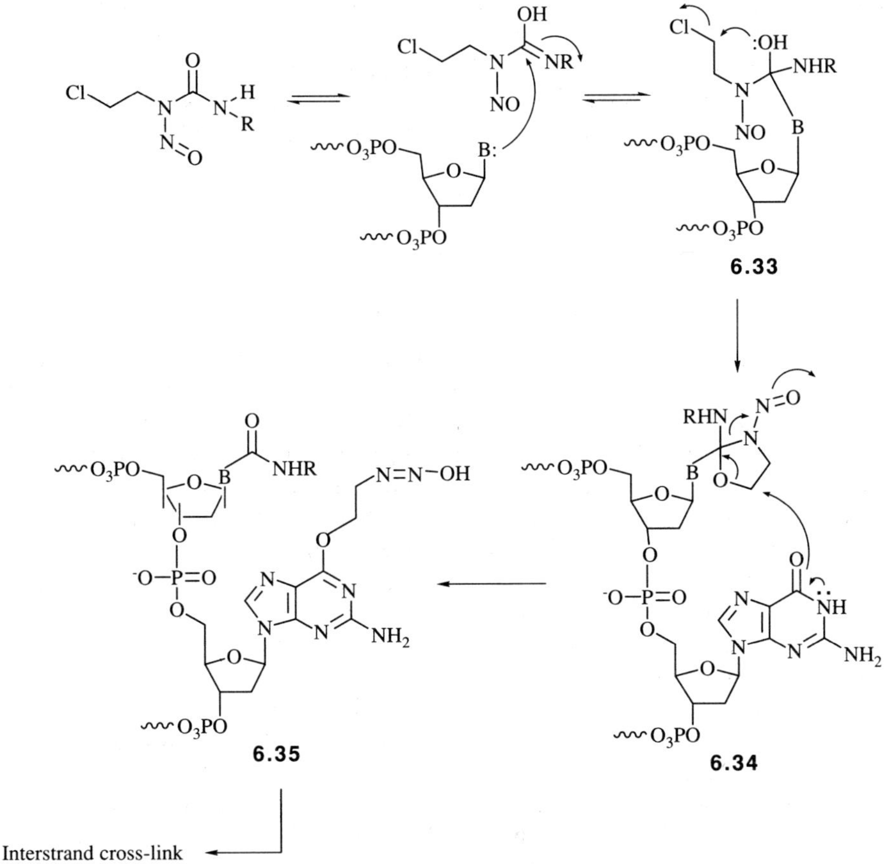

Interstrand cross-link

**Scheme 6.8.** Alternative mechanism for the cross-linking of DNA by (2-chloroethyl)nitrosoureas.

The isocyanate that is generated (e.g., **6.26** in Scheme 6.5) with the chloroethyldiazonium ion does not appear to be directly involved in the antitumor effects of nitrosoureas, but it does react with amines in proteins.[106] More importantly, it inhibits DNA polymerase[107] and other enzymes involved in the repair of DNA lesions,[108] such as $O^6$-alkylguanine–DNA alkyltransferase, DNA nucleotidyltransferase, and DNA glycosylases. It also inhibits RNA synthesis and processing[109] and plays a role in the toxicity of the nitrosoureas.[110]

## 5. Platinum Complexes

The platinum complexes, such as cisplatin (**6.36**) and carboplatin (**6.37**), are neither organic molecules nor alkylating agents, and therefore, by definition, they do not really belong in a text on the organic chemistry of drug action. However, they are part of a very important class of antitumor agents that are related to alkylating agents by virtue of the fact that they form essentially irreversible bonds (albeit Pt–N bonds) to guanine and adenine bases in DNA via (ligand) substitution reactions. Because these compounds are only distant relatives to the alkylating agents, the discussion here is brief.

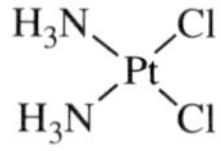

$H_3N$ Cl Pt $H_3N$ Cl

**6.36**

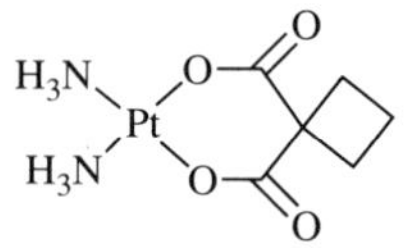

$H_3N$ O O Pt $H_3N$ O O

**6.37**

The discovery of the antitumor activity of cisplatin arose from studies by Rosenberg and co-workers[111] on the effects of electric fields on bacterial cell growth. When alternating current was delivered through platinum electrodes to growing bacteria, the bacterial cells stopped dividing and grew into long filaments, which is the same phenomenon that occurs when bacteria are treated with alkylating agents. It was found that platinum ions were released from the electrodes during electrolysis and, in the presence of ammonium salts and light, *cis*-diamminedichloroplatinum(II) (cisplatin) was generated. Only the cis isomer (two chlorines on the same side) has antitumor activity.

Because cisplatin can form two stable bonds to DNA by displacement of both chloride ligands, it acts in the same way as a bifunctional alkylating agent.[112] Although cisplatin binds to all DNA bases, there is a strong preference for binding to N-7 of guanine.[113] The majority of the DNA damage that is caused by cisplatin is the result of intrastrand cross-links.[114] Three types of intrastrand cross-links have been identified.[114] Of the total cross-linked adducts produced, 65% are platinum complexes at N-7 of two adjacent guanines [d(GpG); **6.38**], 25% of the total complexes are between the N-7 position of a guanine and the N-7 of an adjacent adenine [d(ApG); **6.39**], and 6% occur at the N-7 of two guanines that have another base between them [d(GpNpG); **6.40**]. Adducts at d(ApG) sequences (**6.39**) invariably have the adenine on the 5′-terminus of the dimer. Less than 1% of cross-links that form are interstrand cross-links; cross-links between DNA and proteins comprise only about 0.1% of the total cisplatin adducts.[115] Because *trans*-diamminedichloroplatinum(II), which does not have antitumor activity, forms DNA interstrand and DNA–protein cross-links, it is unlikely that either of these adducts is responsible for the antitumor activity of the cis analog.

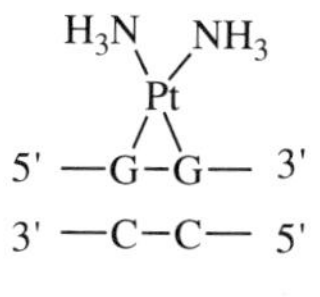

$H_3N$ $NH_3$ / Pt / 5' —G–G— 3' / 3' —C–C— 5'

**6.38**

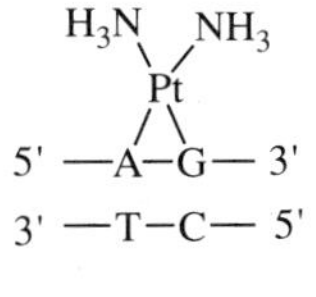

$H_3N$ $NH_3$ / Pt / 5' —A–G— 3' / 3' —T–C— 5'

**6.39**

$H_3N$ $NH_3$ / Pt / 5'—G–N–G— 3' / 3'—C–N–C— 5'

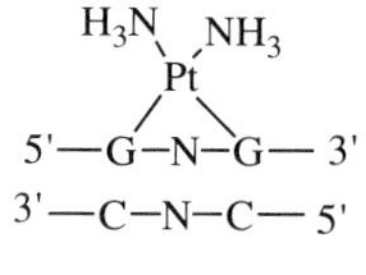

**6.40**

The three principal mechanisms of drug resistance are alterations in transmembrane transport of cisplatin,[116] deactivation as a result of increased sulfhydryl concentrations,[117] and enhanced DNA adduct repair capability.[118]

## C. DNA Strand Breakers

Some DNA-interactive drugs initially intercalate into DNA but then, under certain conditions, react in such a way as to generate radicals. The reaction of these radicals with the sugar moieties leads to DNA strand scission. As examples of this mode of action of DNA-interactive drugs we will consider the anthracycline antitumor antibiotics, bleomycin, and the enediyne antitumor antibiotics. It should be kept in mind that some of the compounds which lead to strand breakage act via the topoisomerase-induced mechanisms discussed in Section III,A,1.

### 1. Anthracycline Antitumor Antibiotics

Doxorubicin (**6.9**, X = OH) and daunorubicin (**6.9**, X = H) are anthracyclines that were discussed in Section III,A,2,c on DNA intercalators; however, these drugs also cause oxygen-dependent DNA damage.[119] Several mechanisms have been proposed to account for this destruction of DNA. Anthracyclines cause protein-associated breaks that correlate with their cytotoxicity,[120] and these breaks may be caused by the reaction of anthracyclines on topoisomerase II, an enzyme that promotes DNA strand cleavage and reannealing.[121]

Another mechanism for DNA damage is electron transfer chemistry. A one-electron reduction of the anthracyclines, probably catalyzed by flavoenzymes such as NADPH–cytochrome-*P*-450 reductase,[122] produces the anthracycline semiquinone radical (**6.41**, Scheme 6.9), which can transfer an electron to oxygen to regenerate the anthracycline and produce superoxide ($O_2^{\dot{-}}$). Both the superoxide and anthracycline semiquinone radical anions can generate the hydroxyl radical (HO·) (Scheme 6.10), which is known to cause DNA strand breaks.[123]

A third possibility for the mechanism of DNA damage by anthracyclines is the formation of a ferric complex, which binds to DNA by a mechanism

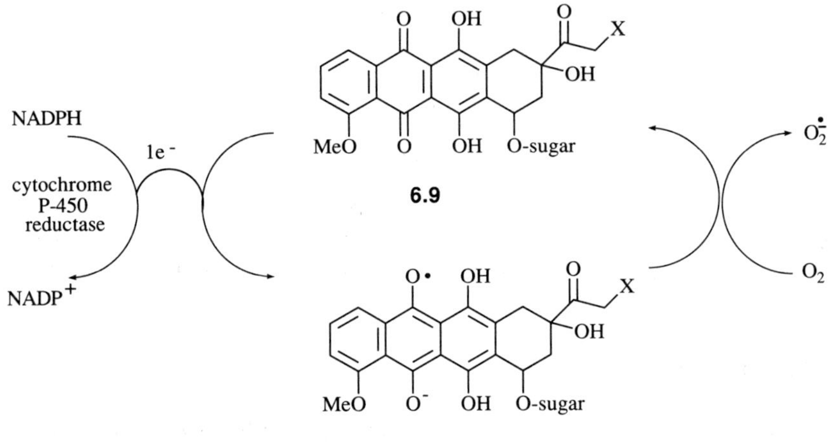

NADPH

cytochrome P-450 reductase

NADP$^+$

1e$^-$

**6.9**

**6.41**

$O_2^{\dot{-}}$

$O_2$

**Scheme 6.9.** Electron transfer mechanism for DNA damage by anthracyclines.

$$2\,O_2^{\dot{-}} + 2\,H^+ \longrightarrow H_2O_2 + O_2$$

$$O_2^{\dot{-}} + H_2O_2 \xrightarrow{\text{slow}} HO\bullet + HO^- + O_2$$

**or**

$$H_2O_2 + \mathbf{6.41} \longrightarrow \mathbf{6.9} + HO\bullet + HO^-$$

**or**

$$O_2^{\dot{-}} + Fe^{3+}\bullet\mathbf{6.9} \longrightarrow O_2 + Fe^{2+}\bullet\mathbf{6.9}$$

$$Fe^{2+}\bullet\mathbf{6.9} + H_2O_2 \longrightarrow Fe^{3+}\bullet\mathbf{6.9} + HO\bullet + HO^-$$

$$HO\bullet + DNA \longrightarrow \text{strand scission}$$

**Scheme 6.10.** Anthracycline semiquinone generation of hydroxyl radicals.

different from intercalation and significantly tighter.[124] The binding constant for the doxorubicin–ferric complex (**6.42**) is $10^{33}$. This ferric complex could react with superoxide to give oxygen and the corresponding ferrous complex (Scheme 6.10). The reaction of ferrous ions with hydrogen peroxide is known as *Fenton's reaction*,[125] which is used in the standard method for the generation of hydroxyl radicals when doing *DNA footprinting*, a technique that indiscriminately cleaves the DNA to determine where protein–DNA interactions occur.[126]

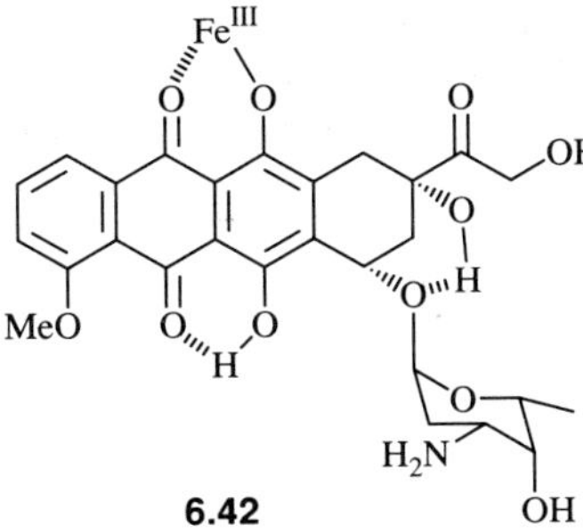

$Fe^{III}$ O O O OH O O H MeO O O H O $H_2N$ OH

**6.42**

As the generation of the hydroxyl radicals occurs adjacent to DNA, it is unlikely that radical scavengers would be an effective method of cell protection, as was shown with anthracycline antibiotic–induced cardiac toxicity. However, an iron chelator (**6.43**) does prevent doxorubicin-induced cardiac toxicity in humans.[127] This iron chelator actually is a prodrug (see Chapter 8) that, because of its nonpolar nature, enters the cell. Once inside the cell it is hydrolyzed to the active iron chelator **6.44** (Scheme 6.11), which is structur-

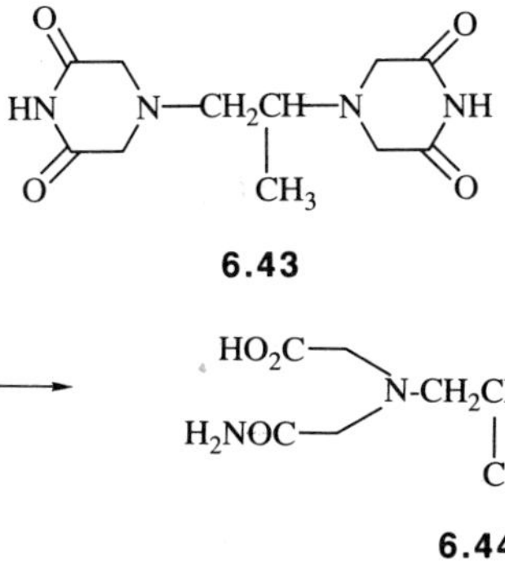

O O HN N—$CH_2CH$—N NH O $CH_3$ O

**6.43**

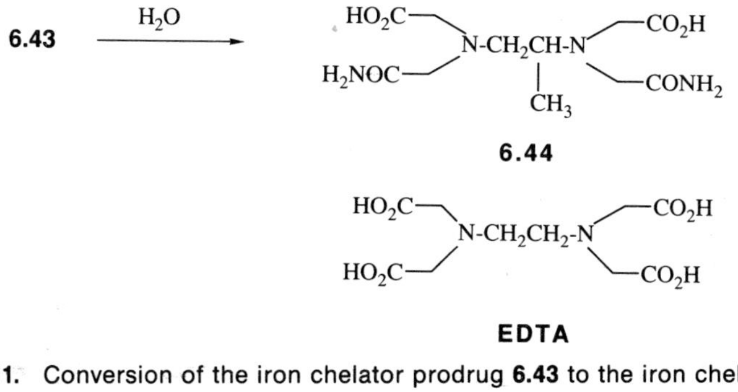

**6.43** $\xrightarrow{H_2O}$ $HO_2C$ N-$CH_2CH$-N $CO_2H$ $H_2NOC$ $CH_3$ $CONH_2$

**6.44**

$HO_2C$ N-$CH_2CH_2$-N $CO_2H$ $HO_2C$ $CO_2H$

**EDTA**

**Scheme 6.11.** Conversion of the iron chelator prodrug **6.43** to the iron chelator **6.44**.

ally related to the well-known iron chelator EDTA. Gianni *et al.*[128] showed that the iron–doxorubicin (**6.9**, X = OH) complex is more reactive than the iron–daunorubicin (**6.9**, X = H) complex because the hydroxymethyl ketone side chain of doxorubicin reacts spontaneously with iron to produce $Fe^{2+}$, $HO\cdot$, and $H_2O_2$.

A fourth mechanism proposed for the action of anthracyclines involves a two-electron reduction to reduced anthracycline (**6.45**), which was thought to cause spontaneous elimination of the sugar. This would generate a Michael acceptor, which could alkylate DNA (Scheme 6.12). However, Sulikowski *et al.*[129] synthesized **6.45** and found that it did not undergo elimination of the sugar. Possibly, elimination of the sugar occurs at the semiquinone oxidation state.

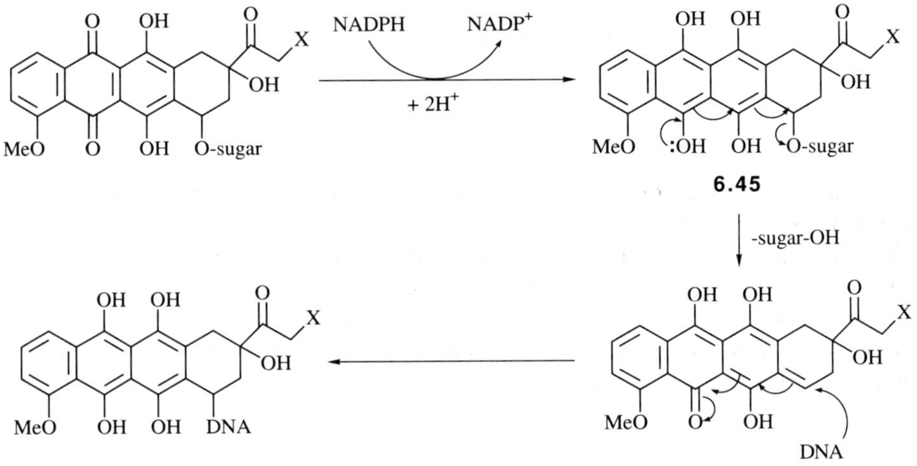

**Scheme 6.12.** Two-electron mechanism for DNA damage by anthracyclines.

Drug resistance is attributed to elevated gene expression of the P170 glycoprotein that pumps drugs out of cells, to decreased activity of the repair enzyme topoisomerase II, and to increased glutathione production, which scavenges free radicals and peroxides.

## 2. Bleomycin

The anticancer drug bleomycin (**6.46**; BLM) is actually a mixture of several glycopeptide antibiotics isolated from a strain of *Streptomyces verticillus*; the major component is bleomycin $A_2$ [**6.46**, R = $NH(CH_2)_3\overset{+}{S}(CH_3)_2$].[130] There are three principal domains in bleomycin.[131] The pyrimidine, the β-aminoalanine, and the β-hydroxyimidazole moieties make up the first domain, which is involved in the formation of a stable complex with iron(II). This complex interacts with $O_2$ to give a ternary complex (**6.47**) which is believed to be responsible for the DNA-cleaving activity.[132] The second domain comprises the bithiazole moiety (the five-membered N and S heterocycles) and the attached sulfonium ion–containing side chain (R). It is thought that the bithiazole intercalates into the double helix; possibly, the sulfonium ion is attracted electrostatically to a phosphate group.[133] The gulose and carbamoylated mannose disaccharide moiety, the third domain, may be responsible for selective accumulation of bleomycin in some cancer cells, but it does not appear to be involved in DNA cleavage.

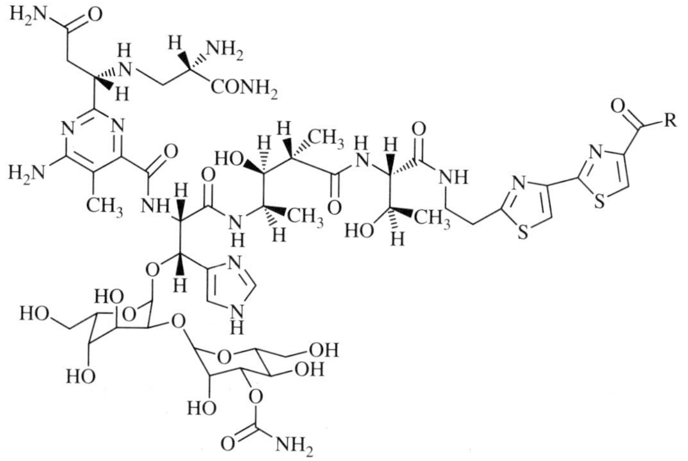

**6.46**

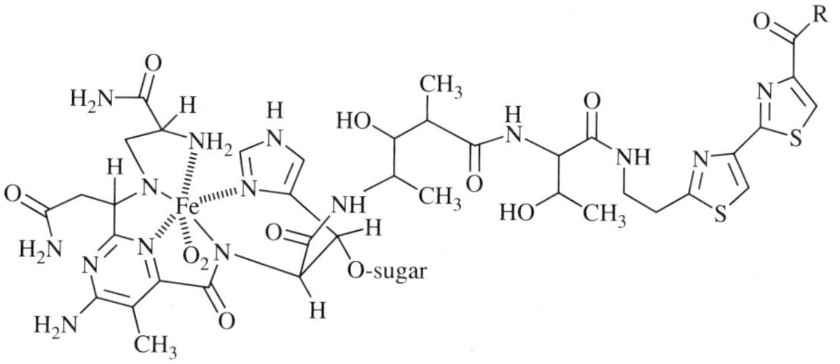

**6.47**

The primary mechanism of action of bleomycin is the generation of single- and double-strand breaks in DNA. This results from the production of radicals by a 1:1:1 ternary complex of bleomycin, Fe(II), and $O_2$ (Scheme 6.13). Activation of the ternary complex may be self-initiated by the transfer of an electron from a second unit of the ternary complex, or activation may be initiated by a microsomal NAD(P)H–cytochrome P-450 reductase–catalyzed

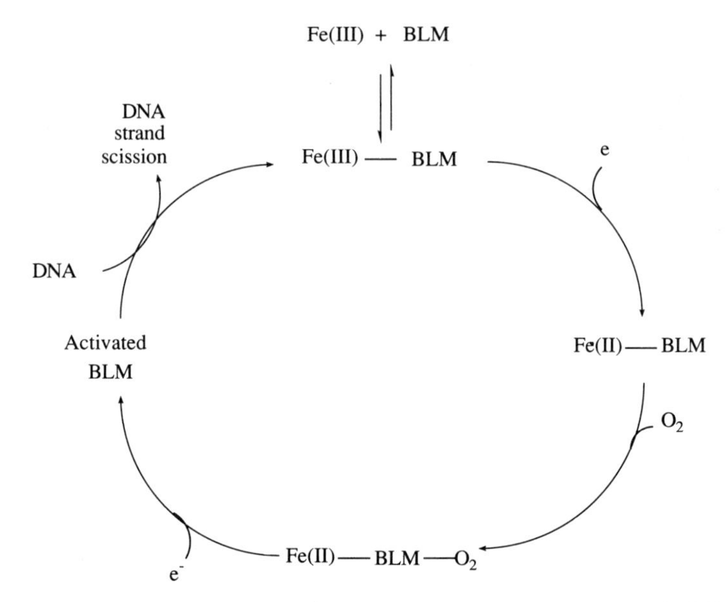


**Scheme 6.13.** Cycle of events involved in DNA cleavage by bleomycin (BLM).

reduction.[134] The activated bleomycin binds tightly to guanine bases in DNA, principally via the amino-terminal tripeptide (called the tripeptide S) containing the bithiazole unit.[135] Binding by the bithiazole to G–T and G–C sequences is favored. Evidence for intercalation comes from when bleomycin, its tripeptide S moiety, or just bithiazole is mixed with DNA, which results in a lengthening of the linear DNA and a relaxation of supercoiled circular DNA.[136]

Activation of the bound ternary complex is believed to occur by a reaction related to that for heme-dependent enzyme activation (see Section III,D of Chapter 4) as depicted in Scheme 6.14. Activated bleomycin appears to be related to the ferryl oxo species of heme-dependent enzymes, as it catalyzes the same reactions that are observed with these enzymes.[131] The two major monomeric products formed when activated bleomycin reacts with DNA are nucleic base propenals (**6.49**, Scheme 6.15) and nucleic acid bases. Base propenal formation consumes 1 equivalent of $O_2$ in addition to that required for bleomycin activation and is accompanied by DNA strand scission with the production of 3′-phosphoglycolate (**6.50**) and 5′-phosphate-modified DNA fragments (**6.48**, Scheme 6.15).[137] DNA base formation does not require additional $O_2$ and results in destabilization of the DNA sugar–phosphate back-

$$\text{BLM–Fe(II)—O}_2 \xrightarrow[2\,\text{H}^+]{e^-} \text{BLM—Fe(V)=O}$$

**Scheme 6.14.** Activation of bleomycin–Fe(II)–$O_2$.

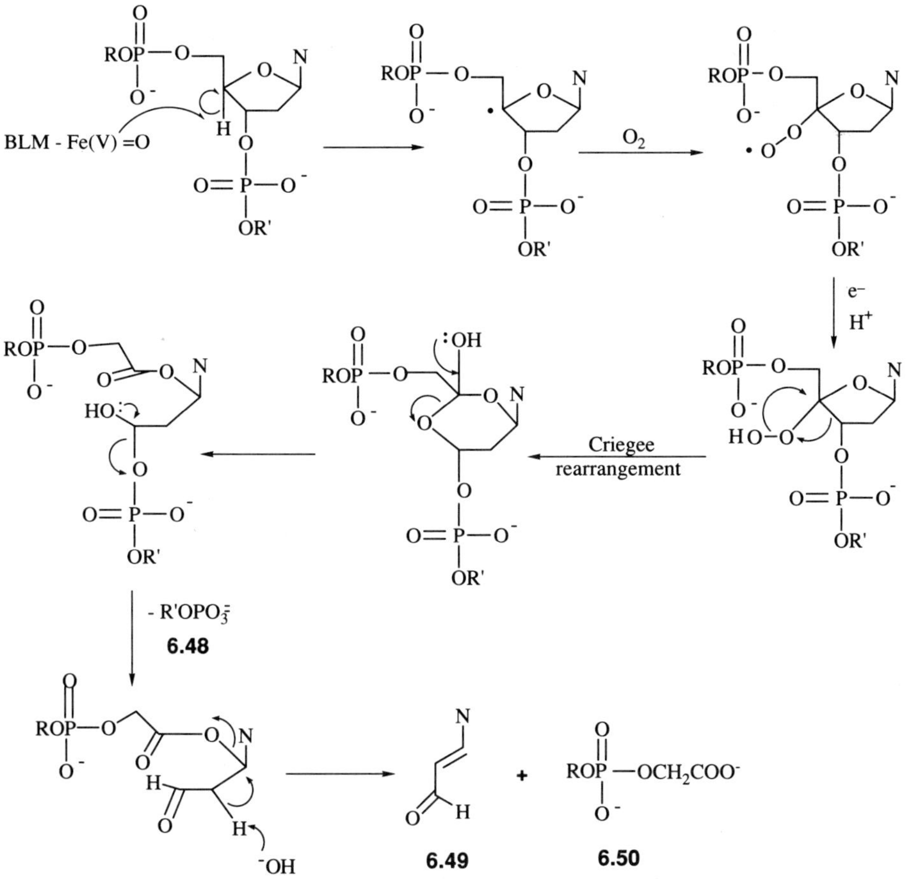

Scheme 6.15. Base propenal formation and DNA strand scission by activated bleomycin.

bone; additional single-strand cleavage of the DNA occurs in the presence of alkali (the *alkali-labile lesion*).[138] The alkali-labile lesion was identified as the 4′-keto aldehyde (**6.53**),[139,140] which could arise from the mechanism shown in Scheme 6.16. After C-4′ hydrogen atom abstraction by activated bleomycin to give **6.51**, an oxidation to **6.52** occurs. Sugar ring opening leads to elimination of the nucleic acid base and gives the alkali-labile product (**6.53**) which is probably in equilibrium with **6.54**.

Sugiyama *et al.*[139] heated the alkali-labile product with hydroxide and were able to isolate the 2-hydroxycyclopentenone analog **6.56**; the mechanism shown in Scheme 6.17 could account for the formation of that product. Under mild conditions Rabow *et al.*[140] were able to isolate the $NaBH_4$-reduced alkali-labile product and show that it is formed in stoichiometric amounts relative to nucleic acid base release, thereby establishing its relevance to the reaction

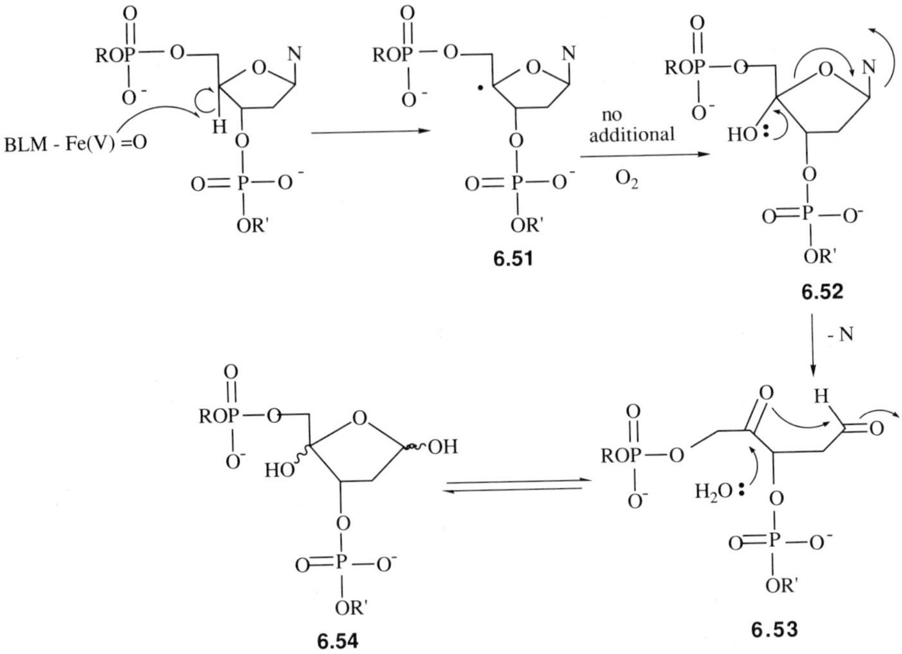

Scheme 6.16. Possible mechanism for the formation of nucleic acid bases and single-strand cleavage of DNA by alkali treatment of activated bleomycin-modified DNA.

with activated bleomycin. When the reaction was carried out with $^{18}O_2$ or $H_2^{18}O$, it was found that the oxygen incorporated at C-4′ comes from $H_2O$, not $O_2$, and that some $H_2O$ also is incorporated at C-1′.[141] The mechanism given in Scheme 6.18 rationalizes that observation. DNA strand scission mediated by bleomycin is sequence selective, occurring most frequently at 5′-GC-3′ and 5′-GT-3′ sequences.[142] The specificity for cleavage of DNA at a residue located at the 3′ side of G appears to be absolute. An important mechanism for resistance of cells to bleomycin is expression of bleomycin hydrolase, an aminopeptidase that hydrolyzes the carboxamide group of the L-aminoalaninecarboxamide substituent in the metal-free antibiotic.[143] The deamido bleomycin is less effective in the activation of oxygen by its Fe(II) complex.[144]

## 3. Enediyne Antitumor Antibiotics

Except for neocarzinostatin (**6.57**),[145] which was isolated in 1965, the other members of the enediyne antibiotic class of antitumor agents, the esperamicins (**6.58**),[146] the calichemicins (**6.59**),[147] and dynemicin A (**6.60**),[148] were isolated from various microorganisms in the mid- to late 1980s. Because their

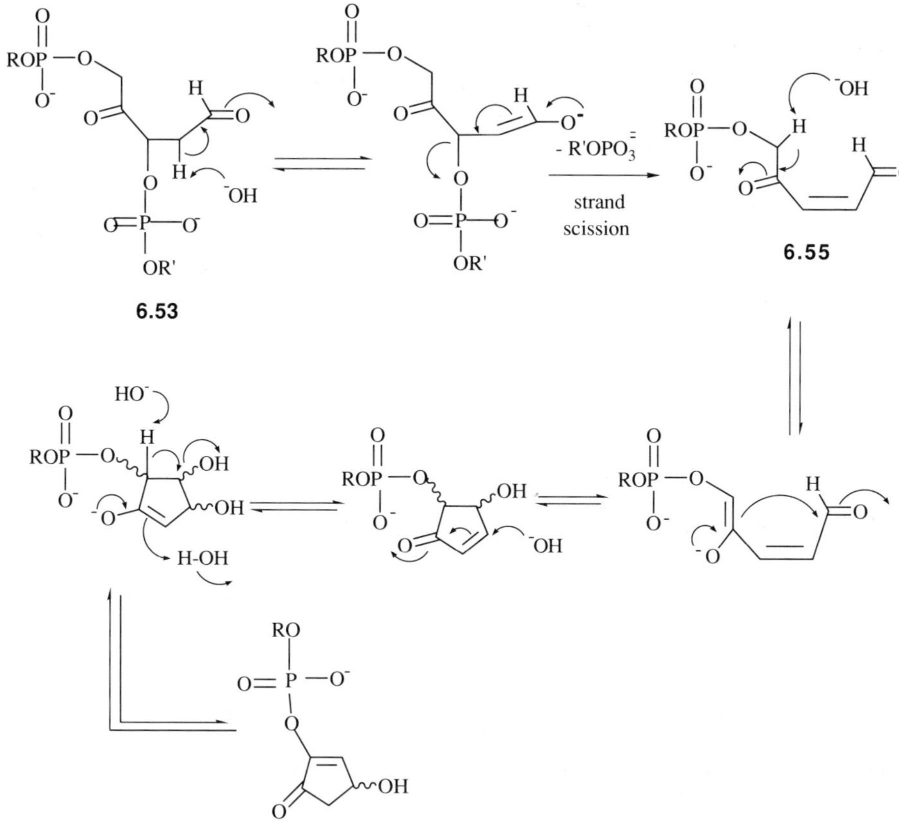

**Scheme 6.17.** Product isolated[129] by hydroxide treatment of the alkali-labile product from the reaction of activated bleomycin with DNA.

common structural feature is a macrocyclic ring containing at least one double bond and two triple bonds, they are referred to as enediyne antitumor antibiotics. All of these compounds appear to share two modes of action. First, there is intercalation of part of the molecule into the minor groove of DNA; then, either thiol or NADPH triggers a reaction that leads to the generation of radicals, which cleave DNA. Although knowledge of the chemistry involved in the latter process is only beginning to surface, we are able to draw reasonable mechanisms for DNA damage by each of these classes of compounds.

Generally, there are two tests that are used to demonstrate minor groove binding to B-DNA. One indication is an asymmetric cleavage pattern on the 3′ side of the opposite strand of the DNA helix.[149] The other test is inhibition of DNA cleavage by the known minor groove binders distamycin A and netrop-

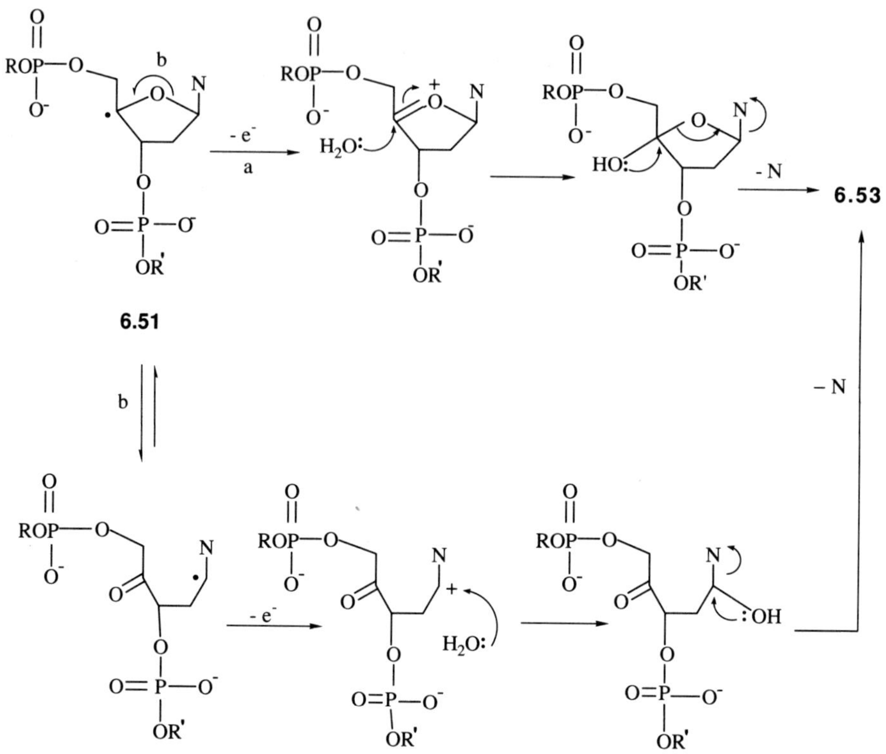

Scheme 6.18. Mechanism to account for incorporation of water into C-4′ and C-1′ of the sugar moiety during reaction of activated bleomycin with DNA.

sin (similar to substrate protection of enzymes from inhibition; see Section IV,A Chapter 5).

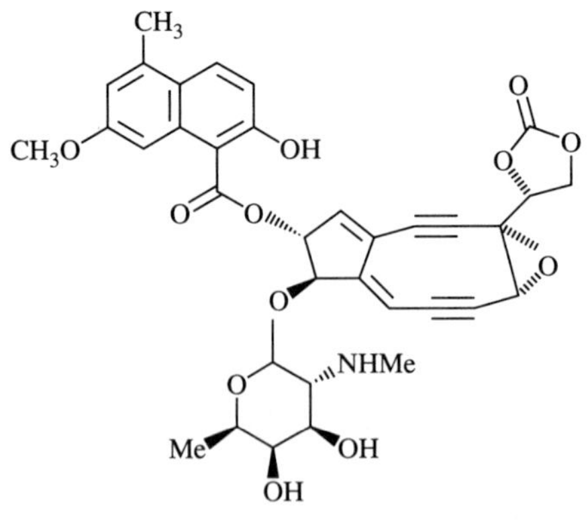

6.57

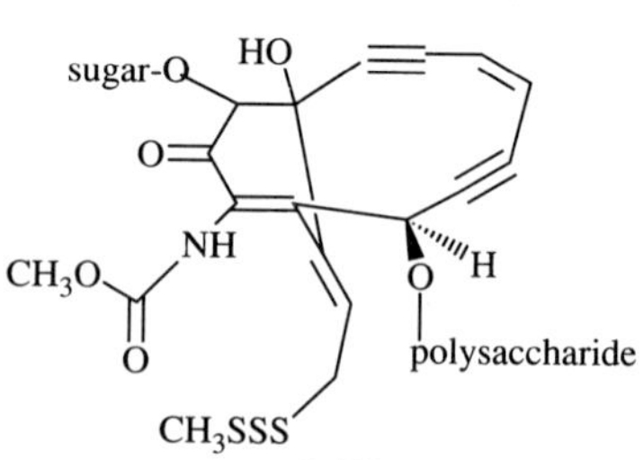

6.58

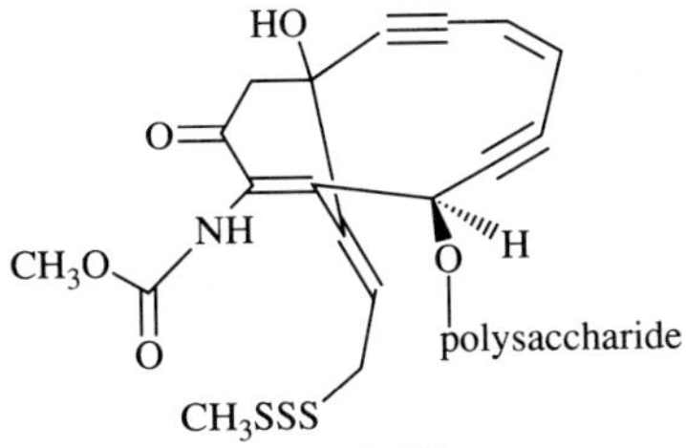

**6.59** 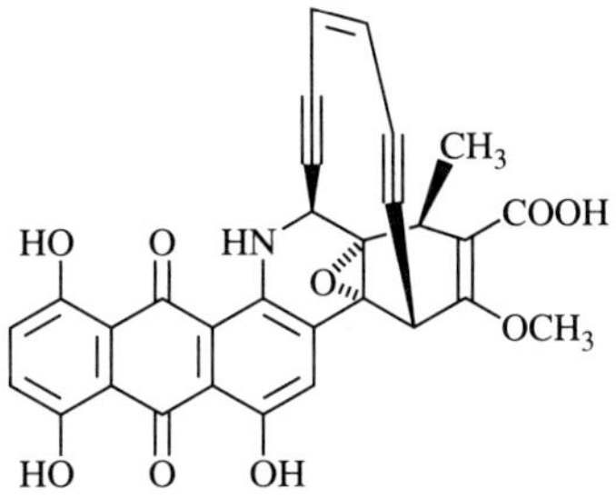 **6.60**

There are two phases to the mechanism of DNA degradation by the enediyne antibiotics. First there is the activation of the antitumor agent, and then there is the action of the activated antitumor agent on DNA. This is much akin to the process of DNA degradation that we discussed for bleomycin (see Section III,C,2). Most of what is now known about the mechanism of the reaction of activated enediynes with DNA comes from studies with neocarzinostatin, which is discussed last. As the chemistry of the activation of esperamicins and calichemicins is virtually identical, we first take a look at these compounds.

***a. Esperamicins and Calichemicins.*** The important structural features of esperamicins (**6.58**) and calichemicins (**6.59**) are a bicyclo[7.3.1] ring system, an allylic trisulfide attached to the bridgehead carbon, a 3-ene-1,5-diyne as part of the macrocycle, and an $\alpha,\beta$-unsaturated ketone in which the double bond is at the bridgehead of the bicyclic system. It is believed that the enediyne moiety partially inserts into the minor groove and then undergoes a reaction with either a thiol or NADPH, which reduces the trisulfide to the corresponding thiolate (**6.61**, Scheme 6.19).[150,151] Michael addition of this thiolate into the $\alpha,\beta$-unsaturated ketone gives the dihydrothiophene (**6.62**) in which the bridgehead carbon hybridization has changed from $sp^2$ in **6.61** to $sp^3$ in **6.62**. This change in geometry at the bridgehead may be sufficient to allow the two triple bonds to interact with each other and to trigger a *Bergman rearrangement*,[152] giving the 1,4-dehydrobenzene diradical (**6.63**), which is the activated esperamicin or calichemicin. The reaction of the diradical with DNA is, presumably, identical to that for the diradical produced from neocarzinostatin, which is discussed in III,C,3,c.

***b. Dynemicin A.*** Another member of the enediyne class of antitumor antibiotics is dynemicin A (**6.60**), which combines the structural features of both the anthracycline antitumor agents and the enediynes. The anthraquinone part of dynemicin A intercalates into the minor groove of DNA; then,

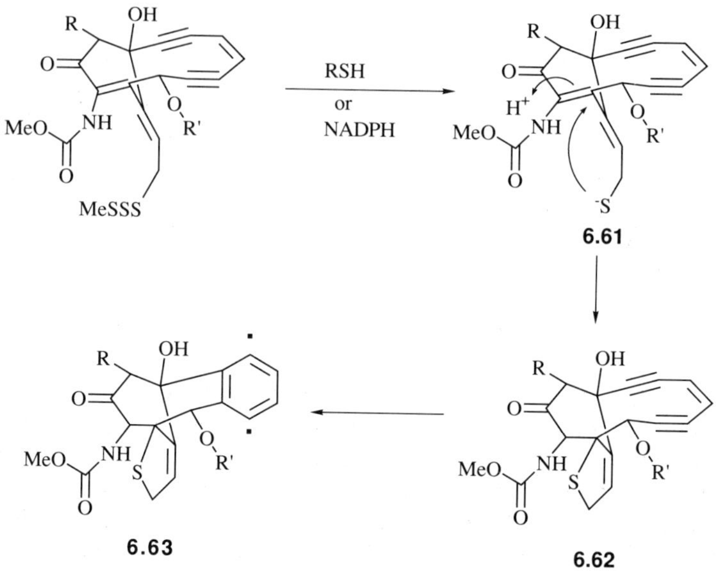

6.61

6.63

6.62

**Scheme 6.19.** Activation of esperamicins and calichemicins.

depending on whether activation is initiated by NADPH or a thiol, a reductive mechanism (Scheme 6.20) or nucleophilic mechanism (Scheme 6.21) of activation, respectively, is possible.[153] In the reductive activation mechanism (Scheme 6.20) either NADPH or a thiol reduces the quinone to a hydroquinone (**6.64**) which leads to opening of the epoxide. Following protonation to **6.65** the geometry of the triple bonds become more favorable for the Bergman reaction[152] to take place, leading to the generation of the 1,4-dehydrobenzene diradical (**6.66**). In the nucleophilic mechanism (Scheme 6.21) reaction of a thiol gives **6.67** which is very similar in structure to the reduced intermediate **6.65** (Scheme 6.20). Bergman reaction[152] of **6.67** gives the 1,4-dehydrobenzene diradical **6.68**. Either diradical (**6.66** or **6.68**) could be responsible for DNA degradation by the mechanism described in detail for neocarzinostatin (see Section III,C,3,c).

***c. Neocarzinostatin (Zinostatin).*** The oldest known member of the enediyne family of antibiotics is neocarzinostatin (now called zinostatin, **6.57**; NCS). More mechanistic studies have been performed with this compound than with any of the more recent additions to the enediyne family. The naphthoate ester moiety is believed to intercalate into DNA, thereby positioning the epoxybicyclo[7.3.0]dodecadiendiyne portion of the chromophore in the minor groove.[154] Activation by a thiol generates an intermediate (**6.69**, Scheme 6.22) capable of undergoing a Bergman rearrangement[152] to diradical **6.70**.[155] This diradical differs from the diradicals generated by activation of

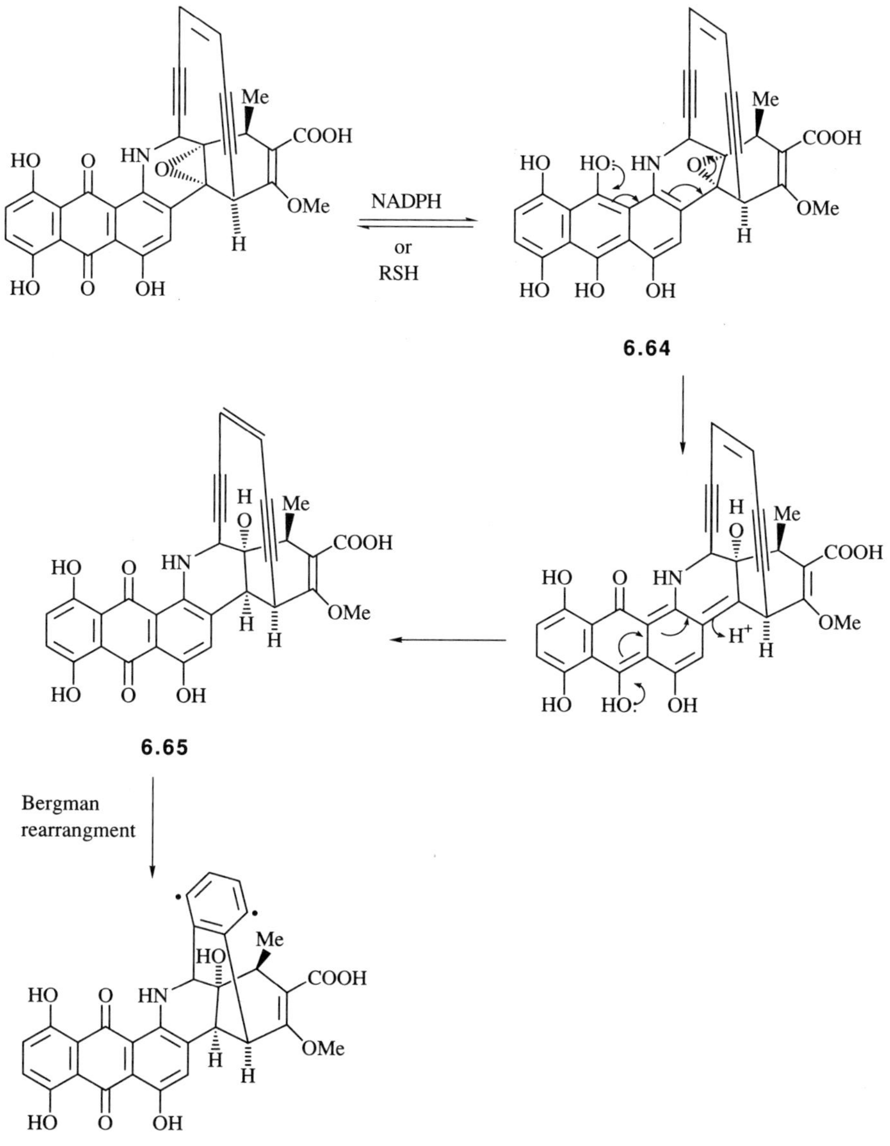

NADPH

or

RSH

6.64

6.65

Bergman rearrangment

6.66

**Scheme 6.20.** Reductive mechanism for activation of dynemicin A.

esperamicin and calichemicin (**6.63**, Scheme 6.19) and dynemicin A (**6.66**, Scheme 6.20, or **6.68**, Scheme 6.21) in that it is not a 1,4-dehydrobenzene diradical, but it is very similar.

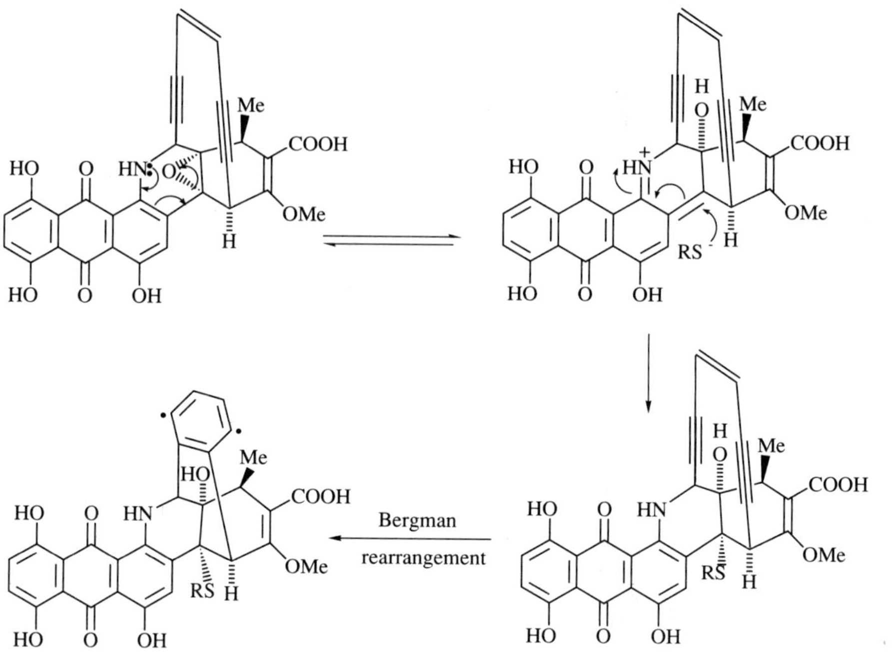

6.68

6.67

Scheme 6.21. Nucleophilic mechanism for activation of dynemicin A.

The highly reactive diradical **6.70** is responsible for DNA strand scission, which consumes 1 equivalent of $O_2$ per strand break. Both the C-4′ and the C-5′ hydrogens of the DNA sugar–phosphate residues are accessible to the diradical within the minor groove, and either can be abstracted. In the presence of $O_2$ two different mechanisms of DNA cleavage can result (Scheme 6.23).[156] About 80% of the cytotoxic lesions are the result of single-strand cleavages, mostly caused by C-5′ hydrogen atom abstraction from thymidine or deoxyadenosine residues (pathway a). This leads to the formation of the nucleoside 5′-aldehyde (**6.71**) and the 3′-phosphate (**6.72**). Abstraction of the C-4′ hydrogen atom (pathway b) gives a radical (**6.73**) that partitions between a modified basic carbohydrate terminus (**6.74**, pathway c, Scheme 6.23) and a 3′-phosphoglycolate terminus (**6.75**, pathway d), as was observed with bleomycin (see Section III,C,2). Also, as in the case of bleomycin, double-strand breaks occur when the NCS-treated DNA is heated in alkali. It appears that single-strand breaks at AGT sequences occur with both C-4′ and C-5′ oxidation but ACT breaks occur only with C-5′ oxidation.

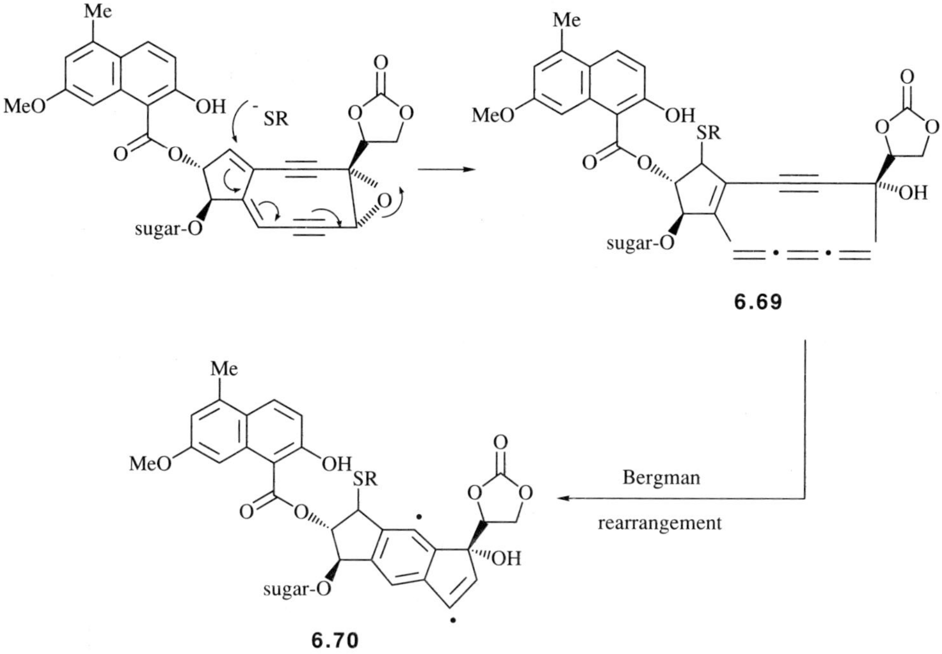

Scheme 6.22. Activation of neocarzinostatin by thiols.

## 4. Sequence Specificity for DNA Strand Scission

The preferential cutting sites for esperamicins are at thymidine residues (T > C > A > G). This base preference differs from that of the calichemicins (C >> T > A = G), neocarzinostatin (T > A > C > G), and bleomycin (T > C > A > G).[132,157] Unlike all of the other DNA strand breakers discussed, dynemicin A prefers to cut at guanine residues.[158] In fact, for esperamicins, calichemicins, and neocarzinostatin, guanine is the least important cleavage site. Whereas bleomycin prefers to cut at 5′-GT-3′ and 5′-GC-3′, esperamicins more often cut at 5′-TG-3′ and 5′-CG-3′ sequences. The preferential cutting site for dynemicin A is on the 3′ side of purine bases, namely, 5′-GC-3′, 5′-GT-3′, 5′-AT-3′, and 5′-AG-3′.

Nucleosomal structure and superhelical density probably influence sequence selectivity; consequently, sequence selectivity data obtained on linear DNA may not be applicable to nuclear DNA. Furthermore, there is no free DNA in the cell; the real target for DNA-interactive agents is the chromatin (DNA packaged by histones to form bundles in the cell). Therefore, it is not clear how informative *in vitro* studies are in the understanding of cellular DNA sequence specificity.

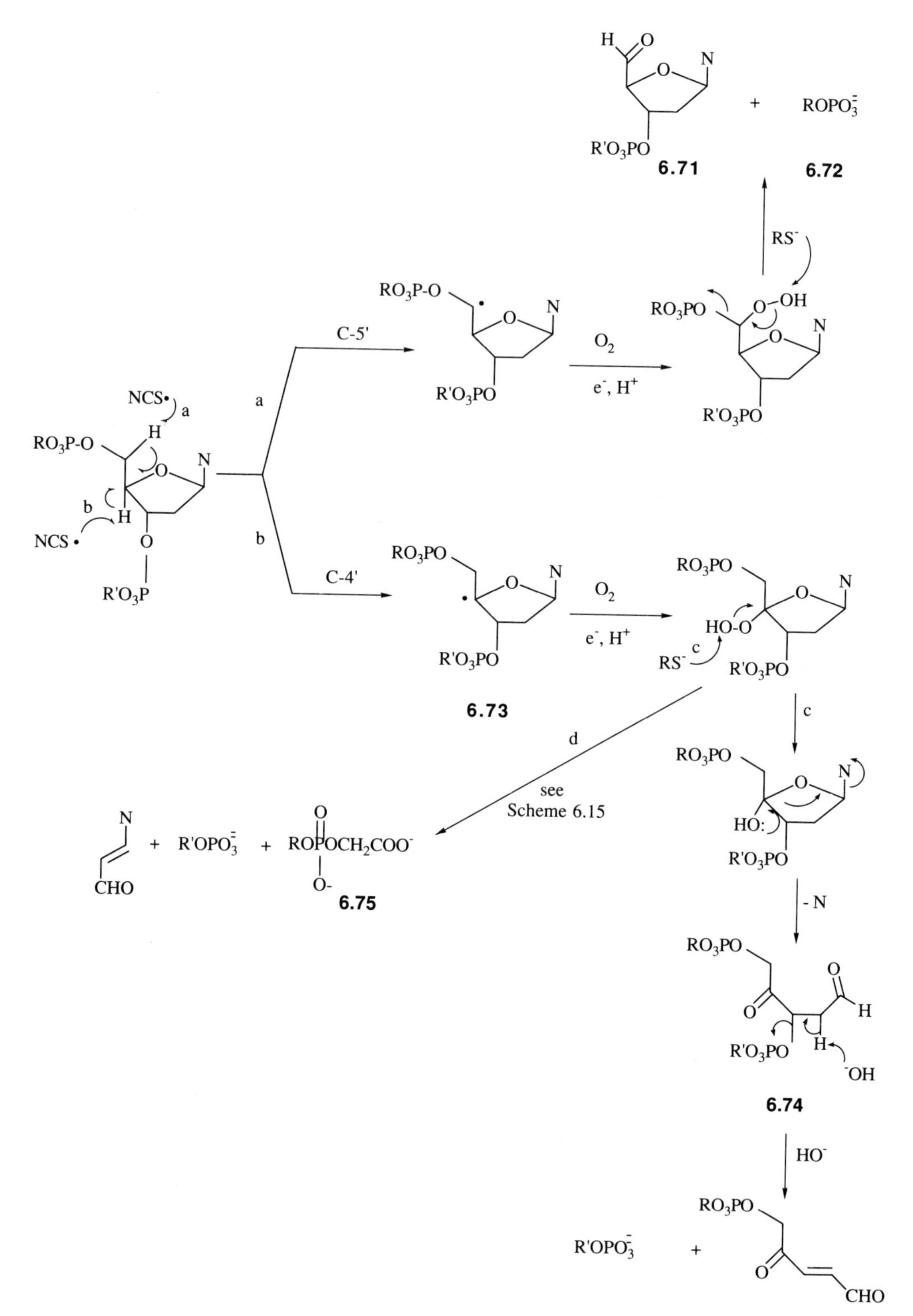


**Scheme 6.23.** DNA strand scission by activated neocarzinostatin (NCS) and other members of the class of enediyne antibiotics.

Although many of the DNA-interactive drugs react with the constituents of the minor groove, it is not yet clear if there is sufficient sequence specificity information in the minor groove of DNA to allow for the design of minor groove-selective agents. Mother Nature has opted (e.g., with protein–DNA interactions) to utilize major groove interactions.

## IV. Epilogue to Receptor-Interactive Agents

Chapters 3–6 have dealt with the structures and functions of various protein and nucleic acid receptors as well as with classes of drugs that interact with these receptors and how the medicinal chemist can begin to design molecules that selectively interact with these receptors. Although these pharmacodynamic aspects are essential starting points in drug design, all of these discussions have ignored an equally important aspect of drug design, namely, pharmacokinetics, the absorption, distribution, metabolism, and excretion of drugs. Absorption and distribution often are accommodated by changes in the lipophilicity and hydrophilicity of the lead molecules. In Chapter 7 we turn our attention to the heroic efforts our bodies make to destroy and excrete xenobiotics, including drugs; then in Chapter 8 we discuss the heroic efforts medicinal chemists make to outwit these metabolic processes by protecting the drugs until they reach the desired sites of action.

## *References*

1. Sancar, A., and Sancar, G. B. 1988. *Annu. Rev. Biochem.* **57**, 29.
2. Friedberg, E. C. 1984. "DNA Repair." Freeman, San Francisco, California.
3. Borison, H. L., Brand, E. D., and Orkand, R. K. 1968. *Am. J. Physiol.* **192**, 410.
4. Collins, A. R. S., Squires, S., and Johnson, R. T. 1982. *Nucleic Acids Res.* **10**, 1203.
5. Cadman, E., Heimer, R., and Davis, L. 1979. *Science* **205**, 1135.
6. Greenberger, L. M., Williams, S. S., and Horwitz, S. B. 1987. *J. Biol. Chem.* **262**, 13685.
7. Riordan, J. R., Deuchars, K., Kartner, N., Alon, N., Trent, J., and Ling, V. 1985. *Nature (London)* **316**, 817.
8. Marchand, D. H., Remmel, R. P., and Abdel-Monem, M. M. 1988. *Drug Metab. Dispos.* **16**, 85.
9. Watson, J. D., and Crick, F. H. C. 1953. *Nature (London)* **171**, 737; Crick, F. H. C., and Watson, J. D. 1954. *Proc. R. Soc., London Ser. A* **223**, 80.
10. Watson, J. D. 1968. "The Double Helix." Weidenfeld and Nicholson, London.
11. Dekker, C. A., Michelson, A. M., and Todd, A. R. 1953. *J. Chem. Soc.* 947.
12. Zamenhof, S., Braverman, G., and Chargaff, E. 1952. *Biochim. Biophys. Acta* **9**, 402.
13. Astbury, W. T. 1947. *Symp. Soc. Exp. Biol.*, 66.
14. Gulland, J. M. 1947. *Cold Spring Harbor Symp. Quant. Biol.* **12**, 95.
15. Wilkins, M. H. F. 1963. *Science* **140**, 941.
16. Furberg, S. 1950. *Acta Crystallogr.* **3**, 325.

17. Hanlon, S. 1966. *Biochem. Biophys. Res. Commun.* **23**, 861.
18. Herskovits, T. T. 1962. *Arch. Biochem. Biophys.* **97**, 474.
19. Beak, P. 1977. *Acc. Chem. Res.* **10**, 186.
20. Wolfenden, R. V. 1969. *J. Mol. Biol.* **40**, 307.
21. Sepiol, J., Kazimierczuk, Z., and Shugar, D. 1976. *Z. Naturforsch. C: Biosci.* **31**, 361.
22. Chenon, M.-T., Pugmire, R. J., Grant, D. M., Panzica, R. P., and Townsend, L. B. 1975. *J. Am. Chem. Soc.* **97**, 4636.
23. Liu, L. F. 1989. *Annu. Rev. Biochem.* **58**, 351.
24. Halligan, B. D., Edwards, K. A., and Liu, L. F. 1985. *J. Biol. Chem.* **260**, 2475.
25. Champoux, J. J. 1981. *J. Biol. Chem.* **256**, 4805.
26. Rowe, T. C., Chen, G. L., Hsiang, Y.-H., and Liu, L. F. 1986. *Cancer Res.* **46**, 2021.
27. Dervan, P. B. 1986. *Science* **232**, 464; Hurley, L. H., and Boyd, F. L. 1987. *Annu. Rep. Med. Chem.* **22**, 259; Hurley, L. H. 1989. *J. Med. Chem.* **32**, 2027.
28. von Hippel, P. H., and Berg, O. G. 1986. *Proc. Natl. Acad. Sci. U.S.A.* **83**, 1608.
29. Neidle, S., and Abraham, Z. 1984. *Crit. Rev. Biochem.* **17**, 73.
30. Lerman, L. S. 1961. *J. Mol. Biol.* **3**, 18.
31. Krugh, T. R., and Reinhardt, C. G. 1975. *J. Mol. Biol.* **97**, 133. Nuss, M. E., Marsh, F. J., and Kollman, P. A. 1979. *J. Am. Chem. Soc.* **101**, 825.
32. Hsiang, Y.-H., Liu, L. F., Wall, M. E., Wani, M. C., Nicholas, A. W., Manikumar, G., Kirschenbaum, S., Silber, R., and Potmesil, M. 1989. *Cancer Res.* **49**, 4385.
33. Nelson, E. M., Tewey, K. M., and Liu, L. F. 1984. *Proc. Natl. Acad. Sci. U.S.A.* **81**, 1361.
34. Zhang, H., D'Arpa, P., and Liu, L. F. 1990. *Cancer Cells* **2**, 23.
35. Bodley, A., Liu, L. F., Israel, M., Seshadri, R., Koseki, Y., Giuliani, F. C., Kirschenbaum, S., Silber, R., and Potmesil, M. 1989. *Cancer Res.* **49**, 5969; D'Arpa, P., and Liu, L. F. 1989. *Biochim. Biophys. Acta* **989**, 163. (c) Zunino F., and Capranico, G. (1990). *Anti-Cancer Drug Design* **5**, 307.
36. Mannaberg, J. 1897. *Arch. Klin. Med.* **59**, 185.
37. Albert, A. 1966. "The Acridines," 2nd Ed. Arnold, London.
38. Cain, B. F., Atwell, G. J., Seelye, R. N. 1971. *J. Med. Chem.* **14**, 311.
39. Goldin, A., Serpick, A. A., and Mantel, N. 1966. *Cancer Chemother. Rep.* **50**, 173.
40. Atwell, G. J., Cain, B. F., and Seelye, R. N. 1972. *J. Med. Chem.* **15**, 611.
41. Denny, W. A., Cain, B. F., Atwell, G. J., Hansch, C., Panthananickal, A., and Leo, A. 1982. *J. Med. Chem.* **25**, 276.
42. Zittoun, R. 1985. *Cancer Treat. Rep. 69*, 1447.
43. Cain, B. F., Atwell, G. J., and Denny, W. A. 1975. *J. Med. Chem.* **18**, 1110.
44. Waring, M. J. 1976. *Eur. J. Cancer* **12**, 995.
45. Braithwaite, A. W., and Baguley, B. C. 1980. *Biochemistry* **19**, 1101.
46. Sakore, T. D., Reddy, B. S., and Sobell, H. M. 1979. *J. Mol. Biol.* **135**, 763.
47. Denny, W. A., Baguley, B. C., Cain, B. F., and Waring, M. J. 1983. *In* "Molecular Aspects of Anticancer Drug Action" (Neidle, S., and Waring, M. J., eds.), p. 1. Verlag Chemie, Weinheim.
48. Denny, W. A., and Wakelin, L. P. G. 1986. *Cancer Res.* **46**, 1717.
49. Baguley, B. C., Denny, W. A., Atwell, G. J., Finlay, G. J., Rewcastle, G. W., Twigden, S. J., and Wilson, W. R. 1984. *Cancer Res.* **44**, 3245.
50. Waksman, S. A., and Woodruff, H. B. 1940. *Proc. Soc. Exp. Biol. Med.* **45**, 609.
51. Schulte, G. 1952. *Z. Krebsforsch.* **58**, 500.
52. Sobell, H. M. 1985. *Proc. Natl. Acad. Sci. U.S.A.* **82**, 5328.
53. Müller, W., Crothers, D. M. 1968, *J. Mol. Biol.* **35**, 251.
54. Cerami, A., Reich, E., Ward, D. C., and Goldberg, I. H. 1967. *Proc. Natl. Acad. Sci. U.S.A.* **57**, 1036.
55. Sobell, H. M., and Jain, S. C. 1972, *J. Mol. Biol.* **68**, 21.

56. Takusagawa, F., Dabrow, M., Neidle, S., and Berman, H. M. 1982. *Nature (London)* **296**, 466.
57. Takusagawa, F., Goldstein, B. M., Youngster, S., Jones, R. A., and Berman, H. M. 1984. *J. Biol. Chem.* **259**, 4714.
58. Takusagawa, F. 1985. *J. Antibiot.* **38**, 1596.
59. Chiao, Y.-C. C., and Krugh, T. R. 1977. *Biochemistry* **16**, 747.
60. Pastan, I., and Gottesman, M. (1987). *N. Engl. J. Med.* **316**, 1388. Polet, H. 1975. *J. Pharmacol. Exp. Ther.* **192**, 270; Bowen, D., and Goldman, I. D. 1975. *Cancer Res.* **35**, 3054.
61. Bosmann, H. B. 1971. *Nature (London)* **233**, 566.
62. Schwarz, H. S. 1974. *Cancer Chemother.* **58**, 55.
63. Quigley, G. J., Wang, A. H.-J., Ughetto, G., van der Marel, G., van Boom, J. H., and Rich, A. 1980. *Proc. Natl. Acad. Sci. U.S.A.* **77**, 7204.
64. Patel, D. J., Kozlowski, S. A., and Rice, J. A. 1981. *Proc. Natl. Acad. Sci. U.S.A.* **78**, 3333.
65. Pigram, W. J., Fuller, W., and Hamilton, L. D. 1972. *Nature (London) New Biol.* **235**, 17.
66. Henry, D. W. 1979. *Cancer Treat. Rep.* **63**, 845.
67. Wright, R. G. McR., Wakelin, L. P. G., Fieldes, A., Acheson, R. M., Waring, M. J. 1980. *Biochemistry* **19**, 5825.
68. Wakelin, L. P. G. 1986. *Med. Res. Rev.* **6**, 275.
69. Krumbhaar, E. B., and Krumbhaar, H. D. 1919. *J. Med. Res.* **40**, 497.
70. Adair, F. E., and Bagg, H. J. 1931. *Ann. Surg. (Chicago)* **93**, 190.
71. Gilman, A., and Philips, F. S. 1946. *Science* **103**, 409.
72. Rhoads, C. P. 1946. *J. Am. Med. Assoc.* **131**, 656; Goodman, L. S., Wintrobe, M. M., Dameshek, W., Goodman, M. J., Gilman, A., and McLennan, M. 1946. *J. Am. Med. Assoc.* **132**, 126.
73. Ross, W. C. J. 1962. "Biological Alkylating Agents." Butterworth, London.
74. Montgomery, J. A., Johnston, T. P., and Shealy, Y. F. 1979. *In* "Burger's Medicinal Chemistry" (Wolff, M. E., ed.) 4th Ed., Part 2, p. 595. Wiley, New York.
75. Lawley, P. D., and Brookes, P. 1963. *Biochem. J.* **89**, 127.
76. Pullman, A., and Pullman, B. 1980. *Int. J. Quantum Chem., Quantum Biol. Symp.* **7**, 245.
77. Price, C. C. 1974. *In* "Handbook of Experimental Pharmacology" (Sartorelli, A. C., and Johns, D. J., eds.), Vol. 38, Part 2, p. 4. Springer-Verlag, Berlin.
78. Kohn, K. W., Spears, C. L., and Doty, P. 1966. *J. Mol. Biol.* **19**, 266; Lawley, P. D., and Brookes, P. 1967. *J. Mol. Biol.* **25**, 143.
79. Colvin, M., and Chabner, B. A. 1990. *In* "Cancer Chemotherapy: Principles and Practice" (Chabner, B. A., and Collins, S. M., eds.), p. 276. Lippincott, Philadelphia, Pennsylvania.
80. Bardos, T. J., Chmielewicz, Z. F., and Hebborn, P. 1969. *Ann. N. Y. Acad. Sci.* **163**, 1006.
81. Niculescu-Duvăz, I., Baracu, I., and Balaban, A. T. 1990. *In* "The Chemistry of Antitumour Agents" (Wilman, D. E. V., ed.), p. 63. Blackie, Glasgow.
82. Kohn, K. W., Erickson, L. C., Laurent, G., Ducore, J. M., Sharkey, N. A., and Ewig, R. A. G. 1981. *In* "Nitrosoureas, Current Status and New Developments" (Prestayo, W., Crooke, S. T., Karter, S. K., and Schein, P. S., eds.), p. 69. Academic Press, New York.
83. Harris, A. L. 1985. *Cancer Surv.* **4**, 601.
84. Prelog, V., and Stepan, V. 1935. *Collect. Czech. Chem. Commun.* **7**, 93.
85. Brookes, P., and Lawley, P. D. 1961. *Biochem. J.* **80**, 496.
86. Haddow, A., Kon, G. A. R., and Ross, W. C. J. 1948. *Nature* **162**, 824.
87. Everett, J. L., Roberts, J. J., and Ross, W. C. J. 1953. *J. Chem. Soc.*, 2386.
88. Schmidt, L. H., Fradkin, R., Sullivan, R., and Flowers, A. 1965. *Cancer Chemother. Rep.* (Suppl. 2), 1.
89. Vistica, D. T. 1983. *Pharmacol. Ther.* **22**, 379.
90. Goldin, A., and Wood, H. B., Jr. 1969. *Ann. N.Y. Acad. Sci.* **163**, 954.

91. Khan, A. H., and Driscoll, J. S. 1976. *J. Med. Chem.* **19**, 313.
92. Haddow, A., and Timmis, G. M. 1953. *Lancet* **1**, 207.
93. Timmis, G. M., and Hudson, R. F. 1958. *Ann. N.Y. Acad. Sci.* **68**, 727.
94. Brookes, P., and Lawley, P. D. 1961. *J. Chem. Soc.*, 3923.
95. Tong, W. P., and Ludlam, D. B. 1980. *Biochim. Biophys. Acta* **608**, 174.
96. Skinner, W. A., Gram, H. F., and Greene, M. O. 1960. *J. Med. Pharm. Chem.* **2**, 299.
97. Schabel, F. M., Jr., Johnston, T. P., McCaleb, G. S., Montgomery, J. A., Laster, W. R., and Skipper, H. E. 1963. *Cancer Res.* **23**, 725.
98. Montgomery, J. A., James, R., McCaleb, G. S., and Johnston, T. P. 1967. *J. Med. Chem.* **10**, 668.
99. Wheeler, G. P. 1974. *In* "Handbook of Experimental Pharmacology" (Sartorelli, A. C., and Johns, D. G., eds.), Vol. 38, Part 2, p. 7. Springer-Verlag, Berlin.
100. Johnston, T. P., and Montgomery, J. A. 1986. *Cancer Treat. Rep.* **70**, 13.
101. Tong, W. P., Kirk, M. C., and Ludlum, D. B. 1982. *Cancer Res.* **42**, 3102; Tong, W. P., Kirk, M. C., and Ludlum, D. B. 1983. *Biochem. Pharmacol.* **32**, 2011; Lown, J. W., McLaughlin, L. W., and Chang, Y.-M. 1978. *Bioorg. Chem.* **7**, 97.
102. Buckley, N., and Brent, T. P. 1988. *J. Am. Chem. Soc.* **110**, 7520.
103. Piper, J. R., Laseter, A. G., Johnston, T. P., and Montgomery, J. A. 1980. *J. Med. Chem.* **23**, 1136.
104. Erickson, L. C., Sharkey, N. A., and Kohn, K. W. 1980. *Nature (London)* **288**, 727; Brent, T. P., Houghton, P. J., and Houghton, J. A. 1985. *Proc. Natl. Acad. Sci. U.S.A.* **82**, 2985.
105. Ludlum, D. B., Mehta, J. R., and Tong, W. P. 1986. *Cancer Res.* **46**, 3353.
106. Schmall, B., Cheng, C. J., Fujimura, S., Gersten, N., Grunberger, D., and Weinstein, I. B. 1973. *Cancer Res.* **33**, 1921.
107. Baril, B. B., Baril, E. F., Laszlo, J., and Wheeler, G. P. 1975. *Cancer Res.* **35**, 1.
108. Kann, H. E., Jr., Blumenstein, B. A., Petkas, A., and Schott, M. A. 1980. *Cancer Res.* **40**, 771; Robins, P., Harris, A. L., Goldsmith, I., and Lindahl, T. 1983. *Nucleic Acids Res.* **11**, 7743.
109. Kann, H. E., Jr., Kohn, K. W., Widerlite, L., and Gullion, D. 1974. *Cancer Res.* **34**, 1982.
110. Panasci, L. C., Green, D., Nagourney, R., Fox, P., and Schein, P. S. 1977. *Cancer Res.* **37**, 2615.
111. Rosenberg, B., Van Kamp, L., and Krigas, T. 1965. *Nature (London)* **205**, 698; Rosenberg, B., Van Camp, L., Trosko, J. E., and Mansour, V. H. 1969. *Nature (London)* **222**, 385.
112. Sherman, S. E., and Lippard, S. J. 1987. *Chem. Rev.* **87**, 1153.
113. Pinto, A. L., and Lippard, S. J. 1985. *Biochim. Biophys. Acta* **780**, 167; Caradonna, J. P., and Lippard, S. J. 1988. *Inorg. Chem.* **27**, 1454.
114. Eastman, A. 1986. *Biochemistry* **25**, 3912.
115. Plooy, A. C. M., van Dijk, M., and Lohman, P. H. M. 1984. *Cancer Res.* **44**, 2043.
116. Waud, W. R. 1987. *Cancer Res.* **47**, 6549.
117. Micetich, K., Zwelling, L. A., and Kohn, K. W. 1983. *Cancer Res.* **43**, 3609; Kelley, S. L., Basu, A., Teicher, B. A., Hacker, M. P., Hamer, D. H., and Lazo, J. S. 1988. *Science* **241**, 1813.
118. Eastman, A., and Schulte, N. 1988. *Biochemistry* **27**, 4730.
119. Lown, J. W., Sim, S.-K., Majumdar, K. C., and Chang, R.-Y. 1977. *Biochem. Biophys. Res. Commun.* **76**, 705; Bachur, N. R., Gordon, S. L., and Gee, M. V. 1977. *Cancer Res.* **38**, 1745.
120. Ross, W. A., Glaubiger, D. L., and Kohn, K. W. 1978. *Biochim. Biophys. Acta* **519**, 23.
121. Tewey, K. M., Chen, G. L., Nelson, E. M., and Liu, L. F. 1984. *J. Biol. Chem.* **259**, 9182.
122. Pan, S.-S., Pedersen, L., and Bachur, N. R. 1981. *Mol. Pharmacol.* **19**, 184.
123. Hertzberg, R. P., and Dervan, P. B. 1984. *Biochemistry* **23**, 3934.
124. Garnier-Suillerot, A. 1988. *In* "Anthracycline and Anthracenedione-Based Anticancer Agents" (Lown, J. W., ed.), p. 129. Elsevier, Amsterdam.

125. Walling, C. 1975. *Acc. Chem. Res.* **8**, 125.
126. Tullius, T. D., Dombrowski, B. A., Churchill, M. E. A., and Kam, L. 1987. *In* "Methods in Enzymology" (Wu, R., ed.), Vol. 155, p. 537. Academic Press, Orlando, Florida.
127. Speyer, J. L., Green, M. D., Kramer, E., Rey, M., Sanger, J., Ward, C., Dubin, N., Ferran, V., Stecy, P., Zeleniuch-Jaquotte, A., Wernz, J., Feit, F., Slater, W., Blum, R., and Mugia, F. 1988. *N. Engl. J. Med.* **319**, 745.
128. Gianni, L., Vigano, L., Lanzi, C., Niggeler, M., and Malatesta, V. 1988. *J. Natl. Cancer Inst.* **80**, 1104.
129. Sulikowski, G. A., Turos, E., Danishefsky, S. J., and Schulte, G. M. 1991. *J. Am. Chem. Soc.* **113**, 1373.
130. Umezawa, H., Suhara, Y., Takita, T., and Maeda, K. 1966. *J. Antibiot., Ser. A.* **19**, 210.
131. Stubbe, J., and Kozarich, J. W. 1987. *Chem. Rev.* **87**, 1107.
132. Sausville, E. A., Stein, R. W., Peisach, J., and Horwitz, S. B. 1978. *Biochemistry* **17**, 2746; Hecht, S. M. 1986. *Acc. Chem. Res.* **19**, 383.
133. Fisher, L. M., Kuroda, R., and Sakai, T. T. 1985. *Biochemistry* **24**, 3199.
134. Ciriolo, M. R., Magliozzo, R. S., and Peisach, J. 1987. *J. Biol. Chem.* **262**, 6290; Mahmutoglu, I., and Kappus, H. 1987. *Biochem. Pharmacol.* **36**, 3677.
135. Umezawa, H., Takita, T., Sugiura, Y., Otsuka, M., Kobayashi, S., and Ohno, M. 1984. *Tetrahedron* **40**, 501.
136. Povirk, L. F., Hogan, M., and Dattagupta, N. 1979. *Biochemistry* **18**, 96.
137. Giloni, L., Takeshita, M., Johnson, F., Iden, C., and Grollman, A. P. 1981. *J. Biol. Chem.* **256**, 8608. Murugesan, N., Xu, C., Ehrenfeld, G. M., Sugiyama, H., Kilkuskie, R. E., Rodriguez, L. O., Chang, L.-H., and Hecht, S. M. 1985. *Biochemistry* **24**, 5735.
138. Ross, S. L., and Moses, R. E. 1978. *Biochemistry* **17**, 581.
139. Sugiyama, H., Xu, C., Murugesan, N., Hecht, S. M., van der Marel, G. A., and van Boom, J. H. 1988. *Biochemistry* **27**, 58.
140. Rabow, L. E., Stubbe, J., and Kozarich, J. W. 1990. *J. Am. Chem. Soc.* **112**, 3196.
141. Rabow, L. E., McGall, G. H., Stubbe, J., and Kozarich, J. W. 1990. *J. Am. Chem. Soc.* **112**, 3203.
142. D'Andrea, A. D., and Haseltine, W. A. 1978. *Proc. Natl. Acad. Sci. U.S.A.* **75**, 3608; Takeshita, M., Grollman, A. P., Ohtsubo, E., and Ohtsubo, H. 1978. *Proc. Natl. Acad. Sci. U.S.A.* **75**, 5983.
143. Umezawa, H., Takeuchi, S., Hori, T., Sawa, T., Ishizuka, T., Ichikawa, T., and Komai, T. 1972. *J. Antibiot.* **25**, 409.
144. Sugiura, Y., Muraoka, Y., Fujii, A., Takita, T., and Umezawa, H. 1979. *J. Antibiot.* **32**, 756.
145. Ishida, N., Miyazaki, K., Kumagai, K., and Rikimaru, M. 1965. *J. Antibiot.* **18**, 68.
146. Konishi, M., Ohkuma, H., Saitoh, K., Kawaguchi, H., Golik, J., Dubay, G., Groenewold, G., Krishnan, B., and Doyle, T. W. 1985. *J. Antibiot.* **38**, 1605.
147. Lee, M. D., Dunne, T. S., Siegel, M. M., Chang, C. C., and Morton, G. O., and Borders, D. B. 1987. *J. Am. Chem. Soc.* **109**, 3464.
148. Konishi, M., Ohkuma, H., Matsumoto, K., Tsuno, T., Kamei, H., Miyaki, T., Oki, T., Kawaguchi, H., Van Duyne, G. D., and Clardy, J. 1989. *J. Antibiot.* **42**, 1449.
149. Sluka, J. P., Horvath, S. J., Bruist, M. F., Simon, M. I., and Dervan, P. B. 1987. *Science* **238**, 1129.
150. Long, B. H., Golik, J., Forenza, S., Ward, B., Rehfuss, R., Dabrowiak, J. C., Catino, J. J., Musial, S. T., Brookshire, K. W., and Doyle, T. W. 1989. *Proc. Natl. Acad. Sci. U.S.A.* **86**, 2.
151. Zein, N., Sinha, A. M., McGahren, W. J., and Ellestad, G. A. 1988. *Science* **240**, 1198.
152. Lockhart, T. P., Comita, P. B., and Bergman, R. G. 1981. *J. Am. Chem. Soc.* **103**, 4082.
153. Sugiura, Y., Arawaka, T., Uesugi, M., Siraki, T., Ohkuma, H., and Konishi, M. 1991. *Biochemistry* **30**, 2989.

154. Dasgupta, D., and Goldberg, I. H. 1985. *Biochemistry* **24**, 6913.
155. Myers, A. G. 1987. *Tetrahedron Lett.* **28**, 4493.
156. Frank, B. L., Worth, L., Jr., Christner, D. F., Kozarich, J. W., Stubbe, J., Kappen, L. S., and Goldberg, I. H. 1991. *J. Am. Chem. Soc.* **113**, 2271; Kappen, L. S., and Goldberg, I. H. 1985. *Nucleic Acids Res.* **13**, 1637; Kappen, L. S., Goldberg, I. H., Frank, B. L., Worth, L., Jr., Christner, D. F., Kozarich, J. W., and Stubbe, J. 1991. *Biochemistry* **30**, 2034.
157. Sugiura, Y., Uesawa, Y., Takahashi, Y., Kuwahara, J., Golik, J., and Doyle, T. W. 1989. *Proc. Natl. Acad. Sci. U.S.A.* **86**, 7672.
158. Sugiura, Y., Shiraki, T., Konishi, M., and Oki, T. 1990. *Proc. Natl. Acad. Sci. U.S.A.* **87**, 3831.

## General References

### DNA Structure and Function

Alberts, B., Bray, D., Lewis, J., Raff, M., Roberts, K., and Watson, J. D. 1989. "Molecular Biology of the Cell," 2nd Ed.; Garland Publ., New York.
Blackburn, G. M., and Gait, M. J. (1990). *Nucleic Acids Chem. Biol.* JRL Press, Oxford.
Palecek, E. 1991. *Crit. Rev. Biochem. Mol. Biol.* **26**, 151.
Saenger, W. 1984. "Principles of Nucleic Acid Structure." Springer-Verlag, New York.
Watson, J. D., Hopkins, N. H., Roberts, J. W., Steitz, J. A., and Weiner, A. M. 1987. "Molecular Biology of the Gene," 4th Ed., Vol. 1. Benjamin/Cummings Publ., Menlo Park, California.

### Topoisomerases

Wang, J. C. 1985. *Annu. Rev. Biochem.* **54**, 665.

### Antitumor Drugs

Baguley, B. C. (1991). *Anti-Cancer Drug Design* **6**, 1.
Chabner, B. A., and Collins, J. M. 1990. "Cancer Chemotherapy: Principles and Practice." Lippincott, Philadelphia, Pennsylvania.
Neidle, S., and Waring, M. J. 1983. "Molecular Aspects of Anti-Cancer Drug Action." Verlag Chemie, Weinheim.
Pratt, W. B., and Ruddon, R. W. 1979. "The Anticancer Drugs." Oxford Univ. Press, New York.
Remers, W. A. 1984. "Antineoplastic Agents." Wiley, New York.
Wilman, D. E. V. 1990. "The Chemistry of Antitumour Agents." Blackie, Glasgow.

### Drug Resistance

Gupta, R. S. 1989. "Drug Resistance in Mammalian Cells," Vols. 1 and 2. CRC Press, Boca Raton, Florida.
Woolley III, P. V., and Tew, K. D. 1988. "Mechanisms of Drug Resistance in Neoplastic Cells." Academic Press, San Diego, California.

CHAPTER 7

# Drug Metabolism



## I. Introduction

When a foreign organism enters the body, the immune system produces antibodies to interact with and destroy it. Small molecules, however, generally do not stimulate an antibody response. So how has the human body evolved to protect itself against low molecular weight environmental toxins? The principal mechanism is the use of nonspecific enzymes that transform the toxins (often highly nonpolar molecules) into polar molecules which are excreted by the normal bodily processes. Although this prophylactic mechanism to rid the body of *xenobiotics* is highly desirable, especially when one considers all of the toxic materials to which we are exposed every day, it can cause problems when the toxin is a drug that we want to enter and be retained in the body long enough for it to be effective. The enzymatic biotransformations of drugs are known as *drug metabolism*. Since many drugs have structures similar to those of endogenous compounds, drugs may be metabolized by specific enzymes for the related natural substrates as well as by nonspecific enzymes.

The principal site of drug metabolism is the liver, but the kidneys, lungs, and gastrointestinal (GI) tract also are important metabolic sites. When a drug is taken orally (the most common and desirable method of administration), it is absorbed through the mucous membrane of the small intestine or from the stomach. Once out of the GI tract it is generally carried via the bloodstream to the liver, where it is usually first metabolized. This drug metabolism by liver enzymes is called the presystemic or *first-pass effect*, which can completely deactivate the drug. If only a fraction of the drug molecules is metabolized, then larger or multiple doses of the drug will be required to get the desired effect. Another significant undesirable effect of drug metabolism is that the metabolites of a drug may be toxic, even though the drug is not.

The first-pass effect can be avoided by changing the route of drug administration. The *sublingual route* (the drug is placed under the tongue) bypasses the liver. After absorption through the buccal cavity, the drug enters the bloodstream directly. This route is used with nitroglycerin (**7.1**; glyceryl trinitrate) for angina pectoris. The *rectal route*, in the form of a solid suppository or in solution as an enema, leads to absorption through the colon mucosa. Ergotamine (**7.2**), a drug used for migraine headaches, is administered this way (who would have guessed?). *Intravenous* (i.v.) *injection* introduces the drug directly into the systemic circulation and is used when a rapid therapeutic response is desired. The effects are almost immediate when drugs are administered by this route because the total blood circulation time in man is about 15 sec. *Intramuscular* (i.m.) *injection* is used when large volumes of drugs need to be administered or if slow absorption is desirable. A *subcutaneous* (s.c.) *injection* delivers the drug through the loose connective tissue of the subcutaneous layer of the skin. Another method of administration, particularly for gaseous or highly volatile drugs such as anesthetics, is by *pulmonary absorption* through the respiratory tract. The asthma drug, isoproterenol (**7.3**), is metabolized in the intestines and liver, but administration by aerosol inhalation is effective at getting the drug directly to the lungs. *Topical application* of the drug to the skin or mucous membrane is used for local effects, because few drugs readily penetrate the intact skin.

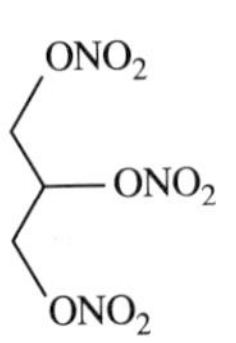

**7.1** 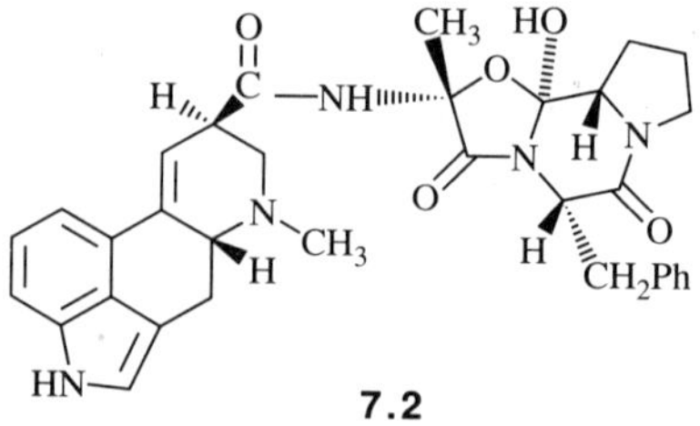 **7.2** 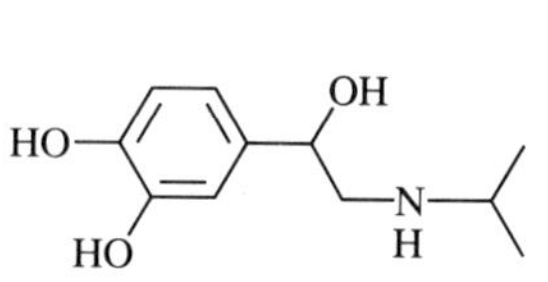 **7.3**

Not all drugs can be administered by the alternate routes, so their structures may have to be altered in order to minimize the first-pass effect or to permit them to be administered by one of the alternate routes. These structural modification approaches in drug design to avoid the first-pass effect are discussed in Chapter 8. Even if the first-pass effect is avoided, there are many enzymes outside of the liver that are capable of catalyzing drug metabolism reactions. Once a drug has reached its site of action and elicited the desired response, it usually is desirable for the drug to be metabolized and eliminated. Otherwise, it may remain in the body and produce the effect longer than desired or become toxic to the cells.

Drug metabolism studies are essential for the determination of the safety of potential drugs. Consequently, prior to approval of a drug for human use the metabolites produced from the drug must be isolated and shown to be nontoxic. These studies also can be a useful lead modification approach. Once the metabolic products are known, it is possible to design a compound that is inactive when administered, but which utilizes the metabolic enzymes to convert it to the active form. These compounds are known as *prodrugs*, and are discussed in Chapter 8. In this chapter we consider the various reactions that are involved in the biotransformations of drugs. Because only very small quantities of drugs generally are required to elicit the appropriate response, it may be difficult to detect all of the metabolic products. In order to increase the sensitivity of the detection process, the drug candidates are radioactively labeled. In the next two sections we look briefly at how radiolabeling is carried out, how metabolites are detected, and how their structures are elucidated.

## II. Synthesis of Radioactive Compounds

Because of the sensitivity of detection of particles of radioactive decay, the most common approach used for detection of metabolites in whole animal studies is the incorporation of a radioactive label (usually a weak $\beta$-emitter such as $^{14}C$ or $^{3}H$) into the drug molecule. When this approach is used, it does not matter how few metabolites or how small the quantities of metabolites are produced, even in the presence of a large number of endogenous compounds. Only the radioactively labeled compounds are isolated from the urine and the feces of the animals, and the structures of these metabolites are elucidated (see Section III).

To incorporate a radioactive label into a compound a synthesis must be designed so that a commercially available radioactively labeled compound or reagent can be used in one of the steps, preferably a step at or near the end of the synthesis. Often the radioactive synthesis is quite different from or longer than the synthesis of the unlabeled compound so that it is possible to utilize a commercially available radioactive material. It is preferable to prepare a $^{14}C$-labeled analog; when tritium is incorporated into the drug, the site of incorporation must be such that loss of the tritium by exchange with the medium does not occur even after an early metabolic step. Generally, only one radioactive label is incorporated into a drug because drug metabolism typically leads to a modified structure with little fragmentation of the molecule. If one goes back far enough in a synthesis, however, it is possible to synthesize a drug with all or almost all carbon atoms radioactively labeled. Radioactive labeling of all of the carbons would permit the identification of any metabolic fragments of the drug.

It must be kept in mind when working with radioactively labeled compounds that commercially available radioactive compounds contain only a trace amount of radioactivity. This means that maybe only 1 in $10^6$ molecules contains the radioactive tag; the remainder of the molecules are unlabeled and are carriers of the relatively few radioactive molecules. In the case of $^{14}C$ there will be no difference in the reactivity of the labeled and unlabeled molecules, so there is the same statistical amount of radioactivity in the products formed as was in the starting materials. The *specific radioactivity* (a measure of the amount of radioactivity per amount of compound) of the metabolites formed during metabolism, then, should be identical to the specific radioactivity of the drug. In the case of tritiated drugs, however, if a carbon–hydrogen bond is broken in a rate-determining step of an enzymatic reaction, and that hydrogen is replaced with a tritium label, there will be a *kinetic isotope effect* on those molecules that are tritiated. This will lead to metabolite formation with a lower specific radioactivity than that of the drug. As a result, quantitation of the various metabolic pathways, where some require C—H bond cleavage and others do not, would require knowledge of the tritium isotope effect. This, then, is another reason it is preferable to use $^{14}C$-labeling of a drug for metabolism studies rather than tritium labeling.

If the drug is a natural product or derivative of a natural product, the easiest procedure for incorporation of a radioactive label could be a biosynthetic approach, namely, to grow the organism that produces the natural product in the presence of a radioactive precursor and let Mother Nature incorporate the radioactivity into the molecule. Because of the volume of media generally involved, and, therefore, the large amount of radioactive precursor required, this could be a very expensive approach; however, in some cases the expense may be compensated by the ease of the method and an attractive yield of product obtained.

An example of a drug class with which this approach could be used is the penicillins, which are biosynthesized by *Penicillium* fungi from valine, cysteine, and various carboxylic acids (Scheme 7.1). Valine is commercially available with a $^{14}C$ label at the carboxylate or may be obtained uniformly

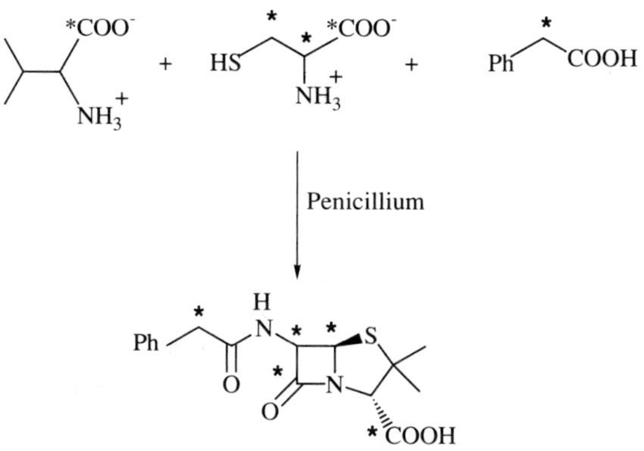

Scheme 7.1. Biosynthesis of penicillins.

labeled, that is, with all of the carbon atoms labeled to some small extent with $^{14}C$ (albeit very few molecules would contain all of the carbon atoms labeled in the same molecule). It also can be purchased with a tritium label at the 2 and 3 positions or at the 3 and 4 positions. Cysteine is available uniformly labeled in $^{14}C$ or with a $^{35}S$ label. Penicillin G could be produced if phenylacetic acid (available with a $^{14}C$ label at either the 1 or 2 position) were inoculated into the *Penicillium* growth medium.

If the drug is not a natural product, which is the case for nearly all drugs, a chemical synthesis must be carried out. For example, the nonradioactive synthesis of the anti-inflammatory drug sulindac (**7.11**) is shown in Scheme 7.2.[1] Friedel–Crafts acylation of fluorobenzene with 2-bromo-2-methylpropionyl bromide (**7.4**) gives **7.5** in greater than 90% yield without isolation of the intermediate $\alpha$-bromo ketone. The mechanism for this reaction was studied in detail.[2] Addition of 4-methylthiobenzylmagnesium chloride (**7.7**), prepared by the reaction of 4-methylthiobenzyl chloride (**7.6**) with magnesium in ether, to **7.5** followed by treatment of the product with sulfuric acid in acetic acid gave 6-fluoro-3-methyl-3-(4-methylthiobenzyl)indene (**7.8**) in an 84% yield. Condensation, catalyzed by Triton B (benzyltrimethylammonium hydroxide), and dehydration of **7.8** with glyoxylic acid gave (*Z*)-6-fluoro-2-methyl-3-(4-methylthiobenzyl)-1-indenylideneacetic acid (**7.9**) in 75% yield. Tautomerization of **7.9** to **7.10** was carried out smoothly in 93% yield in a solution of hydrochloric acid in acetic acid. Oxidation of **7.10** to sulindac (**7.11**) was performed in 92% yield with hydrogen peroxide in acetic acid.

Because of the number of steps involved, there are numerous opportunities to incorporate a radioactive label. As noted above, though, the radioactive label should be incorporated into the drug molecule as late as possible in the synthesis so special handling and microscale reactions can be avoided until near the end of the synthesis. In the synthesis shown in Scheme 7.2 the last opportunity for incorporation of $^{14}C$ into sulindac occurs in the conversion of

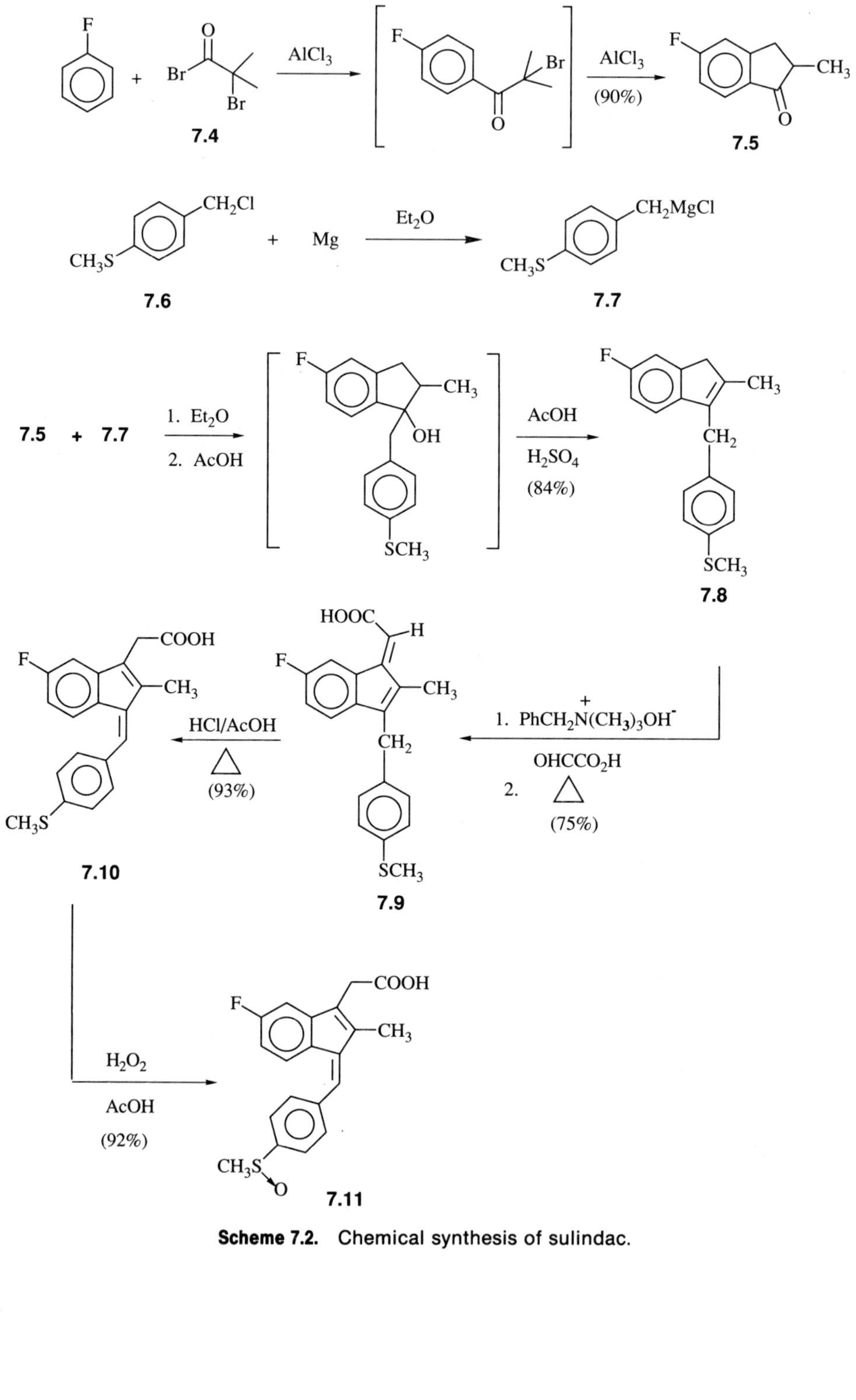

F + Br(C=O)C(Br) → $AlCl_3$ → [F ... Br, O] → $AlCl_3$ (90%) → F ... $CH_3$, O

7.4 7.5

$CH_3S$ ... $CH_2Cl$ + Mg → $Et_2O$ → $CH_3S$ ... $CH_2MgCl$

7.6 7.7

7.5 + 7.7 → 1. $Et_2O$ 2. AcOH → [F ... $CH_3$, OH, $SCH_3$] → AcOH, $H_2SO_4$ (84%) → F ... $CH_3$, $CH_2$, $SCH_3$

7.8

7.8 → 1. $PhCH_2\overset{+}{N}(CH_3)_3OH^-$ 2. $OHCCO_2H$, △ (75%) → 7.9 (HOOC, H, F, $CH_3$, $CH_2$, $SCH_3$)

7.9 → HCl/AcOH, △ (93%) → 7.10 (F, COOH, $CH_3$, $CH_3S$)

7.10 → $H_2O_2$, AcOH (92%) → 7.11 (F, COOH, $CH_3$, $CH_3S$→O)

**Scheme 7.2.** Chemical synthesis of sulindac.

**7.8** to **7.9,** during which step commercially available [$^{14}$C]glyoxylic acid could be used.

In most syntheses it is possible to label almost any carbon atom in the molecule in order to determine the metabolic fates of all parts of the drug. For example, consider the synthesis of sulindac again (Scheme 7.2). [$^{14}$C]Fluorobenzene can be synthesized from commercially available [$^{14}$C]aniline by diazotization ($NaNO_2$/HCl) and fluorination with $HBF_4$.[3] [1-$^{14}$C]2-Bromo-2-methylpropionyl bromide ([1-$^{14}$C]-**7.4**) can be prepared by the reaction of commercially available [1-$^{14}$C]isobutyric acid with bromine using catalytic red phosphorus.[4] This leads to **7.5** with a $^{14}$C label at the carbonyl carbon. If [3-$^{14}$C]isobutyric acid is used (prepared by the method for the synthesis of [3-$^{13}$C]isobutyric acid[5]), then the methyl group and the cyclopentenone ring of **7.5** can be labeled. 4-Methylthiobenzyl chloride (**7.6**) can be synthesized in a 74% yield by the reaction of methylthiobenzene with formaldehyde dimethyl acetal and aluminum chloride.[6] If [$^{14}$C]formaldehyde dimethyl acetal (prepared from commercially available [$^{14}$C]paraformaldehyde and methanol by the procedure carried out for the corresponding $^{13}$C-labeled compound[7]) is used, the chloromethyl methylene group of **7.6** will be labeled. The methylthio methyl group will be labeled if [*methyl*-$^{14}$C]methylthiobenzene (synthesized by the reaction of thiophenol with [$^{14}$C]methyl iodide[8]) is used in the above reaction. The reaction of [$^{14}$C]thiophenol[9] and unlabeled methyl iodide for the synthesis of methylthiobenzene would allow radiolabeling of the benzene ring of **7.6**. It also is possible to tritium-label sulindac in various positions, but, unless incorporation is in the aromatic rings, there is the risk of loss of label by exchange with water during metabolism.

Once the radioactive drug has been synthesized, it is used in metabolism and bioavailability studies first in rats, mice, or guinea pigs, then in dogs or monkeys. Typically, the urine and feces are collected from the animals, and the radioactive compounds are isolated and their structures identified (see Section III). If greater than 95% of the radioactivity can be recovered in the urine and feces, then the radioactive drug can be administered to humans. The U.S. Food and Drug Administration approves a maximum absorbed dose of 3 rem of radioactivity to a specific organ in a healthy adult volunteer for drug metabolism studies.[10] These radioactive levels are estimated from a determination of the absorbed dose in animal models, then 10–100 times lower amounts are used in the human studies. On rare occasion, other fluids such as saliva, cerebrospinal fluid, eye fluids, perspiration, or breath may be examined as well as various organs and tissues.

From the above discussion it appears that drug metabolism studies are straightforward; however, until relatively recently these studies were difficult, at best, to carry out. The ready commercial availability of radioactively labeled precursors made the synthetic work much less tedious. The advent of high-performance liquid chromatography and the advancements in column packing materials permitted the separation of many metabolites very similar in structure. Metabolites that were previously overlooked can now be detected and identified. Structure elucidation by various types of mass spectrometry and by various techniques of nuclear magnetic resonance spectrometry has been relatively routine. As a result of these advances in instrumentation more information can be gleaned from drug metabolism studies than ever before, and this can result in the discovery of new leads (see Section I,B,3 of Chapter 2) or in a basis for prodrug design (see Chapter 8). This, also, means that the U.S. Food and Drug Administration can demand that many more metabolites be identified and their toxicities be determined prior to drug approval (that is good news for consumers but bad news for drug companies). The final step in the process to prove the identity of a metabolite is to synthesize it and demonstrate that its spectral properties are identical to those of the metabolite. In the next section we look briefly at analytical methods in drug metabolism.

## III. Analytical Methods in Drug Metabolism

The four principal steps in drug metabolism studies are isolation (extraction), separation (chromatography), identification (spectrometry), and quantification of the metabolites. Detection systems are sensitive enough to allow the isolation and identification of submicrogram quantities of metabolites.

### A. Isolation

As discussed in Section I, the body usually converts drugs to more polar conjugates for excretion. Enzymatic hydrolysis ($\beta$-glucuronidase and arylsulfatase) of the conjugates releases the less polar drug metabolites for easier extraction and structure identification. A variety of solvents may be used to extract the biological fluid, with the pH being adjusted from basic to neutral to acidic prior to extractions. Ion-pair extraction[11,12] is used to remove hydrophilic ionizable compounds from aqueous solution. Salt–solvent pair extraction[13] separates metabolites into an ethyl acetate–soluble neutral and basic fraction, an ethyl acetate–soluble acidic fraction, and a water–soluble fraction. Various exchange resins such as the anion–exchange resin DEAE-Sephadex,[14] the cation-exchange resin Dowex 50,[15] and the nonionic resin Amberlite XAD-2[16] are used to separate acidic, basic, and neutral metabolites, respectively, from body fluids.

### B. Separation

The two most important techniques for resolving mixtures of metabolites are *high-performance liquid chromatography* (HPLC) and *capillary gas chromatography* (GC). HPLC is more versatile than GC because the metabolites can be charged or uncharged, they can be thermally unstable, and derivation is unnecessary. Normal-phase columns (silica gel) can be used for uncharged metabolites, and reversed-phase columns (silica gel to which $C_4$ to $C_{18}$ alkyl chains are attached to give a hydrophobic environment) can be used for charged metabolites. For GC separation the metabolites must be volatilized. This often requires prior derivatization[17,18] in order for the metabolites to volatilize at lower temperatures. Carboxylic acids can be converted to the corresponding methyl esters with diazomethane; hydroxyl groups can be trimethylsilylated with bis(trimethylsilyl)acetamide or trimethylsilylimidazole in pyridine. Ketone carbonyls can be converted to *O*-substituted oximes. Until recently, the major advantage of GC was that the derivatized sample could be separated and analyzed directly by a GC in tandem with a mass spectrometer. It is now possible, however, to do *tandem HPLC–mass spectrometry*, albeit this technique is still under development.[19]

### C. Identification

The two principal methods of metabolite structure identification are mass spectrometry (MS) and nuclear magnetic resonance (NMR) spectrometry. As indicated above, it is preferable to link the separation and identification steps by running tandem GC–MS or tandem HPLC–MS. In this way there is less chance for metabolite degradation or loss, and work-up procedures for mass spectrometry sample preparation are eliminated. *Mass spectrometers* are operated in several different modes: electron impact (EI), chemical ionization (CI), desorption chemical ionization (DCI), field desorption (FD), secondary-ion mass spectrometry (SIMS), laser desorption, and fast atom bombardment (FAB) mass spectrometry. They are sufficiently sensitive to identify subnanogram amounts of material. Briefly, *electron impact mass spectrometry* (EI–MS) involves the bombardment of the vaporized metabolite by high-energy electrons (0–100 eV), producing a molecular ion ($M^{+}\cdot$) having a mass equivalent to the molecular weight of the compound. The electron bombardment causes bond fission, and the positively charged fragments produced are detected. The mass spectrum is a plot of the percentage of relative abundance of each ion produced versus the mass-to-charge ratio ($m/z$).

*Chemical ionization mass spectrometry* (CI–MS) is important when compounds do not give spectra containing a molecular ion; generally, the molecular ion decomposes to give fragment ions. With CI–MS a reagent gas such as ammonia, isobutane, or methane is ionized in the mass spectrometer and then ion–molecule reactions such as protonation occur instead of electron–molecule reactions. This *soft ionization* process results in little fragmentation. Fragment ions in this case are almost always formed by loss of neutral molecules, and, as a result, much less structural information can be gleaned relative to EI–MS.

A variety of mass spectral techniques for nonvolatile compounds are now available. Included in these is *field desorption mass spectrometry*, which involves the application of a high field to an emitter surface coated with a thin film of the nonvolatile organic compound. *Fast atom bombardment* (FAB) ionization involves the bombardment of a liquid film containing the nonvolatile sample with a beam of energized atoms of xenon or argon. This method also is useful for thermally unstable compounds. *Secondary-ion mass spectrometry* (SIMS) is similar to FAB except that energetic ions ($Xe^+$ and $Ar^+$) instead of atoms are used in SIMS. Both of these techniques are useful for the structure elucidation of peptides and other high molecular weight organic compounds.[20] Another mass spectrometric technique that is most useful for analysis of targeted compounds in complex mixtures or to screen unknown mixtures rapidly for compound classes is *tandem mass spectrometry* or MS/MS.

High-resolution proton and carbon-13 *nuclear magnetic resonance spectroscopies* are very important for structure elucidation of metabolites, although it is a relatively insensitive technique (about 50 $\mu M$ solution is needed) when compared with chromatography or mass spectrometry. However, the content of structural information from NMR spectra is high. Usually, high field strength instruments (400–600 MHz) are required, but, with the use of certain sample preparation techniques, drug metabolism and excretion studies can be carried out in medium field instruments (200–300 MHz).[21] Because this technique is so routinely used in research today, no further discussion is included here. The uninformed reader is directed to the texts listed in the General References for more information.

### D. Quantification

Quantification of drug metabolites is carried out by radioactive labeling, GC, HPLC, and mass fragmentography. For radioactive labeling techniques to be useful, the various radiolabeled metabolites are first separated by chromatography. Each is isolated, and the rate of radioactive disintegration is determined by liquid scintillation counting methods. The amount of the metabolite isolated can be calculated from the specific radioactivity of the drug (see Section II).

GC and HPLC both require the construction of a calibration curve of known quantities of a reference compound, usually a compound of similar structure to that of the metabolite. From the integration of the internal standard chromatography peak, the amount of each metabolite formed can be determined.

*Selected ion monitoring* (SIM) is a highly selective method for detection and quantification of small quantities of metabolites which uses a mass spectrometer as a selective detector of specific components in the effluent from a gas chromatograph. By setting the spectrometer to detect characteristic fragment ions at a single $m/z$ value, other compounds with the same GC retention times that do not produce those fragment ions will go undetected. When a full mass spectrum is recorded repetitively throughout a chromatogram, and a selected ion monitoring profile is reconstructed by computer, it is often called *mass fragmentography*. Subpicogram quantities of metabolites in a mixture can be detected by the selected ion monitoring method.

# IV. Pathways for Drug Deactivation and Elimination

## A. Introduction

The first mammalian drug metabolite that was isolated and characterized was hippuric acid (**7.12**) from benzoic acid in the early nineteenth century.[22] However, not until the late 1940s, when Mueller and Miller[23] demonstrated that the *in vivo* metabolism of 4-dimethylaminoazobenzene could be studied *in vitro* (see Section IV,B,1), was the discipline of drug metabolism established. As a result of the ready commercial availability of radioisotopes and sophisticated separation, detection, and identification techniques that were developed in the latter half of the twentieth century (see Section III), drug metabolism studies burgeoned.

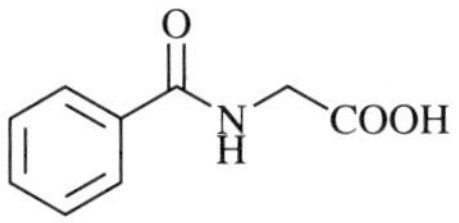

O
N
H
COOH

**7.12**

Drug metabolism reactions have been divided into two general categories,[24a] termed phase I and phase II reactions. *Phase I transformations* involve reactions that introduce or unmask a functional group, such as oxygenation or hydrolysis. *Phase II transformations* mostly generate highly polar derivatives (known as *conjugates*), such as glucuronides and sulfate esters, for excretion through the kidneys into the urine.

The rate and pathway of drug metabolism are affected by species, strain, sex, age, hormones, pregnancy, and liver diseases such as cirrhosis, hepatitis, porphyria, and hepatoma. Drug metabolism can have a variety of profound effects on drugs. It principally causes pharmacological deactivation of a drug by altering its structure so that it no longer interacts appropriately with the target receptor and becomes more susceptible to excretion. Drug metabolism, however, also can activate a drug, as in the case of prodrugs (see Chapter 8). The pharmacological response of a drug may be altered if a metabolite has a new activity; in some cases, the metabolite has the same activity and a similar potency as the drug. A change in drug absorption and drug distribution (i.e., the tissues or organs in which it is concentrated) also can result when it is converted to a much more polar species.

The majority of drug-metabolizing enzymes also catalyze reactions of endogenous compounds. Consequently, the real function of these enzymes may be endogenous metabolism, and it may be fortuitous that they also metabolize drugs and other xenobiotics. The greater affinity of the endogenous substrate over drugs in many cases seems to support this notion; however, many of these enzymes are very broad in specificity, so it is not clear whether at least some of these enzymes have evolved to protect the organism from undesirable substances.

As was discussed in Chapter 3 (Section III,F) and Chapter 4 (Section I,B,1), the interaction of a chiral molecule with a receptor or enzyme produces a diastereomeric complex. Therefore, it is not surprising that drug metabolism also is stereoselective, if not stereospecific.[24b] *Stereoselectivity* can occur with enantiomers of drugs, in which case one enantiomer may be metabolized by one pathway and the other enantiomer by another pathway to give two different metabolites. Another type of stereoselectivity is the conversion of an achiral drug to a chiral metabolite. In both of these cases it may be a rate difference that leads to unequal amounts of metabolites rather than exclusively to one metabolite. In many cases, however, a racemic drug acts as if it were two different xenobiotics, each displaying its own pharmacokinetic and pharmacodynamic profile. In fact, it was concluded by Hignite *et al.*[25] that "warfarin enantiomers should be treated as two drugs," and Silber *et al.*[26] concluded that "*S*-(−)- and *R*-(+)-propranolol are essentially two distinct entities pharmacologically." These are just two of many examples of drugs whose enantiomers are metabolized by different routes. In some cases the inactive enantiomer can produce toxic metabolites which could be avoided by the administration of only the active enantiomer.[27] In other cases, the inactive isomer may inhibit the metabolism of the active isomer.

The metabolism of enantiomers may depend on the route of administration. For example, the antiarrhythmic drug verapamil (**7.13**) is 16 times more potent when administered intravenously than when taken orally because of extensive hepatic presystemic elimination that occurs with oral administration (the first-pass effect; see Section I).[28] The (−)-isomer, which is 10 times more active than the (+)-isomer, is preferentially metabolized during hepatic metabolism, and, therefore, there is much more of the less active (+)-isomer available by the oral route than by the intravenous route.

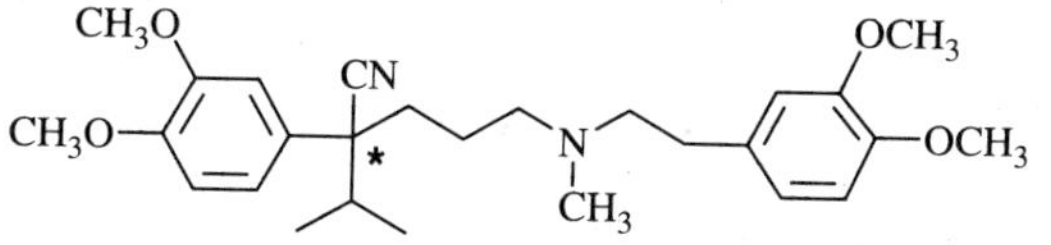

**7.13**

In some cases one enantiomer of a drug can be metabolized to the other enantiomer. The therapeutically inactive isomer of the analgesic ibuprofen (**7.14**) is metabolized in the body to the active (*S*)-isomer.[29] If a racemic mixture is administered, a 70:30 mixture of *S*:*R* is excreted; if a 6:94 mixture of *S*:*R* is administered, an 80:20 mixture of *S*:*R* is excreted.

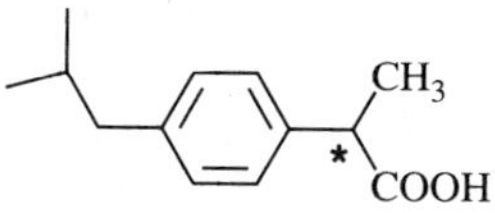

**7.14**

In the next two sections the various types of metabolic transformations are described. Some examples are included to show the effect of stereochemistry on drug metabolism. It should be kept in mind, however, that when a racemic mixture is administered, drug metabolism may be altered from that observed when a pure isomer is utilized.

## B. Phase I Transformations

### 1. Oxidative Reactions

As noted above, Mueller and Miller[23] showed that the *in vivo* metabolism of 4-dimethylaminoazobenzene could be studied *in vitro* using rat liver homogenates. It was demonstrated that the *in vitro* system was functional only if nicotinamide adenine dinucleotide phosphate ($NADP^+$), molecular oxygen, and both the microsomal and soluble fractions from the liver homogenates were included. Later, Brodie and co-workers[30] found that the oxidative activity was in the microsomal fraction and that the soluble fraction could be replaced by either NADPH or a NADPH-generating system. This system was active toward a broad spectrum of structurally diverse compounds. Because it required both $O_2$ and a reducing system, it was classified as a *mixed function oxidase*,[31] that is, one atom of the $O_2$ is transferred to the substrate and the other undergoes a two-electron reduction and is converted to water. This classification was confirmed when it was shown[32a] that aromatic hydroxylation of acetanilide by liver microsomes in the presence of $^{18}O_2$ resulted in incorporation of one atom of $^{18}O$ into the product, and that a heme protein was an essential component for this reaction.[32b,c] When the heme protein was reduced and exposed to carbon monoxide, a strong absorption in the visible spectrum at 450 nm resulted. Because of this observation, these microsomal oxidases were named cytochrome *P*-450.

Cytochrome *P*-450 now represents a family of closely related enzymes that catalyze the same reaction on different substrates, in this case, the oxidation of steroids, fatty acids, and xenobiotics. These related cytochrome *P*-450 enzymes are referred to as *isozymes*; however, strictly speaking, isozymes are related enzymes that catalyze the same reaction with the same substrate. The primary site for these enzymes is the liver, but they also are present in lung, kidney, gut, adrenal cortex, skin, brain, aorta, and other epithelial tissues. The heme is noncovalently bound to the apoprotein. Cytochrome *P*-450 is associated with another enzyme, NADPH–cytochrome *P*-450 reductase, a flavoenzyme that contains one molecule each of flavin adenine dinucleotide (FAD) and flavin mononucleotide (FMN).[33] Heme-dependent oxidation reactions were discussed in Chapter 4 (Section III,D). As shown in Scheme 4.35, the NADPH–cytochrome *P*-450 reductase reduces the flavin which, in turn, transfers an electron to the heme–oxygen complex of cytochrome *P*-450. Actually, there is evidence to suggest that the FAD accepts electrons from the NADPH, the $FADH^-$ then transfers electrons to the FMN, and the $FMNH^-$ donates the electron to the heme or heme–oxygen complex of cytochrome *P*-450.[33]

In general, cytochrome *P*-450 catalyzes either hydroxylation or epoxidation of various substrates (Table 7.1) and is believed to operate via radical intermediates (see Scheme 4.36). When the concentrations of cytochrome *P*-450 and other drug-metabolizing enzymes are modified, drug metabolism becomes altered. Many drugs and environmental chemicals induce either their own metabolism or the metabolism of other drugs in man as a result of their activation of cytochrome *P*-450 and NADPH-cytochrome *P*-450 reductase. Different chemicals induce different isozymes of cytochrome *P*-450. A contrasting problem sometimes arises when multiple drugs are administered and one of the drugs inhibits drug metabolism as a result of its inhibition of cytochrome *P*-450 or other enzymes.

Another important enzyme involved in drug oxidation is the microsomal flavin monooxygenase[34] (see Section III,C of Chapter 4), whose mechanism of oxidation was described in Schemes 4.33 and 4.34. According to these schemes, the flavin peroxide intermediate is an electrophilic species, indicating that the substrates for this enzyme are nucleophiles such as amines and thiols (Table 7.2). The enzyme contains one FAD molecule per subunit and,

**Table 7.1** Classes of Substrates for Cytochrome *P*-450

| Functional group | Product |
|---|---|
| 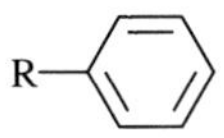 $R-C_6H_5$ | 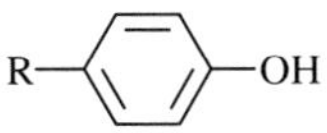 $R-C_6H_4-OH$ (para) |
| 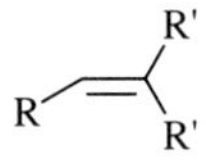 $RCH=CR'_2$ | 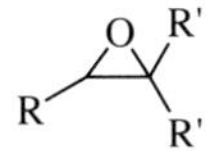 $RCH(O)CR'_2$ (epoxide) |
| $ArCH_2R$ |  $ArCH(OH)R$ |
| 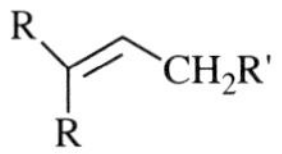 $R_2C=CHCH_2R'$ | 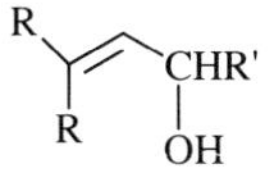 $R_2C=CHCH(OH)R'$ |
| 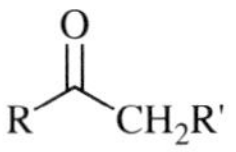 $RC(=O)CH_2R'$ | 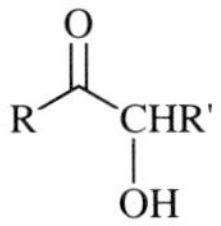 $RC(=O)CH(OH)R'$ |
| $RCH_2R'$ |  $RCH(OH)R'$ |
| $RCH_2$-X-R' ( X = N, O, S, halogen ) | 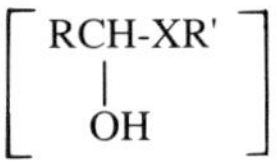 [RCH(OH)-XR'] → RCHO + R'XH |
| R-X-R' (X = NR″, S) |  R-X(→O)-R' |

as in the case of cytochrome *P*-450, requires NADPH to reduce the flavin. It has been found that nucleophilic compounds containing an anionic group are excluded from the active site of this enzyme. As most endogenous nucleophiles contain negatively charged groups, this may be a way that Mother Nature prevents normal cellular components from being oxidized by this enzyme.

Other enzymes involved in oxidative drug metabolism include prostaglandin H synthase (see Section IV,B,1,e), alcohol dehydrogenase, aldehyde dehydrogenase (see Section IV,B,1,i), xanthine oxidase, monoamine oxidase, and aromatase. These enzymes, however, are involved, for the most part, in endogenous compound metabolism.

**Table 7.2** Classes of Substrates for Flavin Monooxygenase

| Functional group | Product |
| --- | --- |
| R-NR'$_2$ | R-N(→O)R'$_2$ |
| R-NHR' | R-N(OH)R' |
| R-N(OH)R' | R=N(→O)R' |
| (cyclic)=NH | (cyclic)-NHOH |
| R-N(R')-NH$_2$ | R-N(→O)(R')-NH$_2$ |
| R-C(=S)NH$_2$ | R-C(=S→O)NH$_2$ |
| RNH-C(SH)=N$^+$HR' | RNH-C($SO_2^-$)=N$^+$HR' |
| 2RSH | RSSR |
| RSSR | 2 $RSO_2^-$ |

***a. Aromatic Hydroxylation.*** In 1950 Boyland[35] hypothesized that aromatic compounds were metabolized initially to the corresponding epoxides (arene oxides). This postulate was confirmed in 1968 by a group at the National Institutes of Health (NIH), who isolated naphthalene 1,2-oxide from the microsomal oxidation of naphthalene[36a,b] (Scheme 7.3). Kinetic isotope effect studies,[36c] however, now indicate that a direct arene epoxidation is a highly unlikely process. Instead, an activated heme iron–oxo species may add to the aromatic ring (Scheme 7.3a), similar to the mechanism described for the addition of the corresponding heme iron–oxo species to alkenes as was discussed in Chapter 4 (Section III,D; Scheme 4.36). This mechanism produces a tetrahedral intermediate that can rearrange via the epoxide (pathway a) or a ketone (pathway b) intermediate to give, ultimately, an arenol. Usually, the arene oxides can undergo rearrangements to arenols, hydration (catalyzed by epoxide hydrolase) to the corresponding *trans*-diol, reaction with glutathione (catalyzed by glutathione *S*-transferase) to the $\beta$-hydroxy sulfide, and reactions with various macromolecular nucleophiles (Scheme 7.4).

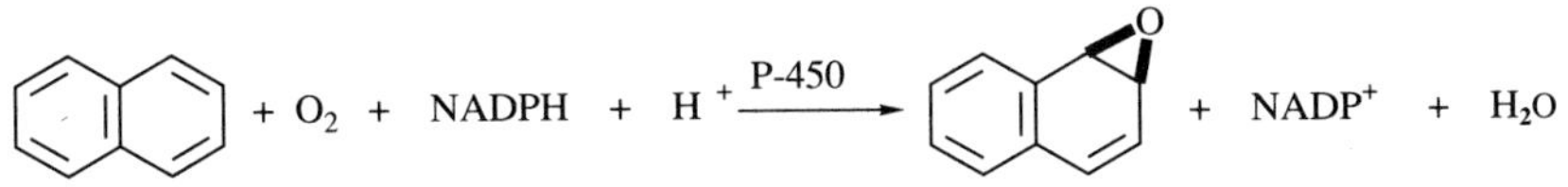

$$C_{10}H_8 + O_2 + NADPH + H^+ \xrightarrow{P\text{-}450} \text{naphthalene 1,2-oxide} + NADP^+ + H_2O$$

**Scheme 7.3.** Cytochrome *P*-450 oxidation of naphthalene.

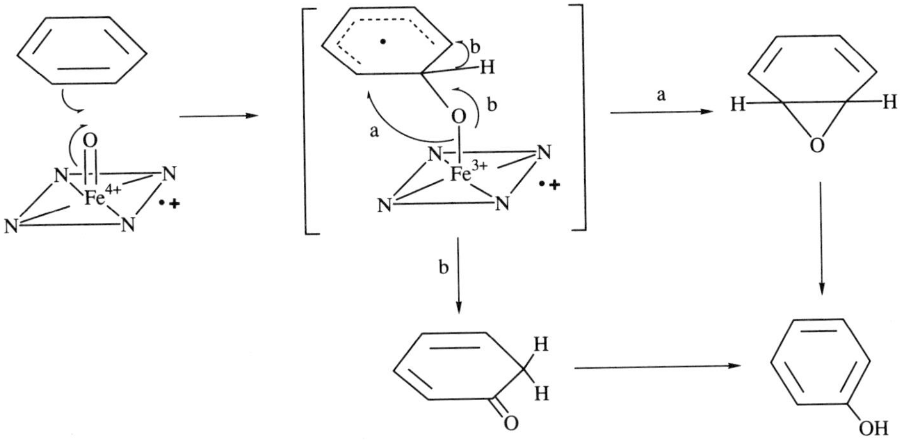

**Scheme 7.3a.** Addition–rearrangement mechanism for arene oxide formation.

The rearrangement of an arene oxide to an arenol is known as the *NIH shift* because a research group at the NIH proposed the mechanism (Scheme 7.5).[37] Ring opening occurs in the direction that gives the more stable carbocation (ortho- or para-hydroxylation when R is electron donating). Also, because of an isotope effect on cleavage of the C—D bond, the proton is preferentially removed, leaving the migrated deuterium. Although there is an isotope effect on the cleavage of the C—D versus the C—H bond, this step is not the rate-determining step in the overall reaction; consequently, there is no overall isotope effect on this oxidation pathway when deuterium is incorporated into the substrate.

In competition with 1,2-hydride (deuteride) shift is deprotonation (dedeuteronation) (Scheme 7.6). The percentage of each pathway depends on the degree of stabilization of the intermediate carbocation by R; the more stabilization, the less need for the hydride shift, and the more deprotonation (dedeuteronation) occurs.[38] For example, when R is $NH_2$, OH, $NHCOCF_3$, or $NHCOCH_3$, only 0–30% of the product phenols retain deuterium (NIH shift),

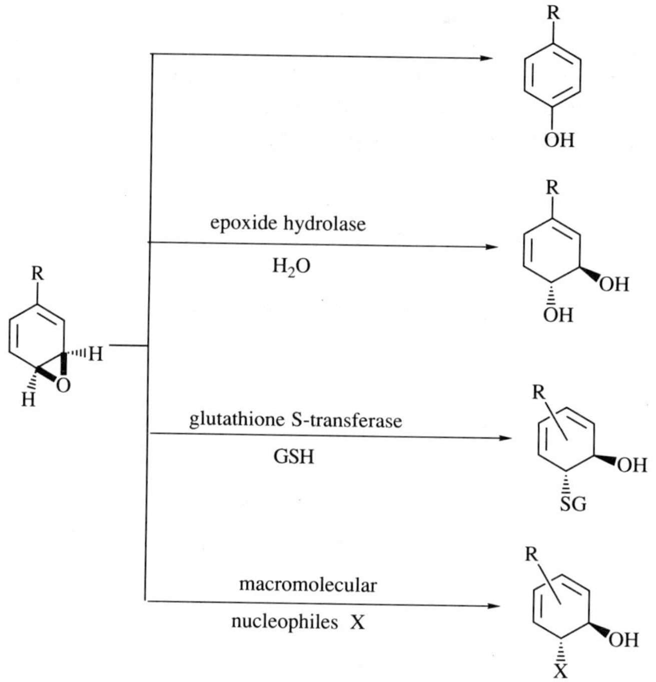
R
OH
epoxide hydrolase
$H_2O$
OH
OH
glutathione S-transferase
GSH
SG
macromolecular
nucleophiles X
X
H
O

**Scheme 7.4.** Possible fates of arene oxides.

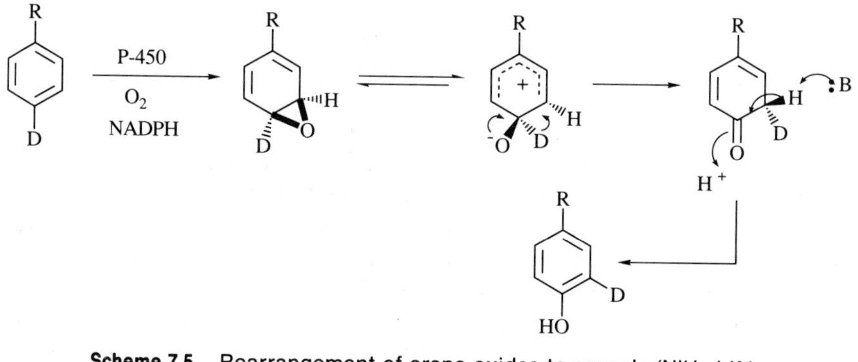
P-450
$O_2$
NADPH
R
D
H
O
B
H+
HO

**Scheme 7.5.** Rearrangement of arene oxides to arenols (NIH shift).

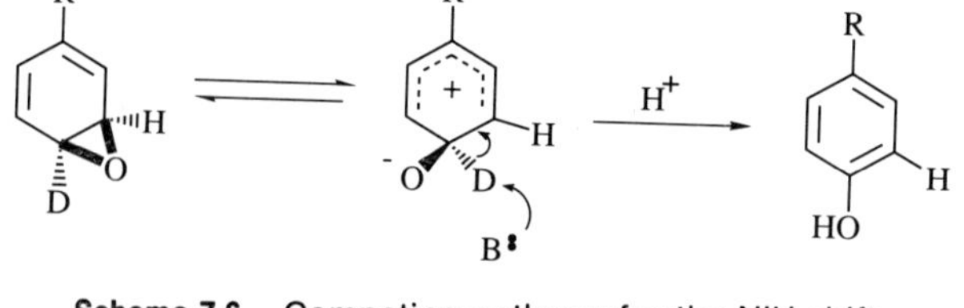
R
H
D
O
H+
B
HO

**Scheme 7.6.** Competing pathway for the NIH shift.

but when R is Br, $CONH_2$, F, CN, or Cl, 40–54% deuterium retention is observed.

The NIH shift also occurs with substituents as well as with hydrogen. For example, rat liver metabolizes *p*-chloroamphetamine (**7.15**) to 3-chloro-4-hydroxyamphetamine[39] (Scheme 7.7). A related substituent NIH shift was observed in the metabolic oxidation of the antiprotozoal agent tinidazole[40] (**7.16**, Scheme 7.8).

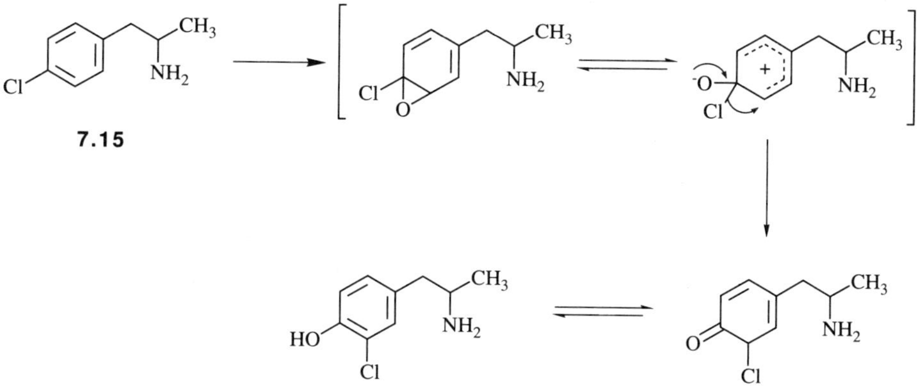

**Scheme 7.7.** NIH shift of chloride ions.

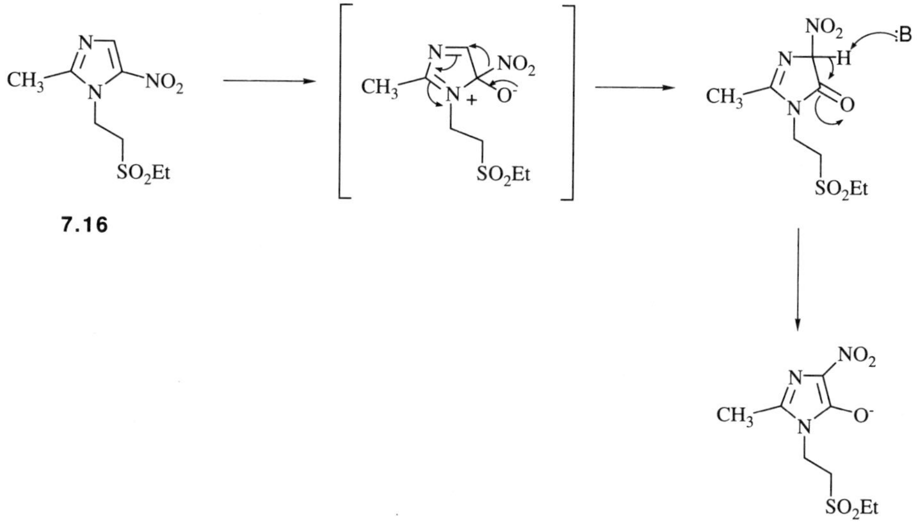

**Scheme 7.8.** NIH shift of nitro groups.

As in the case of electrophilic aromatic substitution reactions, it appears that the more electron rich the aromatic ring (R is electron donating), the faster the microsomal hydroxylation will be.[41] Aniline (electron rich), for example, undergoes extensive ortho- and para-hydroxylation,[42] whereas the strongly electron-poor uricosuric drug probenecid (**7.17**) undergoes no detectable aromatic hydroxylation.[43] In the case of drugs with two aromatic rings, the more electron-rich one, generally, is hydroxylated. The antipsychotic drug chlorpromazine (**7.18**, R = H), for example, undergoes 7-hydroxylation (**7.18**, R = OH).[44] Other factors, such as binding orientation at the active site of the hydroxylases, may play a role in the rate of hydroxylation at different positions on the aromatic ring.

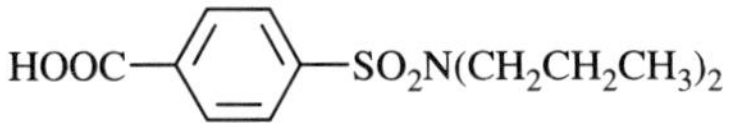

**7.17** 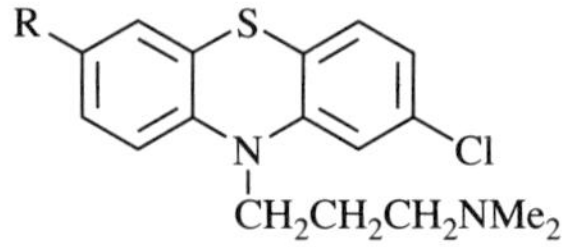 **7.18**

Aromatic hydroxylation, as is the case for all metabolic reactions, is species specific. In man, para-hydroxylation is a major route of metabolism for many phenyl-containing drugs. The site and stereospecificity of hydroxylation depend on the animal studied. In man, the antiepilepsy drug phenytoin (**7.19**, $R^1 = R^2 = R^3 = H$) is para-hydroxylated at the pro-(*S*) phenyl ring (**7.19**, $R^1 = OH$, $R^2 = R^3 = H$) 10 times more often than is the pro-(*R*) ring (**7.19**, $R^3 = OH$, $R^1 = R^2 = H$). In dogs, however, meta-hydroxylation of the pro-(*R*) phenyl ring (**7.19**, $R^2 = OH$, $R^1 = R^3 = H$) is the major pathway. The overall ratio of $R^2:R^1:R^3$ hydroxylation in dogs is 18:2:1.[45] Meta-hydroxylation may be catalyzed by an isozyme of cytochrome *P*-450 which operates by a mechanism different from arene oxide formation.[46] For example, an important metabolite of chlorobenzene is 3-chlorophenol, but it was shown that neither 3- nor 4-chlorophenol oxide gave 3-chlorophenol in the presence of rat liver microsomes, suggesting that a direct oxygen insertion mechanism may be operative in this case. This mechanism also is consistent with the observation of a kinetic isotope effect ($k_H/k_D = 1.3–1.75$) on the *in vivo* 3-hydroxylation.[47]

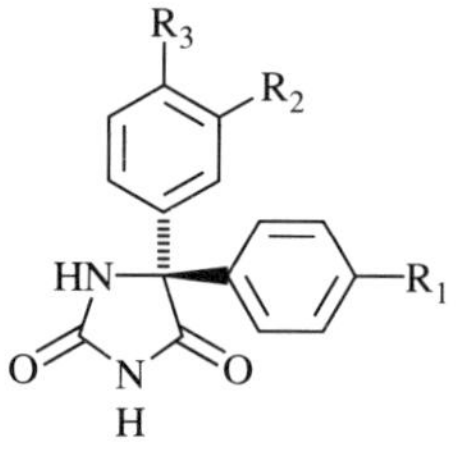

**7.19**

Because of the reactivity of arene oxides, they can undergo rapid reactions with nucleophiles. If cellular nucleophiles react with these compounds, toxicity can result. Consequently, there are enzymes that catalyze deactivation reactions on these reactive species (see Scheme 7.4). Epoxide hydrolase (also called epoxide hydratase or epoxide hydrase) catalyzes the hydration of highly electrophilic arene oxides to give *trans*-dihydrodiols. The mechanism of this reaction is a general base-catalyzed nucleophilic attack of water on one of the two electrophilic carbons of the epoxide ring. As in the case of the corresponding nonenzymatic reaction of hydroxide with epoxides, trans stereochemistry results,[48] and attack occurs predominantly at the less sterically hindered side.[49] The *trans*-dihydrodiol product can be oxidized further to give catechols (Scheme 7.9); catechols also are generated by hydroxylation of arenols (Scheme 7.9). Because of the instability of catechols to oxidation, they may be converted to either *ortho*-quinones or semiquinones (see Scheme 7.9).

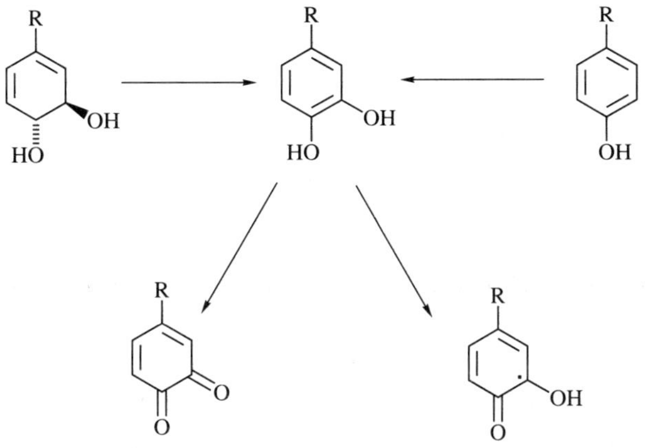

**Scheme 7.9.** Metabolic formation and oxidation of catechols.

Glutathione *S*-transferase is another enzyme that protects the cell from the electrophilic arene oxide metabolites. Glutathione adducts also are known to undergo rearrangement upon dehydration (Scheme 7.10). Other reactions catalyzed by glutathione *S*-transferases are discussed in Section IV,C,5.

When the arene oxide escapes enzymatic destruction, toxicity can result. An important example of this is the metabolism of benzo[*a*]pyrene (**7.20**; Scheme 7.11), a potent carcinogen found in soot and charcoal. The relationship between soot and cancer was noted by Sir Percival Scott, a British surgeon, in 1775 when he observed that chimney sweeps (people who clean out chimneys) frequently developed skin cancer. Metabolic activation of polyaromatic hydrocarbons can lead to the formation of covalent adducts with RNA, DNA, and proteins.[50] Covalent binding of the metabolites to DNA is the initial event that is responsible for malignant cellular transformation.[51] The key reactive intermeuiate responsible for alkylation of nucleic acids is (+)-(7*R*,8*S*)-dihydroxy-(9*R*,10*R*)-oxy-7,8,9,10-tetrahydrobenzo[*a*]pyrene (**7.21**; Scheme 7.11). This reactive metabolite reacts with RNA to form a

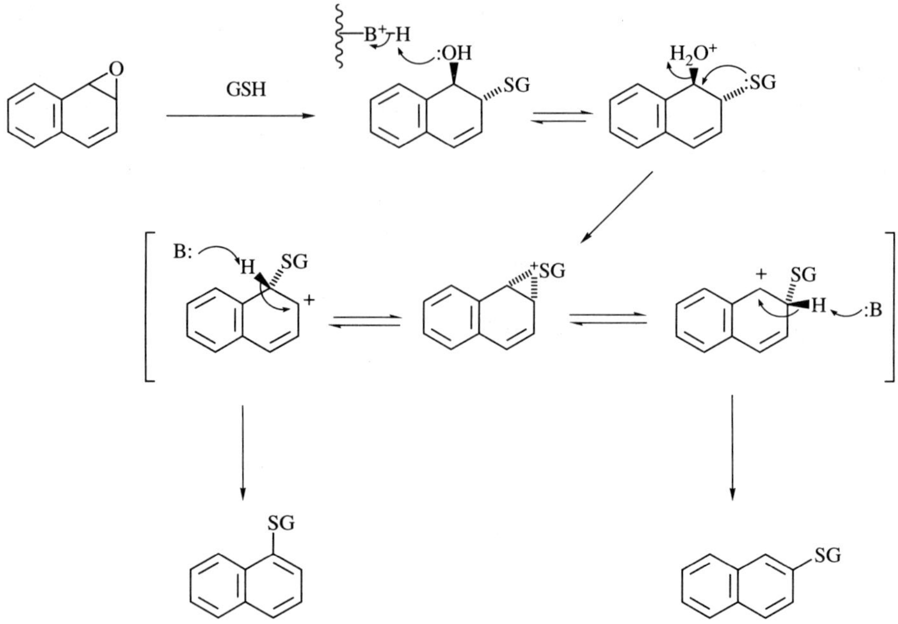

Scheme 7.10. Dehydration and rearrangement of glutathione adducts.

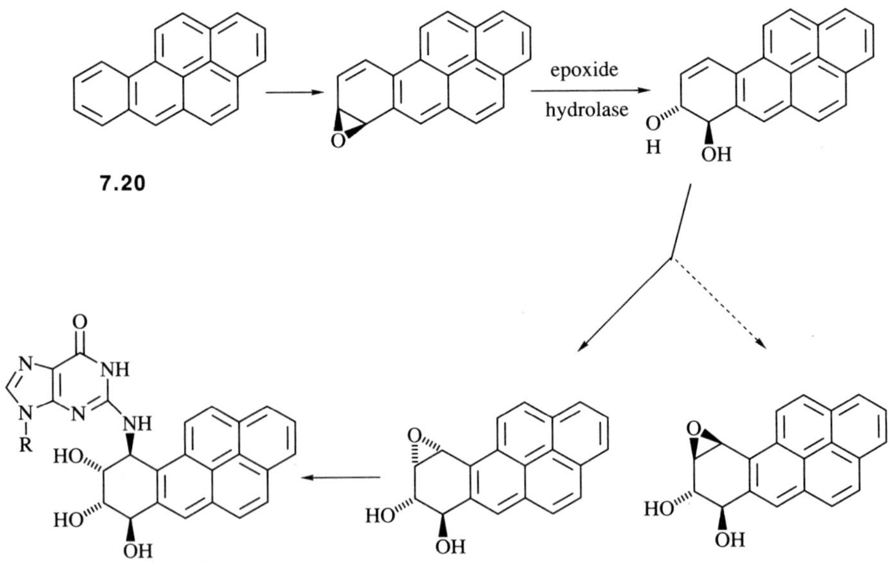

7.20

7.22

7.21

Scheme 7.11. DNA adduct formation with benzo[*a*]pyrene metabolites.

covalent adduct between the C-2 amino group of a guanosine and C-10 of the hydrocarbon[52] (**7.22**). The reactive metabolite (**7.21**) also causes nicks in superhelical DNA of *E. coli*; an adduct between the C-2 amino group of deoxyguanosine and the C-10 position of the diol epoxide was isolated.[53]

***b. Alkene Epoxidation.*** Because alkenes are more reactive than aromatic $\pi$ bonds, it is not surprising that alkenes also are metabolically epoxidized. An example of a drug that is metabolized by alkene epoxidation is the anticonvulsant agent, carbamazepine (**7.23**, Scheme 7.12).[54] Carbamazepine-10,11-epoxide (**7.24**) has been found to be an anticonvulsant agent as well, and this metabolite may actually be responsible for the anticonvulsant activity of carbamazepine.[55] The epoxide is converted stereoselectively to the corresponding (10*S*,11*S*)-diol (**7.25**) by epoxide hydrolase, and this metabolite is stereoselectively conjugated to the glucuronide (**7.26**) with UDP-glucuronosyltransferase.

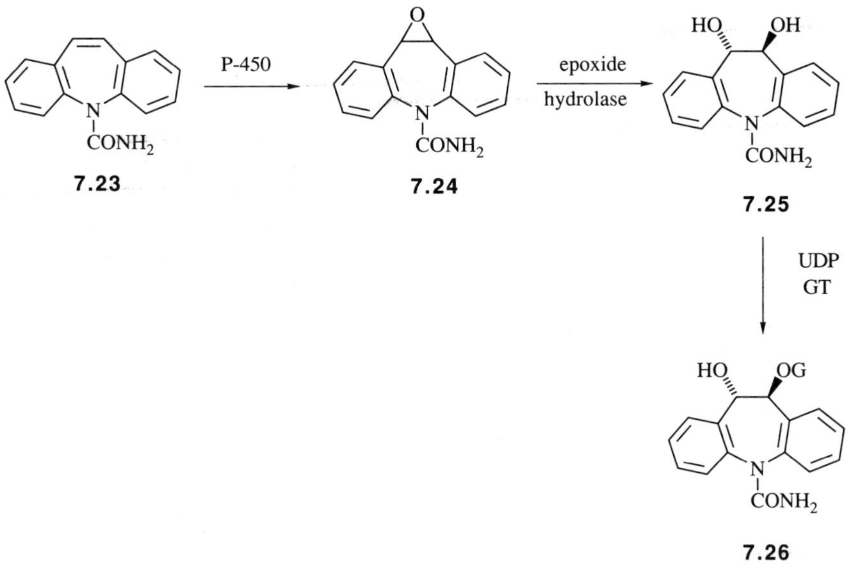

N CONH$_2$ **7.23** —P-450→ O N CONH$_2$ **7.24** —epoxide hydrolase→ HO OH N CONH$_2$ **7.25** —UDP GT→ HO OG N CONH$_2$ **7.26**

**Scheme 7.12.** Metabolism of carbamazepine.

Metabolic epoxidation of an alkene also can lead to the formation of toxic products. The hepatocarcinogen aflatoxin B$_1$ (**7.27**), for example, becomes covalently bound to cellular DNA (Scheme 7.13). With the use of radiolabeled **7.27**,[56] and model studies,[57] it was shown that a covalent bond forms between C-8 of aflatoxin B$_1$ and the N-7 of a guanine residue in DNA (**7.29**). The most likely precursor is the epoxide **7.28**, which has been synthesized and shown to undergo the expected reaction with a guanine-containing oligonucleotide.[57]

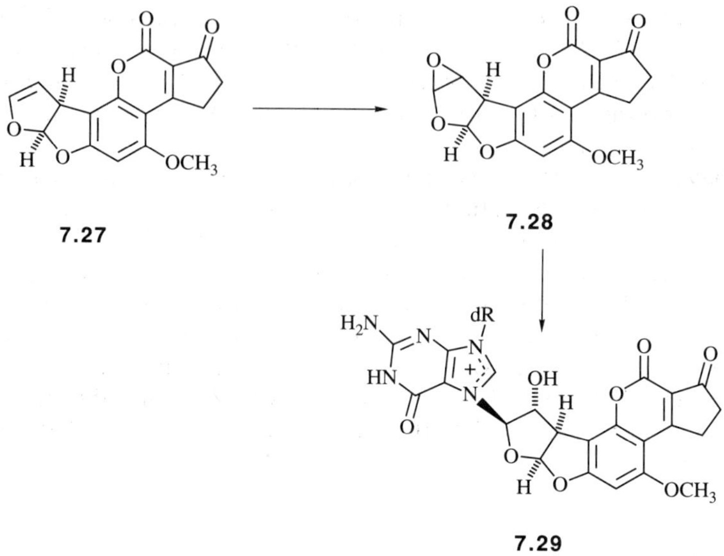

7.27

7.28

7.29

Scheme 7.13. Metabolic reactions of aflatoxin $B_1$.

*c. **Oxidations of Carbons Adjacent to*** $sp^2$ ***Centers.*** Carbons adjacent to aromatic, olefinic, and carbonyl or imine groups undergo metabolic oxidations. The oxidation mechanism is not clearly understood, but as the ease of oxidation parallels the C—H bond dissociation energies, it is likely that a typical cytochrome *P*-450 oxidation is responsible.

Examples of benzylic oxidation are the metabolism of the antidepressent drug amitriptyline (**7.30**, R = R′ = H) which is oxidized to **7.30** (R = H, R′ = OH and R = OH, R′ = H)[58] and the $\beta_1$-adrenoreceptor antagonist ($\beta$-blocker) antihypertensive drug metoprolol (**7.31**).[59] In the case of metoprolol both enantiomers [(1′*R*) and (1′*S*)] of the hydroxylated drug are formed, but (1′*R*)-hydroxylation occurs to a greater extent. Furthermore, the ratio of the (1′*R*)- to (1′*S*)-isomers depends on the stereochemistry at the 2 position. (2*R*)-Metoprolol gives a ratio of (1′*R*,2*R*)/(1′*S*,2*R*) metabolites of 9.4, whereas (2*S*)-metoprolol gives a ratio of (1′*R*,2*S*)/(1′*S*,2*S*) of 26. Therefore, the stereochemistry in the methoxyethyl side chain is influenced by the ste-

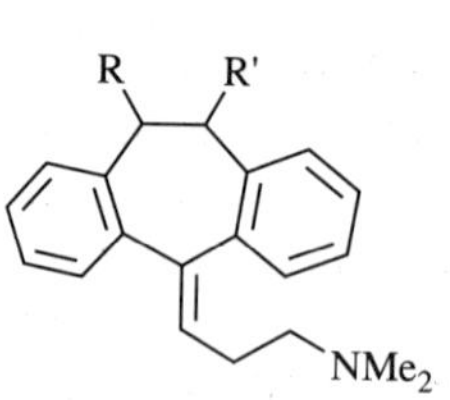

7.30

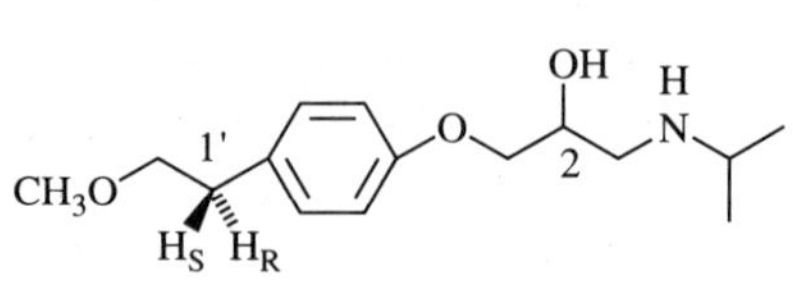

7.31

reochemistry in the para side chain. Hydroxylations generally are highly stereoselective, if not stereospecific.

The cytochrome *P*-450–catalyzed metabolism of the antiarrhythmic drug quinidine (**7.32**, R = H) leads to allylic oxidation (**7.32**, R = OH).[60] The psychoactive constituent of marijuana ($\Delta^9$-tetrahydrocannabinol) (**7.33**, R = R′ = H) is extensively metabolized to all stereoisomers of **7.33** (R = H, R′ = OH; R = OH, R′ = H; and R = R′ = OH).[61] The sedative–hypnotic (+)-glutethimide (**7.34**, R = R′ = H) is converted to 5-hydroxyglutethimide (**7.34**, R = OH, R′ = H).[62] This metabolite is pharmacologically active and may contribute to the comatose state of individuals who have taken toxic overdoses of the parent drug. The (−)-isomer is enantioselectively hydroxylated at the ethyl group to give **7.34** (R = H, R′ = OH).

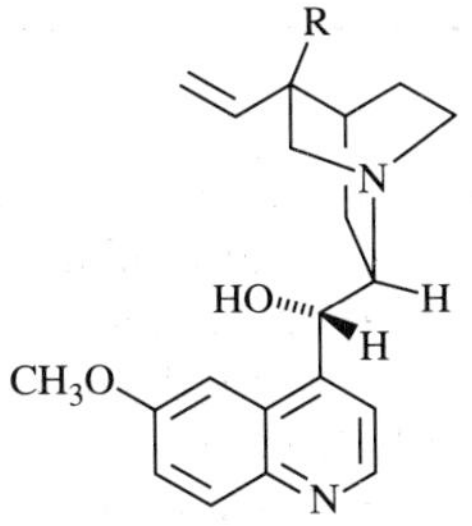

R HO H H $CH_3O$ N N

**7.32**

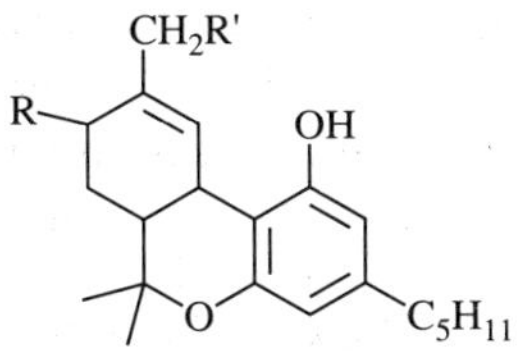

$CH_2R'$ R OH O $C_5H_{11}$

**7.33**

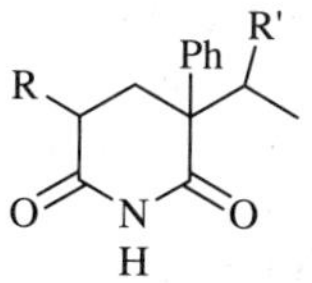

R′ Ph R O N O H

**7.34**

***d. Oxidation at Aliphatic and Alicyclic Carbon Atoms.*** Metabolic oxidation at the terminal methyl group of an aliphatic side chain is referred to as $\omega$-oxidation, and oxidation at the penultimate carbon is $\omega - 1$ oxidation. Because $\omega$-oxidation is a chemically unfavorable process, the active site of the enzyme must be favorably disposed for this particular regiochemistry to proceed. In the case described above of (−)-glutethimide (**7.34**, R = R′ = H), $\omega$-1 oxidation occurred. The anticonvulsant drug ethosuximide (**7.35**) undergoes both $\omega$- and $\omega$-1 oxidations.[63] The coronary vasodilator perhexiline (**7.36**, R = H) is metabolized to the alicyclic alcohol **7.36** (R = OH).[64]

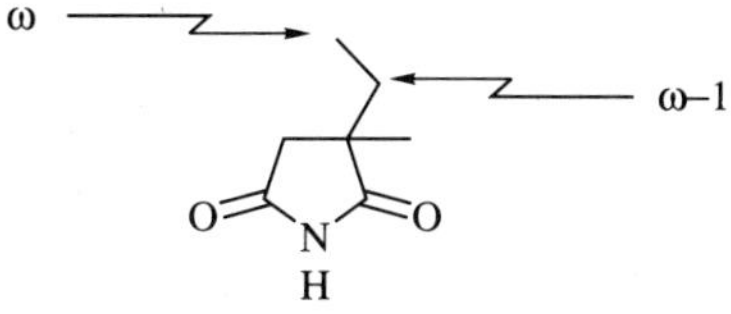

ω ω–1 O N O H

**7.35**

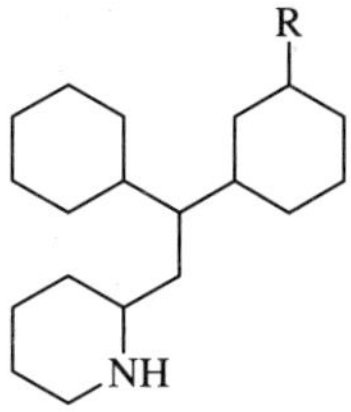

R NH

**7.36**

**Table 7.3** Oxidative Reactions of Primary Amines and Amides

$$RCH(R')NH_2 \longrightarrow RC(=O)R' + NH_4^+$$

$$RCH(R')NH_2 \longrightarrow RCH(R')NHOH \xrightarrow{-H_2O} RC(R'){=}NH \longrightarrow RC(R'){=}NOH \longrightarrow RCH(R')NO_2$$

$$ArNH_2 \longrightarrow ArNHOH \longrightarrow ArN{=}O \longrightarrow ArNO_2$$

$$RC(=O){-}NH_2 \longrightarrow RC(=O){-}NHOH$$

***e. Oxidations of Carbon–Nitrogen Systems.*** The metabolism of organic nitrogen compounds is very complex. Two general classes of oxidation reactions will be considered for primary, secondary, and tertiary amines, namely, carbon- and nitrogen-oxidation reactions that lead to C—N bond cleavage and N-oxidation reactions that do not lead to C—N bond cleavage.

Primary amines and/or amides are metabolized by the oxidation reactions shown in Table 7.3. Primary aliphatic and arylalkyl amines having at least one $\alpha$-carbon–hydrogen bond undergo cytochrome *P*-450–catalyzed hydroxylation at the $\alpha$-carbon (see Scheme 4.37 in Chapter 4 for the mechanism) to give the carbinolamine (**7.37**) which generally breaks down to the aldehyde or ketone and ammonia (Scheme 7.14). This process of oxidative cleavage of ammonia from the primary amine is known as *oxidative deamination*. As is predicted by this mechanism, primary aromatic amines and $\alpha,\alpha$-disubstituted aliphatic amines cannot undergo oxidative deamination. A variety of endogenous arylalkyl amines, such as the neurotransmitters dopamine, norepinephrine, and serotonin, are oxidized by monoamine oxidase by an electron transfer mechanism (see Scheme 4.32 for the mechanism). This enzyme also may be involved in drug metabolism of arylalkyl amines with no $\alpha$-substituents. An example of cytochrome *P*-450–catalyzed primary amine oxidative deamination is the metabolism of amphetamine (**7.38**) to give 1-phenyl-2-propanone and ammonia.[65]

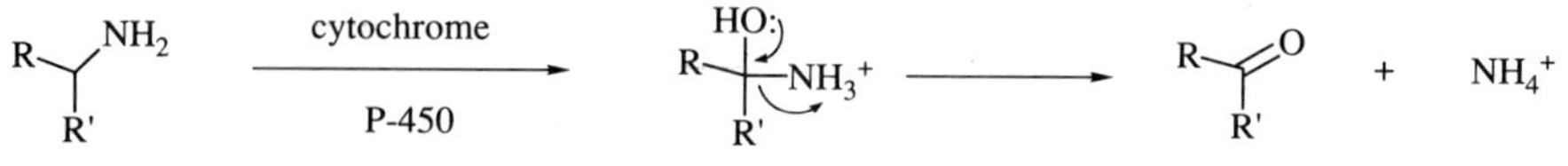

$$RCH(R')NH_2 \xrightarrow[\text{P-450}]{\text{cytochrome}} R{-}C(OH)(R'){-}NH_3^+ \longrightarrow RC(=O)R' + NH_4^+$$

**7.37**

**Scheme 7.14.** Oxidative deamination of primary amines.

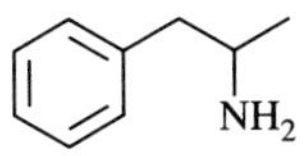

**7.38**

Another important oxidative pathway for primary amines is *N-oxidation* (hydroxylation of the nitrogen atom) to the corresponding hydroxylamine,[66] catalyzed by flavin monooxygenases (see Schemes 4.33 and 4.34 for mechanisms). It has been suggested[67] that the basic amines (p$K_a$ 8–11) are oxidized by the flavoenzymes, nonbasic nitrogen compounds such as amides are oxidized by cytochrome *P*-450 enzymes, and compounds with intermediate basicity, such as aromatic amines, are oxidized by both enzymes. Cytochrome *P*-450 enzymes, however, tend not to catalyze N-oxidation reactions when there are $\alpha$-protons available. Amphetamine undergoes N-oxidation to the corresponding hydroxylamine (**7.39**, Scheme 7.15), the oxime (**7.40**), and the nitro compound (**7.41**). It could be argued that the 1-phenyl-2-propanone stated above as the product of oxidative deamination could be derived from hydrolysis of the oxime **7.40** (see Scheme 7.15). However, metabolism of *N*-hydroxyamphetamine (**7.39**) to the oxime occurs only to a small extent, and

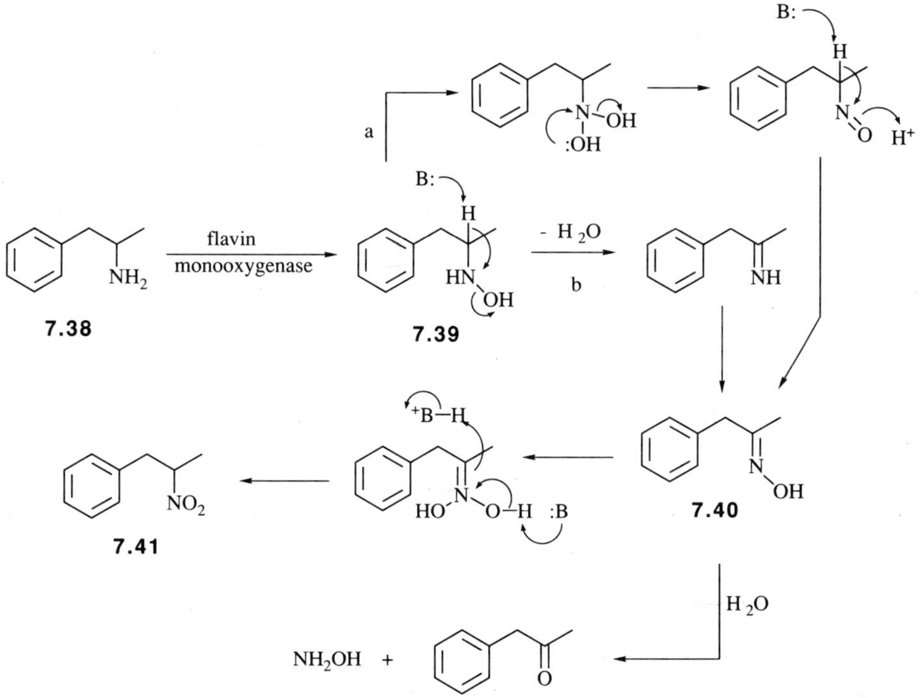

**Scheme 7.15.** Potential alternative oxidation pathway of amphetamine to 1-phenyl-2-propanone.

the oxime is not hydrolyzed to the ketone *in vitro*.[68] Also, *in vitro* metabolism of amphetamine in $^{18}O_2$ leads to substantial enrichment of 1-phenyl-2-propanone with $^{18}O$, indicating the relevance of a cytochrome *P*-450–catalyzed oxidative deamination pathway to the ketone. The amphetamine oxime is derived both from hydroxylation of **7.39** (pathway a) and from dehydration of **7.39** (pathway b) followed by hydroxylation of the imine (see Scheme 7.15). The imine also could be derived from dehydration of the carbinolamine (Scheme 7.16). Hydrolysis of the imine could lead to 1-phenyl-2-propanone.

Secondary aliphatic amines are metabolized by oxidative N-dealkylation, oxidative deamination, and N-oxidation (Table 7.4). When a small alkyl substituent is cleaved from a secondary amine (or tertiary amine; see below) to give a primary amine (or, in the case of a tertiary amine, to give a secondary amine), the process is known as *oxidative N-dealkylation* (Scheme 7.17). On the basis of the reactions shown in Schemes 7.14 and 7.17, it is apparent that oxidative deamination and oxidative N-dealkylation are really the same reaction. The metabolic difference between the two is that the former oxidizes primary and some secondary amines to the aldehyde or ketone and ammonia (from primary amines) or to the aldehyde or ketone and a primary amine (from a secondary amine); the latter converts tertiary and secondary amines to secondary amines (from tertiary amines) and primary amines (from secondary amines). It is mostly a matter of semantics, but oxidative N-dealkylation is a process that cleaves a small alkyl group from the main amine compound, whereas oxidative deamination oxidizes the main amine compound by chopping off ammonia (in the case of a primary amine) or a small primary amine fragment (from a more complex secondary amine). It is not clear if the precise mechanisms of the two processes are the same, but from a reaction standpoint they are the same except viewed from the two sides of the nitrogen atom.

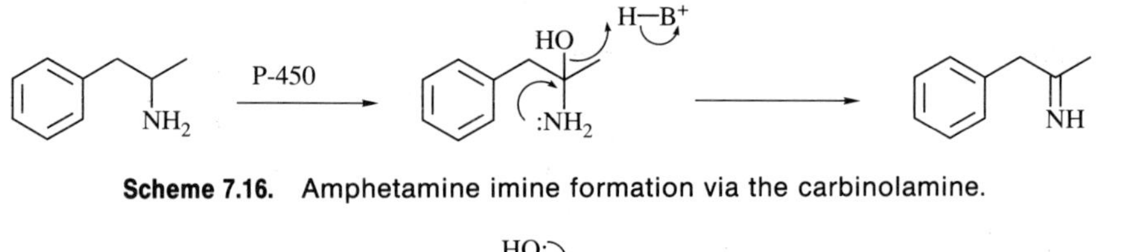

P-450 HO H—B+ :NH2 NH2 NH

**Scheme 7.16.** Amphetamine imine formation via the carbinolamine.

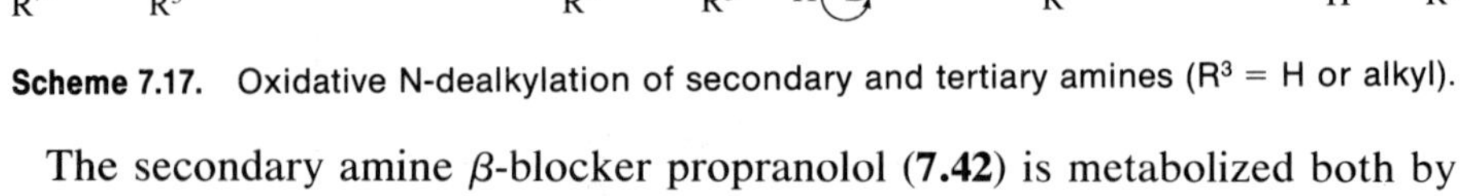

R R' N R2 R3 P-450 HO: H—B+ NHR3 + O H R2

**Scheme 7.17.** Oxidative N-dealkylation of secondary and tertiary amines ($R^3$ = H or alkyl).

The secondary amine β-blocker propranolol (**7.42**) is metabolized both by oxidative N-dealkylation (pathway a, Scheme 7.18) and by oxidative deamination (pathway b, Scheme 7.18).[69] Aldehydic metabolites often are oxidized by soluble aldehyde oxidases or dehydrogenases to the corresponding carbox-

**Table 7.4** Oxidative Reactions of Secondary Amines and Amides

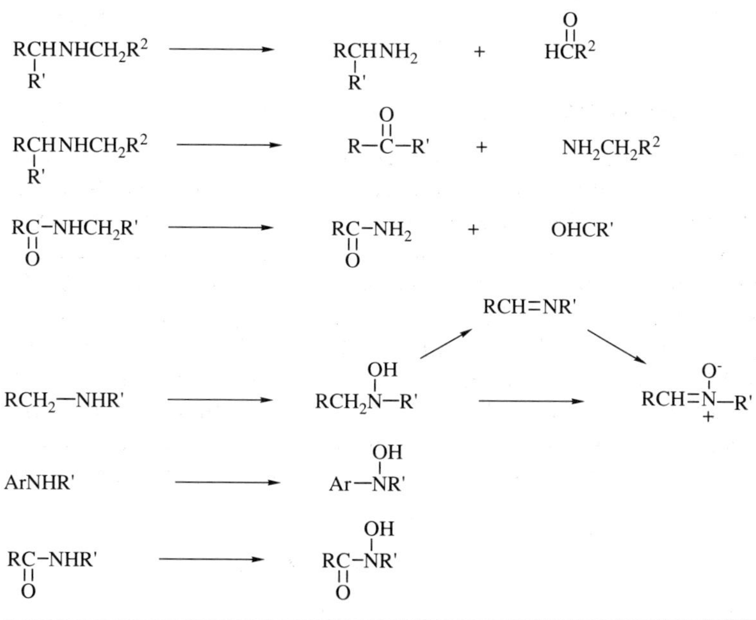

|  | |  | | |
|---|---|---|---|---|
| $RCH(R')NHCH_2R^2$ | → | $RCH(R')NH_2$ | + | $HC(=O)R^2$ |
| $RCH(R')NHCH_2R^2$ | → | $R{-}C(=O){-}R'$ | + | $NH_2CH_2R^2$ |
| $RC(=O){-}NHCH_2R'$ | → | $RC(=O){-}NH_2$ | + | $OHCR'$ |
| $RCH_2{-}NHR'$ | → | $RCH_2N(OH){-}R'$ → ($RCH{=}NR'$) → | | $RCH{=}N^+(O^-){-}R'$ |
| $ArNHR'$ | → | $Ar{-}N(OH)R'$ | | |
| $RC(=O){-}NHR'$ | → | $RC(=O){-}N(OH)R'$ | | |

ylic acids (see pathway b). N-Oxidation of secondary amines leads to a variety of N-oxygenated products. Secondary hydroxylamine formation is common, but these metabolites are susceptible to further oxidation to give nitrones. An example of this is the metabolism of the anorectic drug fenfluramine (**7.43**, Scheme 7.19).

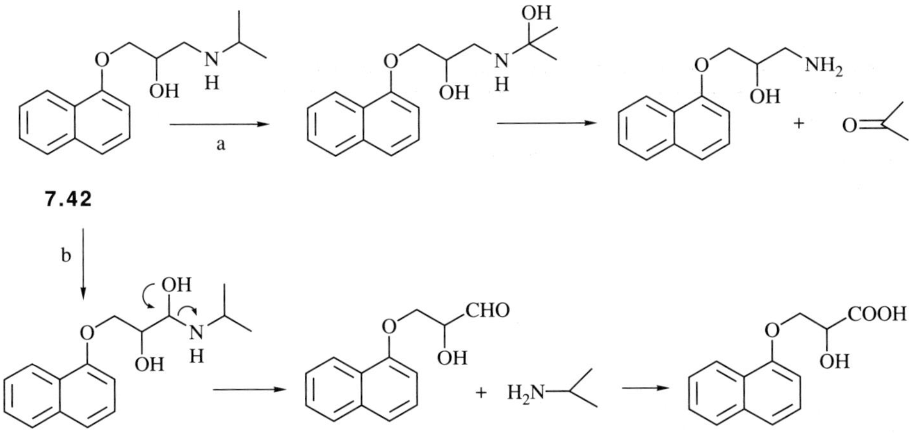

**7.42**

Scheme 7.18. Oxidative metabolism of propranolol.

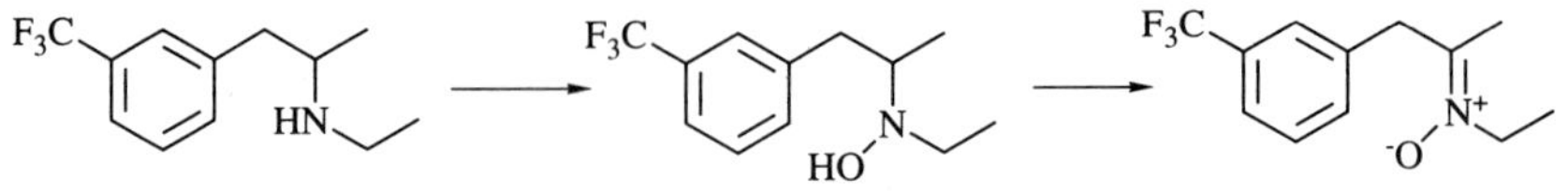

7.43

**Scheme 7.19.** N-Oxidation of fenfluramine.

Tertiary aliphatic amines generally undergo oxidative N-dealkylation and N-oxidation reactions (Table 7.5). Oxidative N-dealkylation (see Scheme 7.17) leads to the formation of secondary amines. In general, this occurs more rapidly than further oxidative N-dealkylation of the secondary amine metabolite to the primary amine. The tricyclic antidepressant drug imipramine (**7.44**, R = $CH_3$), for example, is metabolized to the corresponding secondary amine, desmethylimipramine (desipramine; **7.44**, R = H), which also is an antidepressant drug.[70] Very little oxidative N-demethylation of **7.44** (R = H) occurs. Enantioselective oxidative N-dealkylation can occur with chiral tertiary amines. The (2*S*,3*R*)-(+)-enantiomer of propoxyphene (**7.45**), a narcotic analgesic drug, is N-demethylated more slowly than the (−)-enantiomer.[71]

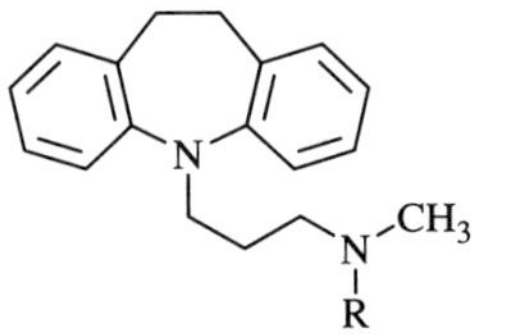

7.44 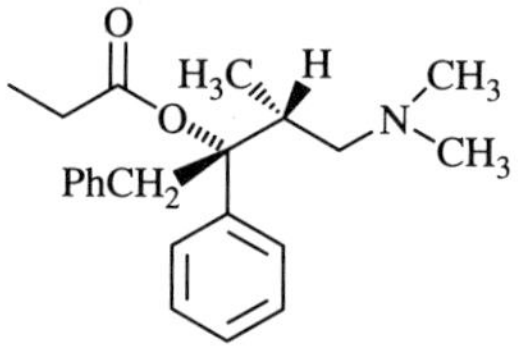 7.45

Monoamine oxidase (MAO; see Section III,C of Chapter 4) also catalyzes the metabolic oxidation of certain tertiary amine drugs. The (*S*)-(+)-isomer of the antiparkinsonian drug deprenyl (**7.46**, Scheme 7.20) is only a weak inhibi-

**Table 7.5** Oxidative Reactions of Tertiary Amines and Amides

| Substrate | | Products |
|---|---|---|
| $RCH(R')N(CH_2R^3)CH_2R^2$ | → | $RCH(R')NHCH_2R^3$ + $HC(=O)-R^2$ |
| $R_3N$ | → | $R_3N^+-O^-$ |
| $ArN(R)R'$ | → | $ArN^+(R)(O^-)-R'$ |
| $RC(=O)-N(R')CH_2R^2$ | → | $RC(=O)-NHR'$ + $HC(=O)-R^2$ |

tor of monoamine oxidase B and is therefore therapeutically ineffective. It is metabolized to (*S*)-(+)-metamphetamine (**7.47**) which is converted to (*S*)-(+)-amphetamine, both of which strongly contribute to the undesirable CNS stimulant side effect. (*R*)-(−)-Deprenyl, however, is a potent MAO B inactivator and is metabolized to (*R*)-(−)-metamphetamine and (*R*)-(−)-amphetamine, which have only weak CNS stimulant side effects.[72] Because of this, only the (−)-enantiomer is utilized in the treatment of Parkinson's disease.

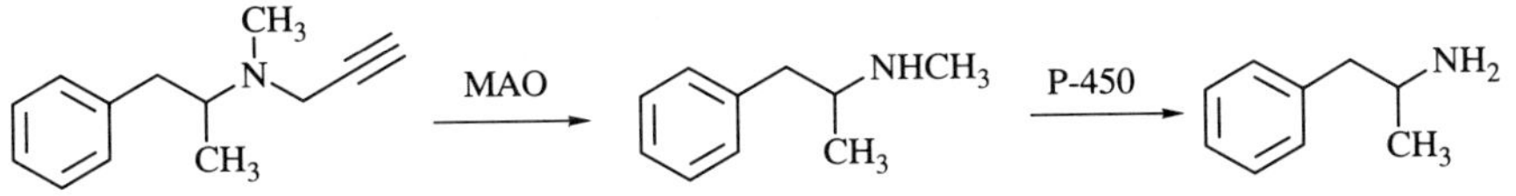

**7.46** **7.47**

**Scheme 7.20.** Metabolism of deprenyl.

When an alicyclic tertiary amine undergoes oxidation of the alicyclic carbon atoms attached to the nitrogen, a variety of oxidation products can result. Nicotine (**7.48**, Scheme 7.21) is $\alpha$-hydroxylated both in the pyrrolidine ring (pathway a) and at the methyl carbon (pathway b). The carbinolamine **7.49** is oxidized further to the major metabolite, cotinine (**7.50**). Cotinine also can undergo hydroxylation, leading to a minor metabolite, $\gamma$-(3-pyridyl)-$\gamma$-oxo-*N*-methylbutyramide (**7.51**).[73] Carbinolamine **7.49** is in equilibrium with the immonium ion **7.52**, which has been trapped with cyanide ion *in vitro* to give **7.53**.[74] The immonium ion is very electrophilic and may be a cause for the toxicity of nicotine. Oxidative N-demethylation of **7.48** to **7.54** is also observed as a metabolic pathway for nicotine. Further evidence for metabolic immonium ion intermediates generated during tertiary amine oxidation is the isolation of the imidazolidinone **7.56** from the metabolism of the local anesthetic lidocaine[75] (**7.55**, Scheme 7.22).

N-Oxidation of tertiary amines gives chemically stable tertiary amine N-oxides which, in contrast to N-oxidation of primary and secondary amines, do not undergo further oxidation. The antihypertensive drug guanethidine (**7.57**) is oxidized at the tertiary cyclic amine atom to give the *N*-oxide **7.58**.[76] The pyrrolidine nitrogen of nicotine (**7.48**) also is N-oxidized to (1′*R*,2′*S*)- and (1′*S*,2′*S*)-nicotine 1′-*N*-oxide. One of the major metabolites of the antihistamine cyproheptadine (**7.59**) in dogs is the $\alpha$-*N*-oxide **7.60**; no $\beta$-*N*-oxide was detected.[77] Hydrogen peroxide oxidation of **7.59**, however, gives both $\alpha$- and $\beta$-*N*-oxides.[78] *N*-Oxides are susceptible to bioreduction, which regenerates the parent amine.

Aromatic amine oxidation is similar to that for aliphatic amines. N-Oxidation of aromatic amines appears to be an important contributor to the carcinogenic and cytotoxic properties of aromatic amines.[79] There are two enzyme systems responsible for N-oxidation of tertiary aromatic amines, a flavopro-

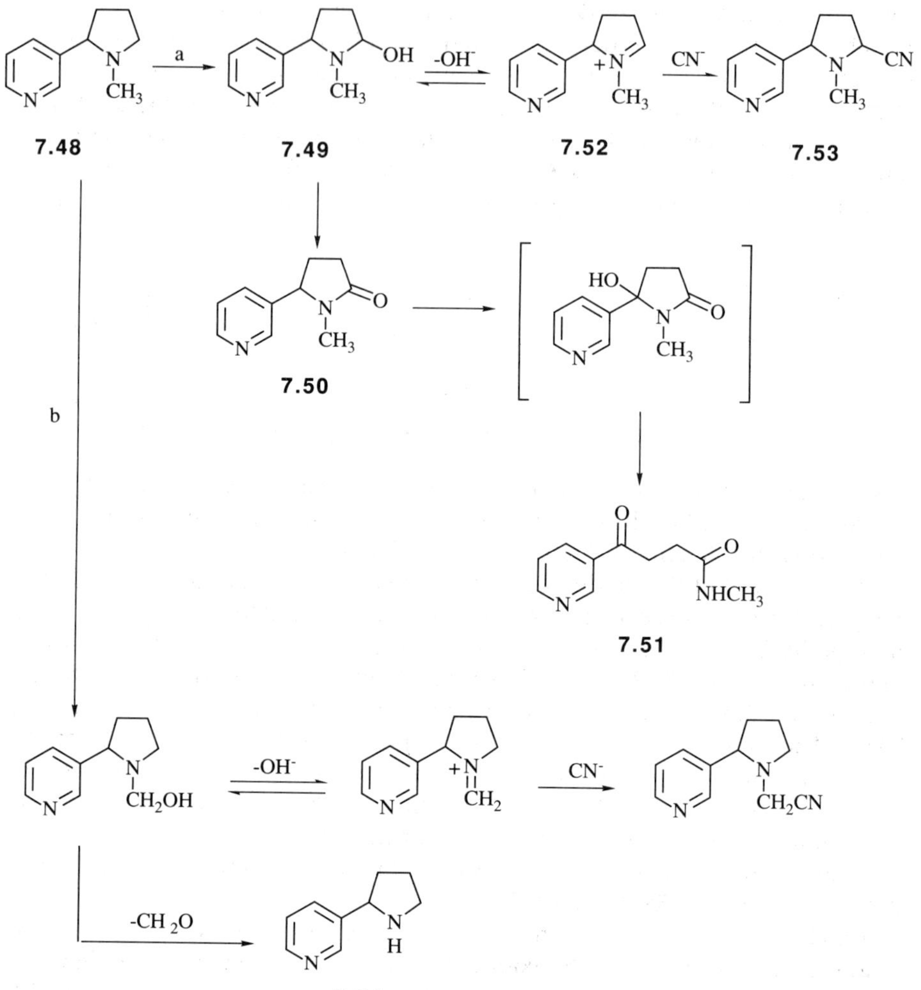

7.48 7.49 7.52 7.53

7.50

7.51

7.54

**Scheme 7.21.** Oxidative metabolism of nicotine leading to C—N bond cleavage.

tein monooxygenase and cytochrome *P*-450. The mechanism for the flavin-dependent reaction was discussed in Section III,C of Chapter 4 (see Schemes 4.33 and 4.34). Cytochrome *P*-450–catalyzed N-oxidation occurs by an electron transfer reaction followed by oxygen rebound (Scheme 7.23). N-Oxidation by cytochrome *P*-450 appears to occur only if there are no $\alpha$-hydrogens available for abstraction, if the iminium radical is stabilized by electron donation, or if Bredt's rule (i.e., that a double bond cannot be generated at a bridgehead carbon of a bicyclic system which does not contain at least one ring having eight or more atoms) prevents $\alpha$-hydrogen abstraction.[80a]

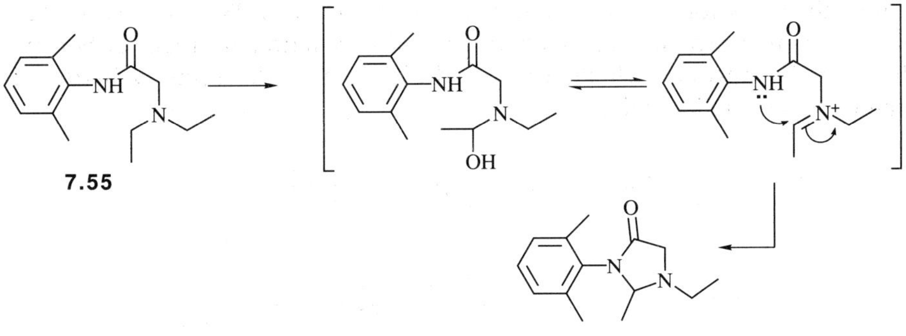

7.55

7.56

Scheme 7.22. Oxidation of lidocaine.

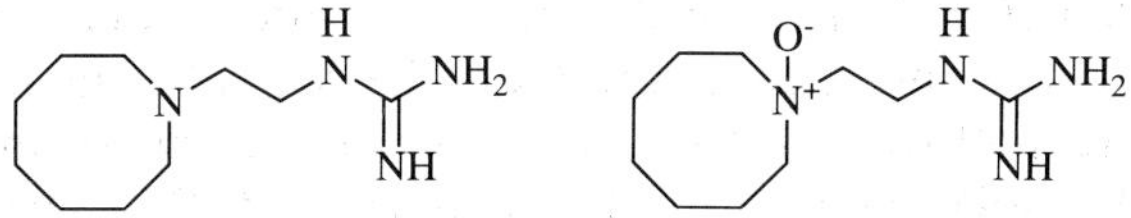

7.57

7.58

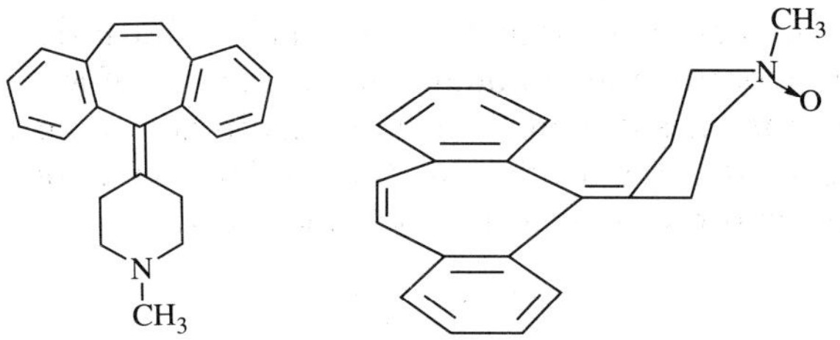

7.59

7.60

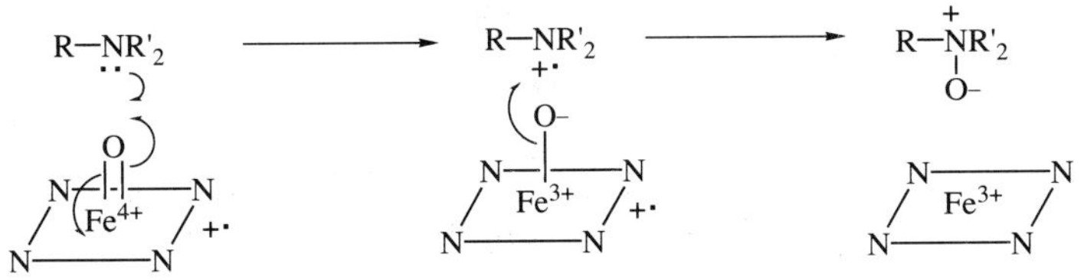

Scheme 7.23. Mechanism for cytochrome *P*-450–catalyzed N-oxidation.

The mechanism for oxidation of primary arylamines may be different from that of secondary and tertiary arylamines. Although purified flavin monooxygenases are inactive toward most primary arylamines, they readily N-oxygenate secondary and tertiary arylamines. Ziegler *et al.*[80b] showed that primary arylamines could be N-methylated by *S*-adenosylmethionine-dependent amine *N*-methyltransferases, and that the secondary arylamines produced were substrates for flavin monooxygenases (Scheme 7.23a). The secondary hydroxylamine products of this N-oxygenation reaction are then further oxidized to the nitrones which, upon hydrolysis, give the primary hydroxylamines and formaldehyde.

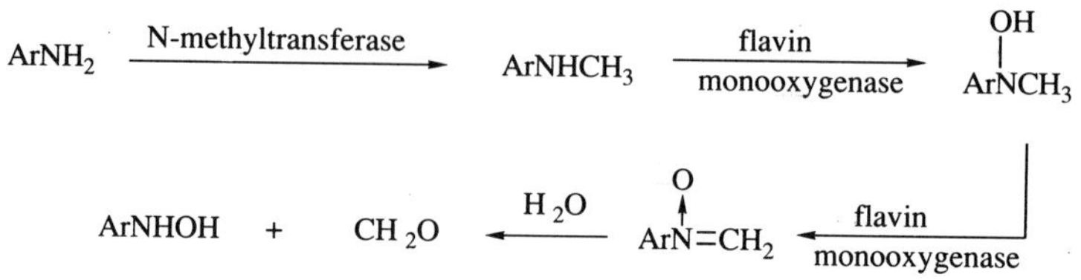

$$ArNH_2 \xrightarrow{\text{N-methyltransferase}} ArNHCH_3 \xrightarrow[\text{monooxygenase}]{\text{flavin}} ArN(OH)CH_3$$

$$ArNHOH + CH_2O \xleftarrow{H_2O} ArN(\rightarrow O){=}CH_2 \xleftarrow[\text{monooxygenase}]{\text{flavin}}$$

**Scheme 7.23a.** Possible mechanism for N-oxidation of primary arylamines.

N-Demethylation of tertiary aromatic amines is believed to proceed by two mechanisms[81] (Scheme 7.24), one that involves intermediate carbinolamine formation (pathway a) and one that produces the *N*-oxide which rearranges to the carbinolamine (pathway b). Evidence to support carbinolamine formation is provided by the isolation of the carbinolamine (**7.61**, R = OH) during cytochrome *P*-450–catalyzed oxidation of *N*-methylcarbazole[82] (**7.61**, R = H). Intrinsic isotope effects on the cytochrome *P*-450–catalyzed oxidation of tertiary aromatic amines were found to be small.[83] Intrinsic isotope effects associated with deprotonation of aminium radicals are of low magnitude ($k_H/k_D < 3.6$), but those associated with direct hydrogen atom abstraction are large; this suggests that the mechanism for carbinolamine formation involves electron transfer and proton transfer (Scheme 7.25).

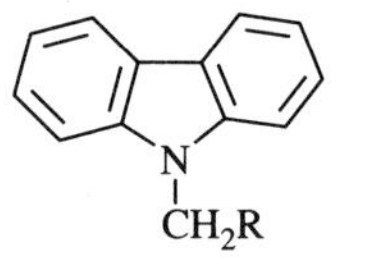

N
$CH_2R$

**7.61**

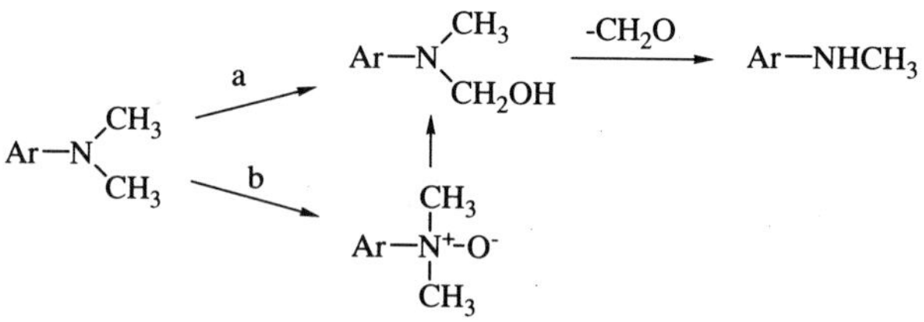

$$Ar{-}N(CH_3)_2 \xrightarrow{a} Ar{-}N(CH_3)CH_2OH \xrightarrow{-CH_2O} Ar{-}NHCH_3$$

$$Ar{-}N(CH_3)_2 \xrightarrow{b} Ar{-}N^{+}(CH_3)_2{-}O^{-} \rightarrow Ar{-}N(CH_3)CH_2OH$$

**Scheme 7.24.** Two pathways to N-demethylation of tertiary aromatic amines.

N-Oxidation of primary and secondary aromatic amines leads to the generation of reactive electrophilic species that form covalent bonds to cellular macromolecules (Scheme 7.26).[79,84] The $X^+$ in Scheme 7.26 represents acety-

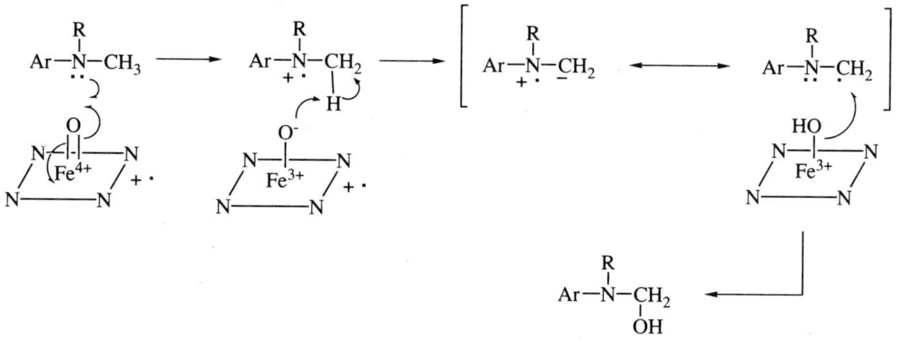

Scheme 7.25. Mechanism of carbinolamine formation during oxidation of tertiary aromatic amines.

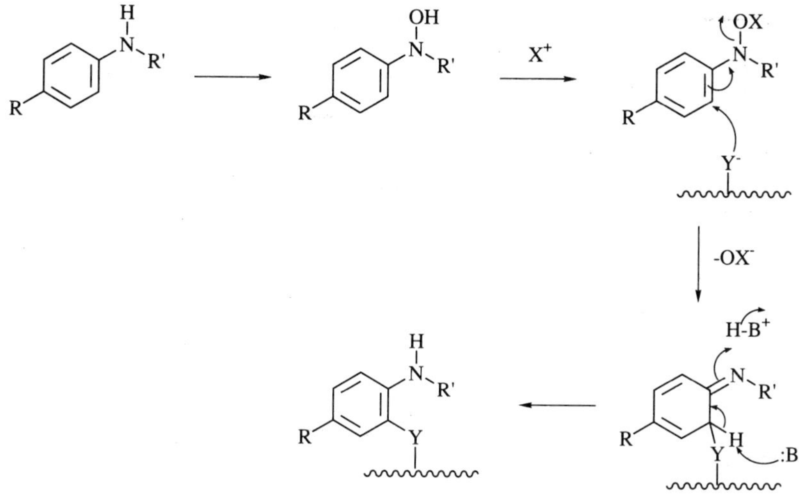

Scheme 7.26. Metabolic activation of primary and secondary aromatic amines.

lation or sulfation of the hydroxylamine to give a good leaving group ($OX^-$). Attachment of aromatic amines to proteins, DNA, and RNA is known.[85] Although C—N bond cleavage of heterocyclic aromatic nitrogen-containing compounds does not occur, N-oxidation to the aromatic N-oxide is prevalent.

Amides also are metabolized both by oxidative N-dealkylation and N-oxidation (see Tables 7.3–7.5). The sedative diazepam (**7.62**, R = $CH_3$) undergoes extensive oxidative N-demethylation to **7.62** (R = H). As in the case of primary and secondary aromatic amines, which are activated to cytotoxic and carcinogenic metabolites, N-oxidation of aromatic amides also leads to electrophilic intermediates. The carcinogenic agent 2-acetylaminofluorene (**7.63**, R = H) undergoes cytochrome *P*-450–catalyzed oxidation to the N-hydroxy analog (**7.63**, R = OH). Activation of the hydroxyl group as in Scheme 7.26 leads to an electrophilic species capable of undergoing attack by cellular nucleophiles.

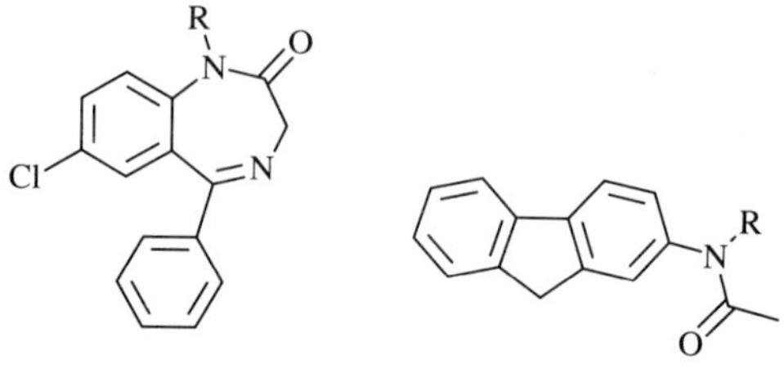

**7.62** **7.63**

Liver arylhydroxamic acid *N*,*O*-acyltransferase has been shown to catalyze the rearrangement of **7.63** (R = OH) to the corresponding *O*-acetyl hydroxylamine (**7.64**) which is activated for nucleophilic attack by the acyltransferase or by other cellular macromolecules (Scheme 7.27).[86] When covalent bond formation occurs to arylhydroxamic acid *N*,*O*-acyltransferase (prior to release of the activated species), then it is known as mechanism-based enzyme inactivation (see Section V,C of Chapter 5). Release of **7.64** into the cell can lead to covalent bond formation with other cellular macromolecules, resulting in cytotoxicity and carcinogenicity.

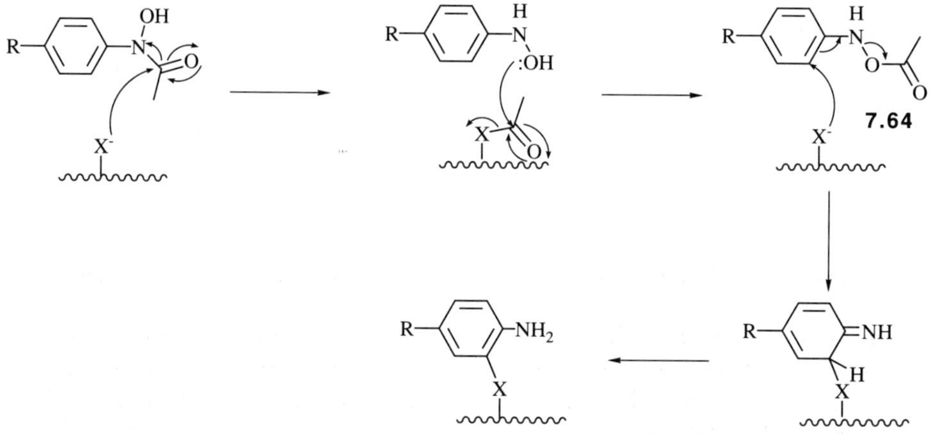

**Scheme 7.27.** Arylhydroxamic acid *N*,*O*-acyltransferase-catalyzed activation of *N*-hydroxy-2-acetylaminoarenes.

The analgesic agent acetaminophen (**7.65**, Scheme 7.28) is relatively nontoxic at therapeutic doses, but in large doses it causes severe liver necrosis.[87] The hepatotoxicity arises from the depletion of liver glutathione levels as a result of the reaction of glutathione with an electrophilic metabolite of acetaminophen. When the glutathione levels drop by 80% or greater, then hepatic macromolecules react with the electrophilic metabolite, leading to the observed liver necrosis.[87] The original proposal for the mechanism of formation of the electrophilic metabolite (**7.66**) involves N-oxidation of the drug (Scheme 7.28, pathway a).[87] The $X^-$ in Scheme 7.28 represents either glutathione or a cellular macromolecular nucleophile. No evidence, however, for the formation of *N*-hydroxyacetaminophen could be found.[88] Another mechanism for the formation of the electrophilic metabolite **7.66** would involve acetaminophen epoxidation (Scheme 7.28, pathway b). If this were correct, incubation of acetaminophen in the presence of $^{18}O_2$ should result in the incorporation of $^{18}O$ into metabolites; however, none was incorporated.[89]

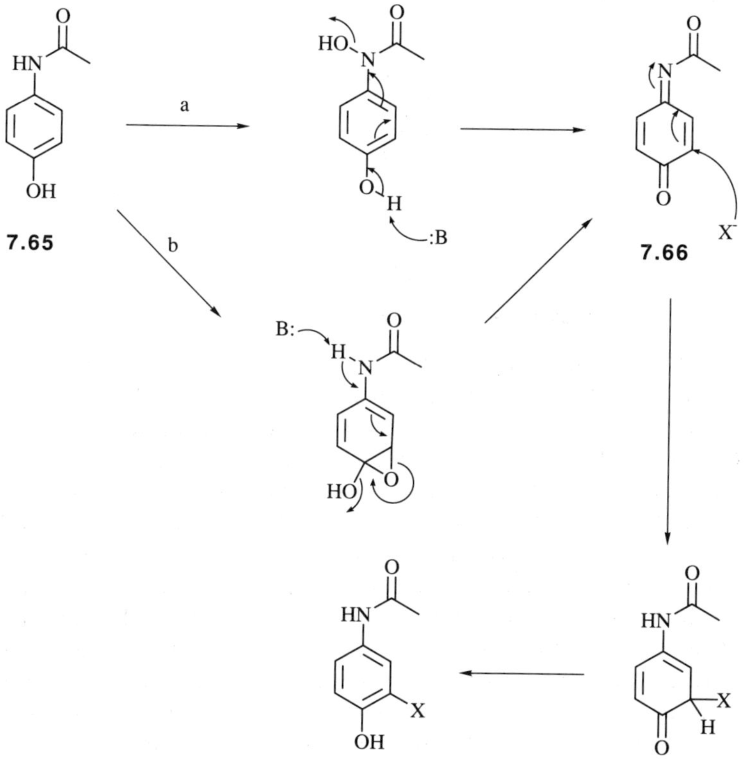

**Scheme 7.28.** Early proposals for bioactivation of acetaminophen.

A third mechanistic possibility is a one-electron transfer mechanism via the acetaminophen radical (**7.67**, Scheme 7.29).[90] The acceptor Y could be $O_2$ or oxygen bound to the heme of cytochrome *P*-450. Electron paramagnetic resonance (EPR) evidence was obtained[91] for the formation of acetaminophen radicals, which in the presence of $O_2$ produced superoxide. Because ethanol induces a cytochrome *P*-450, the hepatotoxicity of acetaminophen in ethanol-fed animals was compared with normal animals. It was found that the alcoholic animals had increased hepatotoxicity and that radical scavengers protected the animals. These results support a mechanism that involves cytochrome *P*-450–dependent acetaminophen radical formation followed by second electron transfer to oxygen (Scheme 7.29) as a mechanism for the generation of the electrophilic metabolite responsible for acetaminophen hepatotoxicity. Because of its reactivity, it was suggested that **7.67** may be responsible for the hepatotoxicity[92]; however, the benzoquinone imine (**7.66**) has been detected as a metabolite of the oxidation of acetaminophen by purified cytochrome *P*-450[93] and by microsomes, NADPH, and $O_2$.[94] Furthermore, **7.66** reacts rapidly with glutathione *in vitro* to give the same conjugate as that found *in vivo*.[93,94]

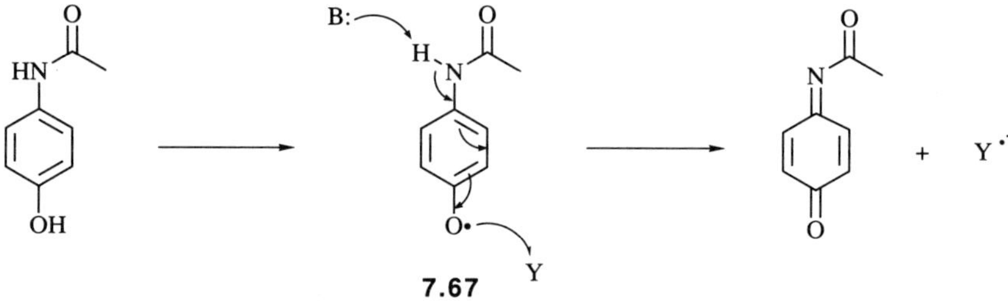

**7.67**

**Scheme 7.29.** Bioactivation of acetaminophen.

High doses of acetaminophen also cause renal damage in humans;[87] however, cytochrome *P*-450 activity is low in the kidneys, and, therefore, it may not be the major enzyme responsible for acetaminophen toxicity there. Prostaglandin H synthase (also called prostaglandin synthetase or cyclooxygenase), the enzyme that catalyzes the cyclooxygenation of arachidonic acid to prostaglandin $G_2$ ($PGG_2$) followed by the reduction of $PGG_2$ to $PGH_2$ (see Scheme 5.14 in Section V,B,2,b of Chapter 5), is present in high concentrations in the kidneys, and it may be important in promoting acetaminophen toxicity in that organ. During the reduction of $PGG_2$ to $PGH_2$, prostaglandin H synthase can simultaneously cooxidize a variety of substrates including certain drugs. Acetaminophen is metabolized to an intermediate that reacts with glutathione to form the glutathione conjugate.[95] By comparison of one- and two-electron reactions in the presence and absence of glutathione, it was found that prostaglandin H synthase also can catalyze the bioactivation of acetaminophen by one- and two-electron mechanisms[96] (Scheme 7.30). Prostaglandin H synthase may be another important drug-metabolizing enzyme in tissues that are rich in this enzyme and low in the concentration of cytochrome *P*-450.

***f. Oxidations of Carbon–Oxygen Systems.*** *Oxidative O-dealkylation* is a common biotransformation which, as in the case of oxidative N-dealkylation (see Section IV,B,1,e), is catalyzed by microsomal mixed function oxidases.

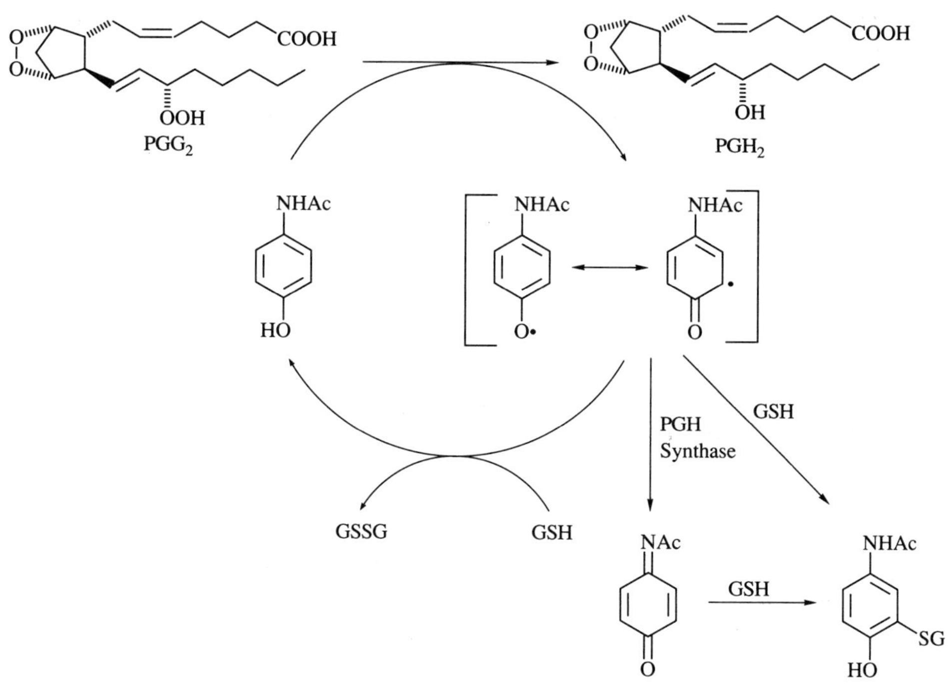

**Scheme 7.30.** Proposed bioactivation of acetaminophen by prostaglandin H synthase.

The mechanism appears to be the same as for oxidative N-dealkylation (see Scheme 7.17) and involves hydroxylation on the carbon attached to the oxygen followed by C—O bond cleavage to give the alcohol and the aldehyde or ketone. Although O-demethylation is rapid, dealkylation of longer chain *n*-alkyl substituents is slow, and often ω-1 hydroxylation (see Section IV,B,1,d) competes with O-dealkylation.

A major metabolite of the anti-inflammatory drug indomethacin (**7.68**, R = $CH_3$) is the O-demethylated compound (**7.68**, R = H).[97] Sometimes O-dealkylated metabolites also are pharmacologically active, as in the case of the narcotic analgesic codeine (**7.69**, R = $CH_3$) which is O-demethylated to mor-

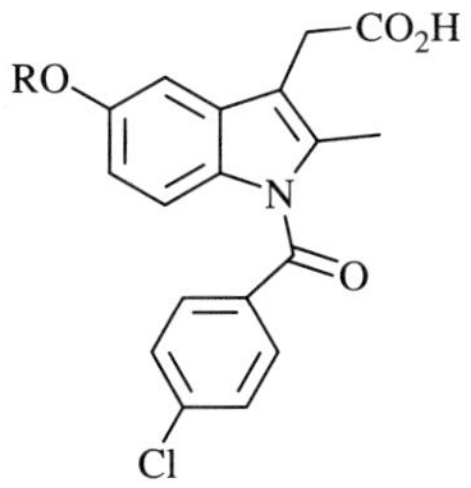

**7.68** 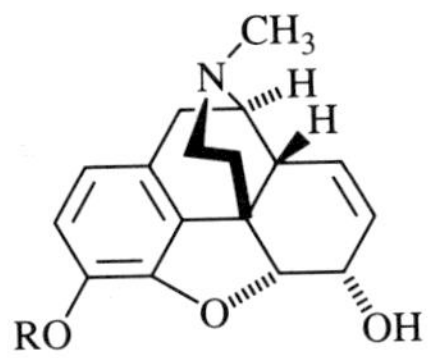 **7.69**

phine (**7.69**, R = H).[98] Nonequivalent methyl groups in drugs can be regioselectively O-demethylated; the blood pressure maintenance drug methoxamine (**7.70**, R = $CH_3$) gives exclusive 2′-O-demethylation (**7.70**, R = H) in dogs.[99] Stereoselective O-dealkylation can occur in the metabolism of chiral ethers.

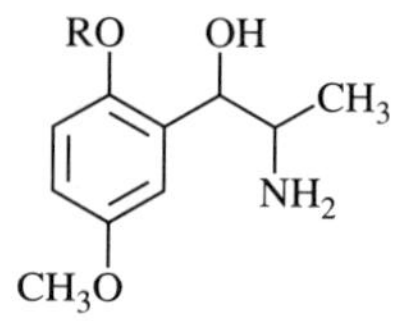

RO OH $CH_3$ $NH_2$ $CH_3O$

**7.70**

*g. **Oxidations of Carbon–Sulfur Systems.*** Fewer drugs contain sulfur than oxygen or nitrogen. The three principal types of biotransformations of carbon–sulfur systems are oxidative S-dealkylation, desulfuration, and S-oxidation. *Oxidative S-dealkylation* is not nearly as prevalent as the corresponding oxidative N- or O-dealkylations discussed above. The sedative methitural (**7.71**, R = $CH_3$) is metabolized by S-demethylation to **7.71** (R = H).[100]

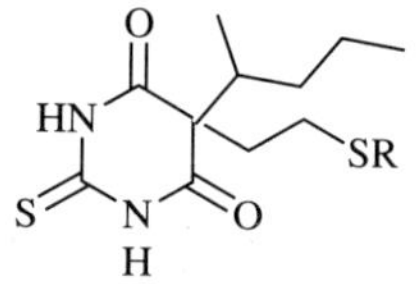

O HN SR S N O H

**7.71**

*Desulfuration* is the conversion of a carbon-sulfur double bond (C=S) to a carbon–oxygen double bond (C=O). The anesthetic thiopental (**7.72**, X = S) undergoes desulfuration to pentobarbital (**7.72**, X = O).[101]

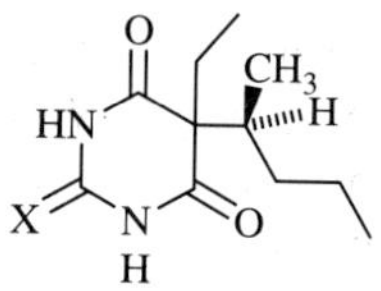

O $CH_3$ HN H X N O H

**7.72**

*S-Oxidation* of sulfur-containing drugs to the corresponding sulfoxide is a common metabolic transformation which is catalyzed both by flavin monooxygenase and by cytochrome *P*-450.[102] Flavin monooxygenase produces exclusively the sulfoxide by a mechanism discussed in Chapter 4 (see Scheme 4.33 or 4.34), but cytochrome *P*-450 metabolizes sulfides to S-dealkylation products and to sulfoxides.[103] The common intermediate for these two metabolites is probably the sulfenium cation radical (**7.73**, Scheme 7.31); the mecha-

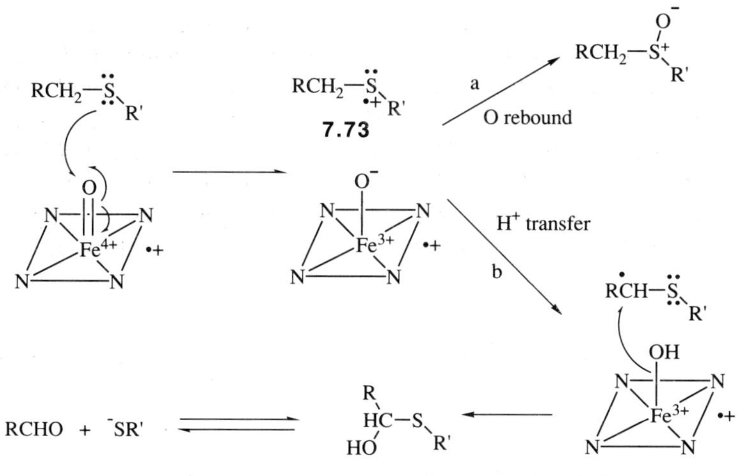

Scheme 7.31. Cytochrome *P*-450–catalyzed oxidation of sulfides.

nism for S-dealkylation may be related to that discussed in Chapter 4 for cytochrome *P*-450–catalyzed amine oxidations (see Scheme 4.37). The relative contribution of these two pathways was examined for the S-oxidation of the anthelmintic agent albendazole (**7.74**).[104] Another example of S-oxidation is the oxidation of the antipsychotic drug thioridazine (**7.75**, R = $SCH_3$), in which case both sulfur atoms in the drug are metabolized to the corresponding sulfoxides.[105] The metabolite with only the methylthio substituent converted to the sulfoxide [**7.75**, R = $S(O)CH_3$] is twice as potent as thioridazine and also is used as an antipsychotic drug (mesoridazine). Further oxidation of sulfoxides to sulfones also occurs; for example, the immunosuppressive drug oxisuran (**7.76**) is metabolized to the corresponding sulfone.[106]

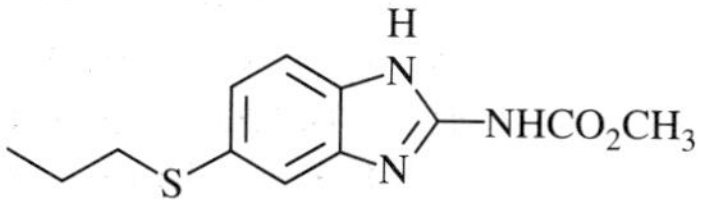

**7.74** 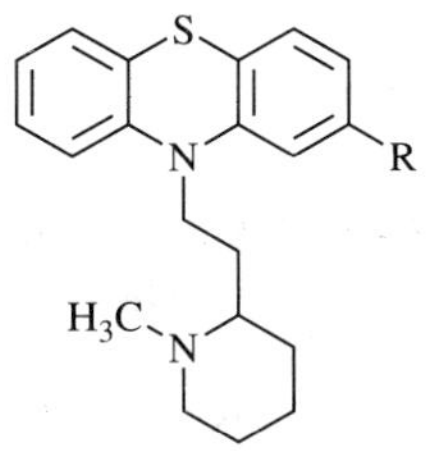 **7.75** 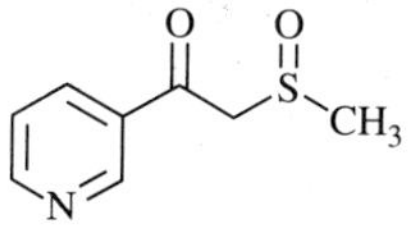 **7.76**

***h. Other Oxidative Reactions.*** Oxidative dehalogenation, oxidative aromatization, and oxidation of arenols to quinones are other important drug metabolism pathways that are catalyzed by cytochrome *P*-450 enzymes. *Oxidative dehalogenation* occurs with the volatile anesthetic halothane (**7.77**, Scheme 7.32) which is metabolized to trifluoroacetic acid.[107] The acid chloride may be responsible for the covalent binding of halothane to liver microsomes. This oxidation mechanism is the same as that for the mechanism-based inactivation of cytochrome *P*-450 by the antibacterial drug chloramphenicol (see Section V,C,3,i in Chapter 5). *Oxidative aromatization* of the A ring in the antifertility drug norgestrel (**7.78**, R = Et) by a cytochrome *P*-450 isozyme gives the corresponding phenol (**7.79**, R = Et).[108] Catechols may be converted enzymatically to electrophilic *ortho*-quinones. Morphine, for example, is metabolized to its 2,3-catechol (**7.80**), which is oxidized to the *ortho*-quinone (**7.81**).[109]

**7.77**

**Scheme 7.32.** Oxidative dehalogenation of halothane.

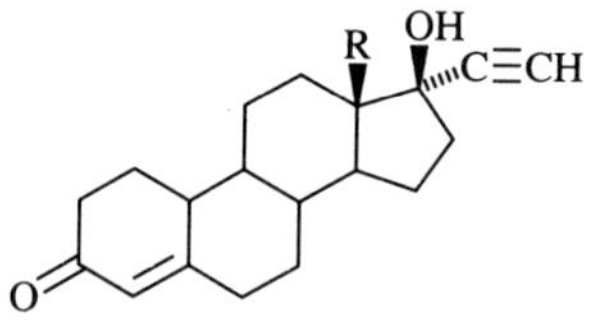

**7.78** 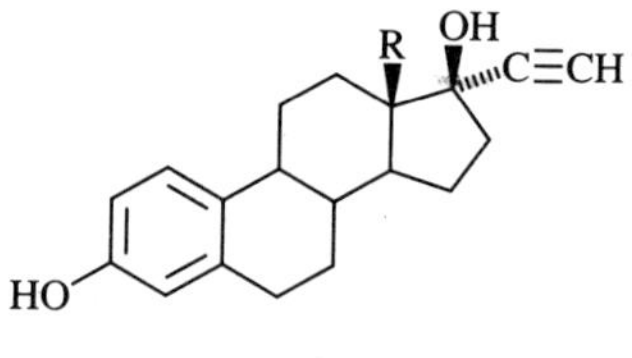 **7.79**

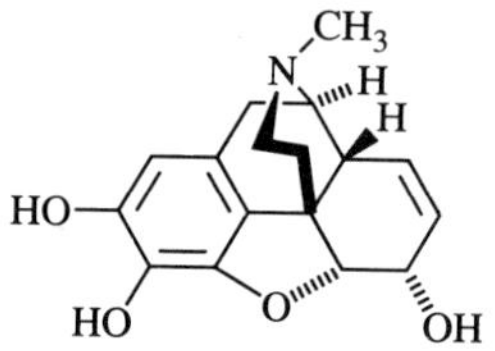

**7.80** 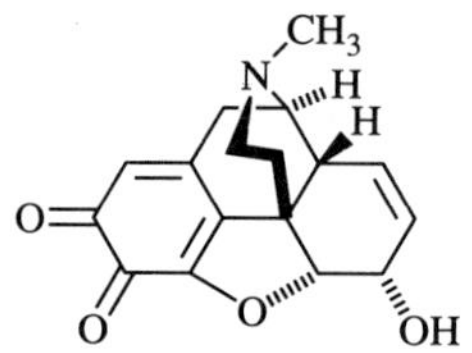 **7.81**

***i. Alcohol and Aldehyde Oxidations.*** The oxidation of alcohols to aldehydes and of aldehydes to carboxylic acids, in general, is catalyzed by alcohol dehydrogenase and aldehyde dehydrogenase, respectively, soluble enzymes that require $NAD^+$ or $NADP^+$ as the cofactor (see Section III,B of Chapter 4). They are found in highest concentrations in the liver but are also present in virtually every other organ. There also is a flavin-dependent aldehyde oxidase that is involved in aldehyde oxidation. The reaction catalyzed by alcohol dehydrogenase is the oxidation of primary alcohols to the aldehyde as well as the reverse reaction (Scheme 7.33). As is apparent from this equation, the equilibrium is pH dependent. The oxidation of alcohols is favored at higher pH (~pH 10), and the reduction of aldehydes is favored at lower pH (~pH 7). Therefore, it would be predicted that at physiological pH reduction would be favored. However, this is not observed because further oxidation of the aldehyde to the carboxylic acid (catalyzed by aldehyde dehydrogenase) is a lower energy pathway. Aldehydes are almost always metabolized to the acid; only a few examples of aldehyde reduction *in vivo* are known.

$$RCH_2OH + NAD^+ \rightleftharpoons RCHO + NADH + H^+$$

**Scheme 7.33.** Reaction catalyzed by alcohol dehydrogenase.

Actually, very few drugs contain an aldehyde group. The main exposure to aldehydes (which generally are toxic) is through ingestion of the metabolic precursors such as primary alcohols or various amines (see Section IV,B,1,e). Aromatic methyl groups also are oxidized to primary alcohols (see Section V,B,1,c), which then are metabolized to the aldehyde and the carboxylic acid. The anti-inflammatory and analgesic drug mefenamic acid (**7.82**, R = $CH_3$) is metabolized to the corresponding carboxylic acid (**7.82**, R = COOH).

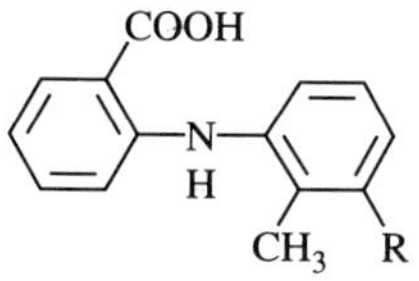

COOH
N
H
$CH_3$ R

**7.82**

## 2. Reductive Reactions

Oxidative processes are, by far, the major pathways of drug metabolism, but reductive reactions are important for the biotransformations of the functional groups listed in Table 7.6. Reductive reactions are important for the formation of hydroxyl and amino groups that render the drug more hydrophilic and set it up for Phase II conjugation (see Section IV,C).

***a. Carbonyl Reduction.*** Carbonyl reduction typically is catalyzed by aldo–keto reductases that require NADPH or NADH as the coenzyme. As described in the previous section, alcohol dehydrogenase catalyzes the reduction of aldehydes as well as the oxidation of alcohols. It is rare, however, to observe reduction of aldehydes to alcohols. A large variety of aliphatic and aromatic ketones also are reduced to alcohols by NADPH-dependent ketone reductases.[110] As discussed in Section III,B of Chapter 4, the NADPH carbonyl reductases are stereospecific with regard to the pyridine nucleotide cofactor. In general, the aldehyde reductases exhibit A-(pro-4*R*)-hydrogen specificity, and ketone reductases exhibit B-(pro-4*S*)-hydrogen specificity.[111] Stereoselectivity for enantiomer substrates as well as stereospecific reduction of the ketone carbonyl also are typical.

**Table 7.6** Classes of Substrates for Reductive Reactions

| Functional group | Product |
|---|---|
| 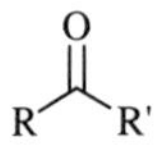 $RC(=O)R'$ | 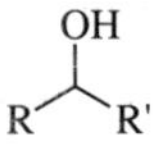 $RCH(OH)R'$ |
| $RNO_2$ | RNHOH |
| RNO | RNHOH |
| RNHOH | $RNH_2$ |
| RN=NR' | $RNH_2 + R'NH_2$ |
| $R_3N \rightarrow O$ | $R_3N$ |
| 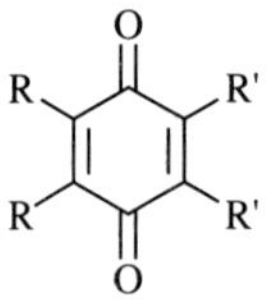 2,3,5,6-tetrasubstituted (R, R, R', R') 1,4-benzoquinone | 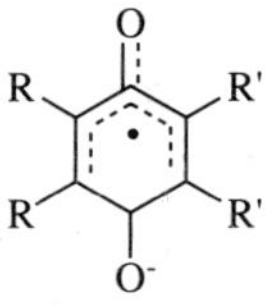 semiquinone radical anion ($O^-$) |
| R—X | $R\cdot + X^-$ |

The reduction of the anticoagulant drug warfarin (**7.83**, R = H) is selective for the (*R*)-(+)-enantiomer; reduction of the (*S*)-(−)-isomer occurs only at high substrate concentrations.[112] (*R*)-Warfarin is reduced in humans principally to the (*R*,*S*)-warfarin alcohol, whereas (*S*)-warfarin is metabolized mainly to 7-hydroxywarfarin (**7.83**, R = OH) but also is reduced to a 4:1

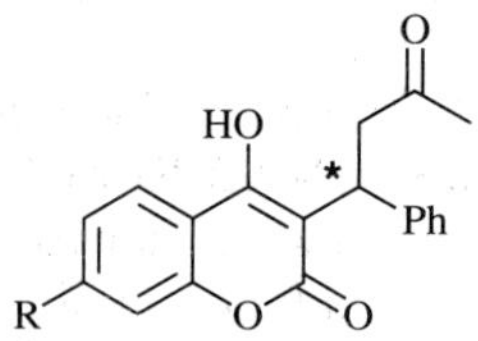

HO, Ph, R, O, O, *

**7.83**

mixture of the (*S*,*S*)-alcohol and the (*S*,*R*)-alcohol.[112] These studies were carried out by administration of the enantiomers separately to human volunteers. When warfarin was administered to human volunteers as the racemic mixture, rather than as separate enantiomers, a slightly different picture emerged.[113] Of the 50% of the drug recovered as metabolites, 19% arose from (*R*)-isomer metabolism and 31% from the (*S*)-isomer. The difference in the enantiomeric selectivity is a result of the difference in the rate of clearance from the body of the (*S*)-isomer relative to that of the (*R*)-isomer.[114] The major metabolite was (*S*)-7-hydroxywarfarin (22%); (*S*)-6-hydroxywarfarin (6%) also was obtained. In this study the main (*R*)-metabolite was not the (*R*,*S*)-warfarin alcohol (6%), but rather (*R*)-6-hydroxywarfarin (9%); (*R*)-7-hydroxywarfarin (3%) also was obtained. By comparison of the results from the two warfarin studies,[112,113] it is apparent that one warfarin enantiomer can have an effect on the metabolism of the other. This is yet another reason to administer pure enantiomers, rather than racemic mixtures, as drugs. Enzymatic reduction of ketones in general produces the (*S*)-alcohol as the major metabolite,[115] even in the presence of other chiral centers.

Species variation in the stereochemistry of the reduction of ketones is not uncommon. Naltrexone (**7.84**), a nonaddictive oral opioid antagonist used to treat narcotics addicts who have been withdrawn from opiates in rehabilitation programs, is reduced to the 6$\alpha$-alcohol in the chicken[116] and to the 6$\beta$-alcohol in rabbit and man.[117]

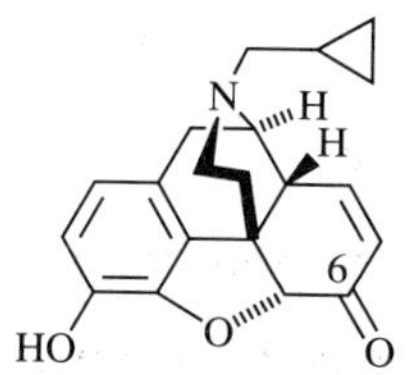

**7.84**

$\alpha,\beta$-Unsaturated ketones can be metabolized to saturated alcohols (reduction of both the carbon–carbon double bond and the carbonyl group). The antifertility drugs norgestrel (**7.78**, R = Et) and norethindrone (**7.78**, R = Me) are reduced in women to the 5$\beta$-*H*-3$\alpha$,17$\beta$-diol (**7.85**; $R^1$ = H, $R^2$ = OH, $R^3$ = Et) and 5$\beta$-*H*-3$\beta$,17$\beta$-diol derivatives (**7.85**, $R^1$ = OH, $R^2$ = H, $R^3$ = Me), respectively.[118] The $\Delta^4$-double bond of both drugs is reduced to give the 5$\beta$-configuration, but the 3-keto group is reduced to the 3$\alpha$-epimer in the case of norgestrel and to the 3$\beta$-epimer with norethindrone, even though the only difference in the two molecules is a 13$\beta$-ethyl group versus a 13$\beta$-methyl group, respectively.

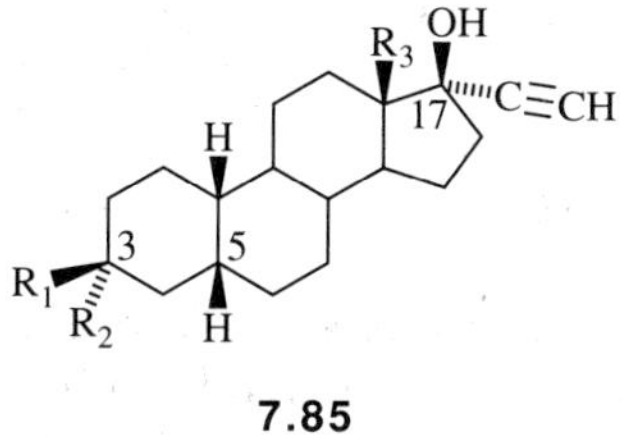

**7.85**

***b. Nitro Reduction.*** Aromatic nitro reduction, catalyzed by cytochrome *P*-450 in the presence of NADPH, except under anaerobic conditions ($O_2$ inhibits the reaction), and by the flavin-dependent NADPH–cytochrome-*P*-450 reductase (see Section IV,B,1), is a multistep process (Scheme 7.34); the reduction of the nitro group to the nitroso group (**7.87**) is the rate-determining step.[119] On the basis of EPR spectra and the correlation of rates of radical formation with product formation, it has been proposed[120] that the nitro anion radical (**7.86**) is the first intermediate in the reduction of the nitro group. The reoxidation of this radical by oxygen to give the nitro compound back and a superoxide radical anion may explain the inhibition of this metabolic pathway by oxygen.[121] Other enzymes that catalyze nitro group reduction are the bacterial nitro reductase in the gastrointestinal tract,[122] xanthine oxidase,[123] aldehyde oxidase,[124] and quinone reductase [NAD(P)H dehydrogenase (quinone); DT-diaphorase].[125] Metabolic oxidation–reduction cycling of drugs may occur; the balance between the oxidative and reductive pathways is important in determining the pharmacological and toxicological profile of a drug.

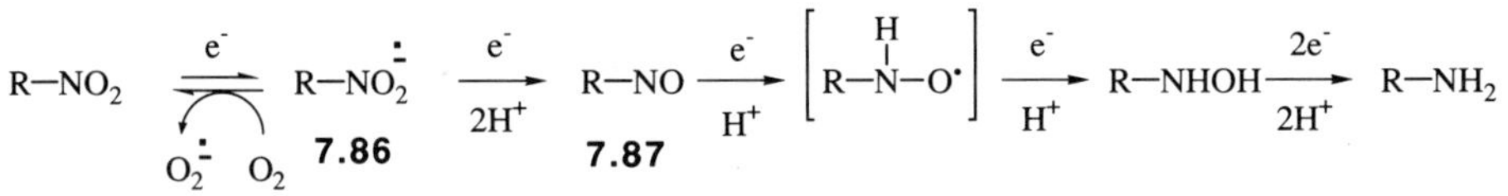

$$R{-}NO_2 \underset{O_2^{\cdot-}\ \ O_2}{\overset{e^-}{\rightleftharpoons}} \underset{\mathbf{7.86}}{R{-}NO_2^{\cdot-}} \xrightarrow[2H^+]{e^-} \underset{\mathbf{7.87}}{R{-}NO} \xrightarrow[H^+]{e^-} \left[R{-}\overset{H}{N}{-}O^{\cdot}\right] \xrightarrow[H^+]{e^-} R{-}NHOH \xrightarrow[2H^+]{2e^-} R{-}NH_2$$

**Scheme 7.34.** Nitro group reduction.

An example of nitro reduction is the metabolism of the anticonvulsant drug clonazepam (**7.88**, R = $NO_2$) to its corresponding amine (**7.88**, R = $NH_2$).[126] In some cases the reduced metabolite is not observed because it is easily air oxidized back to the parent compound. For example, the antiparasitic agent niridazole (**7.89**, R = $NO_2$) is reduced to the hydroxylamine metabolite (**7.89**, R = NHOH), which is reoxidized in air to **7.89** (R = $NO_2$).[123,127] The antibacterial drug nitrofurazone (**7.90**, Scheme 7.35) is reduced both to the corresponding 5-hydroxylamino derivative (**7.91**) and the 5-amino derivative (**7.92**). The latter is unstable and isomerizes to **7.93**.[128]

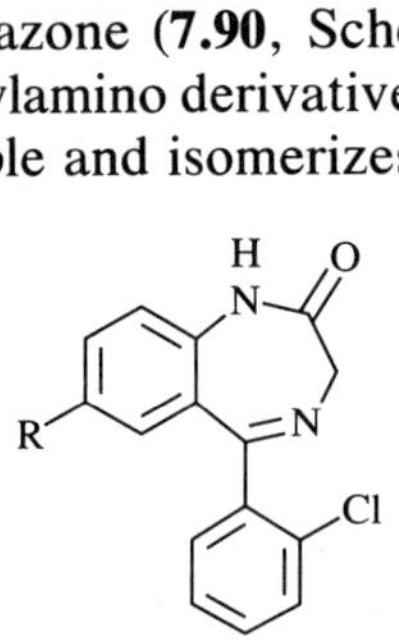

H N O N R Cl

**7.88**

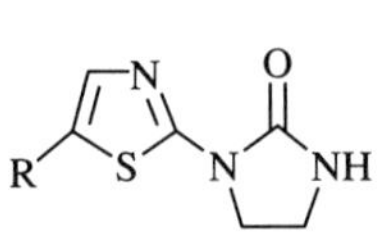

N O R S N NH

**7.89**

***c. Azo Reduction.*** Azo group (RN=NR) reduction is similar to nitro reduction in many ways. It, too, is mediated both by cytochrome *P*-450 and by NADPH–cytochrome-*P*-450 reductase (see Section IV,B,1), and oxygen of-

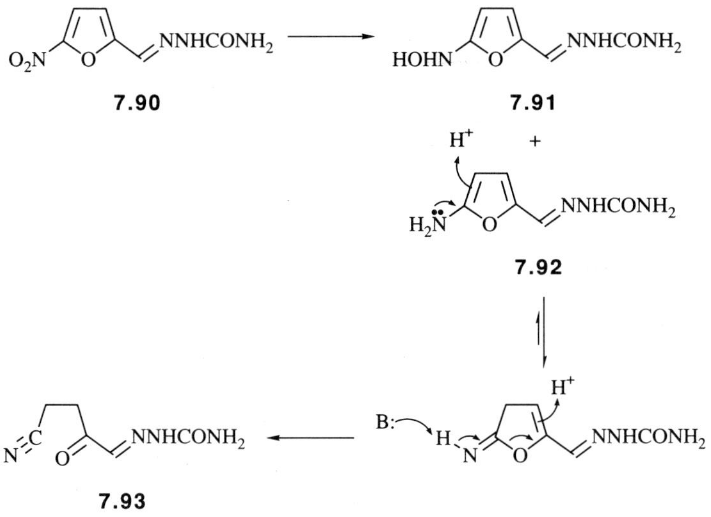

$O_2N$–(furan)–$CH{=}NNHCONH_2$ (**7.90**) → HOHN–(furan)–$CH{=}NNHCONH_2$ (**7.91**) → $H^+$ + $H_2N$–(furan)–$CH{=}NNHCONH_2$ (**7.92**) ⇌ B: H–N=(ring)–$CH{=}NNHCONH_2$ → $N{\equiv}C$–$CH_2CH_2$–C(=O)–$CH{=}NNHCONH_2$ (**7.93**)

**Scheme 7.35.** Reductive metabolism of nitrofurazone.

ten inhibits the reaction. The initial reduction in the oxygen-sensitive metabolism appears to proceed via the azo anion radical (**7.94**, Scheme 7.36)[129]; the oxygen apparently reverses this reduction, with concomitant conversion of the oxygen to the superoxide anion radical.[130] Oxygen-insensitive azoreductases presumably proceed by a two-electron reduction of the azo compound directly to the hydrazo intermediate.

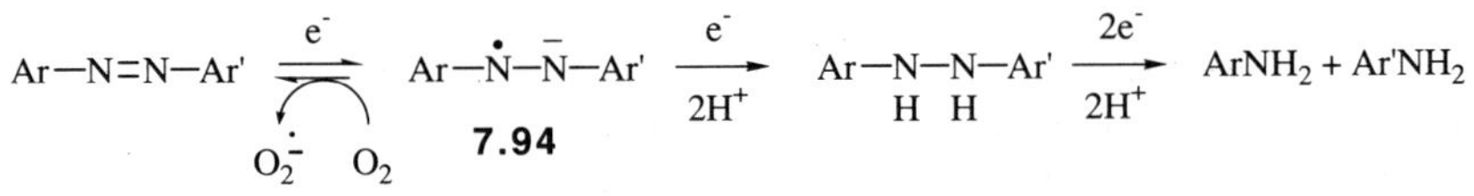

$$\mathrm{Ar{-}N{=}N{-}Ar'} \underset{O_2^{\cdot -}\ \ O_2}{\overset{e^-}{\rightleftharpoons}} \underset{\mathbf{7.94}}{\mathrm{Ar{-}\dot{N}{-}\bar{N}{-}Ar'}} \xrightarrow[2H^+]{e^-} \mathrm{Ar{-}NH{-}NH{-}Ar'} \xrightarrow[2H^+]{2e^-} \mathrm{ArNH_2 + Ar'NH_2}$$

**Scheme 7.36.** Azo group reduction.

Bacteria in the gastrointestinal tract also are important in azo reduction.[122] Reduction of sulfasalazine (**7.95**, Scheme 7.37), used in the treatment of ulcerative colitis, to sulfapyridine (**7.96**) and 5-aminosalicylic acid (**7.97**) occurs primarily in the colon via the action of intestinal bacteria.[131]

***d. Tertiary Amine Oxide Reduction.*** A wide variety of aliphatic and aromatic tertiary amine oxides, such as imipramine *N*-oxide (**7.98**), are reduced by cytochrome *P*-450 in the absence of oxygen.[132]

***e. Reductive Dehalogenation.*** Under hypoxic or anaerobic conditions the volatile anesthetic halothane (**7.99**) is reductively dehalogenated by cytochrome *P*-450[133] (Scheme 7.38). The first electron is transferred to halothane from cytochrome *P*-450, which is reduced by NADPH–cytochrome-*P*-450 reductase. This electron transfer ejects the bromide ion and produces the

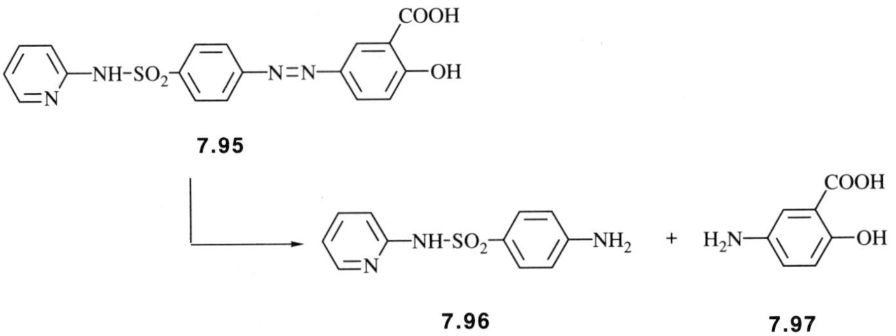

7.95

7.96 7.97

Scheme 7.37. Reductive metabolism of sulfasalazine.

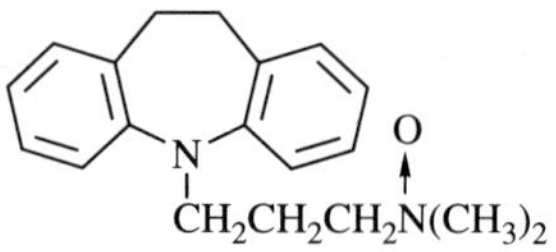

$CH_2CH_2CH_2N(CH_3)_2$

7.98

cytochrome *P*-450–bound 1-chloro-2,2,2-trifluoroethyl radical. If this radical escapes from the active site (pathway a), it either can be reduced by hydrogen atom transfer (pathway c) to give 2-chloro-1,1,1-trifluoroethane (**7.100**) or can form a covalent bond to cellular proteins (pathway d). A second electron reduction from **7.99** (pathway b) produces the carbanion; $\beta$-elimination of fluoride ion gives chlorodifluoroethylene (**7.101**). Pathway d, resulting in covalent attachment to proteins, has been proposed[133] to be the cause for halothane hepatitis, a toxic reaction to halothane exposure in the liver. The second electron transfer (pathway b) is thought to be derived from cytochrome $b_5$; this leads to nontoxic products and competes with pathway a, thereby aiding detoxification in the metabolic process.

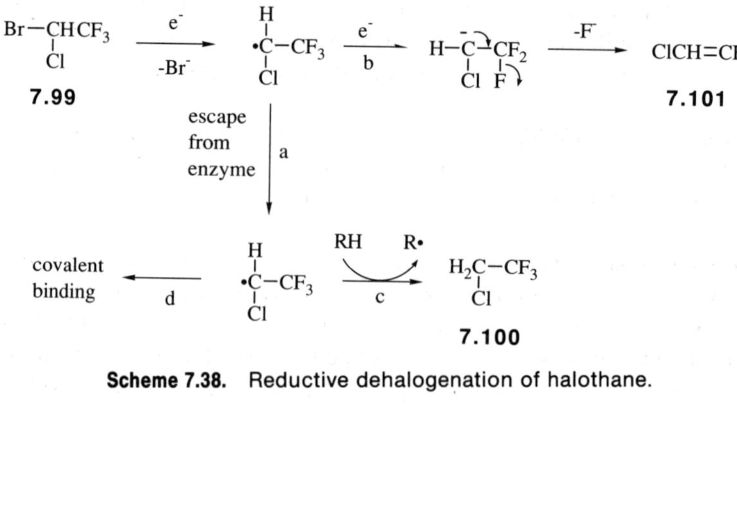

7.99 7.101

escape from enzyme a

covalent binding d c RH R•

7.100

Scheme 7.38. Reductive dehalogenation of halothane.

### 3. Hydrolytic Reactions

The hydrolytic metabolism of esters and amides leads to the formation of carboxylic acids, alcohols, and amines, all of which are quite susceptible to phase II conjugation reactions and excretion (see Section IV,C). As described in Section II,C of Chapter 4 enzyme-catalyzed hydrolysis can be acid and/or base catalyzed. Base-catalyzed hydrolysis is accelerated nonenzymatically when electron-withdrawing groups are substituted on either side of the ester or amide bond. When the carbonyl is in conjugation with a $\pi$-system, nonenzymatic base hydrolysis is decelerated.

A wide variety of nonspecific esterases and amidases involved in drug metabolism are found in plasma, liver, kidney, and intestine.[134] All mammalian tissues may contribute to the hydrolysis of a drug; however, the liver, the gastrointestinal tract, and the blood have the greatest hydrolytic capacity. Aspirin (**7.102**) is an example of a drug that is hydrolyzed by all human tissues.[135] The hydrolysis of xenobiotics is very similar in all mammals; however, there are some exceptions, and large species differences can be observed. Some esterases catalyze hydrolysis of aliphatic esters and others aromatic esters. For example, only the benzoyl ester in cocaine (**7.103**, R = $CH_3$) is hydrolyzed by human liver *in vitro*, not the alicyclic ester[136]; however, *in vivo*, the major metabolite of cocaine is the alicyclic ester hydrolysis product, benzoylecgonine (**7.103**, R = H).[137]

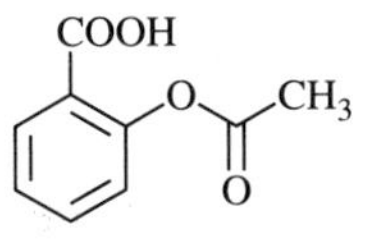

**7.102** 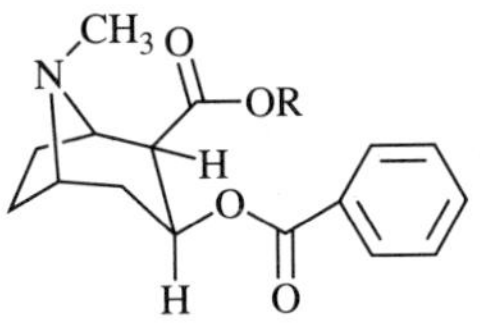 **7.103**

Generally, amides are more slowly hydrolyzed than esters. For example, the enzymatic hydrolysis of the antiarrhythmic drug procainamide (**7.104**, X = NH) is slow relative to that of the local anesthetic procaine (**7.104**, X = O).[138] The ester group, not the amide, is hydrolyzed in the anesthetic propanidid (**7.105**).[139] However, the amide bond of the local anesthetic butanilicaine (**7.106**) is hydrolyzed by human liver at rates comparable to those of good ester substrates.[139]

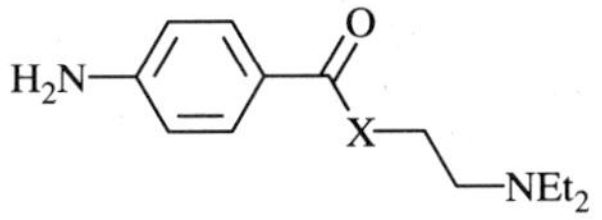

**7.104**

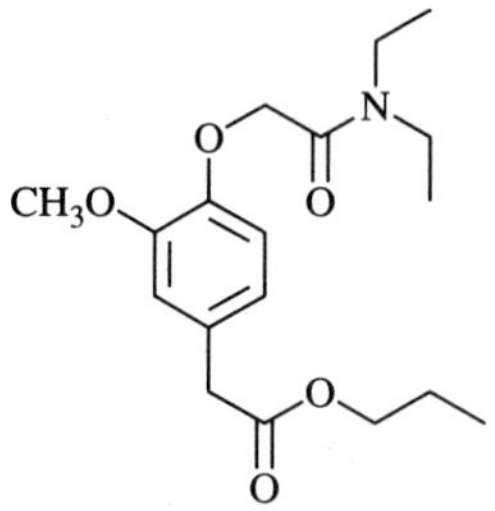

7.105 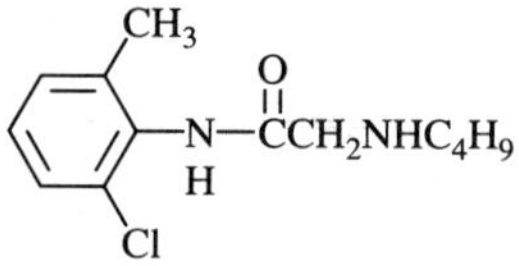 7.106

In some cases the hydrolysis of an ester or amide bond produces a toxic compound. Aromatic amines generated upon hydrolysis of *N*-acylanilides become methemoglobin-forming agents after N-oxidation. Phenacetin (**7.107**) causes methemoglobinemia in rats which is reduced drastically when the carboxylesterase inhibitor bis(4-nitrophenyl)phosphate is coadministered.[140] The danger of using a racemic mixture as a drug is further exemplified by the observation that although both isomers of prilocaine (**7.108**) have local anesthetic action, only the (*R*)-(−)-isomer is hydrolyzed to toluidine, which causes methemoglobinemia. The (*S*)-(+)-isomer, which is not hydrolyzed, does not cause this side effect.[141]

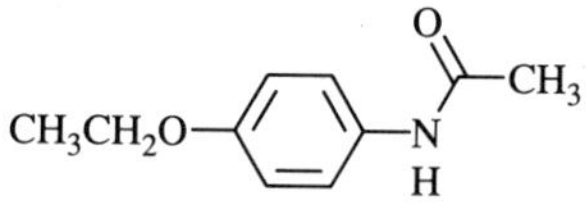

7.107 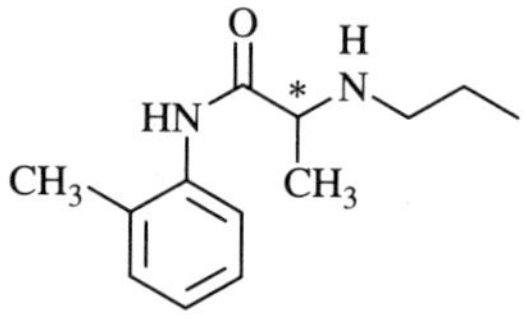 7.108

As shown in the above example, enzymatic hydrolysis often exhibits enantiomeric specificity. A racemic mixture of the anticonvulsant drug phensuximide (**7.109**) is enzymatically hydrolyzed stereospecifically to (*R*)-(−)-2-phenylsuccinamic acid (**7.110**).[142] Enantiomeric specificity may be organ selective as is found with the metabolic hydrolysis of the tranquilizer prodrug oxazepam acetate (**7.111**). Preferential hydrolysis of the (*R*)-(−)-ester occurs in the liver, but the opposite is found in brain.[143]

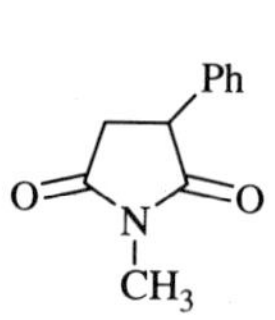

7.109 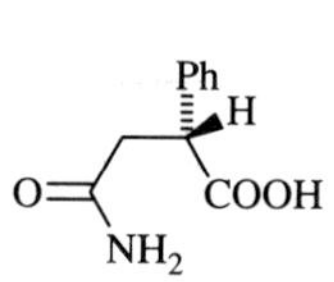 7.110 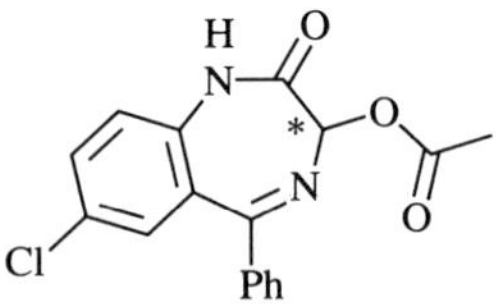 7.111

Because of the stereoselectivity of various enzymes, one enantiomer may be a preferential substrate for one enzyme and the other enantiomer a substrate for a different enzyme. Both enantiomers of the hypnotic drug etomidate (**7.112**, Scheme 7.39) are metabolized, but by different routes.[144,145] The active (*R*)-(+)-isomer is a better substrate for hydrolysis (pathway a) than is the (*S*)-(−)-isomer, but it appears that the (*S*)-(−)-isomer is a better substrate for hydroxylation (pathway b) than is the (*R*)-(+)-isomer. *In vitro*, only the (*S*)-(−)-isomer produces acetophenone (**7.113**).

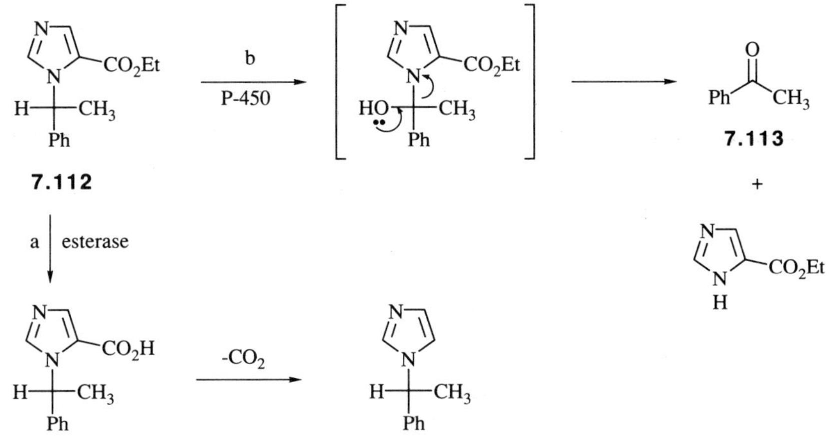

**Scheme 7.39.** Competitive metabolism of (*R*)- and (*S*)-etomidate.

In addition to carboxylesterases and amidases, hydrolytic reactions are also carried out by various other mammalian enzymes such as phosphatases, $\beta$-glucuronidases, sulfatases, and deacetylases. We now turn our attention to the next phase of drug metabolism, that of transforming the phase I metabolites into conjugates for excretion.

## C. Phase II Transformations: Conjugation Reactions

### 1. Introduction

Phase II or *conjugating enzymes*, in general, catalyze the attachment of small polar endogenous molecules such as glucuronic acid, sulfate, and amino acids to drugs or, more often, to metabolites of phase I enzymes. This further deactivates the drug and produces water-soluble metabolites that are readily excreted in the urine or bile. Phase II reactions such as methylation and acetylation do not yield more polar metabolites, but instead serve to terminate or attenuate the biological activity. Metabolic reactions with the potent nucleophile glutathione serve to trap highly electrophilic metabolites before they

**Table 7.7** Mammalian Phase II Conjugating Agents[a]

| Conjugate | Coenzyme Form | Groups Conjugated | Transferase Enzyme |
|---|---|---|---|
| Glucuronide | Uridine-5'-diphospho-α-D-glucuronic acid (UDPGA) 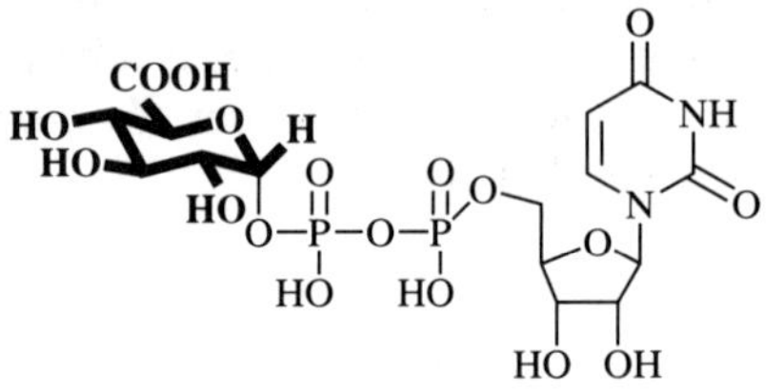 | -OH, -COOH, $-NH_2$, $-NR_2$, -SH, C-H | UDP-Glucuronosyl-transferase |
| Sulfate | 3'-Phosphoadenosine-5'-phosphosulfate (PAPS) 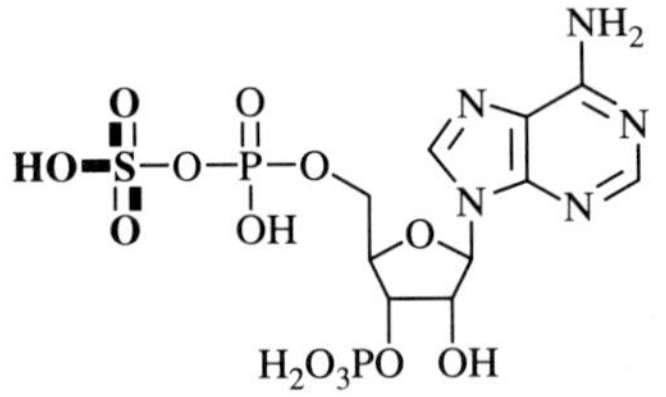 | -OH, $-NH_2$ | Sulfotransferase |
| Glycine and glutamine | Activated acyl or aroyl coenzyme A cosubstrate 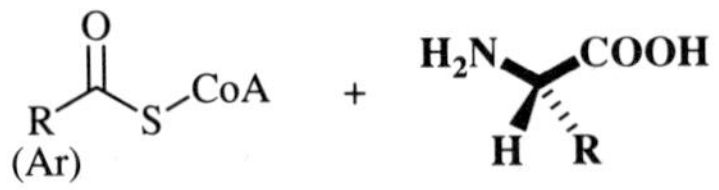 | -COOH | Glycine *N*-acyltransferase<br>Glutamine *N*-acyltransferase |
| Glutathione | Glutathione (GSH) 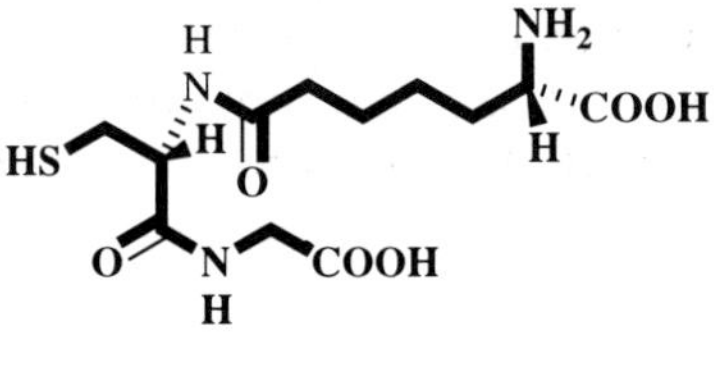 | Ar-X, arene oxide, epoxide, carbocation or related | Glutathione *S*-transferase |
| Acetyl | Acetyl coenzyme A 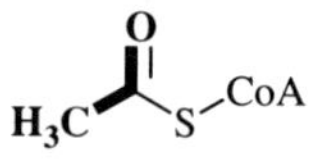 | -OH, $-NH_2$ | Acetyl-transferase |
| Methyl | *S*-Adenosyl methionine (SAM) 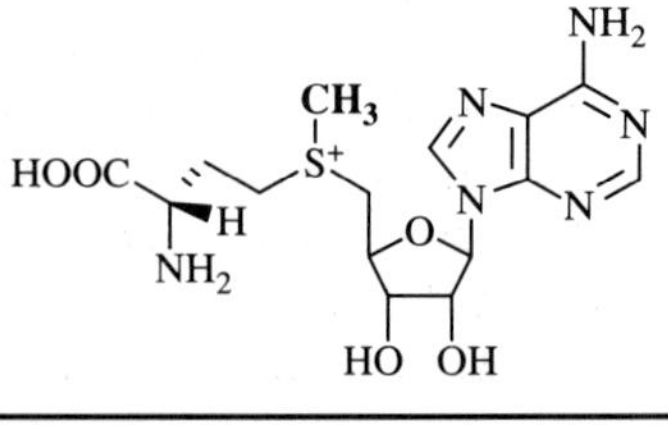 | -OH, $-NH_2$, -SH, heterocyclic N | Methyl-transferase |

[a] The bold-faced parts are transferred to the drug or metabolite.

damage biologically important macromolecules such as proteins, RNA, and DNA.

Conjugation reactions take place primarily with hydroxyl, carboxyl, amino, heterocyclic nitrogen, and thiol groups. If these groups are not present in the drug, they are introduced or unmasked by the phase I reactions. For the most part, the conjugating group is an endogenous molecule that is first activated in a coenzyme form prior to its transfer to the acceptor group. The enzymes that catalyze these reactions are known as transferases (Table 7.7).

## 2. Glucuronic Acid Conjugation

*Glucuronidation* is the most common mammalian conjugation pathway; it occurs in tissues of all mammals except the cat. As shown in Table 7.7, the coenzyme form of glucuronic acid, namely uridine 5′-diphospho-α-D-glucuronic acid (UDP-glucuronic acid, **7.116**, Scheme 7.40) is biosynthesized from α-D-glucose 1-phosphate (**7.114**) by phosphorylase-catalyzed conversion to the nucleotide sugar **7.115**, followed by UDP glucose dehydrogenase-catalyzed oxidation. UDP-glucuronic acid (**7.116**) contains D-glucuronic acid in the α-configuration, but glucuronic acid conjugates (**7.117**) are β-glycosides, suggesting that the glucuronidation reaction involves an inversion of stereo-

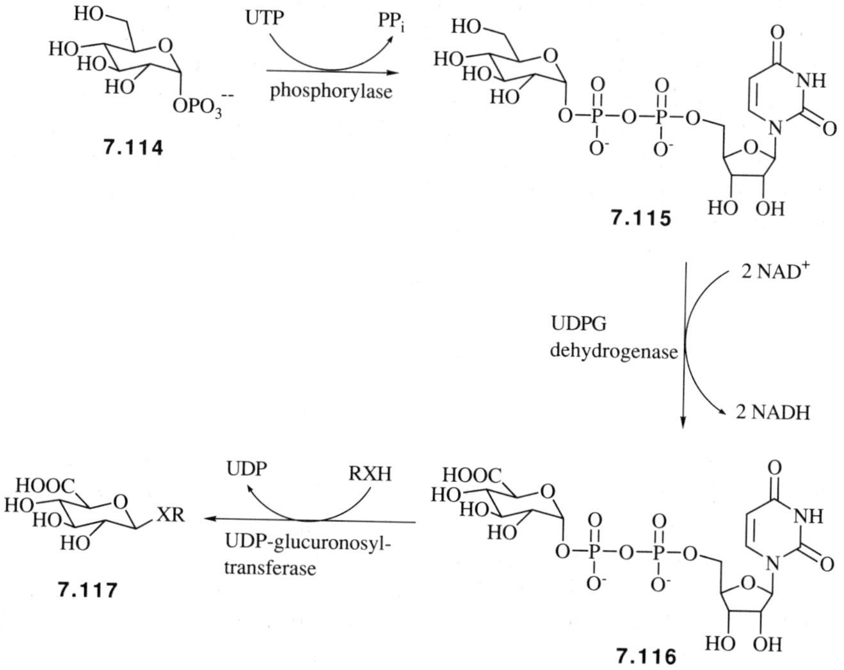

Scheme 7.40. Biosynthesis and reactions of UDP-glucuronic acid.

chemistry at the anomeric carbon. Owing to the carboxylate and hydroxyl groups of the glucuronyl moiety, glucuronides are very water soluble and, therefore, are set up for excretion. Glucuronides generally are excreted in the urine, but when the molecular weight of the conjugate exceeds 300, excretion in the bile becomes significant. There is some evidence that UDP-glucuronosyltransferase is closely associated with cytochrome *P*-450, so that as drugs become oxidized by the phase I cytochrome *P*-450 reactions, the metabolites are efficiently conjugated.[146]

Four general classes of glucuronides have been established, the *O*-, *N*-, *S*-, and *C*-glucuronides; a small sampling of examples is given in Table 7.8[147–155]; arrows point to the sites of glucuronidation.

There are certain disease states (inborn errors of metabolism) that are associated with defective glucuronide formation or attachment of the glucuronide to bilirubin, for example, Crigler–Najjar syndrome and Gilbert's disease, both characterized by a deficiency of UDP-glucuronosyltransferase activity.[156] Neonates, which have undeveloped liver UDP-glucuronosyltransferase activity, may exhibit similar metabolic problems. In these cases there is a greater susceptibility to adverse effects caused by the accumulation of drugs that normally are glucuronidated. An example is the inability of neonates to conjugate the antibacterial drug chloramphenicol, thereby leading to "gray baby syndrome" from the accumulation of toxic levels of the drug.[157]

As in the case of phase I metabolism, phase II reactions also can be species specific, regioselective, and stereoselective. The antibacterial drug sulfadimethoxine (see Table 7.8) is glucuronidated in man but not in rat, guinea pig, or rabbit. The bronchodilator fenoterol (**7.118**) is conjugated as two different glucuronides, a *para*-glucuronide and a *meta*-glucuronide, as there are both *para*- and *meta*-hydroxyl groups.[158] The (*R*,*R*′)-(−)-isomer is conjugated with higher affinity but with lower velocity than is the (*S*,*S*′)-(+)-isomer.

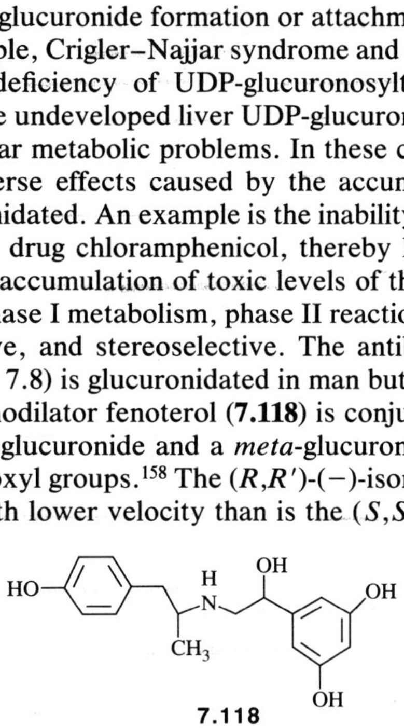

HO, H, N, OH, OH, $CH_3$, OH

**7.118**

The antidepressant drug nortriptyline (**7.119**, R = H) is metabolized (cytochrome *P*-450) predominantly to the *E*-(−)-hydroxy analog [**7.119**, R = OH; the absolute configurations of the (+)- and (−)-enantiomers are not known, so the stereochemistry is not specified]. This metabolite is converted stereospecifically to the corresponding *O*-glucuronide, but the stereospecificity is organ dependent.[159] The liver and kidneys convert only the *E*-(+)-isomer of **7.119** (R = OH) to the *O*-glucuronide, whereas the intestine metabolizes only the *E*-(−)-isomer. The enantiomer that is not glucuronidated inhibits the glucuronidation of its antipode.

**Table 7.8** Classes of Compounds Forming Glucuronides

| Type | Example | Structure (arrow indicates site of glucuronidation) | Ref. |
|---|---|---|---|
| ***O*-Glucuronide** | | | |
| *Hydroxyl* | | | |
| phenol | **acetaminophen** | 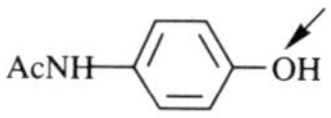 AcNH–C₆H₄–OH | 147 |
| alcohol | **chloramphenicol** | 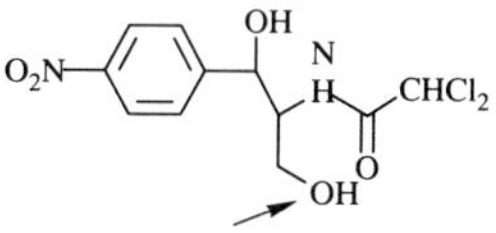 $O_2N$–C₆H₄–CH(OH)–CH(NHCOCHCl₂)–CH₂OH | 148 |
| *Carboxyl* | **fenoprofen** | 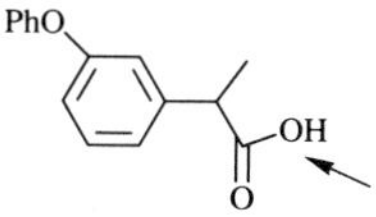 PhO–C₆H₄–CH(CH₃)–COOH | 149 |
| ***N*-Glucuronide** | | | |
| *Amine* | **desipramine** | 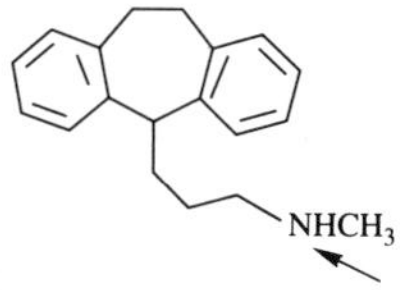 NHCH₃ | 150 |
| *Amide* | | | |
| carbamate | **meprobamate** | 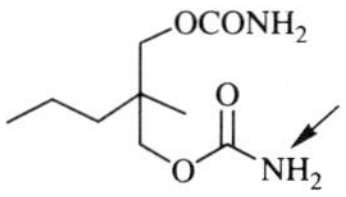 OCONH₂; O–C(=O)–NH₂ | 151 |
| *Sulfonamide* | **sulfadimethoxine** | 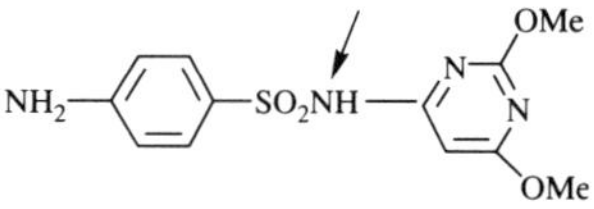 NH₂–C₆H₄–SO₂NH–(dimethoxypyrimidine, OMe, OMe) | 152 |
| ***S*-Glucuronide** | | | |
| *Sulfhydryl* | **methimazole** | 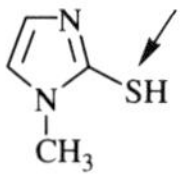 N, N–CH₃, SH | 153 |
| *Carbodithioic acid* | **disulfiram** | 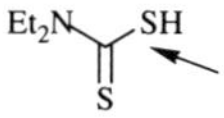 Et₂N–C(=S)–SH (reduced metabolite) | 154 |
| ***C*-Glucuronide** | | | |
| | **phenylbutazone** | 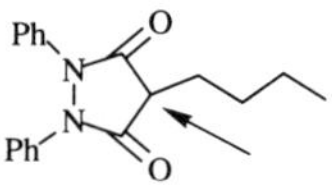 Ph, Ph, N–N, O, O | 155 |

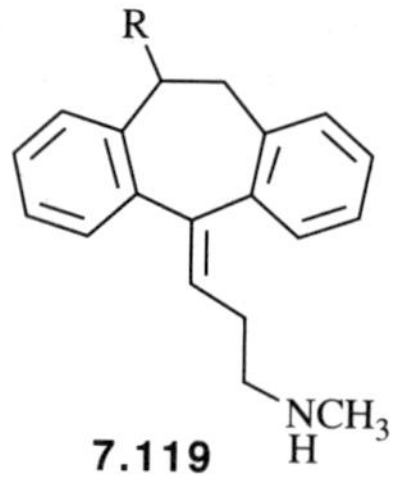

7.119

## 3. Sulfate Conjugation

*Sulfate conjugation* occurs less frequently than does glucuronidation, presumably because of the limited availability of inorganic sulfate in mammals and the fewer number of functional groups (phenols, alcohols, arylamines, and *N*-hydroxy compounds) that undergo sulfate conjugation.[160] There are three enzyme-catalyzed reactions involved in sulfate conjugation (Scheme 7.41). Inorganic sulfate is activated by the ATP-sulfurylase (sulfate adenylyltransferase)-catalyzed reaction with ATP to give adenosine 5′-phosphosulfate (APS, **7.120**), which is phosphorylated in an APS phosphokinase (adenylylsulfate kinase)-catalyzed reaction to 3′-phosphoadenosine 5′-phosphosulfate (PAPS, **7.121**), the coenzyme form used for sulfation. The acceptor molecule (RXH) undergoes sulfotransferase-catalyzed sulfation to **7.122** with release of 3′-phosphoadenosine 5′-phosphate (PAP).

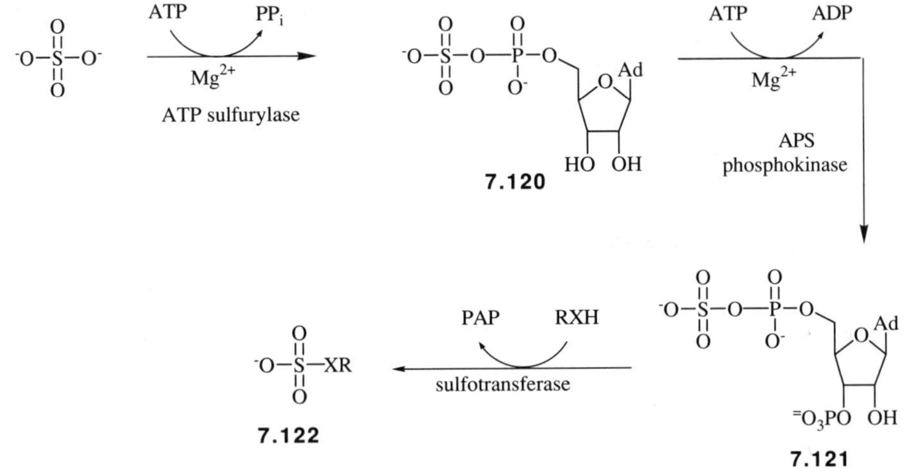

ATP $PP_i$ $Mg^{2+}$ ATP sulfurylase

7.120

ATP ADP $Mg^{2+}$ APS phosphokinase

PAP RXH sulfotransferase

7.122

7.121

**Scheme 7.41.** Sulfate conjugation.

There are a variety of sulfotransferases in the liver and other tissues.[160] The main substrates for these enzymes are phenols, but aliphatic alcohols, amines, and, to a much lesser extent, thiols also are active. Often both glucuronidation and sulfation occur on the same substrates, but the $K_m$ for sulfation is usually lower, so it predominates.[161] In addition to substrate binding differences, sulfotransferases are cytoplasmic (soluble) enzymes and glucuronosyltransferases are microsomal (membrane) enzymes. Therefore, sulfate conjugation tends to predominate at low doses, when there is less to diffuse into membranes,[162] and with smaller, less lipid-soluble molecules.[163] However, high lipid solubility does not necessarily mean that a compound will be glucuronidated rather than sulfated, because subcellular distances are small, diffusion out of membranes generally is rapid, and the $K_m$ values for sulfotransferases are generally lower than those for glucuronosyltransferases. The bronchodilator albuterol (**7.123**, R = H) is metabolized to the corresponding sulfate ester (**7.123**, R = $SO_3^-$).[164] Note that although there are three hydroxyl groups in albuterol, phenolic sulfation predominates.

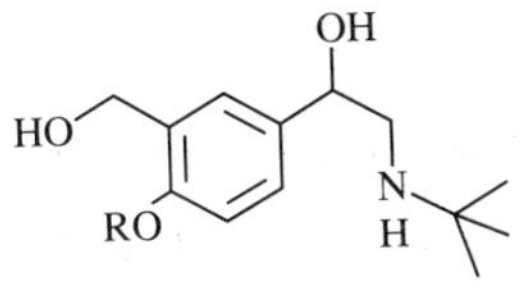

**7.123**

Phenolic *O*-glucuronidation often competes favorably with sulfation because of limited sulfate availability. In some cases the reverse situation occurs. Acetaminophen is metabolized in adults mainly to the *O*-glucuronide, although some sulfate ester can be detected.[165] However, neonates and children 3–9 years old excrete primarily the acetaminophen sulfate conjugate because they have a limited capacity to conjugate with glucuronic acid.[166] Sulfation of aliphatic alcohols and arylamines occurs, but these are minor metabolic pathways. Sulfate conjugates can be hydrolyzed back to the parent compound by various sulfatases.

Sulfoconjugation plays an important role in the hepatotoxicity and carcinogenicity of *N*-hydroxyarylamides.[167] As described in Section IV,B,1,e (see Scheme 7.27), activated *N*-hydroxyarylamines are quite electrophilic and can react with protein and DNA nucleophiles. N-Hydroxysulfation also activates these compounds as highly electrophilic nitrenium-like species. For example, sulfoconjugation of the N-hydroxylation metabolite of the analgesic phenacetin (**7.124**, Scheme 7.42) produces a reactive metabolite (**7.125**) that may be responsible for hepatotoxicity and nephrotoxicity of that compound.[168]

### 4. Amino Acid Conjugation

The first mammalian drug metabolite isolated, hippuric acid (see **7.12**),[22] was the product of glycine conjugation of benzoic acid. *Amino acid conjugation* of a variety of carboxylic acids, particularly aromatic, arylacetic, and heterocy-

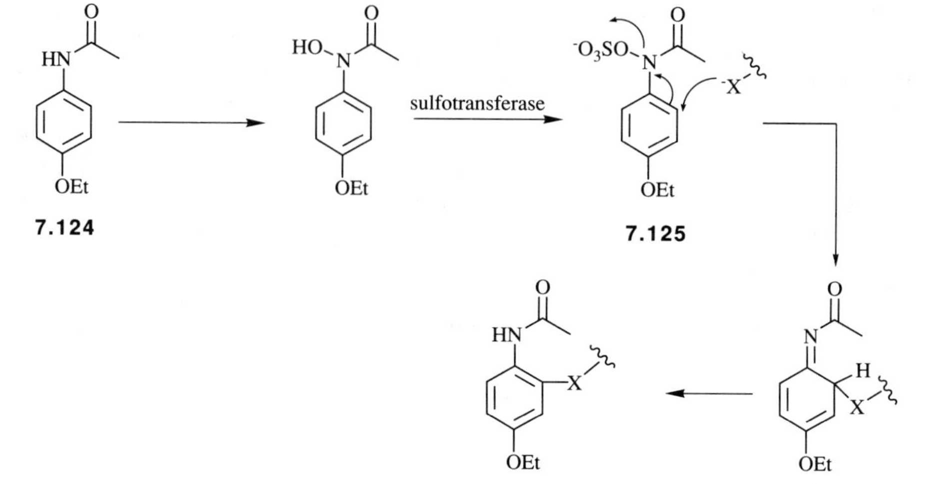

7.124

sulfotransferase

7.125

Scheme 7.42. Bioactivation of phenacetin.

clic carboxylic acids, leads to amide bond formation. The specific amino acid involved in conjugation within a class of animals usually depends on the bioavailability of that amino acid from endogenous and dietary sources. Glycine conjugates are the most common amino acid conjugates in animals; glycine conjugation in mammals follows the order herbivores > omnivores > carnivores. Conjugation with L-glutamine is most common in primate drug metabolism[169]; it does not occur to any significant extent in nonprimates. Taurine, arginine, asparagine, histidine, lysine, glutamate, aspartate, alanine, and serine conjugates also have been found in mammals.[170]

The mechanism of amino acid conjugation involves three steps (Scheme 7.43). The carboxylic acid is first activated by ATP to the AMP ester (**7.126**) which is converted to the corresponding coenzyme A thioester (**7.127**) with CoASH; these first two steps are catalyzed by acyl-CoA synthetases (long-chain-fatty-acid–CoA ligases). The appropriate amino acid *N*-acyltransferase

ATP $PP_i$ acyl CoA synthetase

7.126

CoASH AMP

7.127

CoASH

amino acid N-acyltransferase

7.128

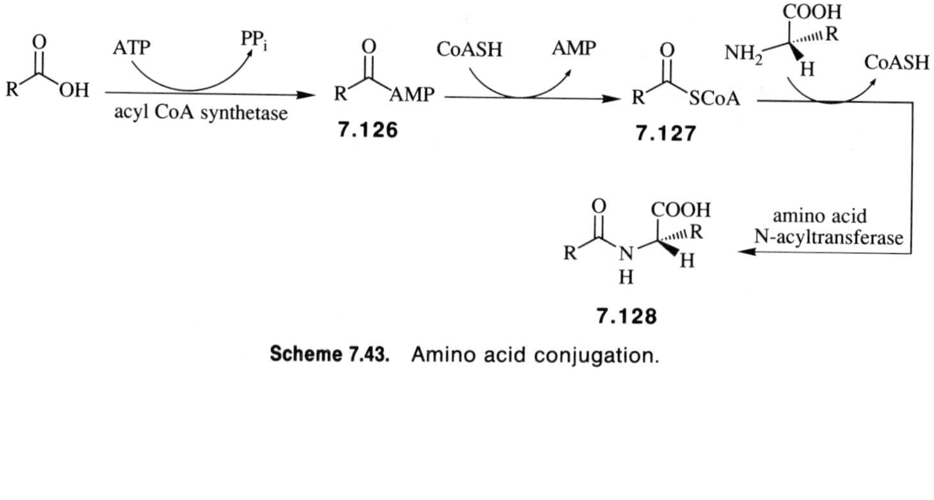

Scheme 7.43. Amino acid conjugation.

then catalyzes the condensation of the amino acid and coenzyme A thioester to give the amino acid conjugate (**7.128**). Conjugation does not take place with the AMP ester (**7.126**) directly because the AMP ester hydrolyzes readily; conversion to the CoA thioester produces a more hydrolytically stable product (**7.127**) that is still quite reactive toward amine nucleophiles.

The antihistamine brompheniramine (**7.129**) undergoes phase I metabolism in both dogs and man (N-dealkylation, oxidative deamination, and aldehyde oxidation; see Section IV,B) to the carboxylic acid (**7.130**), which then is glycine conjugated to give **7.131** (Scheme 7.44).[171] All of the metabolites shown in Scheme 7.44 (including **7.132**), except the aldehyde, were isolated from dog urine, and all except the aldehyde and the *N*-oxide (**7.132**) were isolated from human urine. The related antihistamine diphenhydramine (**7.133**) undergoes similar phase I oxidation but is glutamine conjugated.[172]

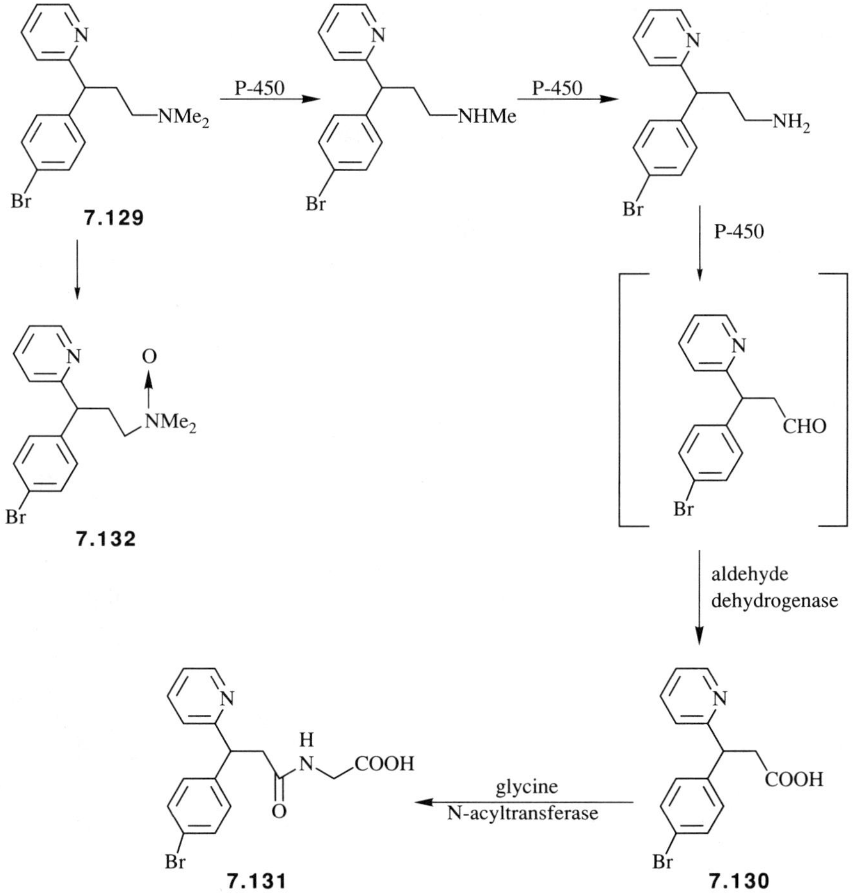

**Scheme 7.44.** Metabolism of brompheniramine.

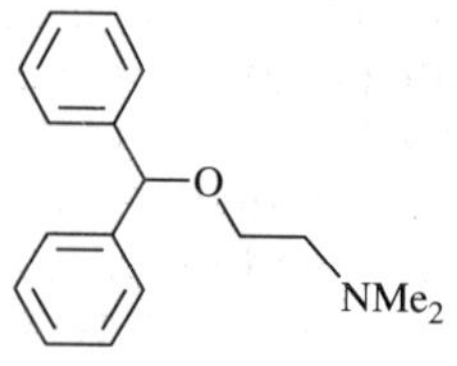

**7.133**

## 5. Glutathione Conjugation

The tripeptide glutathione (**7.134**) is found in virtually all mammalian tissues. It contains a potent nucleophilic thiol group, and its function appears to be as a scavenger of harmful electrophilic compounds ingested or produced by metabolism. Xenobiotics that are conjugated with glutathione are either highly electrophilic as such or are first metabolized to an electrophilic product prior to conjugation. Drug toxicity can result from the reaction of cellular nucleophiles with electrophilic metabolites (see Scheme 7.28, Section IV,B,1,e) if glutathione does not first intercept these reactive compounds. Electrophilic species include any group capable of undergoing $S_N2$- or $S_NAr$-like reactions (e.g., alkyl halides, epoxides, and aryl halides), acylation reactions (e.g., anhydrides and sulfonate esters), Michael additions (addition to a double or triple bond in conjugation with a carbonyl or related group), and reductions (e.g., disulfides and radicals). All of the reactions catalyzed by glutathione *S*-transferase occur nonenzymatically, but at a slower rate.

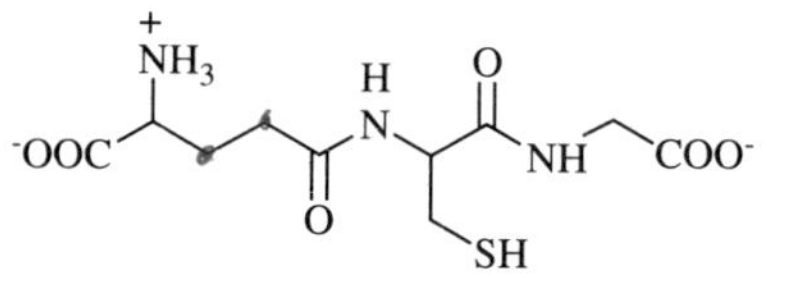

**7.134**

A few examples of glutathione conjugation are given in Scheme 7.45. Examples of $S_N2$ reactions are glutathione conjugation of the leukemia drug busulfan (**7.135**)[173] and of the coronary vasodilator nitroglycerin (**7.136**).[174] The reaction of glutathione with the immunosuppressive drug azathioprine (**7.137**)[175] is an example of an $S_NAr$ reaction. These reactions are direct deactivations of the drugs. Morphine (**7.138**) has been reported to undergo oxidation by two different pathways, both of which lead to potent Michael acceptors that undergo subsequent glutathione conjugation. Pathway a, catalyzed by morphine 6-dehydrogenase, gives morphinone (**7.139**), which undergoes Michael addition with glutathione to **7.140**.[176] Pathway b is a cytochrome *P*-450–catalyzed route that produces the strongly electrophilic quinone methide (**7.141**). Stereospecific glutathione conjugation gives **7.142**.[177]

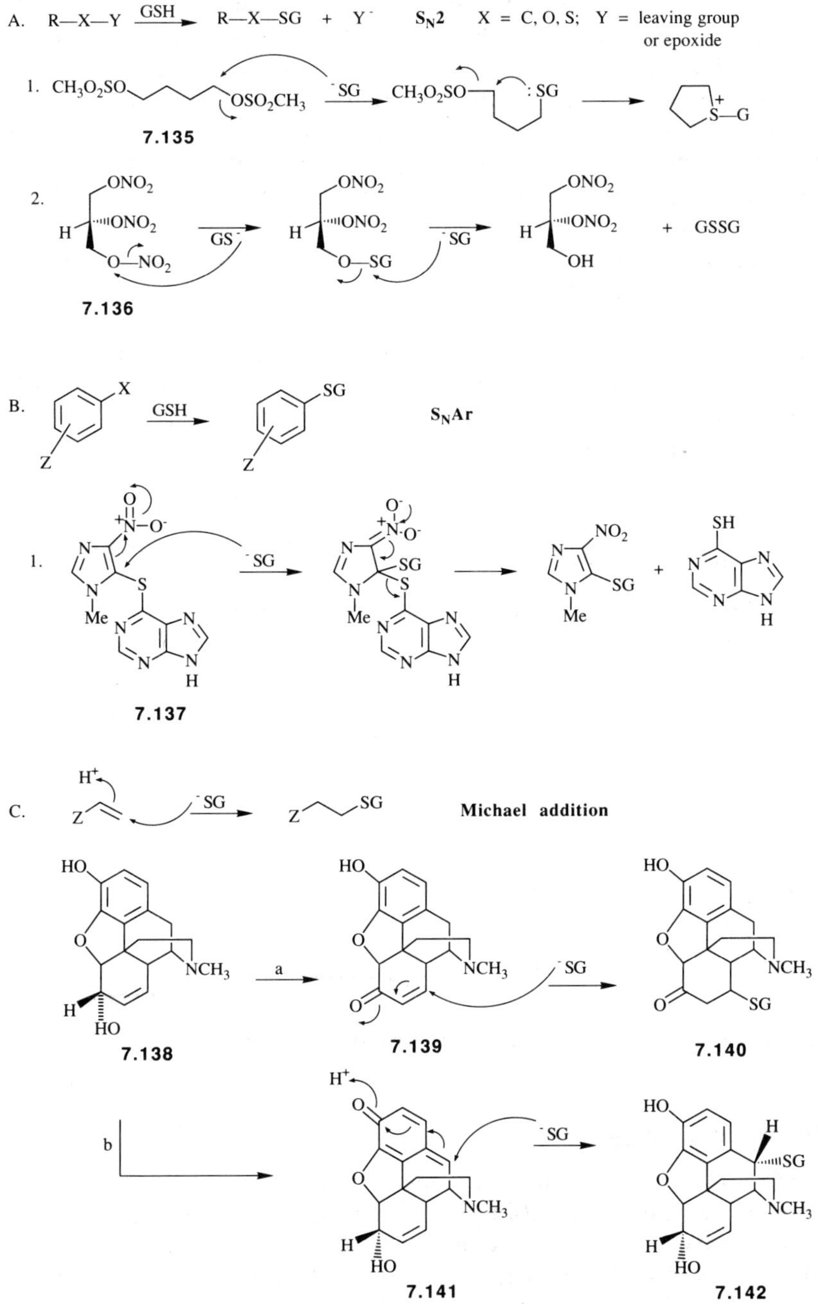

A. R—X—Y $\xrightarrow{GSH}$ R—X—SG + $Y^-$ $S_N2$ X = C, O, S; Y = leaving group or epoxide

1. $CH_3O_2SO$ ... $OSO_2CH_3$ $\xrightarrow{^-SG}$ $CH_3O_2SO$ ... :SG ⟶ S⁺—G

7.135

2. $ONO_2$, $ONO_2$, O—$NO_2$ $\xrightarrow{GS^-}$ $ONO_2$, $ONO_2$, O—SG $\xrightarrow{^-SG}$ $ONO_2$, $ONO_2$, OH + GSSG

7.136

B. X $\xrightarrow{GSH}$ SG $S_NAr$

1. $\xrightarrow{^-SG}$ ⟶ $NO_2$, SG + SH

7.137

C. $H^+$ Z $\xrightarrow{^-SG}$ Z SG Michael addition

7.138 $\xrightarrow{a}$ 7.139 $\xrightarrow{^-SG}$ 7.140

b ⟶ 7.141 $\xrightarrow{^-SG}$ 7.142

**Scheme 7.45.** Examples of glutathione conjugation.

Glutathione conjugates are rarely excreted in the urine; because of their high molecular weight and amphiphilic character, when they are eliminated it is in the bile. Most typically, however, glutathione conjugates are not excreted; instead they are metabolized further and are excreted ultimately as *N*-acetyl-L-cysteine (also known as mercapturic acid) conjugates (**7.146**, Scheme 7.46).[178a] The mercapturic acid pathway begins from the glutathione conjugate (**7.143**). The γ-glutamyl residue is hydrolyzed to glutamate and the cysteinyl-glycine conjugate **7.144** in a reaction catalyzed by γ-glutamyltransferase. Cysteinyl-glycine dipeptidase–catalyzed hydrolysis of **7.144** leads to the release of glycine and the formation of the cysteine conjugate **7.145**, which is N-acetylated by acetyl-CoA in a reaction catalyzed by cysteine-*S*-conjugate *N*-acetyltransferase. Some workers in the field consider this further metabolism of glutathione conjugates as phase III metabolism. Conjugation with glutathione occurs in the cytoplasm of most cells, especially in the liver and kidney where the glutathione concentration is 5–10 m*M*.

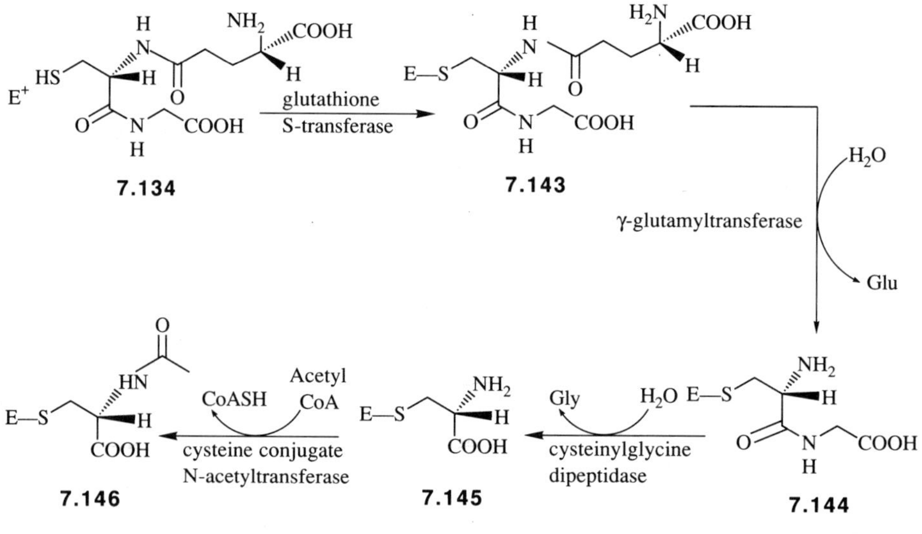

$E^+$ + 7.134 —glutathione S-transferase→ 7.143 —$H_2O$, γ-glutamyltransferase, −Glu→ 7.144 —$H_2O$, cysteinylglycine dipeptidase, −Gly→ 7.145 —Acetyl CoA, cysteine conjugate N-acetyltransferase, −CoASH→ 7.146

**Scheme 7.46.** Metabolism of glutathione conjugates to mercapturic acid conjugates.

## 6. Water Conjugation

Epoxide hydrolase (also called epoxide hydratase or epoxide hydrase),[178b] the enzyme principally involved in *water conjugation* (i.e., *hydration*), was discussed already in the context of hydration of arene oxides (see Section IV,B,1,a). Because this enzyme catalyzes the hydration of endogenous epoxides, such as androstene oxide, at much faster rates than exogenous epoxides, it probably has an important role in endogenous metabolism.

### 7. Acetyl Conjugation

*Acetylation* is an important route of metabolism for xenobiotics containing a primary amino group, including aliphatic and aromatic amines, amino acids, sulfonamides, hydrazines, and hydrazides. In all of the previously discussed conjugation reactions a more hydrophilic metabolite is formed. Acetylation, however, converts the ionized primary ammonium group (the amine is protonated at physiological pH) to an uncharged amide which is less water soluble. The physicochemical consequences of N-acetylation, therefore, are different from those of the other conjugation reactions. The function of acetylation may be to deactivate metabolites further, although *N*-acetylprocainamide is as active as the parent antiarrhythmic drug procainamide.[179]

Acetylation occurs widely in the animal kingdom; however, the dog and fox are unable to N-acetylate arylamines or the $N^4$-amino group of sulfonamides.[180] The extent of N-acetylation of a number of drugs in humans is a genetically determined individual characteristic. The three phenotypes are homozygous fast, homozygous slow, and heterozygous (intermediate) acetylators.[181] The distribution of fast and slow acetylator phenotypes depends on the population studied, varying from mostly slow acetylators (Egyptians) to 50% fast/50% slow in the United States to 90% fast/10% slow in Orientals.[182a–c] Because of the differences in the rates of N-acetylation of certain drugs, there are significant individual variations in the therapeutic and toxicological responses to drugs exhibiting *acetylation polymorphism*. Slow acetylators in general often develop adverse reactions as a result of toxic buildup of the drugs; however, this also may result in longer drug effectiveness. Fast acetylators are more likely to show an inadequate therapeutic response to standard doses of the drug. Examples of drugs exhibiting acetylation polymorphism are the antibacterial drug sulfamethazine (**7.147**),[181] the antituberculosis drug isoniazid (**7.148**),[182c,183] and dapsone (**7.149**),[184] used in the treatment of leprosy. Many of the other enzymes involved in drug metabolism also exhibit polymorphisms.

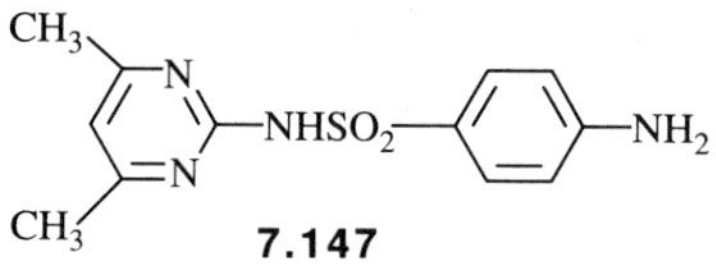

**7.147** 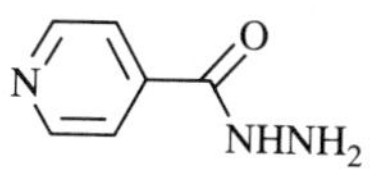 **7.148**

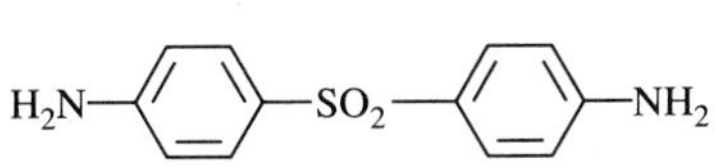

**7.149**

Acetylation is a two-step process (Scheme 7.47). First, acetyl-CoA acetylates an active site amino acid residue of the soluble hepatic *N*-acetyltransferase (**7.150**), then the acetyl group is transferred to the substrate amino

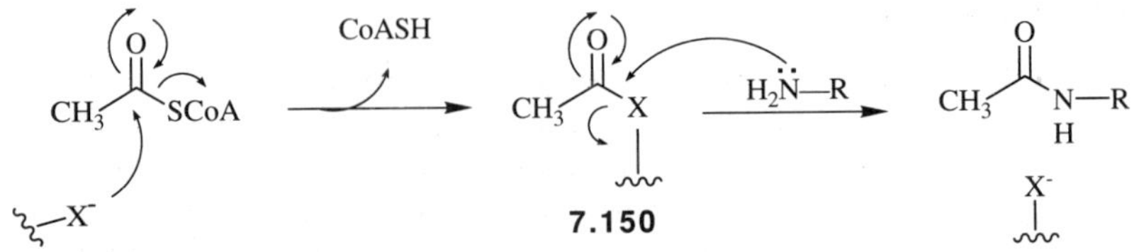

7.150

Scheme 7.47. N-Acetylation of amines.

group. Presumably, this two-step process allows the enzyme to have better control over the catalytic process.

One of the few examples of an aliphatic amine drug that is acetylated is cilastatin (**7.151**, R = H), which is metabolized to **7.151** (R = Ac).[185] Actually, cilastatin is not a drug but is used in combination (Primaxin®) with the antibacterial drug imipenem (**7.152**). When imipenem is administered alone, it is rapidly hydrolyzed in the kidneys by dehydropeptidase I.[186] Cilastatin is a potent inhibitor of this enzyme, and it effectively prevents renal metabolism of imipenem when the two compounds are administered in combination.

Aromatic amines resulting from phase I reduction of aromatic nitro compounds, such as the amine produced from the anticonvulsant drug clonazepam (see **7.88**, R = $NO_2$), also may be N-acetylated (**7.88**, R = NHAc).[187]

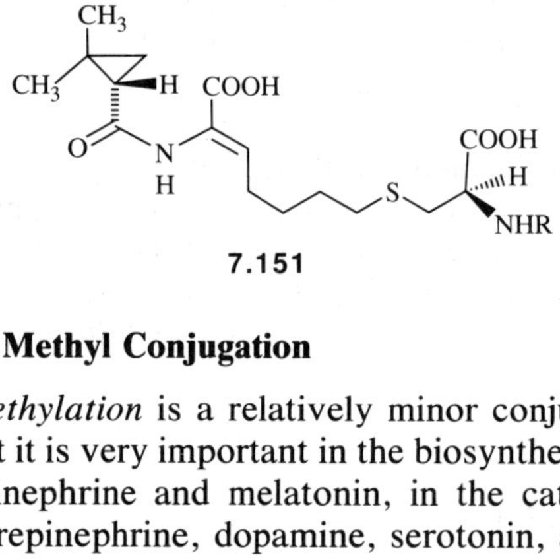

7.151 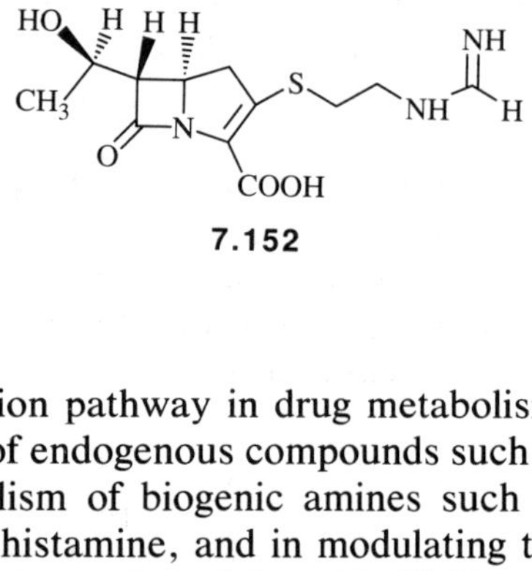 7.152

## 8. Methyl Conjugation

*Methylation* is a relatively minor conjugation pathway in drug metabolism, but it is very important in the biosynthesis of endogenous compounds such as epinephrine and melatonin, in the catabolism of biogenic amines such as norepinephrine, dopamine, serotonin, and histamine, and in modulating the activities of macromolecules such as proteins and nucleic acids.[188] Except when tertiary amines are converted to quaternary ammonium salts, methylation differs from almost all other conjugation reactions (excluding acetylation) in that it reduces the polarity and hydrophilicity of the substrates. The purpose of this conjugation is to deactivate the biological activity. In general, xenobiotics that undergo methylation do so because of their marked structure similarities to endogenous substrates that are methylated.

Methylation is a two-step process (Scheme 7.48). First the methyl-transferring coenzyme, *S*-adenosylmethionine (SAM) (**7.154**), is biosynthesized mostly from methionine (**7.153**) in a reaction catalyzed by methionine adenosyltransferase. Some *S*-adenosylmethionine also is produced by the donation of the methyl group of $N^5$-methyltetrahydrofolate (see Section III,B of Chapter 4) to *S*-adenosyl-L-homocysteine. Then the *S*-adenosylmethionine is utilized in the transfer of the activated methyl group (Mother Nature's methyl iodide) to the acceptor molecules (RXH), which include catechols and phenols, amines, and thiols. A variety of methyltransferases, such as catechol *O*-methyltransferase (COMT), phenol *O*-methyltransferase, phenylethanolamine *N*-methyltransferase, and nonspecific amine *N*-methyltransferases and thiol *S*-methyltransferases, are responsible for catalyzing the transfer of the methyl group from *S*-adenosylmethionine to RXH, which produces $RXCH_3$ and *S*-adenosylhomocysteine (**7.155**).[189]

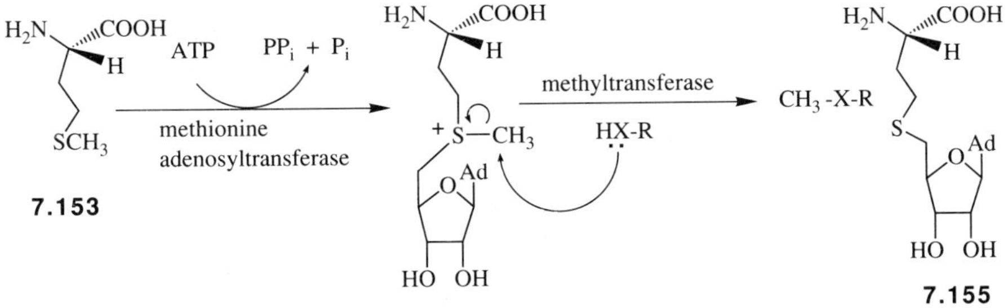

7.153 7.154 7.155

**Scheme 7.48.** Methylation of xenobiotics.

COMT-catalyzed O-methylation of xenobiotic catechols leads to O-monomethylated catechol metabolites. Unlike the free catechols, these metabolites are not oxidized to reactive *ortho*-quinonoid species that can produce toxic effects (see Section IV,B,1,h). An example of this reaction is the metabolism of the $\beta_2$-adrenergic bronchodilator isoproterenol (**7.156**), which is regioselectively methylated at the C-3 catechol hydroxyl group.[190] Compounds that are methylated by COMT must contain an aromatic 1,2-dihydroxy group. Terbutaline (**7.157**), another $\beta_2$-adrenergic bronchodilator related in structure to isoproterenol except that it contains a *meta*-dihydroxy arrangement of hydroxyl groups, does not undergo methylation.[191]

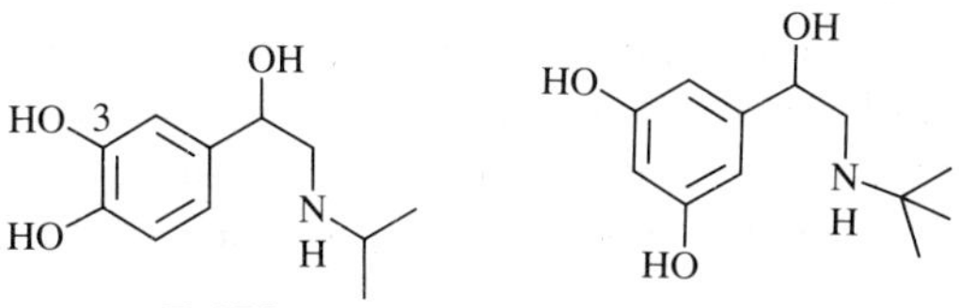

7.156 7.157

Phenol hydroxyl groups also undergo methylation. Morphine (**7.158**, R = H), for example, is metabolized in man to significant amounts of codeine

(**7.158**, R = $CH_3$).[192] N-Methylation of xenobiotics is less common but does occur. The N-dealkylated primary amine metabolite (**7.159**, R = H) of the coronary vasodilator ($\beta$-adrenergic blocker) oxprenolol [**7.159**, R = $CH(CH_3)_2$] is N-methylated to **7.159** (R = $CH_3$).[193] Heterocyclic nitrogen atoms also are susceptible to N-methylation, as in the case of the sedative clomethiazole (**7.160**).[194] Methylation of sulfhydryl groups in xenobiotics also is known. The thiol group of the antihypertensive drug captopril (**7.161**)[195] and the antithyroid drug propylthiouracil (**7.162**)[196] undergo metabolic S-methylation.

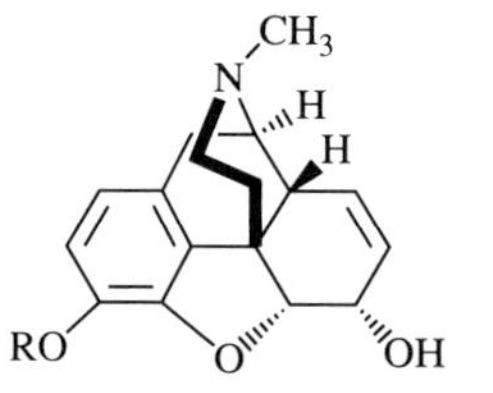

7.158

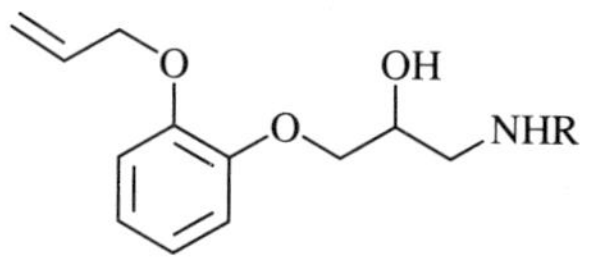

7.159

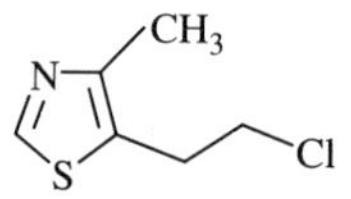

7.160

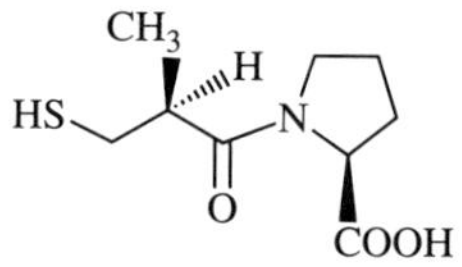

7.161

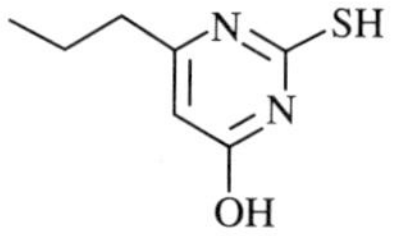

7.162

### D. Hard and Soft Drugs

Bodor[197] divided drugs into two classes, hard drugs and soft drugs. *Hard drugs* are defined as nonmetabolizable compounds, characterized either by high lipid solubility and accumulation in adipose tissues and organelles or high water solubility. They are poor substrates for the metabolizing enzymes; the potential metabolically sensitive parts of these drugs are either sterically hindered or the hydrogen atoms are substituted with halogens to block oxidation.

*Soft drugs* are biologically active drugs characterized by a predictable and controllable metabolism to nontoxic products after they have achieved their desired pharmacological effect. The advantages of soft drugs are significant: (1) elimination of toxic metabolites, thereby increasing the therapeutic index of the drug; (2) avoidance of pharmacologically active metabolites; (3) elimination of drug interactions resulting from metabolite inhibition of enzymes; and (4) simplification of pharmacokinetic problems caused by multicomponent systems.

An example of this concept in drug design is the isosteric soft analog (**7.163**) of the hard antifungal drug cetylpyridinium chloride (**7.164**).[198] The soft analog has a metabolically soft spot (the ester group) built into the structure for detoxification (Scheme 7.49). The important features of this class of drugs are (1) a close structural analogy to a known hard drug; (2) a metabolically soft spot (in this case, an ester group) built in a noncritical part of the drug; (3) a predicted metabolism (in this case, hydrolysis) that is the major or only metabolic route; (4) a rate of metabolism that can be controlled by structural modification (e.g., by substitution on either side of the ester group); (5) metabolic products (in this case, a fatty acid, formaldehyde, and pyridine) that are of very low toxicity and activity; and (6) a planned metabolic deactivation that does not require enzymatic reactions (e.g., cytochrome *P*-450), which may lead to reactive intermediates.

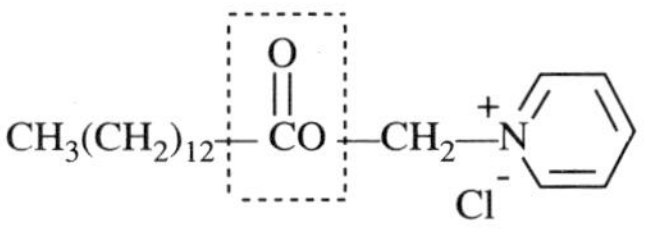

$CH_3(CH_2)_{12}$—CO—$CH_2$—$N^+$(pyridinium) $Cl^-$

**7.163**

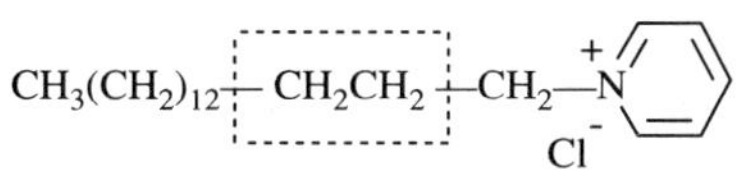

$CH_3(CH_2)_{12}$—$CH_2CH_2$—$CH_2$—$N^+$(pyridinium) $Cl^-$

**7.164**

$CH_3(CH_2)_{12}C(=O)$—O—$CH_2$—$N^+$(pyridinium) $Cl^-$ $\xrightarrow{\text{esterase}}$ $CH_3(CH_2)_{12}COOH$ + O=$CH_2$ + pyridine

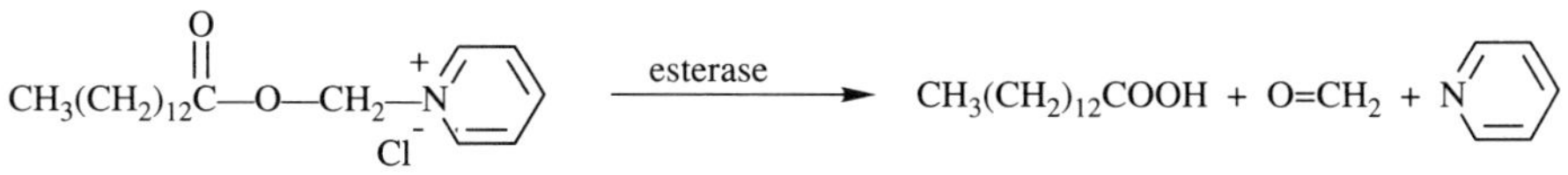

**Scheme 7.49.** Example of a soft drug analog.

A soft drug, then, is an active drug that is destroyed by metabolism after it carries out its therapeutic role. A more common way of utilizing metabolism in drug design is to modify a drug so that it is not active but, when it reaches the site of action, is metabolically transformed into the active drug. These compounds are known as *prodrugs* and are the topic of Chapter 8.

## References


1. Shuman, R. F., Pines, S. H., Shearin, W. E., Czaja, R. F., Abramson, N. L., and Tull, R. 1977. *J. Org. Chem.* **42,** 1914.
2. Pines, S. H., and Douglas, A. W. 1976. *J. Am. Chem. Soc.* **98,** 8119.
3. Anjaneyulu, B., Maller, R. K., Nagarajan, K., Kueng, W., and Wirz, B. 1985. *J. Labelled Compd. Radiopharm.* **22,** 313.
4. Smith, C. W., and Norton, D. G. 1953. *Org. Synth.* **33,** 29.
5. Baretz, B. H., Lollo, C. P., and Tanaka, K. 1978. *J. Labelled Compd. Radiopharm.* **15,** 369.
6. Pines, S. H., Czaja, R. F., and Abramson, N. L. 1974. *J. Org. Chem.* **40,** 1920.
7. Baxter, R. L., and Abbot, E. M. 1985. *J. Labelled Compd. Radiopharm.* **22,** 1211.
8. Gabriel, J., and Seebach, D. 1984. *Helv. Chim. Acta* **67,** 1070.
9. Saito, T., Kudo, A., and Morikawa, N. 1973. *Radioisotopes* **22,** 25.
10. U.S. Code of Federal Regulations, Title 21, 1986. "Drugs Used in Research," Part 361.1, p. 160. U.S. Government Printing Office, Washington, D.C.
11. Schill, G., Borg, K. O., Modin, R., and Persson, B. A. 1977. *In* "Progress in Drug Metabolism" (Bridges, J. W., and Chasseaud, L. F., eds.), Vol. 2, p. 219. Wiley, New York.

12. Schill, G., Modin, R., Borg, K. O., and Persson, B. A. 1977. *In* "Drug Fate and Metabolism: Methods and Techniques" (Garrett, E. R., and Hirtz, J. L., eds.), Vol. 1, p. 135. Dekker, New York.
13. Horning, M. G., Gregory, P., Nowlin, J., Stafford, M., Letratanangkoon, K., Butler, C., Stillwell, W. G., and Hill, R. M. 1974. *Clin. Chem.* **20,** 282.
14. Thompson, J. A., and Markey, S. P. 1975. *Anal. Chem.* **47,** 1313.
15. Brodie, B. B., Cho, A. K., and Gessa, G. L. 1970. *In* "Amphetamines and Related Compounds" (Costa, E., and Garattini, S., eds.), p. 217. Raven, New York.
16. Stolman, A., and Pranitis, P. A. F. 1977. *Clin. Toxicol.* **10,** 49.
17. Crippen, R. C., and Smith, C. E. 1965. *J. Gas Chromatogr.* **3,** 37.
18. Ahuja, S. 1976. *J. Pharm. Sci.* **65,** 163.
19. Rudewicz, P., and Straub, K. 1987. *In* "Drug Metabolism—From Molecules to Man" (Benford, D. J., Bridges, J. W., and Gibson, G. G., eds.), p. 208. Taylor & Francis, London.
20. Horning, E. C., Carroll, D. I., Stilwell, R. N., Horning, M. G., Nowlin, J. G., Hughes, H., and Mitchell, J. R. 1984. *In* "Drug Metabolism and Drug Toxicity" (Mitchell, J. R., and Horning, M. G., eds.), p. 383. Raven, New York.
21. Nicholson, J. K., and Wilson, I. D. 1987. *In* "Drug Metabolism—From Molecules to Man" (Benford, D. J., Bridges, J. W., and Gibson, G. G., eds.), p. 189. Taylor & Francis, London.
22. Liberg, J. 1829. *Poggendorff's Ann. Phys. Chem.* **17,** 389; Lehmann, C. G. 1835. *J. Prakt. Chem.* **6,** 113; Ure, A. 1841. *Pharm. J. Trans.* **1,** 24.
23. Mueller, G. C., and Miller, J. A. 1948. *J. Biol. Chem.* **176,** 535; Mueller, G. C., and Miller, J. A. 1949. *J. Biol. Chem.* **180,** 1125; Mueller, G. C., and Miller, J. A. 1953. *J. Biol. Chem.* **202,** 579.
24a. Williams, R. T. 1959. "Detoxification Mechanisms," 2nd Ed. Chapman & Hall, London.
24b. Jamali, F., Mehvar, R., and Pasutto, F. M. 1989. *J. Pharm. Sci.* **78,** 695.
25. Hignite, C., Utrecht, J., Tschang, C., and Azarnoff, D. 1980. *Clin. Pharmacol. Ther.* (*St. Louis*) **28,** 99.
26. Silber, B., Holford, N. H. G., and Riegelman, S. 1982. *J. Pharm. Sci.* **71,** 699.
27. Ariëns, E. J. 1986. *Med. Res. Rev.* **6,** 451; Simonyi, M. 1984. *Med. Res. Rev.* **4,** 359.
28. Eichelbaum, M. 1988. *Biochem. Pharmacol.* **37,** 93.
29. Kaiser, D. G., Vangeissen, G. J., Reischer, R. J., and Wechter, W. J. 1976. *J. Pharm. Sci.* **65,** 269; Lee, E., Williams, K., Day, R., Graham, G., and Champion, D. 1985. *Br. J. Clin. Pharmacol.* **19,** 669.
30. Brodie, B. B., Axelrod, J., Cooper, J. R., Gaudette, L., LaDu, B. N., Mitoma, C., and Udenfriend, S. 1953. *Science* **121,** 603.
31. Mason, H. S. 1957. *Science* **125,** 1185.
32a. Posner, H. S., Mitoma, C., Rothberg, S., and Udenfriend, S. 1961. *Arch. Biochem. Biophys.* **94,** 280.
32b. Klingenberg, M. 1958. *Arch. Biochem. Biophys.* **75,** 376.
32c. Garfinkel, D. 1958. *Arch. Biochem. Biophys.* **77,** 493.
33. Vermilion, J. L., Ballou, D. P., Massey, V., and Coon, M. J. 1981. *J. Biol. Chem.* **256,** 266; Oprian, D. D., and Coon, M. J. 1982. *J. Biol. Chem.* **257,** 8935; Strobel, H. W., Dignam, J. D., and Gum, J. R. 1982. *Int. Encycl. Pharmacol. Ther.* **108,** 361.
34. Ziegler, D. M. 1988. *Drug Metab. Rev.* **6,** 1.
35. Boyland, E. 1950. *Biochem. Soc. Symp.* **5,** 40.
36a. Jerina, D. M., Daly, J. W., Witkop, B., Zaltzman-Nirenberg, P., and Udenfriend, S. 1968. *J. Am. Chem. Soc.* **90,** 6525.
36b. Jerina, D. M., Daly, J. W., Witkop, B., Zaltzman-Nirenberg, P., and Udenfriend, S. 1970. *Biochemistry* **9,** 147.
36c. Korzekwa, K. R., Swinney, D. C., and Trager, W. F. 1989. *Biochemistry* **28,** 9019.
37. Guroff, G., Daly, J. W., Jerina, D. M., Renson, J., Witkop, B., and Udenfriend, S. 1967. *Science* **157,** 1524; Jerina, D. 1973. *Chem. Technol.* **4,** 120.

38. Daly, J., Jerina, D., and Witkop, B. 1968. *Arch. Biochem. Biophys.* **128,** 517; Daly, J. W., Jerina, D. M., and Witkop, B. 1972. *Experientia* **28,** 1129.
39. Parli, C. J., and Schmidt, B. 1975. *Res. Commun. Chem. Pathol. Pharmacol.* **10,** 601.
40. Wood, S. G., Scott, P. W., Chasseaud, L. F., Faulkner, J. K., Matthews, R. W., and Henrick, K. 1985. *Xenobiotica* **15,** 107.
41. Daly, J. 1971. *In* "Concepts in Biochemical Pharmacology" (Brodie, B. B., and Gillette, J. R., eds.), Part 2, p. 285. Springer-Verlag, Berlin.
42. Parke, D. V. 1960. *Biochem. J.* **77,** 493.
43. Dayton, P. G., Perel, J. M., Cummingham, R. F., Israeli, Z. H., and Weiner, I. M. 1973. *Drug Metab. Dispos.* **1,** 742.
44. Perry. T. L., Culling, C. F. A., Berry, K., and Hansen, S. 1964. *Science* **146,** 81.
45. Butler, T. C., Dudley, K. H., Johnson, D., and Roberts, S. B. 1976. *J. Pharmacol. Exp. Ther.* **199,** 82.
46. Selander, H. G., Jerina, D. M., Piccolo, D. E., and Berchtold, G. A. 1975. *J. Am. Chem. Soc.* **97,** 4428; Billings, R. E., and McMahon, R. E. 1978. *Mol. Pharmacol.* **14,** 145.
47. Tomaszewski, J. E., Jerina, D. M., and Daly, J. W. 1975. *Biochemistry* **14,** 2024.
48. DuBois, G. C., Appella, E., Levin, W., Lu, A. Y. H., and Jerina, D. M. 1978. *J. Biol. Chem.* **253,** 2932; Oesch, F. 1973. *Xenobiotica* **3,** 305.
49. Hanzlik, R. P., Edelman, M., Michaely, W. J., and Scott, G. 1976. *J. Am. Chem. Soc.* **98,** 1952; Hanzlik, R. P., Hiedeman, S., and Smith, D. 1978. *Biochem. Biophys. Res. Commun.* **82,** 310.
50. Nebert, D. W., Boobis, A. R., Yagi, H., Jerina, D. M, and Khouri, R. E. 1977. *In* "Biological Reactive Intermediates" (Jollow, D. J., Kocsis, J. J., Snyder, R., and Vainio, H., eds.), p. 125. Plenum, New York; Yamamoto, J., Subramaniam, R., Wolfe, A. R., and Meehan, T. 1990. *Biochemistry* **29,** 3966.
51. Heidelberger, C. 1975. *Annu. Rev. Biochem.* **44,** 79.
52. Weinstein, I. B., Jeffrey, A. M., Jennette, K. W., Blobstein, S. H., Harvey, R. G., Harris, C., Autrup, H., Kasai, H., and Nakanishi, K. 1976. *Science* **193,** 592; Koreeda, M., Moore, P. D., Yagi, H., Yeh, H. J. C., and Jerina, D. M. 1976. *J. Am. Chem. Soc.* **98,** 6720.
53. Straub, K. M., Meehan, T., Burlingame, A. L., and Calvin, M. 1977. *Proc. Natl. Acad. Sci. U.S.A.* **74,** 5285.
54. Bellucci, G., Berti, G., Chiappe, C., Lippi, A., and Marioni, F. 1987. *J. Med. Chem.* **30,** 768.
55. Johannessen, S. I., Gerna, N. M., Bakke, J., Strandjord, R. E., and Morselli, P. L. 1976. *Br. J. Clin. Pharmacol.* **3,** 575.
56. Essigmann, J. M., Croy, R. G., Nadzan, A. M., Busby, W. F., Jr., Reinhold, V. N., Büchi, G., and Wogan, G. N. 1977. *Proc. Natl. Acad. Sci. U.S.A.* **74,** 1870; Croy, R. G., Essigmann, J. M., Reinhold, V. N., and Wogan, G. N. 1978. *Proc. Natl. Acad. Sci. U.S.A.* **75,** 1745.
57. Gopalakrishnan, S., Stone, M. P., and Harris, T. M. 1989. *J. Am. Chem. Soc.* **111,** 7232.
58. Hucker, H. B. 1962. *Pharmacologist* **4,** 171.
59. Shetty, H. U., and Nelson, W. L. 1988. *J. Med. Chem.* **31,** 55.
60. Guengerich, F. P., Müller-Enoch, D., and Blair, I. A. 1986. *Mol. Pharmacol.* **30,** 287.
61. Nakahara, Y., and Cook, C. E. 1988. *J. Chromatogr.* **434,** 247.
62. Keberle, H., Reiss, W., and Hoffman, K. 1963. *Arch. Int. Pharmacodyn.* **142,** 117.
63. Horning, M. G., Stratton, C., Nowlin, J., Harvey, D. J., and Hill, R. M. 1973. *Drug Metab. Dispos.* **1,** 569.
64. Cooper, R. G., Evans, D. A. P., and Whibley, E. J. 1984. *J. Med. Genet.* **21,** 27.
65. Wright, J., Cho, A. K., and Gal, J. 1977. *Life Sci.* **20,** 467.
66. Coutts, R. T., and Beckett, A. H. 1977. *Drug Metab. Rev.* **6,** 51.
67. Gorrod, J. W. 1973. *Chem.–Biol. Interact.* **7,** 289.
68. Parli, C. H., and McMahon, R. E. 1973. *Drug Metab. Dispos.* **1,** 337.

69. Bakke, O. M., Davies, D. S., Davies, L., and Dollery, C. T. 1973. *Life Sci.* **13,** 1665.
70. Gram, T. E., Wilson, J. T., and Fouts, J. R. 1968. *J. Pharmacol. Exp. Ther.* **159,** 172.
71. Anders, M. W., Cooper, M. J., and Takemori, A. E. 1973. *Drug Metab. Dispos.* **1,** 642.
72. Reynolds, G. 1978. *Br. J. Clin. Pharmacol.* **6,** 543.
73. Langone, J. J., and Van Vunakis, H. 1982. *In* "Methods in Enzymology" (Langone, J. J., and Vunakis, H. V., eds.), Vol. 84, p. 628. Academic Press, New York. Schievelbein, H. 1984. *Int. Encyl. Pharmacol. Ther.* **114,** 1.
74. Nguyen, T.-L., Gruenke, L. D., and Castagnoli, N., Jr. 1976. *J. Med. Chem.* **19,** 1168.
75. Nelson, S. D., Garland, W. A., Breck, G. D., and Trager, W. F. 1977. *J. Pharm. Sci.* **66,** 1180.
76. McMartin, C., and Simpson, P. 1971. *Clin. Pharmacol. Ther.* (*St Louis*) **12,** 73.
77. Hucker, H. B., Balletto, A. J., Stauffer, S. C., Zacchei, A. G., and Arison, B. H. 1974. *Drug Metab. Dispos.* **2,** 406.
78. Christy, M. E., Anderson, P. S., Arison, B. H., Cochran, D. W., and Engelhardt, E. L. 1977. *J. Org. Chem.* **42,** 378.
79. Weisburger, J. H., and Weisburger, E. K. 1973. *Pharmacol. Rev.* **25,** 1; Miller, J. A. 1970. *Cancer Res.* **30,** 559.
80a. Bondon, A., Macdonald, T. L., Harris, T. M., and Guengerich, F. P. 1989. *J. Biol. Chem.* **264,** 1988.
80b. Ziegler, D. M., Ansher, S. S., Nagata, T., Kadlubar, F. F., and Jakoby, W. B. 1988. *Proc. Natl. Acad. Sci. U.S.A.* **85,** 2514.
81. Barker, E. A., and Smuckler, E. A. 1972. *Mol. Pharmacol.* **8,** 318; Willi, P., and Bickel, M. H. 1973. *Arch. Biochem. Biophys.* **156,** 772.
82. Gorrod, J. W., and Temple, D. J. 1976. *Xenobiotica* **6,** 265.
83. Miwa, G. T., Walsh, J. S., Kedderis, G. L., and Hollenberg, P. F. 1983. *J. Biol. Chem.* **258,** 14445.
84. Weisburger, E. K. 1978. *Annu. Rev. Pharmacol. Toxicol.* **18,** 395.
85. Lin, J.-K., Miller, J. A., and Miller, E. C. 1969. *Biochemistry* **8,** 1573; Lin, J.-K., Miller, J. A., and Miller, E. C. 1975. *Cancer Res.* **35,** 844.
86. Wick, M. J., Jantan, I., and Hanna, P. E. 1988. *Biochem. Pharmacol.* **37,** 1225.
87. Mitchell, J. R., Jollow, D. J., Potter, W. Z., Gillette, J. R., and Brodie, B. B. 1973. *J. Pharmacol. Exp. Ther.* **187,** 211; Potter, W. Z., Davis, D. C., Mitchell, J. R., Jollow, D. J., Gillette, J. R., and Brodie, B. B. 1973. *J. Pharmacol. Exp. Ther.* **187,** 203.
88. Hinson, J. A., Pohl, L. R., and Gillette, J. R. 1979. *Life Sci.* **24,** 2133; Nelson, S. D., Forte, A. J., and Dahlin, D. C. 1980. *Biochem. Pharmacol.* **29,** 1617.
89. Nelson, S. D., McMurty, R. J., Mitchell, J. R. 1978. *In* "Biological Oxidation of Nitrogen" (Gorrod, J. W., ed.), p. 319. Elsevier/North-Holland, Amsterdam; Hinson, J. A., Pohl, L. R., Monks, T. J., Gillette, J. R., and Guengerich, F. P. 1980. *Drug Metab. Dispos.* **8,** 289.
90. Nelson, S. D., Dahlin, D. C., Rauckman, E. J., and Rosen, G. M. 1981. *Mol. Pharmacol.* **20,** 195.
91. Rosen, G. M., Singletary, W. V., Jr., Rauckman, E. J., and Killenberg, P. G. 1983. *Biochem. Pharmacol.* **32,** 2053.
92. West, P. R., Harman, L. S., Josephy, P. D., and Mason, R. P. 1984. *Biochem. Pharmacol.* **33,** 2933.
93. Dahlin, D. C., Miwa, G. T., Lu, A. Y. H., and Nelson, S. D. 1984. *Proc. Natl. Acad. Sci. U.S.A.* **81,** 1327.
94. Potter, D. W., and Hinson, J. A. 1987. *J. Biol. Chem.* **262,** 966.
95. Moldéus, P., Andersson, B., Rahimtula, A., and Berggren, M. 1982. *Biochem. Pharmacol.* **31,** 1363.
96. Potter, D. W., and Hinson, J. A. 1987. *J. Biol. Chem.* **262,** 974.
97. Duggan, D. E., Hogans, A. F., Kwan, K. C., and McMahon, F. G. 1972. *J. Pharmacol. Exp. Ther.* **181,** 563.

98. Adler, T. K., Fujimoto, J. M., Way, E. L., and Baker, E. M. 1955. *J. Pharmacol. Exp. Ther.* **114,** 251.
99. Klutch, A., and Bordun, M. 1967. *J. Med. Chem.* **10,** 860.
100. Mazel, P., Henderson, J. F., and Axelrod, J. 1964. *J. Pharmacol. Exp. Ther.* **143,** 1.
101. Spector, E., and Shideman, F. E. 1959. *Biochem. Pharmacol.* **2,** 182.
102. Mitchell, S. C., and Waring, R. H. 1986. *Drug Metab. Rev.* **16,** 255.
103. Oae, S., Mikami, A., Matsuura, T., Ogawa-Asada, K., Watanabe, Y., Fujimori, K., and Iyanagi, T. 1985. *Biochem. Biophys. Res. Commun.* **131,** 567.
104. Souhaili el Amri, H., Fargetton, X., Delatour, P., and Batt, A. M. 1987. *Xenobiotica* **17,** 1159.
105. Gruenke, L. D., Craig, J. C., Dinovo, E. C., Gottschalk, L. A., Noble, E. P., and Biener, R. 1975. *Res. Commun. Chem. Pathol. Pharmacol.* **10,** 221.
106. Crew, M. C., Melgar, M. D., Haynes, L. J., Gala, R. L., and DiCarlo, F. J. 1972. *Xenobiotica* **2,** 431.
107. Cohen, E. N., and Van Dyke, R. A. 1977. "Metabolism of Volatile Anesthetics" Addison-Wesley, Reading, Massachusetts.
108. Sisenwine, S. F., Kimmel, H. B., Lin, A. L., and Ruelius, H. W. 1975. *Drug Metab. Dispos.* **3,** 180.
109. Misra, A. L., Vadlamani, N. L., Pontani, R. B., and Mulé, S. J. 1973. *Biochem. Pharmacol.* **22,** 2129.
110. Bachur, N. R. 1976. *Science* **193,** 595; Wermuth, B. 1982. *Prog. Clin. Biol. Res.* **114,** 261.
111. Felsted, R. L., Richter, D. R., Jones, D. M., and Bachur, N. R. 1980. *Biochem. Pharmacol.* **29,** 1503.
112. Chan, K. K., Lewis, R. J., and Trager, W. F. 1972. *J. Med. Chem.* **15,** 1265; Lewis, R. J., Trager, W. F., Chan, K. K., Breckenridge, A., Orme, M., Roland, M., and Schary, W. 1974. *J. Clin. Invest.* **53,** 1607; Moreland, T. A., and Hewick, D. S. 1975. *Biochem. Pharmacol.* **24,** 1953.
113. Toon, S., Lon, L. K., Gibaldi, M., Trager, W. F., O'Reilly, R. A., Mottey, C. H., and Goulart, D. A. 1986. *Clin. Pharmacol. Ther.* (*St. Louis*) **39,** 15.
114. Holford, N. H. G. 1986. *Clin. Pharmacokinet.* **11,** 483.
115. Prelog, V. 1964. *Pure Appl. Chem.* **9,** 119; Horjales, E., and Brändén, C.-I. 1985. *J. Biol. Chem.* **260,** 15445.
116. Roerig, S., Fujimoto, J. M., Wang, R. I. H., and Lange, D. 1976. *Drug Metab. Dispos.* **4,** 53.
117. Dayton, H. E., and Inturrisi, C. E. 1976. *Drug Metab. Dispos.* **4,** 474.
118. Gerhards, E., Hecker, W., Hitze, H., Nieuweboer, B., and Bellmann, O. 1971. *Acta Endocrinol.* **68,** 219.
119. Uehleke, H. 1963. *Naturwissenschaften* **50,** 335; Gillette, J. R. 1963. *Arzneim-Forsch.* **6,** 11.
120. Mason, R. P., and Josephy, P. D. 1985. *In* "Toxicity of Nitroaromatic Compounds" (Rickert, D., ed.), p. 121. Hemisphere, New York; Mason, R. P., and Holtzman, J. L. 1975. *Biochemistry* **14,** 1626; Moreno, S. N. J. 1988. *Comp. Biochem. Physiol.* **91C,** 321.
121. Mason, R. P., and Holtzman, J. L. 1975. *Biochem. Biophys. Res. Commun.* **67,** 1267.
122. Scheline, R. R. 1973. *Pharmacol. Rev.* **25,** 451; Wheeler, L. A., Soderberg, F. B., and Goldman, P. 1975. *J. Pharmacol. Exp. Ther.* **194,** 135.
123. Morita, M., Feller, D. R., and Gillette, J. R. 1971. *Biochem. Pharmacol.* **20,** 217.
124. Wolpert, M. K., Althaus, J. R., and Johns, D. G. 1973. *J. Pharmacol. Exp. Ther.* **185,** 202.
125. Poirier, L. A., and Weisburger, J. H. 1974. *Biochem. Pharmacol.* **23,** 661.
126. Garattini, S., Marcucci, F., and Mussini, E. 1977. *In* "Psychotherapeutic Drugs" (Usdin, E., and Forrest, I. S., eds.), Part 2, p. 1039. Dekker, New York.
127. Feller, D. R., Morita, M., and Gillette, J. R. 1971. *Biochem. Pharmacol.* **20,** 203.
128. Tatsumi, K., Kitamura, S., and Yoshimura, H. 1976. *Arch. Biochem. Biophys.* **175,** 131.
129. Mason, R. P., Peterson, F. J., Holtzman, J. L. 1977. *Biochem. Biophys. Res. Commun.* **75,**

532; Peterson, F. J., Holtzman, J. L., Crankshaw, D., and Mason, R. P. 1988. *Mol. Pharmacol.* **34,** 597.
130. Mason, R. P., Peterson, F. J., and Holtzman, J. L. 1978. *Mol. Pharmacol.* **14,** 665.
131. Peppercorn, M. A., and Goldman, P. 1972. *J. Pharmacol. Exp. Ther.* **181,** 555; Schröder, H., and Gustafsson, B. E. 1973. *Xenobiotica* **3,** 225.
132. Kato, R., Iwasaki, K., and Noguchi, H. 1978. *Mol. Pharmacol.* **14,** 654.
133. Tamura, S., Kawata, S., Sugiyama, T., and Tarui, S. 1987. *Biochim. Biophys. Acta* **926,** 231.
134. Heymann, E. 1980. *In* "Enzymatic Basis of Detoxification" (Jakoby, W. B., ed.), Vol. 2, p. 291. Academic Press, New York.
135. Puetter, J. 1979. *Eur. J. Drug Metab. Pharmacokinet.* **4,** 1.
136. Steward, D. J., Inaba, T., Lucassen, M., and Kalow, W. 1979. *Clin. Pharmacol. Ther. (St. Louis)* **25,** 464.
137. Kogan, M. J., Verebey, K. G., DePace, A. C., Resnick, R. B., and Mulé, S. J. 1977. *Anal. Chem.* **49,** 1965.
138. Mark, L. C., Kayden, H. J., Steele, J. M., Cooper, J. R., Berlin, I., Rovenstein, E. A., and Brodie, B. B. 1951. *J. Pharmacol. Exp. Ther.* **102,** 5.
139. Junge, W., and Krisch, K. 1975. *Crit. Rev. Toxicol.* **3,** 371.
140. Heymann, E., Krisch, K., Büch, H., and Buzello, W. 1969. *Biochem. Pharmacol.* **18,** 801.
141. Akerman, B., and Ross, S. 1970. *Acta Pharmacol. Toxicol.* **28,** 445.
142. Dudley, K. H., and Roberts, S. B. 1978. *Drug Metab. Dispos.* **6,** 133.
143. Maksay, G., Tegyey, Z., and Ötvös, L. 1978. *J. Pharm. Sci.* **67,** 1208.
144. Heykants, J. J. P., Meuldermans, W. E. G., Michiels, L. J. M., Lewi, P. J., and Janssen, P. A. J. 1975. *Arch. Int. Pharmacodyn. Ther.* **216,** 113.
145. Meuldermans, W. E. G., Lauwers, W. F. J., and Heykants, J. J. P. 1976. *Arch. Int. Pharmacodyn. Ther.* **221,** 140.
146. Vainio, H. 1976. *In* "Mechanisms of Toxicity and Metabolism, Proceedings of the 6th International Congress of Pharmacology" (Karki, N. T., ed.), Vol. 6, p. 53. Pergamon, Oxford.
147. Cummings, A. J., King, M. L., and Martin, B. K. 1967. *Br. J. Pharmacol. Chemother.* **29,** 150.
148. Nakagawa, T., Masada, M., and Uno, T. 1975. *J. Chromatogr.* **111,** 355.
149. Rubin, A., Warrick, P., Wolen, R. L., Chernish, S. M., Ridolfo, A. S., and Gruber, C. M., Jr. 1972. *J. Pharmacol. Exp. Ther.* **183,** 449.
150. Bickel, M. H., Minder, R., and diFrancesco, C. 1973. *Experientia* **29,** 960.
151. Tsukamoto, H., Yoshimura, H., and Tatsumi, K. 1963. *Chem. Pharm. Bull.* **11,** 421.
152. Adamson, R. H., Bridges, J. W., Kibby, M. R., Walker, S. R., and Williams, R. T. 1970. *Biochem. J.* **118,** 41.
153. Sitar, D. S., and Thornhill, D. P. 1973. *J. Pharmacol. Exp. Ther.* **184,** 432.
154. Dutton, G. J., and Illing, H. P. A. 1972. *Biochem. J.* **129,** 539.
155. Dieterle, W., Faigle, J. W., Frueh, F., Mory, H., Theobald, W., Alt, K. O., and Richter, W. J. 1976. *Arzneim-Forsch.* **26,** 572.
156. Burchell, B., Coughtrie, M. W. H., Jackson, M. R., Shepherd, S. R. P., Harding, D., and Hume, R. 1987. *Mol. Aspects Med.* **9,** 429; Atlas, S. A., and Nebert, D. W. 1977. *In* "Drug Metabolism: From Microbe to Man" (Parke, D. V., and Smith, R. L., eds.), p. 394. Taylor & Francis, London.
157. Weiss, C. F., Glazko, A. J., and Weston, J. K. 1960. *N. Engl. J. Med.* **262,** 787.
158. Koster, A. S., Frankhuijzen-Sierevogel, A. C., and Mentrup, A. 1986. *Biochem. Pharmacol.* **35,** 1981.
159. Dahl-Puustinen, M.-L., Dumont, E., and Bertilsson, L. 1989. *Drug Metab. Dispos.* **17,** 433.
160. Mulder, G. J., and Jakoby, W. B. 1990. *In* "Conjugation Reactions in Drug Metabolism" (Mulder, G. J., ed.), p. 107. Taylor & Francis, London; Mulder, G. J., ed. 1981. "Sulfation

of Drugs and Related Compounds." CRC Press, Boca Raton, Florida; Dodgson, K. S. 1977. *In* "Drug Metabolism: From Microbe to Man" (Parke, D. V., and Smith, R. L., eds.), p. 91. Taylor & Francis, London.

161. Pang, K. S. 1990. *In* "Conjugation Reactions in Drug Metabolism" (Mulder, G. J., ed.), p. 5. Taylor & Francis, London.
162. Capel, I. D., French, M. R., Milburn, P., Smith, R. I., and Williams, R. T. 1972. *Xenobiotica* **2,** 25.
163. Mulder, G. J. 1990. *In* "Conjugation Reactions in Drug Metabolism" (Mulder, G. J., ed.), p. 41. Taylor & Francis, London; Whitmer, D. I., Ziurys, J. C., and Gollan, J. L. 1984. *J. Biol. Chem.* **259,** 11969.
164. Lin, C., Li, Y., McGlotten, J., Morton, J. B., and Symchowicz, S. 1977. *Drug Metab. Dispos.* **5,** 234.
165. Albert, K. S., Sedman, A. J., and Wagner, J. G. 1974. *J. Pharmacokinet Biopharm.* **2,** 381.
166. Miller, R. P., Roberts, R. J., and Fischer, L. J. 1976. *Clin. Pharmacol. Ther. (St. Louis)* **19,** 284; Levy, G., Khana, N. N., Soda, D. M., Tsuzuki, O., and Stern, L. 1975. *Pediatrics* **55,** 818.
167. Mulder, G. J., Meerman, J. H. H., and van den Goorbergh, A. M. 1986. *In* "Xenobiotic Conjugation Chemistry" (Paulson, G. D., Caldwell, J., Hutson, D. H., and Menn, J. J., eds.), p. 282. American Chemical Society, Washington, D.C.
168. Mulder, G. J., Hinson, J. A., and Gillette, J. R. 1977. *Biochem. Pharmacol.* **26,** 189.
169. Smith, R. L., and Caldwell, J. 1977. *In* "Drug Metabolism: From Microbe to Man" (Parke, D. V., and Smith, R. L., eds.), p. 331. Taylor & Francis, London.
170. Killenberg, P. G., and Webster, L. T., Jr. 1980. *In* "Enzymatic Basis of Detoxification" (Jakoby, W. B., ed.), Vol. 2, p. 141. Academic Press, New York.
171. Bruce, R. B., Turnbull, L. B., Newman, J. H., and Pitts, J. E. 1968. *J. Med. Chem.* **11,** 1031.
172. Drach, J. C., Howell, J. P., Borondy, P. E., and Glazko, A. J. 1970. *Proc. Soc. Exp. Biol. Med.* **135,** 849.
173. Marchand, D. H., Remmel, R. P., and Abdel-Monem, M. M. 1988. *Drug Metab. Dispos.* **16,** 85.
174. Needleman, P. 1975. *In* "Organic Nitrates" (Needleman, P., ed.), p. 57. Springer-Verlag, Berlin.
175. de Miranda, P., Beacham, III, L. M., Creagh, T. H., and Elion, G. B. 1973. *J. Pharmacol. Exp. Ther.* **187,** 588.
176. Ishida, T., Kumagai, Y., Ikeda, Y., Ito, K., Yano, M., Toki, S., Mihashi, K., Fujioka, T., Iwase, Y., and Hachiyama, S. 1989. *Drug Metab. Dispos.* **17,** 77.
177. Correia, M. A., Krowech, G., Caldera-Munoz, P., Yee, S. L., Straub, K., and Castagnoli, N., Jr. 1984. *Chem.–Biol. Interact.* **51,** 13.

178a. Stevens, J. L., and Jones, D. P. 1989. *In* "Glutathione" (Dolphin, D., Poulson, R., and Avramović, O., eds.), Part B, p. 45. Wiley, New York.

178b. Guenthner, T. M. 1990. *In* "Conjugation Reactions in Drug Metabolism" (Mulder, G. J., ed.), p. 365. Taylor & Francis, London.

179. Elson, J., Strong, J. M., and Atkinson, A. J., Jr. 1975. *Clin. Pharmacol. Ther. (St. Louis)* **17,** 134.
180. Williams, R. T. 1967. *In* "Biogenesis of Natural Compounds" (Bernfeld, P., ed.), 2nd Ed., p. 589. Pergamon, Oxford.
181. Drayer, D. E., and Reidenberg, M. M. 1977. *Clin. Pharmacol. Ther. (St. Louis)* **22,** 251.

182a. Kalow, W. 1962. "Pharmacogenetics, Heredity and the Response to Drugs." Saunders, Philadelphia, Pennsylvania.

182b. Weber, W. W. 1973. *In* "Metabolic Conjugation and Metabolic Hydrolysis" (Fishman, W. H., ed.), Vol. 3, p. 250. Academic Press, New York.

182c. Lunde, P. K. M., Frislid, K., and Hansteen, V. 1977. *Clin. Pharmacokinet.* **2,** 182.

183. Weber, W. W. 1986. *In* "Therapeutic Drugs" (Dollery, C. T., ed.), Churchill Livingstone, New York.
184. Patterson, E., Radtke, H. E., and Weber, W. W. 1980. *Mol. Pharmacol.* **17,** 367.
185. Lin, J. H., Chen, I.-W., and Ulm, E. H. 1989. *Drug Metab. Dispos.* **17,** 426.
186. Kropp, H., Sundelof, J. G., Hajdu, R., and Kahan, F. M. 1982. *Antimicrob. Agents Chemother.* **22,** 62.
187. Eschenhof, E. 1973. *Arzneim.-Forsch.* **23,** 390.
188. Thakker, D. R., and Creveling, C. R. 1990. *In* "Conjugation Reactions in Drug Metabolism" (Mulder, G. J., ed.), p. 193. Taylor & Francis, London; Ansher, S. S., and Jakoby, W. B. 1990. *In* "Conjugation Reactions in Drug Metabolism" (Mulder, G. J., ed.), p. 233. Taylor & Francis, London; Stevens, J. L., and Bakke, J. E. 1990. *In* "Conjugation Reactions in Drug Metabolism" (Mulder, G. J., ed.), p. 251. Taylor & Francis, London.
189. Axelrod, J. 1971. *In* "Concepts in Biochemical Pharmacology" (Brodie, B. B., and Gilette, J. R., eds.), Part 2, p. 609. Springer-Verlag, Berlin; Usdin, E., Borchardt, R. T., and Creveling, C. R., eds. 1982. "The Biochemistry of *S*-Adenosylmethionine and Related Compounds." Macmillan, London.
190. Morgan, C. D., Sandler, M., Davies, D. S., Connolly, M., Paterson, J. W., and Dollery, C. T. 1969. *Biochem. J.* **114,** 8P.
191. Persson, K., and Persson, K. 1972. *Xenobiotica* **2,** 375.
192. Börner, U., and Abbott, S. 1973. *Experientia* **29,** 180.
193. Leeson, G. A., Garteiz, D. A., Knapp, W. C., and Wright, G. J. 1973. *Drug Metab. Dispos.* **1,** 565.
194. Herbertz, G., Metz, T., Reinauer, H., and Staib, W. 1973. *Biochem. Pharmacol.* **22,** 1541.
195. Drummer, O. H., Miach, P., and Jarrott, B. 1983. *Biochem. Pharmacol.* **32,** 1557.
196. Lindsay, R. H., Hulsey, B. S., and Aboul-Enein, H. Y. 1975. *Biochem. Pharmacol.* **24,** 463.
197. Bodor, N. 1977. *In* "Design of Biopharmaceutical Properties of Prodrugs and Analogs" (Roche, E. B., ed.), Chap. 7. Academy of Pharmaceutical Sciences, Washington, D.C.; Bodor, N. 1984. *Adv. Drug Res.* **13,** 255; Bodor, N. 1984. *Med. Res. Rev.* **4,** 449.
198. Bodor, N., Kaminski, J. J., and Selk, S. 1980. *J. Med. Chem.* **23,** 469.

## General References

### Synthesis of Radioactive Compounds

Muccino, R. R., ed. 1986. "Synthesis and Applications of Radioactive Compounds." Elsevier, Amsterdam.

### Analytical Methods in Drug Metabolism

Abraham, R. J., Fisher, J., and Loftus, P. 1988. "Introduction to NMR Spectroscopy." Wiley, Chichester.
Bovey, F. A. 1988. "Nuclear Magnetic Resonance Spectroscopy," 2nd Ed. Academic Press, San Diego, California.
Busch, K. L., Glish, G. L., and McLuckey, S. A. 1988. "Mass Spectrometry/Mass Spectrometry: Techniques and Applications of Tandem Mass Spectrometry." VCH, New York.
Coutts, R. T., and Jones, G. R. 1980. *In* "Concepts in Drug Metabolism" (Jenner, P., and Testa, B., eds.), p. 1. Dekker, New York.
Derome, A. E. 1987. "Modern NMR Techniques for Chemistry Research." Pergamon, Oxford.

Horning, M. G., Sheng, L.-S., and Lertratanangkoon, K. 1984. *In* "Drug Metabolism and Drug Toxicity" (Mitchell, J. R., and Horning, M. G., eds.), p. 353. Raven, New York.
McEwen, C. N., and Larsen, B. S., eds. 1990. "Mass Spectrometry of Biological Materials." Dekker, New York.
Martin, G. E., and Zektzer, A. S. 1988. "Two-Dimensional NMR Methods for Establishing Molecular Connectivity." VCH, New York.
Watson, J. T. 1985. "Introduction to Mass Spectrometry," 2nd Ed. Raven, New York.

## Pathways for Drug Deactivation and Elimination

Armstrong, R. N. 1987. *Crit. Rev. Biochem.* **22,** 39.
Caldwell, J. 1980. *In* "Concepts in Drug Metabolism" (Jenner, P., and Testa, B., eds.), Part A, p. 211. Dekker, New York.
Dolphin, D., Poulson, R., and Avramović, O. 1989. "Glutathione," Vols. 1 and 2. Wiley, New York.
Gibson, G. G., and Skett, P. 1986. "Introduction to Drug Metabolism." Chapman & Hall, London.
Guengerich, F. P., ed. 1987. "Mammalian Cytochromes *P*-450," Vols. 1 and 2. CRC Press, Boca Raton, Florida.
Guengerich, F. P. 1990. *Crit. Rev. Biochem. Mol. Biol.* **25,** 97.
Jakoby, W. B., ed. 1980. "Enzymatic Basis of Detoxification," Vols. 1 and 2. Academic Press, New York.
Low, L. K., and Castagnoli, N., Jr. 1980. *In* "Burger's Medicinal Chemistry" (Wolff, M. E., ed.), Part 1, p. 107. Wiley, New York.
Mulder, G. J., ed. 1990. "Conjugation Reactions in Drug Metabolism." Taylor & Francis, London.
Sies, H., and Ketterer, B., eds. 1988. "Glutathione Conjugation: Mechanisms and Biological Significance." Academic Press, London.
Usdin, E., Borchardt, R. T., and Creveling, C. R., eds. 1982. "The Biochemistry of *S*-Adenosylmethionine and Related Compounds." Macmillan, London.
Weber, W. W. 1987. "The Acetylator Genes and Drug Response." Oxford Univ. Press, New York.

CHAPTER 8

# Prodrugs and Drug Delivery Systems



## I. Enzyme Activation of Drugs

The term *prodrug*, which was used initially by Albert,[1] refers to a pharmacologically inactive compound that is converted to an active drug by a metabolic biotransformation. A prodrug also can be activated by a nonenzymatic process such as hydrolysis, but in this case the compounds usually are inherently unstable and may cause stability problems. The prodrug-to-drug conversion can occur before absorption, during absorption, after absorption, or at a specific site in the body. In the ideal case a prodrug is converted to the drug as soon as the desired goal for designing the prodrug has been achieved. It should be noted that although the compounds discussed in this chapter are illustrative of the approaches taken for the design of prodrugs, many of them have not been approved for medical use.

### A. Utility of Prodrugs

There are numerous reasons why one may wish to utilize a prodrug strategy in drug design. Specific examples of each of these categories are given in Section II,A,2.

### 1. Solubility

Consider an active drug that is insufficiently soluble in water so that it cannot be injected in a small dose. A water-soluble group could be attached which could be metabolically released after drug administration.

### 2. Absorption and Distribution

If the drug is not absorbed and transported to the target site in sufficient concentration, it can be made more water soluble or lipid soluble, depending on the desired site of action. Once absorption has occurred or when the drug is at the appropriate site of action, the water- or lipid-soluble group is removed enzymatically.

### 3. Site Specificity

Specificity for a particular organ or tissue can be made if there are high concentrations of or uniqueness of enzymes present at that site which can cleave the appropriate appendages from the prodrug and unmask the drug.

### 4. Instability

A drug may be rapidly metabolized and rendered inactive prior to when it reaches the site of action. The structure may be modified to block that metabolism until the drug is at the desired site.

### 5. Prolonged Release

It may be desirable to have a steady low concentration of a drug released over a long period of time. The drug may be altered so that it is metabolically converted to the active form slowly.

### 6. Toxicity

A drug may be toxic in its active form and would have a greater therapeutic index if it were administered in a nontoxic, inactive form that was converted to the active form only at the site of action.

### 7. Poor Patient Acceptability

An active drug may have an unpleasant taste or odor, produce gastric irritation, or cause pain when administered (e.g., when injected). The structure of the drug can be modified to alleviate these problems, but once administered, the altered drug can be metabolized to the active drug.

### 8. Formulation Problems

If the drug is a volatile liquid, it would be more desirable to prepare it in a solid form so that it could be formulated as a tablet. An inactive solid derivative could be prepared which would be converted in the body to the active drug.

## B. Types of Prodrugs

There are several classifications of prodrugs. Some prodrugs are not designed as such; the biotransformations are fortuitous, and it is discovered after isolation and testing of the metabolites that activation of the drug had occurred. In most cases a specific modification in a drug has been made on the basis of known metabolic transformations. It is expected that after administration it will be appropriately metabolized to the active form. This has been termed *drug latentiation* to signify the rational design approach rather than serendipity.[2] The term drug latentiation has been refined even further by Wermuth[3] into two classes which he called carrier-linked prodrugs and bioprecursors.

A *carrier-linked prodrug* is a compound that contains an active drug linked to a carrier group that can be removed enzymatically, such as an ester which is hydrolyzed to an active carboxylic acid-containing drug. The bond to the carrier group must be labile enough to allow the active drug to be released efficiently *in vivo*, and the carrier group must be nontoxic and biologically inactive when detached from the drug. Carrier-linked prodrugs can be subdivided even further into bipartate, tripartate, and mutual prodrugs. A *bipartate prodrug* is a prodrug comprised of one carrier attached to the drug. When a carrier is connected to a linker arm which is connected to the drug, the term *tripartate prodrug* is used. A *mutual prodrug* consists of two, usually synergistic, drugs attached to each other (one drug is the carrier for the other and vice versa).

A *bioprecursor* is a compound that is metabolized by molecular modification into a new compound which is the active principle or which can be metabolized further to the active drug. For example, if the drug contains a carboxylic acid group, the bioprecursor may be a primary amine which is metabolized by oxidation to the aldehyde which is further metabolized to the carboxylic acid drug (see Section IV,B,1,e of Chapter 7). Unlike a carrier-linked prodrug, which is the active drug linked to a carrier that generally is released by a hydrolytic reaction, a bioprecursor contains a different structure that cannot be converted to the active drug by simple cleavage of a group from the prodrug.

The concept of prodrugs can be analogized to the use of protecting groups in organic synthesis.[4] If, for example, you wanted to carry out a reaction on a compound that contained a carboxylic acid group, it may be necessary first to protect the carboxylic acid as, say, an ester, so that the acidic proton of the carboxylic acid does not interfere with the desired reaction. After the desired synthetic transformation is completed, the carboxylic acid analog could be unmasked by deprotection, that is, hydrolysis of the ester (Scheme 8.1A). This is analogous to a carrier-linked prodrug; an ester functionality can be used to give the drug more desirable properties until it reaches the appropriate biological site where it is "deprotected." Another type of protecting group in organic synthesis is one which has no resemblance to the desired functional group. For example, a terminal alkene can be oxidized with ozone to an aldehyde,[5] and the aldehyde can be oxidized to a carboxylic acid with hydrogen peroxide (Scheme 8.1B). As in the case of a bioprecursor, a drastic structural change is required to unmask the desired group. Oxidation is a common metabolic biotransformation for bioprecursors.

$$\text{A.}\quad RCO_2H \xrightarrow[\text{HCl},\ \Delta]{\text{EtOH}} RCO_2Et \xrightarrow[\text{on R}]{\text{reaction}} R'CO_2Et \xrightarrow[\Delta]{H_3O^+} R'CO_2H$$

$$\text{B.}\quad RCH{=}CH_2 \xrightarrow[\text{on R}]{\text{reaction}} R'CH{=}CH_2 \xrightarrow[\text{2. } H_2O_2]{\text{1. } O_3} R'CO_2H$$

**Scheme 8.1.** Protecting group analogy for a prodrug.

When designing a prodrug, you should keep in mind that a particular metabolic transformation may be species specific (see Chapter 7). Therefore, a prodrug designed on the basis of rat metabolism studies may not necessarily be effective in humans.

## II. Mechanisms of Prodrug Activation

### A. Carrier-Linked Prodrugs

The most common reaction for activation of carrier-linked prodrugs is hydrolysis. First, we consider the general functional groups involved, then specific examples for different types of prodrugs will be given.

#### 1. Carrier Linkages for Various Functional Groups

***a. Alcohols and Carboxylic Acids.*** There are several reasons why the most common prodrug form for drugs containing alcohol or carboxylic acid functional groups is an ester. First, esterases are ubiquitous, so metabolic regeneration of the drug is a facile process. Also, it is possible to prepare ester derivatives with virtually any degree of hydrophilicity or lipophilicity. Finally, a variety of stabilities of esters can be obtained by appropriate manipulation of electronic and steric factors. Therefore, a multitude of ester prodrugs can be prepared to accommodate a wide variety of problems that require the prodrug approach.

Alcohol-containing drugs can be acylated with aliphatic or aromatic carboxylic acids to decrease water solubility (increase lipophilicity) or with carboxylic acids containing amino or additional carboxylate groups to increase water solubility (Table 8.1).[6] Conversion to phosphate or sulfate esters also increases water solubility. By using these approaches a wide range of solubilities can be achieved that will affect the absorption and distribution properties of the drug. These derivatives also can have an important effect on the dosage form, that is, whether used in tablet form or in aqueous solution. One problem with the use of this prodrug approach is that in some cases the esters are not very good substrates for the endogenous esterases, sulfatases, or phosphatases, and they may not be hydrolyzed at a rapid enough rate. When that occurs, however, a different ester can be tried. Another approach to accelerate the hydrolysis rate could be to attach electron-withdrawing groups (if a

**Table 8.1** Ester Analogs of Alcohols as Prodrugs

Drug—OH ⟶ Drug—OX

| X | Effect on water solubility |
|---|---|
| $C(=O)—R$ | (R = aliphatic or aromatic) decreases |
| $C(=O)—CH_2\overset{+}{N}HMe_2$ | increases (p$K_a$ ~ 8) |
| $C(=O)—CH_2CH_2COO^-$ | increases (p$K_a$ ~ 5) |
| 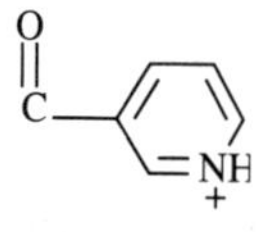 $C(=O)$—(3-pyridinium, $\overset{+}{N}H$) | increases (p$K_a$ ~ 4) |
| $PO_3^{=}$ | increases (p$K_a$ ~ 2 and ~ 6) |
| $C(=O)CH_2SO_3^-$ | increases (p$K_a$ ~ 1) |

base hydrolysis mechanism is relevant) or electron-donating groups (if an acid hydrolysis mechanism is important)[7] to the carboxylate side of the ester. Succinate esters can be used to accelerate the rate of hydrolysis by intramolecular catalysis (Scheme 8.2). If the ester is too reactive, substituents can be appended that cause steric hindrance to hydrolysis. Alcohol-containing drugs also can be converted to the corresponding acetals or ketals for rapid hydrolysis in the acidic medium of the gastrointestinal tract.

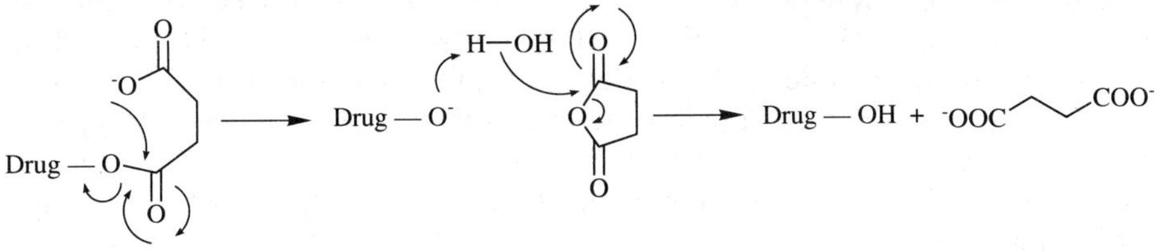

Drug — O⁻ ... H—OH ... Drug — OH + ⁻OOC⌄COO⁻

**Scheme 8.2.** Intramolecular hydrolysis of succinate esters.

Carboxylic acid-containing drugs also can be esterified; the reactivity of the derivatized drug can be adjusted by appropriate structural manipulations. If a slower rate of ester hydrolysis is desired, long-chain aliphatic or sterically hindered esters can be used. If hydrolysis is too slow, addition of electron-withdrawing groups on the alcohol part of the ester can increase the rate. The $pK_a$ of a carboxylic acid can be raised by conversion to a choline ester (**8.1**, R = R′ = Me; $pK_a \sim 7$) or an amino ester (**8.1**, R = H, R′ = H or Me; $pK_a \sim 9$).

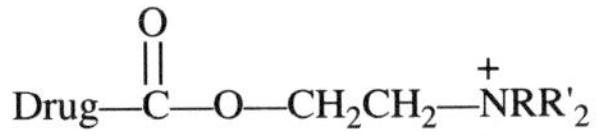

$$\text{Drug—}\overset{\displaystyle O}{\overset{\|}{C}}\text{—O—CH}_2\text{CH}_2\text{—}\overset{+}{N}\text{RR}'_2$$

**8.1**

***b. Amines.*** N-Acylation of amines to give amide prodrugs is not commonly used, in general, because of the stability of amides toward metabolic hydrolysis. Activated amides, generally of low basicity amines, or amides of amino acids are more susceptible to enzymatic cleavage (Table 8.2). Although

**Table 8.2** Prodrug Analogs of Amines

$$\text{Drug—NH}_2 \longrightarrow \text{Drug—NHX}$$

| X | | | | | |
|---|---|---|---|---|---|
| $-\overset{O}{\overset{\|}{C}}R$ | $-\overset{O}{\overset{\|}{C}}CH(R)\overset{+}{N}H_3$ | $-\overset{O}{\overset{\|}{C}}-OPh$ | $-CH_2NH\overset{O}{\overset{\|}{C}}Ar$ | $=CHAr$ | $=NAr$ |

carbamates in general are too stable, phenyl carbamates ($RNHCO_2Ph$) are rapidly cleaved by plasma enzymes,[8] and, therefore, they can be used as prodrugs.

The p$K_a$ values of amines can be lowered by approximately 3 units by conversion to their *N-Mannich bases* (Table 8.2, X = $CH_2NHCOAr$). This lowers the basicity of the amine so that at physiological pH few of the prodrug molecules are protonated, thereby increasing its lipophilicity. For example, the partition coefficient (see Chapter 2, Section II,E,2,b) between octanol and phosphate buffer, pH 7.4, for the *N*-Mannich base (**8.2**, R = $CH_2NHCOPh$) derived from benzamide and the decongestant phenylpropanolamine (**8.2**, R = H) is almost 100 times greater than that for the parent amine.[9] However, the rate of hydrolysis of *N*-Mannich bases depends on the amide carrier group; salicylamide and succinimide are more susceptible to hydrolysis than is benzamide.[10]

Another approach for lowering the p$K_a$ values of amines and, thereby, making them more lipophilic, is to convert them to imines (*Schiff bases*); however, imines often are too labile in aqueous solution. The anticonvulsant agent progabide (**8.3**) is a prodrug form of γ-aminobutyric acid, an important inhibitory neurotransmitter (see Chapter 5, Section V,C,3,a). The lipophilicity of **8.3** allows the compound to cross the blood–brain barrier; once inside the brain it is hydrolyzed to γ-aminobutyric acid.[11]

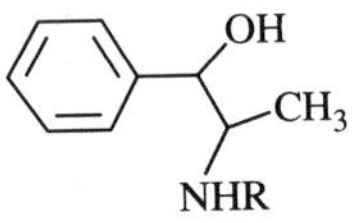

OH CH3 NHR

**8.2**

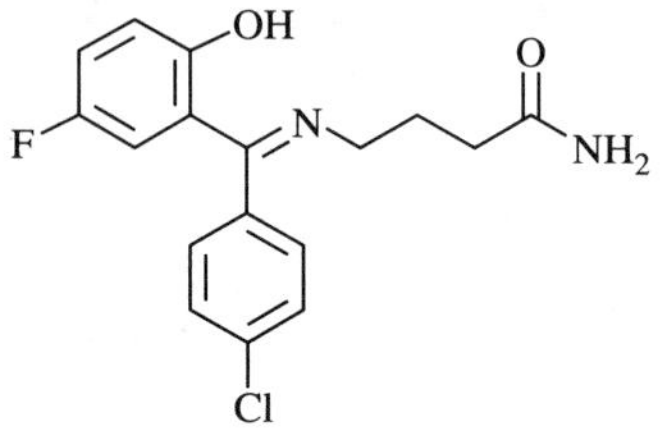

OH O F N NH2 Cl

**8.3**

***c. Carbonyl Compounds.*** The most important prodrug forms of aldehydes and ketones are Schiff bases, oximes, acetals (ketals), enol esters, oxazolidines, and thiazolidines (Table 8.3). A more complete review of bioreversible derivatives of the functional groups was written by Bundgaard.[6]

### 2. Examples of Carrier-Linked Bipartate Prodrugs

***a. Prodrugs for Increased Water Solubility.*** Prednisolone (**8.4**; R = R′ = H) and methylprednisolone (**8.4**; R = $CH_3$, R′ = H) are poorly water-soluble corticosteroid drugs. In order to permit aqueous injection or ophthalmic delivery of these drugs, they must be converted to water-soluble forms

**Table 8.3** Prodrug Analogs of Carbonyl Compounds

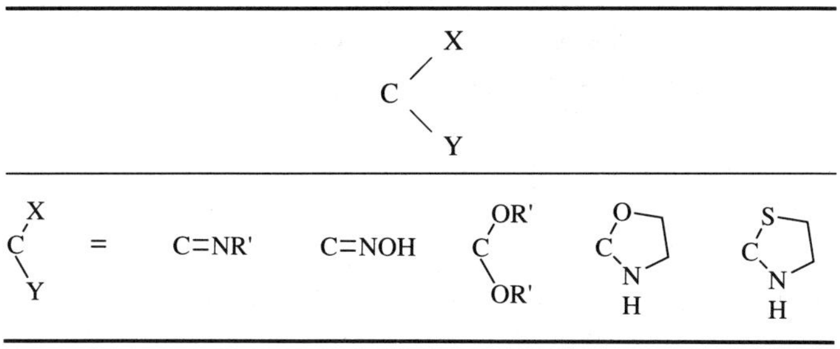

Drug(R)C=O ⟶ Drug(R)C(X)(Y)

| C(X)(Y) | | | | | |
|---|---|---|---|---|---|
| C(X)(Y) = | C=NR' | C=NOH | C(OR')(OR') | C(O–N(H)) ring | C(S–N(H)) ring |

such as one of the ionic esters described in Section II,A,1,a. However, there are two considerations in the choice of a solubilizing group: the ester must be stable enough in aqueous solution so that a ready-to-inject solution has a reasonably long shelf life (greater than 2 years; half-life about 13 years), but it must be hydrolyzed *in vivo* with a reasonably short half-life after administration (less than 10 min). For this optimal situation to occur the *in vivo*/*in vitro* lability ratio would have to be on the order of $10^6$. This is possible when the biotransformation is enzyme catalyzed.

The water-soluble prodrug form of methylprednisolone that is in medical use is methylprednisolone sodium succinate (**8.4**, R = $CH_3$, R′ = $COCH_2$-$CH_2CO_2Na$). However, the *in vitro* stability is low; consequently, it is distributed as a lyophilized (freeze-dried) powder that must be reconstituted with water and then used within 48 hr. The lyophilization process adds to the cost of the drug and makes its use less convenient. On the basis of physical–organic chemical rationalizations, a series of more stable water-soluble methylprednisolone esters was synthesized, and several of the analogs were shown to have shelf lives in solution of greater than 2 years at room temperature.[12] Ester hydrolysis studies of these compounds in human and monkey serum indicated that derivatives having an anionic solubilizing moiety such as carboxylate or sulfonate are poorly or not hydrolyzed, but compounds with a

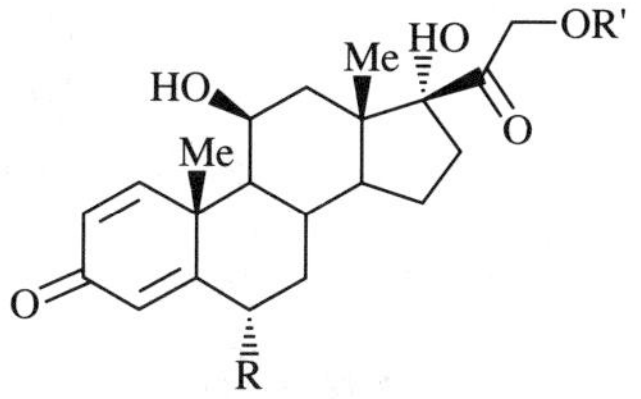

**8.4**

cationic (tertiary amino) solubilizing moiety are hydrolyzed rapidly by serum esterases.[13] Prednisolone phosphate (**8.4**; R = H, R′ = $PO_3Na_2$) is prescribed as a water-soluble prodrug for prednisolone that is activated *in vivo* by phosphatases.

The local anesthetic benzocaine (**8.5**, R = H) has been converted to water-soluble amide prodrug forms with various amino acids (**8.5**, R = $\overset{+}{N}H_3$-CHR′CO); amidase-catalyzed hydrolysis in human serum occurs rapidly.[14]

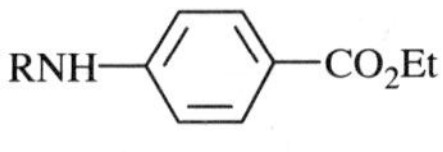

RNH—C₆H₄—CO₂Et

**8.5**

Another prodrug approach is to design acyclic derivatives that are enzymatically hydrolyzed to a product that spontaneously cyclizes to the desired drug. The benzodiazepine tranquilizer diazepam (**8.8**, Scheme 8.3) is very sparingly water soluble, but the open chain amino ketone coupled to an amino acid or peptide is a stable, freely water-soluble prodrug (**8.6**); *in vivo* peptidases hydrolyze the peptide bond, and the resulting 2-aminoacetamidobenzophenone analog (**8.7**) spontaneously cyclizes to give the benzodiazepine.[15,16] The rate of *in vivo* hydrolysis of the peptide bond depends on which L-amino acid is attached; peptides derived from Phe and Lys are cleaved much faster than those from Gly and Glu. The rate of cyclization depends on the substituents in the phenyl ring and on the amide nitrogen. Although the cyclization of **8.7** to **8.8** occurs with a half-life of 73 sec, that for the corresponding *N*-desmethyl analog is 15 min. As an example of how effective this approach is for increasing the water solubility, the benzodiazepine triazolam (**8.9**, R = Cl) has a solubility of 0.015 mg/ml at 25° C, but the corresponding open-chain glycylaminobenzophenone derivative (HCl salt) has a solubility of 109 mg/ml.[17] A similar prodrug approach was taken for the benzodiazepine alprazolam (**8.9**, R = H).[18]

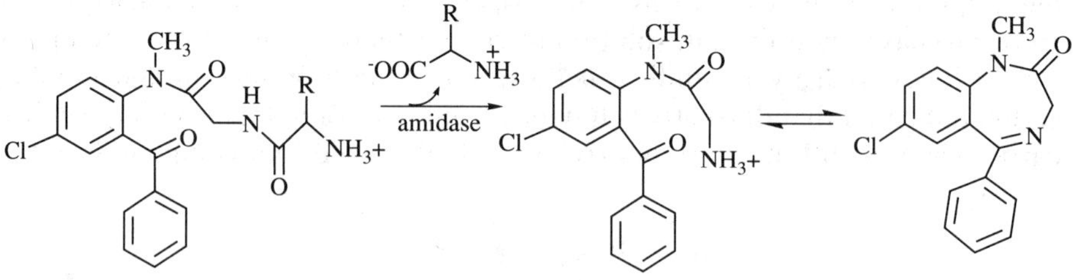

**8.6** —amidase ($^{-}OOC\text{-}CHR\text{-}\overset{+}{N}H_3$)→ **8.7** ⇌ **8.8**

**Scheme 8.3.** Benzodiazepine prodrug activation.

***b. Prodrugs for Improved Absorption and Distribution.*** The skin is designed to maintain the body fluids and prevent absorption of xenobiotics into the general circulation. Consequently, drugs applied to the skin are poorly absorbed.[19] Even steroids have low dermal permeability, particularly if they contain hydroxyl groups which can interact with the skin or binding sites in the keratin. Corticosteroids for the topical treatment of inflammatory, allergic, and pruritic skin conditions can be made more suitable for topical absorption by esterification or acetonidation. For example, both fluocinolone acetonide (**8.10**, R = H) and fluocinonide (**8.10**, R = $COCH_3$) are prodrugs used for inflammatory and pruritic manifestations. Once absorbed through the skin an esterase releases the drug.

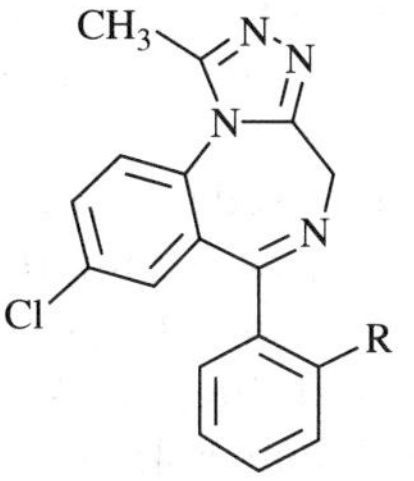

**8.9** 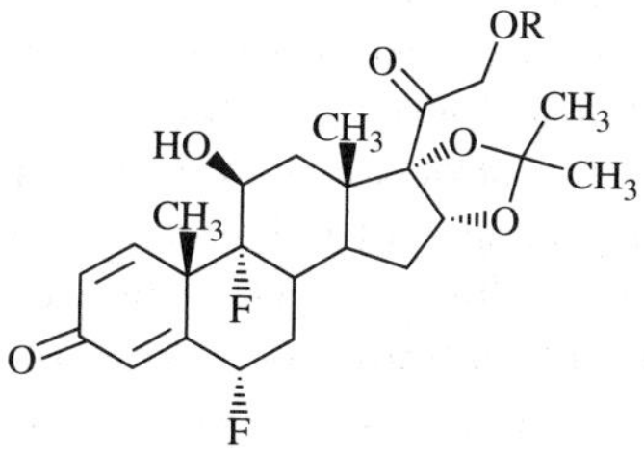 **8.10**

Dipivaloylepinephrine (dipivefrin; **8.11**, R = $Me_3CCO$), a prodrug for the antiglaucoma drug epinephrine (**8.11**, R = H), is better able to penetrate the cornea than is epinephrine. The cornea and aqueous humor have significant esterase activity.[20]

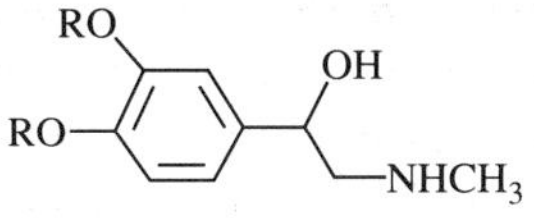

**8.11**

***c. Prodrugs for Site Specificity.*** The targeting of drugs for a specific site in the body by conversion to a prodrug is plausible when the physicochemical properties of the parent drug and prodrug are optimal for the target site. It should be kept in mind, however, that when the lipophilicity of a drug is increased, it will improve passive transport of the drug nonspecifically to all tissues.

Oxyphenisatin (**8.12**, R = H) is a bowel sterilant that is active only when administered rectally. However, when the hydroxyl groups are acetylated (**8.12**, R = Ac), the prodrug can be administered orally, and it is hydrolyzed at the site of action in the intestines to oxyphenisatin.

One important membrane that often is targeted for drug delivery is the *blood–brain barrier*, a unique lipidlike protective barrier that prevents hydrophilic compounds from entering the brain unless they are actively transported.[21] The blood–brain barrier also contains active enzyme systems to

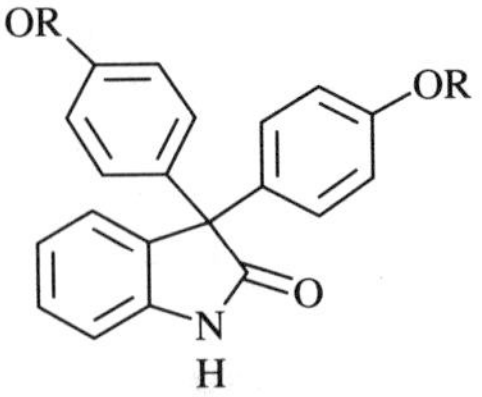

**8.12**

protect the central nervous system even further. Consequently, molecular size and lipophilicity are often necessary, not sufficient, criteria for gaining entry into the brain.[22] Also, once the drug has entered the brain, it must be modified so that it does not escape.

Bodor and co-workers have devised a reversible redox drug delivery system for getting drugs into the central nervous system and then, once in, preventing their efflux.[22,23] The approach is based on the attachment of a hydrophilic drug to a lipophilic carrier (a dihydropyridine, **8.13**) thereby making the bipartate prodrug overall lipophilic (Scheme 8.4). Once inside the brain, the lipophilic carrier is converted enzymatically to a highly hydrophilic species (**8.14**), which is then enzymatically hydrolyzed back to the drug and *N*-methylnicotinic acid (**8.15**) which is eliminated from the brain. The XH group on the drug is an amino, hydroxyl, or carboxyl group. When it is a carboxylic acid, the linkage is an acyloxymethyl ester (**8.16**), which decomposes by the reaction shown in Scheme 8.5. The oxidation of the dihydropyridine (**8.13**) to the pyridinium ion (**8.14**) (half-life generally 20–50 min) prevents the drug from escaping out of the brain because it becomes charged. This drives the equilibrium of the lipophilic precursor (**8.13**) throughout all of the tissues of the body to favor the brain. Any oxidation occurring outside of the brain produces a hydrophilic species that can be rapidly eliminated from the body (see Chapter 7). The released oxidized carrier (**8.15**) is relatively non-

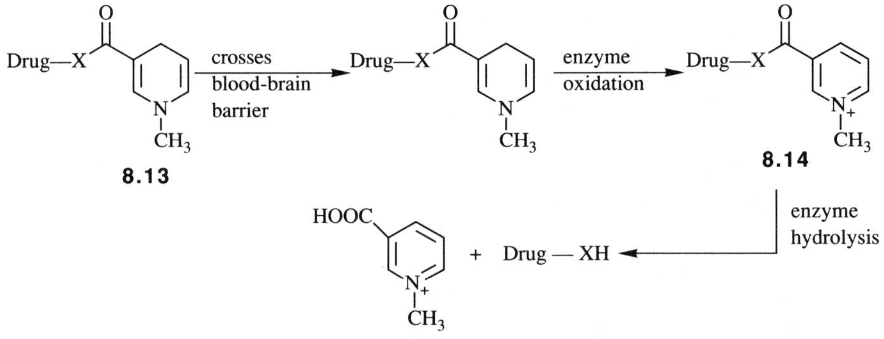

Drug—X (8.13) —crosses blood-brain barrier→ Drug—X —enzyme oxidation→ Drug—X (8.14) —enzyme hydrolysis→ HOOC-pyridinium $CH_3$ (8.15) + Drug—XH

**Scheme 8.4.** Redox drug delivery system.

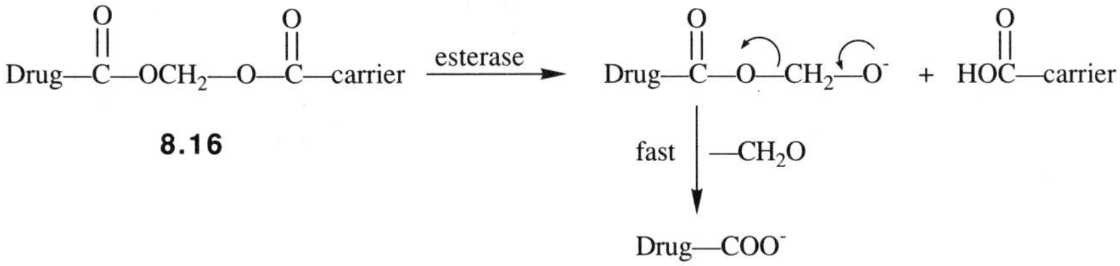

$$\text{Drug—}\overset{\text{O}}{\overset{\|}{\text{C}}}\text{—OCH}_2\text{—O—}\overset{\text{O}}{\overset{\|}{\text{C}}}\text{—carrier} \xrightarrow{\text{esterase}} \text{Drug—}\overset{\text{O}}{\overset{\|}{\text{C}}}\text{—O—CH}_2\text{—O}^- + \text{HO}\overset{\text{O}}{\overset{\|}{\text{C}}}\text{—carrier}$$

**8.16**

$$\xrightarrow[\text{—CH}_2\text{O}]{\text{fast}} \text{Drug—COO}^-$$

**Scheme 8.5.** Hydrolysis of acyloxymethyl esters.

toxic and easily eliminated from the brain. Although this is a carrier-linked prodrug, it requires enzymatic oxidation to target the drug to the brain. The oxidation reaction is a bioprecursor reaction (see Section II,B,2,c).

An example of this approach is the brain delivery of β-lactam antibiotics for the possible treatment of bacterial meningitis. The difficulty in purging the central nervous system of infections arises from the fact that the cerebrospinal fluid contains less than 0.1% of the number of immunocompetent leukocytes found in the blood and almost no immunoglobins; consequently, antibody generation to these foreign organisms is not significant. Since the β-lactam antibiotics are hydrophilic, they enter the brain very slowly, but they are actively transported back into the blood. Therefore, they are not as effective in the treatment of brain infections as elsewhere. Bodor and co-workers[24] prepared a variety of penicillin prodrugs attached to the dihydropyridine carrier through various linkers (**8.17**) and showed that β-lactam antibiotics could be delivered in high concentrations into the brain.

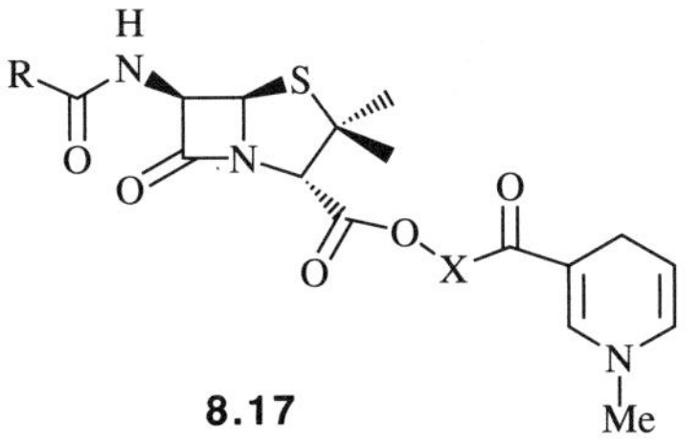

**8.17**

As was discussed in Section V,C,3,a of Chapter 5, increasing the brain concentration of the inhibitory neurotransmitter β-aminobutyric acid (GABA) results in anticonvulsant activity. However, GABA is too polar to cross the blood–brain barrier, so it is not an effective anticonvulsant drug. In order to increase the lipophilicity of GABA, a series of γ-aminobutyric acid and γ-aminobutyric Schiff bases were synthesized.[11] Progabide (**8.3**) emerged as an effective lipophilic analog of GABA that crosses the blood–brain barrier, releases GABA inside the brain, and shows anticonvulsant activity.[25]

Another related approach for anticonvulsant drug design was the synthesis of a glyceryl lipid (**8.18**, R = linolenoyl) containing one GABA molecule and one vigabatrin molecule, a mechanism-based inactivator of GABA aminotransferase and anticonvulsant drug (see Section V,C,3,a of Chapter 5).[26] This compound inactivates GABA aminotransferase *in vitro* only if brain esterases are added to cleave the vigabatrin from the glyceryl lipid. It also is 300 times more potent than vigabatrin, *in vivo*, presumably because of its increased ability to enter the brain.

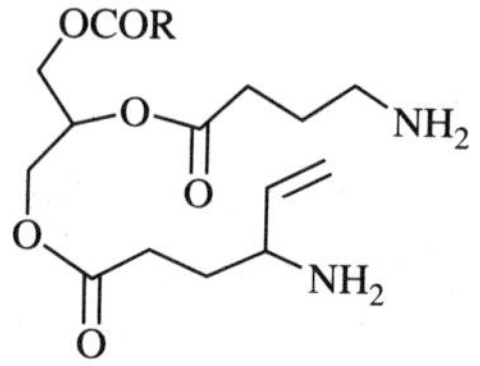

OCOR

O

NH$_2$

O

O

NH$_2$

O

**8.18**

In the above examples, the lipophilicity of the drugs was increased so that they could diffuse through various membranes. Another approach for site-specific drug delivery is to design a prodrug that requires activation by an enzyme found predominantly at the desired site of action. For example, tumor cells contain a higher concentration of phosphatases and amidase than do normal cells. Consequently, a prodrug of a cytotoxic agent could be directed to tumor cells if either of these enzymes were important to the prodrug activation process. Diethylstilbestrol diphosphate (**8.19**, R = $PO_3^{2-}$) was designed for site-specific delivery of diethylstilbestrol (**8.19**, R = H) to prostatic carcinoma tissue.[2,27] In general, though, this tumor-selective approach has not been very successful because the appropriate prodrugs are too polar to reach the enzyme site, the relative enzymatic selectivity is insufficient, and the tumor cell perfusion rate is too poor.

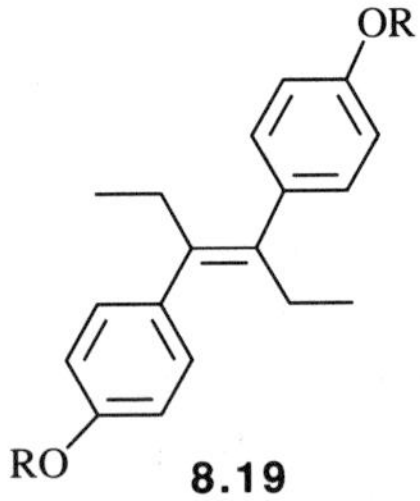

OR

RO

**8.19**

***d. Prodrugs for Stability.*** Some prodrugs protect the drug from the first-pass effect (see Section I of Chapter 7). Propranolol (**8.20**, R = R′ = H) is a widely used antihypertensive drug, but because of first-pass elimination an oral dose has a much lower bioavailability than does an intravenous injection. The major metabolites (see Chapter 7) are propranolol *O*-glucuronide (**8.20**, R = H, OR′ = glucuronide), *p*-hydroxypropranolol (**8.20**, R = OH, R′ = H), and its *O*-glucuronide (**8.20**, R = OH, OR′ = glucuronide). The hemisuccinate ester of propranolol (**8.20**, R = H, R′ = $COCH_2CH_2COOH$)

was prepared to block glucuronide formation; following oral administration of propranolol hemisuccinate, the plasma levels of propranolol were 8 times greater than when propranolol was used.[28]

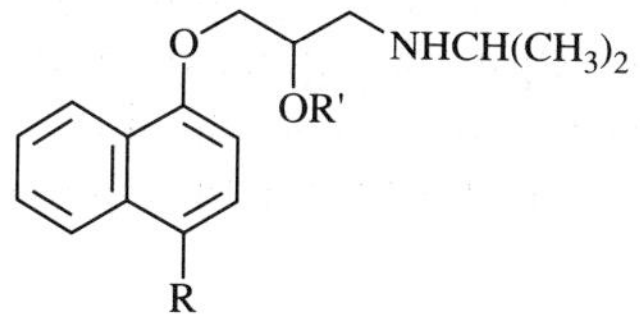

O OR' $NHCH(CH_3)_2$ R

**8.20**

Naltrexone (**8.21**, R = H), used in the treatment of opioid addiction, is nonaddicting and is well absorbed from the gastrointestinal tract. However, it undergoes extensive first-pass metabolism when given orally. Ester prodrugs, namely, the anthranilate (**8.21**, R = CO-*o*-$NO_2$Ph) and the acetylsalicylate (**8.21**, R = CO-*o*-AcOPh), enhanced the bioavailability 45- and 28-fold, respectively, relative to **8.21** (R = H).[29]

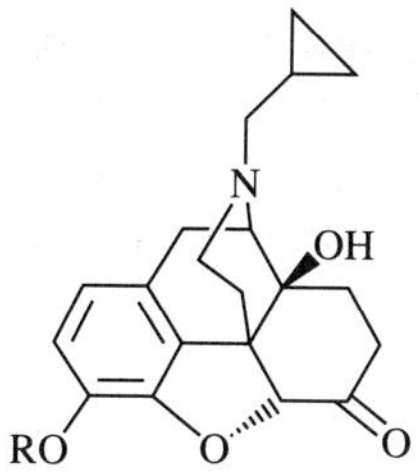

N OH RO O O

**8.21**

***e. Prodrugs for Slow and Prolonged Release.*** The utility of slow and prolonged release of drugs is severalfold. (1) It reduces the number and frequency of doses required. (2) It eliminates nighttime administration of drugs. (3) Because the drug is taken less frequently, slow, prolonged release minimizes patient noncompliance. (4) When a fast released drug is taken, there is a rapid surge of the drug throughout the body. As metabolism of the drug proceeds, the concentration of the drug diminishes. A slow release drug would eliminate the peaks and valleys of fast released drugs which are a strain on cells. (5) Because a constant lower concentration of the drug is being released, it reduces the possibility of toxic levels of drugs. (6) It reduces gastrointestinal side effects. A common strategy in the design of slow release prodrugs is to make a long chain aliphatic ester because these esters hydrolyze slowly.

Prolonged release drugs are quite important in the treatment of psychoses because these patients require medication for extended periods of time and often show high patient noncompliance rates. Haloperidol (**8.22**, R = H) is a potent, orally active central nervous system depressant, sedative, and tranquilizer. However, peak plasma levels are observed between 2 and 6 hr after administration. Haloperidol decanoate [**8.22**, R = $CO(CH_2)_8CH_3$], however, is injected intramuscularly as a solution in sesame oil and its antipsychotic activity lasts for about 1 month.[30] The antipsychotic fluphenazine (**8.23**, R = H) also has a short duration of activity (6–8 hr). Fluphenazine enanthate [**8.23**, R = $CO(CH_2)_5CH_3$] and fluphenazine decanoate [**8.23**, R = $CO(CH_2)_8CH_3$], however, have durations of activity of about a month.[31]

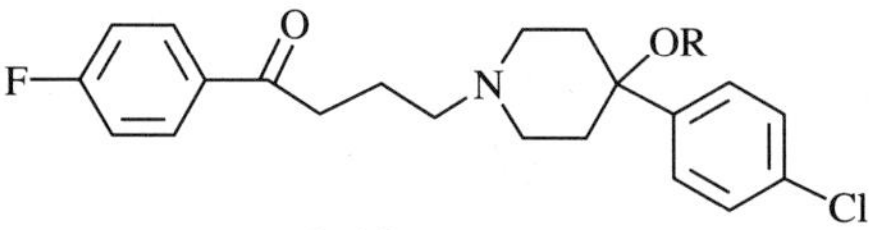

F O N OR Cl

**8.22**

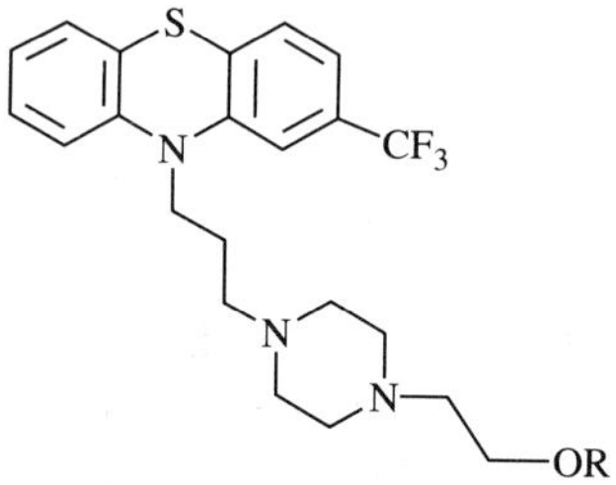

S N $CF_3$ N N OR

**8.23**

Conversion of the nonsteroidal anti-inflammatory (antiarthritis) drug tolmetin sodium (**8.24**, R = $O^-$ $Na^+$) to the corresponding glycine conjugate (**8.24**, R = $NHCH_2COOH$) increases the potency and extends the peak concentration of tolmetin from 1 to about 9 hrs because of the slow hydrolysis of the prodrug amide linkage.[32]

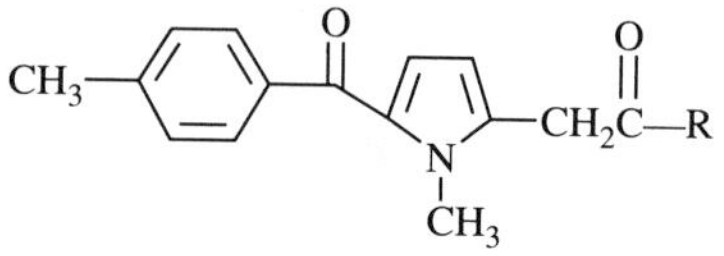

$CH_3$ O N $CH_3$ $CH_2C$—R O

**8.24**

*f. **Prodrugs to Minimize Toxicity.*** The prodrugs that were designed for improved absorption (Section II,A,2,b), for site specificity (Section II,A,2,c), for stability (Section II,A,2,d), and for slow release (Section II,A,2,e) also lowered the toxicity of the drug. For example, epinephrine (**8.11**, R = H) (see Section II,A,2,b), used in the treatment of glaucoma, has a number of ocular and systemic side effects associated with its use. The prodrug dipivaloylepinephrine (**8.11**, R = $Me_3CCO$), has been shown to be more potent than epinephrine in dogs and rabbits and nearly as effective in humans[20] with a significantly improved toxicological profile compared with epinephrine.

Another example of the utility of the prodrug approach to lower the toxicity of a drug can be found in the design of aspirin (**8.25**, R = H) analogs.[33] Side effects associated with the use of aspirin are gastric irritation and bleeding. The gastric irritation and ulcerogenicity associated with aspirin use may result from an accumulation of the acid in the gastric mucosal cells. Esterification of aspirin (**8.25**, R = alkyl) and other nonsteroidal anti-inflammatory agents greatly suppresses gastric ulcerogenic activity. However, esterification also renders the acetyl ester of aspirin extremely susceptible to enzymatic hydrolysis (the $t_{1/2}$ for deacetylation of aspirin in human plasma is about 2 hrs, but that for deacetylation of aspirin esters is 1–3 min). Esters of certain N,N-disubstituted 2-hydroxyacetamides (**8.25**, R = $CH_2CONR_1R_2$) were found to be chemically highly stable but were hydrolyzed very rapidly by pseudocholinesterase (cholinesterase) in plasma[33]; therefore, they are well suited as aspirin prodrugs to lower the gastric irritation effects of aspirin.

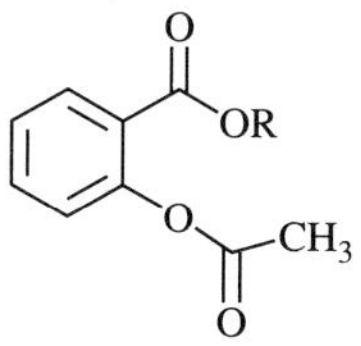

**8.25**

***g. Prodrugs to Encourage Patient Acceptance.*** A fundamental tenet in medicine is that in order for a drug to be effective, the patient has to take it. Painful injections and unpleasant taste or odor are the most common reasons for the lack of patient acceptance of a drug. An excellent example of how a prodrug can increase the potential for patient acceptance is related to the antibacterial drug clindamycin (**8.26**, R = H). Whereas clindamycin causes pain on injection, the prodrug clindamycin phosphate (**8.26**, R = $PO_3H_2$) is well tolerated; hydrolysis of the prodrug *in vivo* occurs with a $t_{1/2}$ of approximately 10 min.[34] Also, clindamycin has a bitter taste, so it is not well accepted by children who do not take pills. However, it was found that by increasing the chain length of 2-acyl esters of clindamycin the taste improved from bitter (acetate ester) to no bitter taste (palmitate ester).[35] Of course, when dealing with young children, it is not sufficient for a drug to be just tasteless; consequently, clindamycin palmitate [**8.26**, R = $CO(CH_2)_{14}CH_3$] is sold for pediatric use in a cherry-flavored syrup. Bitter taste results from a compound dissolving in the saliva and interacting with a bitter taste receptor in the mouth. Esterification with long-chain fatty acids makes the drug less water soluble

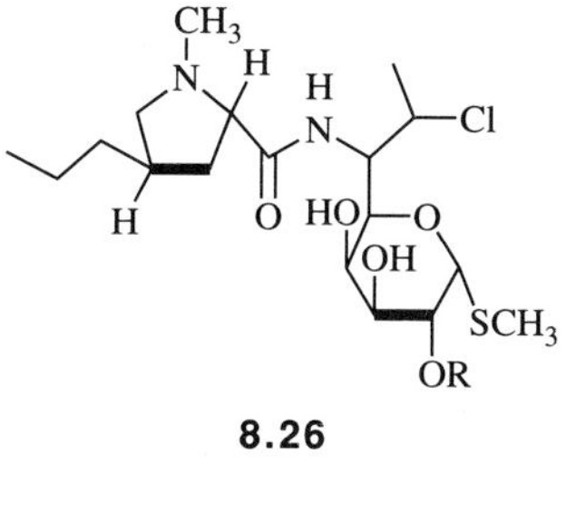

**8.26**

and unable to dissolve in the saliva. It also may alter the interaction of the compound with the taste receptor.

The antibacterial sulfa drug sulfisoxazole (**8.27**, R = H) also is bitter tasting, but sulfisoxazole acetyl (**8.27**, R = $COCH_3$) is tasteless. For pediatric use this drug is combined with the tasteless prodrug form of erythromycin, namely, erythromycin ethylsuccinate, in a strawberry–banana-flavored suspension (my 6-year-old loves it).

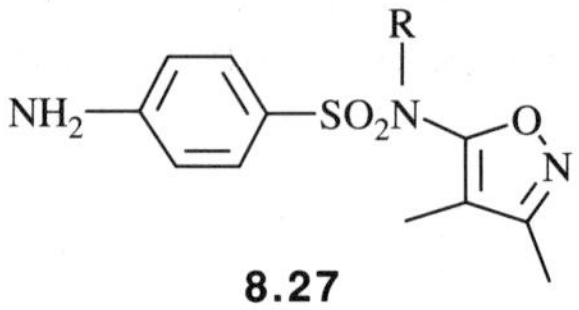

**8.27**

***h. Prodrugs to Eliminate Formulation Problems.*** Formaldehyde ($CH_2O$) is a flammable, colorless gas with a pungent odor that is used as a disinfectant. Solutions of high concentrations of formaldehyde are toxic. Consequently, it cannot be used directly in medicine. However, the reaction of formaldehyde with ammonia produces a stable adamantane-like solid compound, methenamine (**8.28**). In media of acidic pH, methenamine hydrolyzes to formaldehyde and ammonium ions. Since the pH of urine in the bladder can be made acidic, methenamine is used as a urinary tract antiseptic.[36] To prevent hydrolysis of this prodrug in the acidic environment of the stomach, the tablets are enteric coated.

The topical fungistatic prodrug triacetin (**8.29**) owes its activity to acetic acid, the product of skin esterase hydrolysis of triacetin.

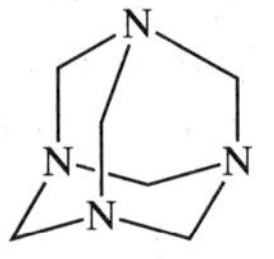

**8.28** 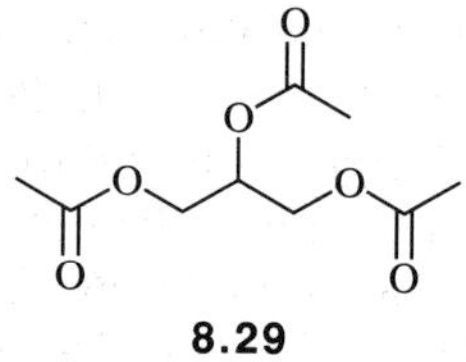 **8.29**

### 3. Macromolecular Drug Carrier Systems

***a. General Strategy.*** Although the prodrug approach has been very fruitful in general, there are three areas that need improvement: site specificity, protection of the drug from biodegradation, and minimization of side effects. Another carrier-linked bipartate prodrug approach that has been utilized to address these shortcomings is *macromolecular drug delivery*. This is a drug carrier system in which the drug is covalently attached to a macromolecule, such as a synthetic polymer, a glycoprotein, a lipoprotein, a lectin, a hormone, albumin, a liposome, DNA, dextran, an antibody, or a cell. Because the absorption and distribution of the drug depend on the physicochemical properties of the macromolecular carrier, not the drug, these parameters can be altered by manipulation of the properties of the carrier. This approach has the potential advantage of targeting drugs for a specific site and improving the therapeutic index by minimizing interactions with nontarget tissues (i.e., lowering the toxicity) as well as reducing premature drug metabolism and excretion. However, it has the disadvantages that the macromolecules may not be well absorbed after oral administration, requiring alternative means of administration, and may be immunogenic. Although polymer conjugates generally cannot pass through membranes, they can gain access to the interior of a cell by *pinocytosis*, the process by which the cell membrane invaginates the particle and then pinches itself off to form an intracellular vesicle which moves into the cell and eventually fuses with lysosomes. Because the breakdown of proteins and other macromolecules is believed to occur in the lysosomes,[37] and because this breakdown then liberates the drug, the design of a macromolecular drug carrier system should be a fruitful approach to deliver the drug inside a cell.

An ideal drug carrier (macromolecular or otherwise) must (1) protect the drug until it is at the site of action, (2) localize the drug at the site of action, (3) allow for release of the drug chemically or enzymatically, (4) minimize host toxicity, (5) be biodegradable, biochemically inert, and nonimmunogenic, (6) be easily prepared inexpensively, and (7) be chemically and biochemically stable in its dosage form. Some of the macromolecular drug carrier systems exert their effects while the drug is still attached to the carrier, but these are not prodrugs. Several examples of macromolecular drug carrier systems follow.

***b. Synthetic Polymers.*** Aspirin linked to poly(vinyl alcohol) (**8.30**) was shown to have the same potency as aspirin but was less toxic. Another anti-inflammatory agent, ibuprofen (the carboxylic acid of **8.31**) was attached as a poly(oxyethylene) diester (**8.31**).[38] This macromolecular carrier system resulted in a sustained release of ibuprofen, giving prolonged anti-inflammatory activity and a higher plasma half-life relative to the free drug.

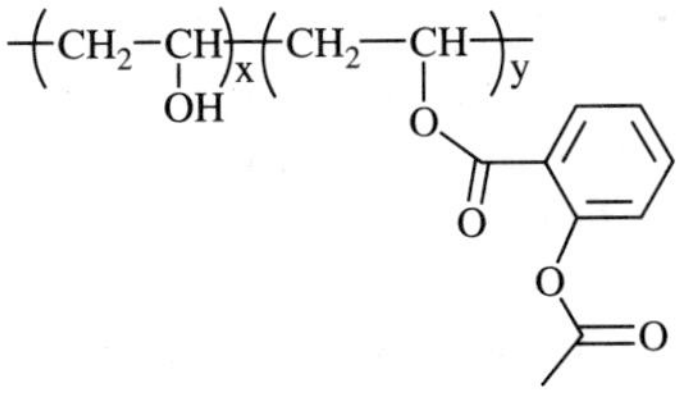

**8.30**

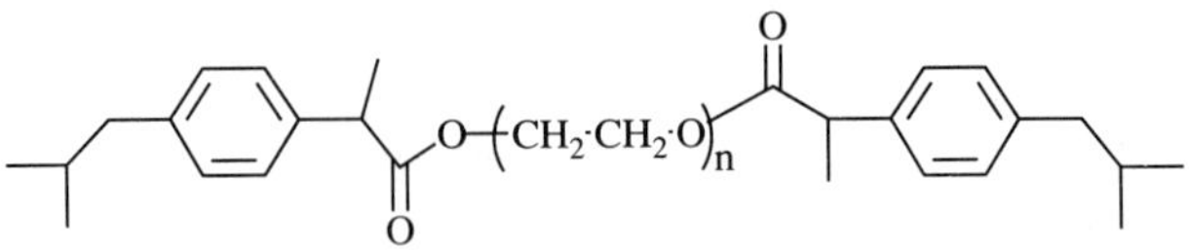

**8.31**

Because it is necessary for a drug to be released from the polymer backbone, steric hindrance by the polymer to chemical or enzymatic hydrolysis may cause problems. For example, when the steroid hormone testosterone is linked to poly(methacrylate) (**8.32**), no androgenic effect is observed, however, when a spacer arm is inserted between the polymer and the testosterone (**8.33**), the macromolecular drug carrier was as effective as testosterone. The 3-thiabutyl oxide chain was attached to the polymer to enhance water solubility.

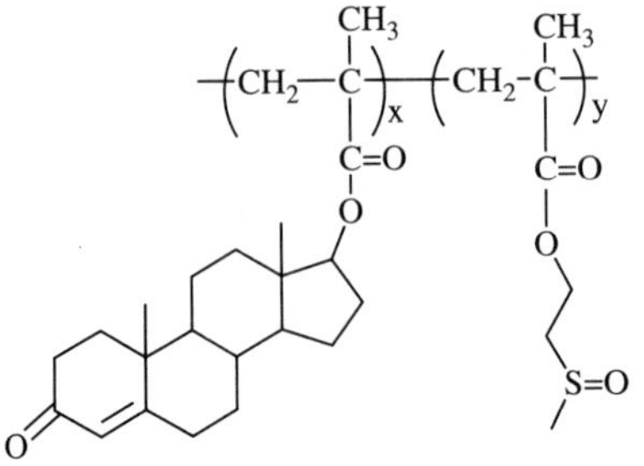

**8.32** 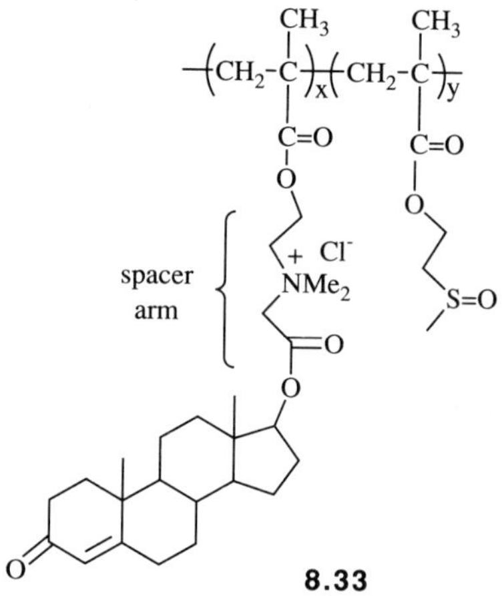 **8.33**

***c. Poly(α-Amino Acids).*** The disadvantage of using synthetic polymers is that they are generally not biodegradable and can take 5–12 months to be eliminated from the body. Poly(α-amino acids) are biodegradable (at least the L-isomers are), with the rate of biodegradability depending on the choice of amino acid.

Conjugation of the antitumor drug methotrexate to poly(L-lysine) (**8.34**; attachment of the polymer also may be to the α-carboxyl group) markedly increased the cellular uptake of the drug and provided a new way to overcome drug resistance related to deficient drug transport.[39] As the activity of methotrexate is a function of its ability to inhibit dihydrofolate reductase (see Section III,B of Chapter 4) and **8.34** is a poor inhibitor of this enzyme *in vitro*, the methotrexate must become detached from the polymer backbone inside the cell. Furthermore, attachment of methotrexate to poly(D-lysine), which, unlike poly(L-lysine) does not undergo proteolytic digestion inside the cell, gave a conjugate devoid of activity with resistant or normal cell lines. Methotrexate attached to poly(L-lysine) also is more inhibitory to the growth of human solid tumor cell lines than to the growth of human lymphocytes; free methotrexate is equally toxic to both kinds of cells.[40]

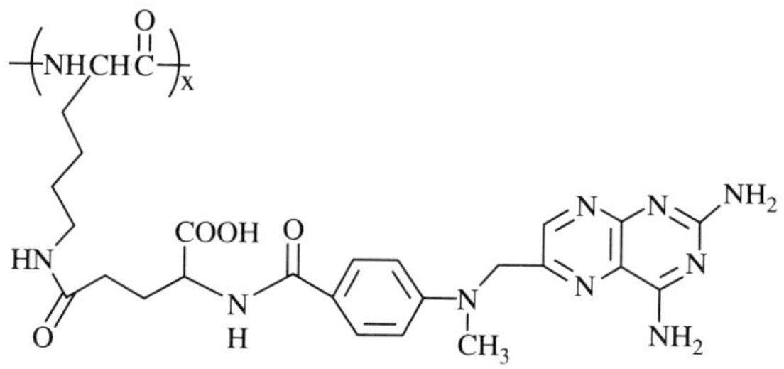

**8.34**

Research directed at a sustained release contraceptive resulted in the macromolecular drug delivery system **8.35**.[41] The contraceptive norethindrone was attached via a 17-carbonate linkage to poly-$N^5$-(3-hydroxypropyl)-L-glutamine. In rats the contraceptive agent was slowly released over a 9-month period.

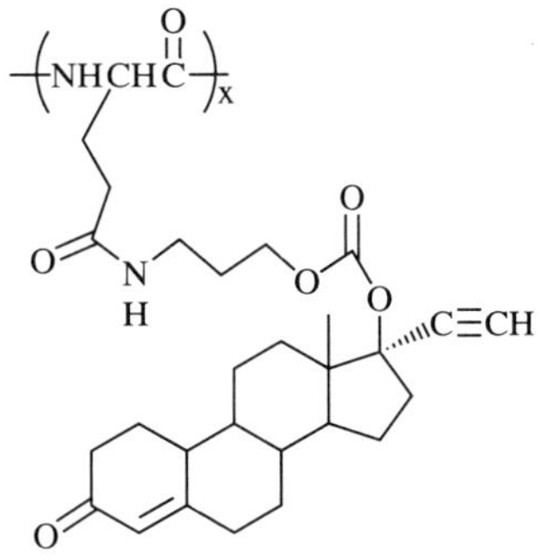

**8.35**

A general scheme for the design of a site-specific macromolecular drug delivery system was described by Ringsdorf[42] (**8.36**). A drug is attached to the polymer backbone, usually through a spacer so that it can be cleaved hydrolytically or enzymatically without steric hindrance. The desired solubility of the drug–polymer conjugate can be adjusted by attachment of an appropriate hydrophilic or hydrophobic ligand. Finally, site specificity, for example, to a particular cancer cell line, can be manipulated by attachment of a "homing device" such as an antibody raised against that cell line.

An elegant example of this approach in which a nitrogen mustard was delivered to tumor cells is shown in **8.37**.[43] Poly(L-glutamate) was used as the

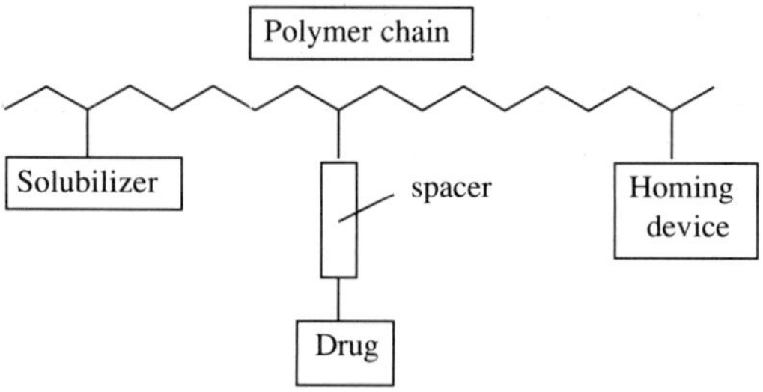

Polymer chain

Solubilizer

spacer

Homing device

Drug

**8.36**

polymeric backbone so that the side-chain carboxylic acid groups could be functionalized appropriately. The water-solubilizing groups were the unsubstituted glutamate side-chain carboxylate groups, the antitumor alkylating agent (the *p*-phenylenediamine mustard) was attached to the built-in spacer arm, that is, to the glutamate side chain, and the homing device was an immunoglobulin (Ig) derived from a rabbit antiserum against mouse lymphoma cells. This macromolecular drug delivery system was much more effective than the individual components or a mixture of the components. Whereas none of the five control mice was alive and tumor free after 60 days, all five of the polymer prodrug-treated mice were. Also, the therapeutic index of *p*-phenylenediamine mustard is greatly enhanced (40-fold) when it is attached to the polymer system, because it is less toxic to normal proliferating cells. Similar results were obtained when the neutral and water-soluble polymer dextran was used.[44]

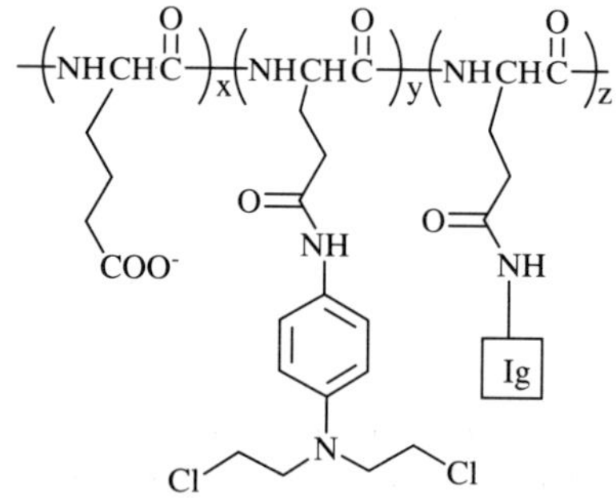

**8.37**

***d. Other Macromolecular Supports.*** Because inhibitors of DNA synthesis generally are toxic to normal rapidly proliferating cells as well, a targeted macromolecular approach to the delivery of the antitumor agents floxuridine (**8.38**, R = H) and cytosine arabinoside (cytarabine; **8.39**, R = H) was taken to decrease their toxicity.[45] The drugs were conjugated to albumin because once proteins enter cells, they are rapidly broken down by lysosomal enzymes, and this would release the drugs from the albumin inside the cells. As certain neoplastic proliferating cells are highly endocytic (high protein uptake) and normal cells with high protein uptake do not proliferate, selective toxicity to neoplastic cells or to DNA viruses that replicate in cells with high protein uptake could be accomplished. Both conjugates (**8.38** and **8.39**, R = albumin–CO) were shown to inhibit the growth of *Ectromelia* virus in mouse liver, whereas the free inhibitors were ineffective. The conjugates exert their antiviral activity in liver macrophages (cells with high protein uptake), suggesting that the drugs are concentrated in these cells.

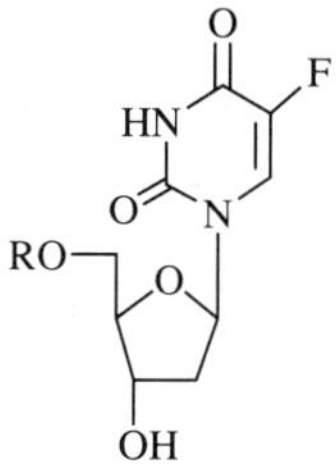

**8.38** 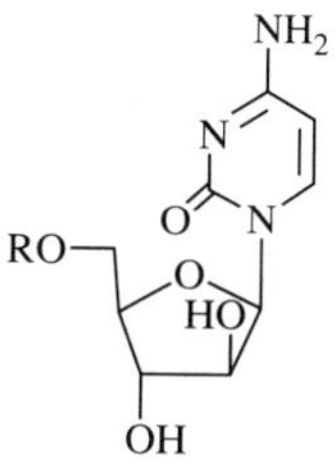 **8.39**

A clever strategy for the delivery of cytotoxic agents to solid tumors was devised.[46] Monoclonal antibodies raised against tumor cells are used as carriers of enzymes that are capable of converting nontoxic prodrugs of antitumor agents to active drugs. The prodrugs are converted to drugs at the desired site of action where they then can penetrate into the tumor cell and destroy it. Alkaline phosphatase (which catalyzes the hydrolysis of phosphate esters) conjugated with a monoclonal antibody to human carcinomas was injected into mice, then phosphates of etoposide (**8.40**, R = $PO_3^{2-}$) and other antitumor agents were administered. Pronounced antitumor activities were observed in animals treated with the antibody–enzyme conjugate prior to the phosphates. Similar results were obtained when carboxypeptidases were attached to antibodies and injected into mice prior to administration of antitumor agent pep-

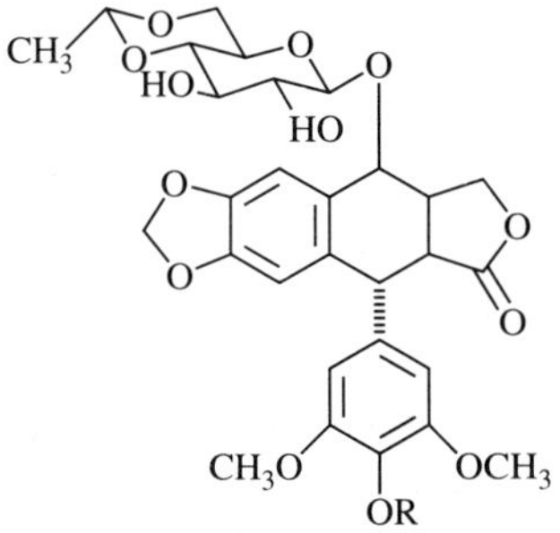

**8.40**

tides or when penicillin V amidase–antibody conjugates were used with amide prodrugs of antitumor agents.

## 4. Tripartate Prodrugs

Bipartate prodrugs may be ineffective because the prodrug linkage is too labile (e.g., certain esters) or too stable (because of steric hindrance to hydrolysis). In a tripartate prodrug the carrier is not connected directly to the drug, but rather to a linker arm which is attached to the drug (Scheme 8.6).[47] This allows for different kinds of functional groups to be incorporated for varying stabilities, and it also displaces the drug farther from the hydrolysis site, which decreases the steric interference by the carrier [as was suggested by Ringsdorf[42] for macromolecular drug delivery systems (see **8.36**)]. The drug–linker connection, however, must be designed so that it cleaves spontaneously after the carrier has been detached. One approach to accomplish this has been termed the *double prodrug* or, in the case where X is COO, the *double ester* concept, generalized in Scheme 8.7[8] (X = COO, O, NH; the double ester strategy was shown earlier in Scheme 8.5).

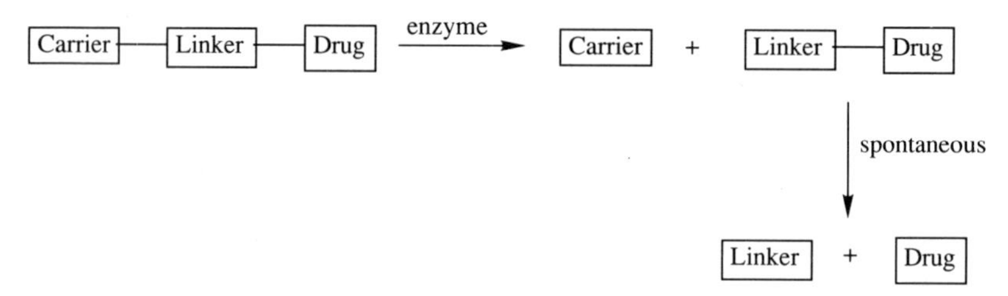


**Scheme 8.6.** Tripartate prodrugs.

$$\text{Drug—X—CH}_2\text{—O—}\overset{\text{O}}{\overset{\|}{\text{C}}}\text{R} \xrightarrow{\text{esterase}} \text{Drug—X—CH}_2\text{—O}^- + \text{RCOOH}$$

$$\downarrow \text{fast}$$

$$\text{Drug—X}^- + \text{CH}_2\text{O}$$

**Scheme 8.7.** Double prodrug concept.

This strategy was employed in the design of prodrugs of ampicillin (**8.41**), a $\beta$-lactam antibiotic that is poorly absorbed when administered orally. As only 40% of the drug is absorbed, 2.5 times more must be administered orally than by injection. Furthermore, the nonabsorbed antibiotic may destroy important intestinal bacteria. A lipid-soluble prodrug of ampicillin would be a useful approach to increase absorption of the drug. However, although various simple alkyl and aryl esters of the thiazolidine carboxyl group are hydrolyzed

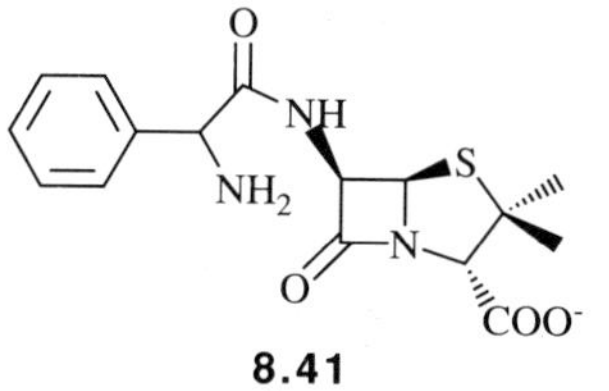

**8.41**

rapidly to ampicillin in rodents, they are too stable in man to be therapeutically useful. This suggests that the esterases in rodents and man are different and that, most likely, steric hindrance of the ester carbonyl by the thiazolidine ring is important in the human esterase. A solution to the problem was the construction of a "double ester," an acyloxymethyl ester[48] such as **8.42** (R = $CH_3$, R′ = OEt; bacampicillin)[49] or **8.42** (R = H, R′ = *tert*-Bu; pivampicillin)[50] (Scheme 8.8), which would extend the terminal ester carbonyl away from the thiazolidine ring and eliminate the inherent steric hindrance with the enzyme. Hydrolysis of the terminal ester (or carbonate, in the case of bacampicillin) gives an unstable hydroxymethyl ester (**8.43**) which spontaneously decomposes to ampicillin and either acetaldehyde (bacampacillin) or formaldehyde (pivampicillin). Bacampicillin is a nontoxic prodrug because it decomposes to ampicillin and compounds which are all natural metabolites in the body, namely, $CO_2$, acetaldehyde, and ethanol (as the usual recommended dose of bacampicillin is 400 mg twice a day, only about 50 $\mu$l of ethanol would be released with each dose, so do not expect to get high). Unlike ampicillin, bacampicillin is absorbed to the extent of 98–99%, and ampicillin is liberated into the bloodstream in less than 15 min. Because of the excellent absorption properties of bacampicillin, only one-half to one-third of the ampicillin dose is required orally.

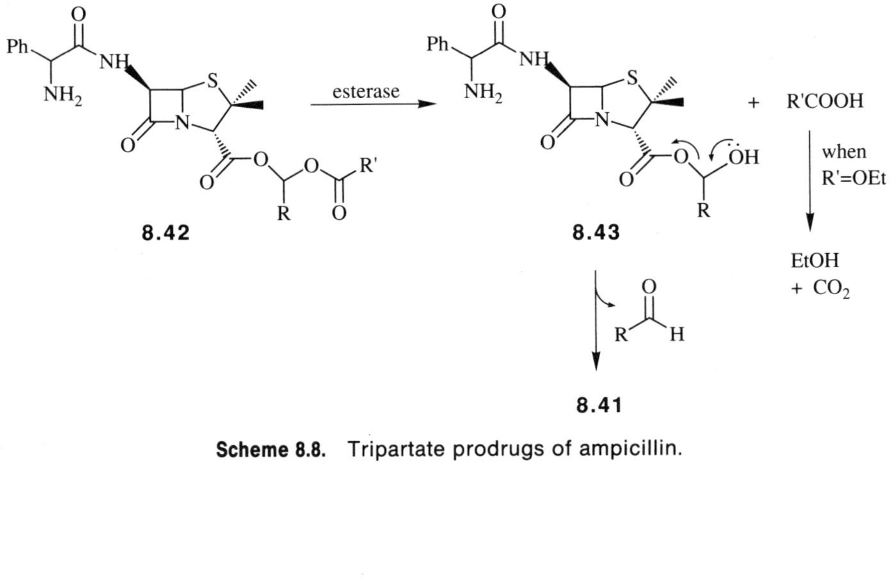

**8.42** —esterase→ **8.43** + R'COOH (when R'=OEt → EtOH + $CO_2$)

**8.43** → **8.41** + RCHO

**Scheme 8.8.** Tripartate prodrugs of ampicillin.

The antitumor agent 5-fluorouracil (**8.44**, R = H) has also been used in the treatment of certain skin diseases. However, because of its low lipophilicity, it does not produce optimal topical bioavailability. *N*-1-Acyloxymethyl derivatives (**8.44**, R = $CH_2OCOR'$) were prepared for increased lipophilicity. The prodrugs were shown to penetrate the skin about 5 times faster than **8.44** (R = H) and to be metabolized to **8.44** (R = H) rapidly.[51] The mechanism for conversion of **8.44** (R = $CH_2OCOR'$) to **8.44** (R = H) is the same as that shown in Scheme 8.8 for ampicillin derivatives.

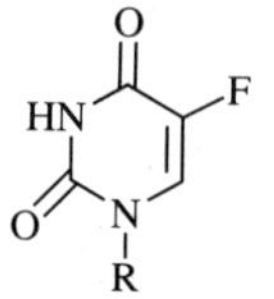

**8.44**

Microorganisms have specialized transport systems for the uptake of peptides (*permeases*), and these transport systems generally have little side-chain specificity. Consequently, peptidyl derivatives of 5-fluorouracil (**8.45**) were designed as potential antifungal and antibacterial agents that would be substrates for both microbial permeases and peptidases.[52] In accord with the known stereochemical selectivity of peptide permeases, only the peptidyl prodrug with the L,L-configuration was active. The mechanism for release of 5-fluorouracil after peptidase action is shown in Scheme 8.9.

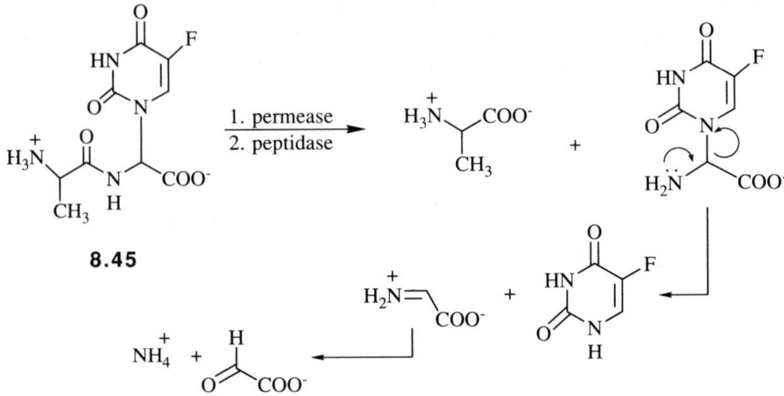

1. permease
2. peptidase

**8.45**

**Scheme 8.9.** Activation of peptidyl derivatives of 5-fluorouracil.

## 5. Mutual Prodrugs

When it is necessary for two synergistic drugs to be at the same site at the same time, a mutual prodrug approach should be considered. A *mutual pro-*

*drug* is a bipartate or tripartate prodrug in which the carrier is a synergistic drug of the drug to which it is linked. In Chapter 5 (Sections V,B,2,a and V,C,3,g) a form of resistance to β-lactam antibacterial drugs was discussed in which these bacteria have a high concentration of the enzyme β-lactamase. For resistant bacteria, compounds that inhibit β-lactamase are given in combination with a β-lactam antibacterial drug. For example, the combination of the penicillin derivative amoxicillin (**8.46**) and the β-lactamase inactivator potassium clavulanate (**8.47**) is used for oral treatment of infections caused by β-lactamase–producing bacteria. Another combination used is the ampicillin prodrug pivampicillin (**8.42**, R = H; R′ = *tert*-Bu) plus the double ester (**8.48**, R = $CH_2OCOCMe_3$) of the β-lactamase inactivator penicillanic acid sulfone (**8.48**, R = H). However, if the two prodrugs are given separately, it is not clear that they are absorbed and transported to the site of action at the same time and in equivalent amounts. An example of a tripartate mutual prodrug is sultamacillin (**8.49**), which upon hydrolysis by an esterase produces ampicillin, penicillanic acid sulfone, and formaldehyde in a reaction like that shown in Scheme 8.8.[53] A mutual prodrug would have a high probability of success provided it is well absorbed, both components are released concomitantly and quantitatively after absorption, the maximal effect of the combination of the two drugs occurs at a 1:1 ratio, and the distribution and elimination of the two components are similar.

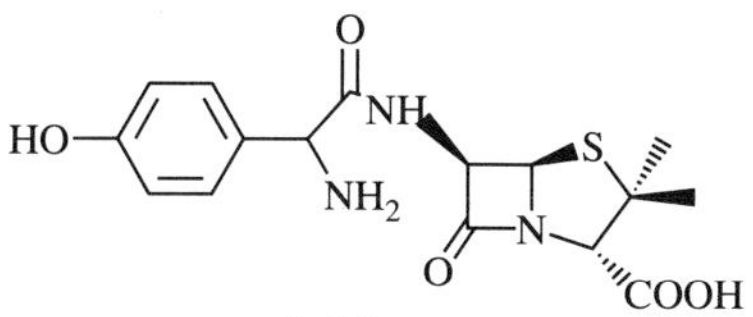

**8.46** 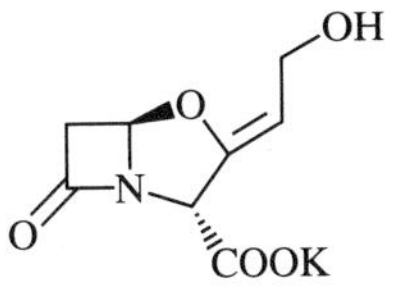 **8.47**

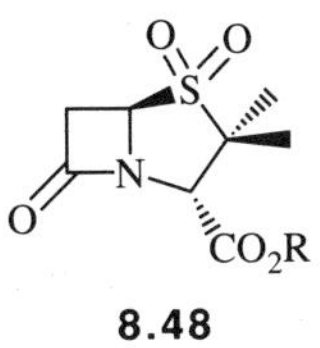

**8.48** 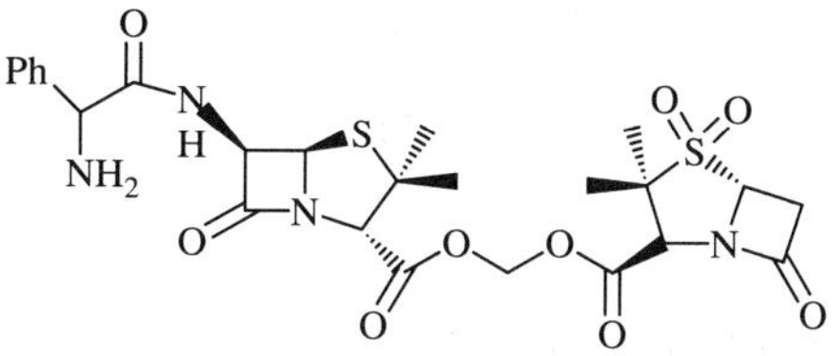 **8.49**

## *B. Bioprecursor Prodrugs*

### 1. Origins

The birth of bioprecursor prodrugs occurred when it was demonstrated that the antibacterial agent prontosil was active *in vivo* only because it was metabolized to the actual drug sulfanilamide (see Section IV,B,1 of Chapter 5). In this case the azo prodrug prontosil was reduced to the amine sulfa drug. This exemplifies the bioprecursor strategy. Whereas carrier-linked prodrugs rely largely on hydrolysis reactions for their effectiveness, bioprecursor prodrugs mostly utilize either oxidative or reductive activation reactions. The examples given below are arranged according to the type of metabolic activation reaction involved.

## 2. Oxidative Activation

***a. N- and O-Dealkylations.*** Open-ring analogs of benzodiazepines, such as the anxiolytic drug alprazolam (**8.50**, X = H) and the sedative triazolam (**8.50**, X = Cl), undergo metabolic N-dealkylation and spontaneous cyclization (Scheme 8.10).[54]

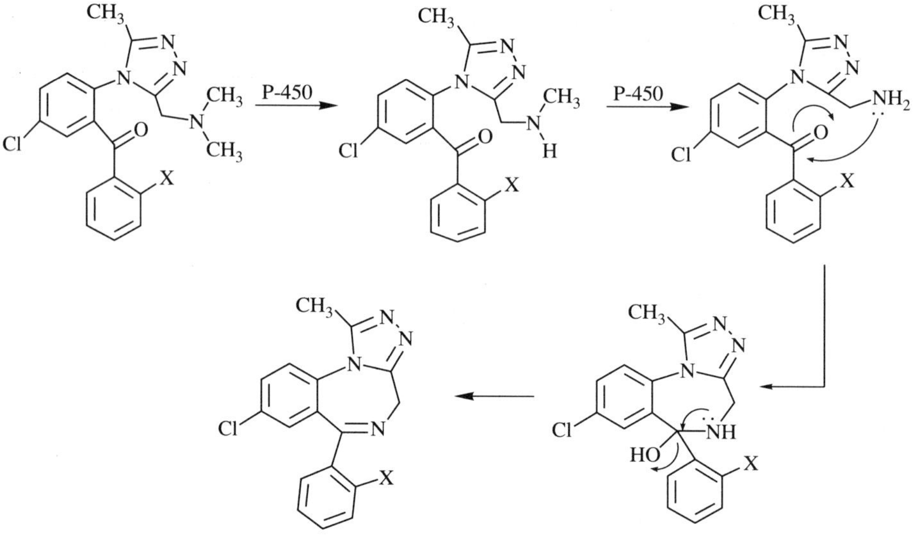

**8.50**

**Scheme 8.10.** Bioprecursor prodrugs for alprazolam and triazolam.

The triazene antitumor drugs are also activated by N-dealkylation.[55] One important class of analogs is the triazenoimidazoles, such as 5-(3,3-dimethyl-1-triazenyl)-1*H*-imidazole-4-carboxamide (**8.51**, dacarbazine) which is active against a broad range of cancers but is used preferably for the treatment of melanotic melanoma.[56] Although dacarbazine is a structural analog of 5-aminoimidazole-4-carboxamide, an intermediate in purine biosynthesis, the cytotoxicity of **8.51** is a result of its conversion to an alkylating agent, not its structural similarity to the metabolic intermediate. With the use of [*methyl-*

$^{14}$C]dacarbazine, it was shown that formaldehyde is generated and that the DNA becomes methylated at the 7 position of guanine.[57] A mechanism that rationalizes these results is shown in Scheme 8.11.

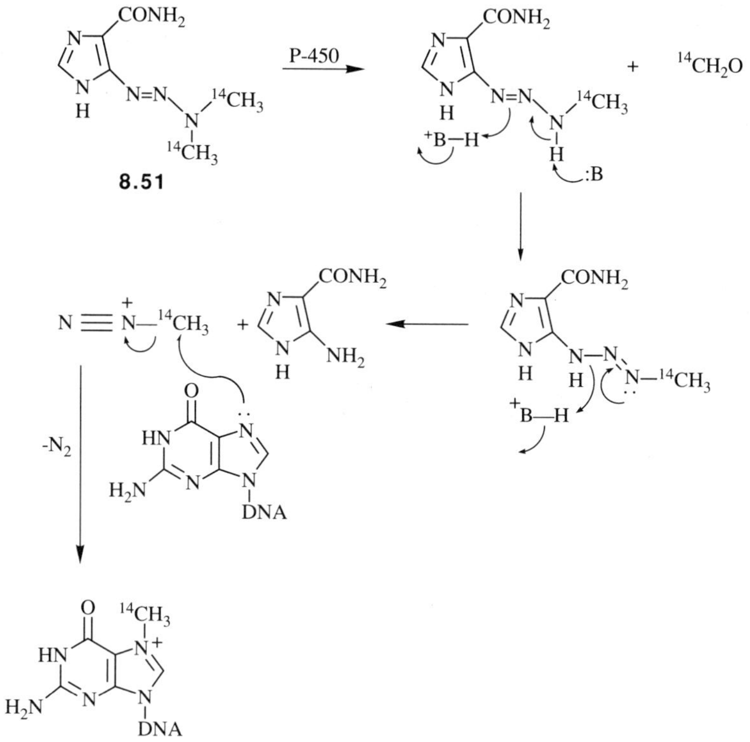

**Scheme 8.11.** Methylation of DNA by dacarbazine.

An example of a bioprecursor prodrug that is activated by O-dealkylation is the analgesic and antipyretic agent phenacetin (**8.52**, R = $CH_2CH_3$). Phenacetin owes its activity to its conversion by O-dealkylative metabolism to acetaminophen (**8.52**, R = H).[58]

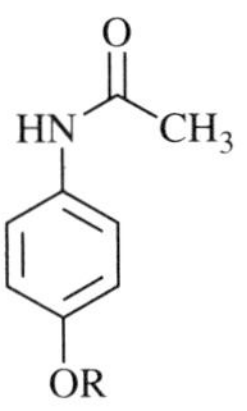

***b. Oxidative Deamination.*** Because of the high concentration of phosphoramidases in neoplastic cells, hundreds of phosphamide analogs of nitrogen mustards were synthesized and tested as carrier-linked antitumor prodrugs. Cyclophosphamide (**8.53**, Scheme 8.12) emerged as an important drug for the treatment of a wide variety of malignant diseases; however, it was later found that it was inactive in tissue culture. Preincubation of the compound with liver homogenates activated it, suggesting that cyclophosphamide is a prodrug requiring an oxidative mechanism (see Section IV,B,1 of Chapter 7).[59a–c] The activation mechanism is believed to be that shown in Scheme 8.12 (there are other metabolites that are not shown in Scheme 8.12 derived from each of the intermediates). It is not clear which of the toxic metabolites, the phosphoramide mustard (**8.56**) or the parent nitrogen mustard (**8.58**), is responsible for the therapeutic action; the major adduct isolated by high-performance liquid chromatography (HPLC) from *in vitro* and *in vivo* studies in rat is *N*-(2-hydroxyethyl)-*N*-[2-(7-guaninyl)ethyl]amine (**8.58b**),[59d] the hydrolysis product of **8.58a**. The reaction of nitrogen mustards with DNA was discussed in Chapter 6 (Section III,B,1). Acrolein (**8.57**) is a potent Michael acceptor that may be responsible for the hemorrhagic cystitis side effect; administra-

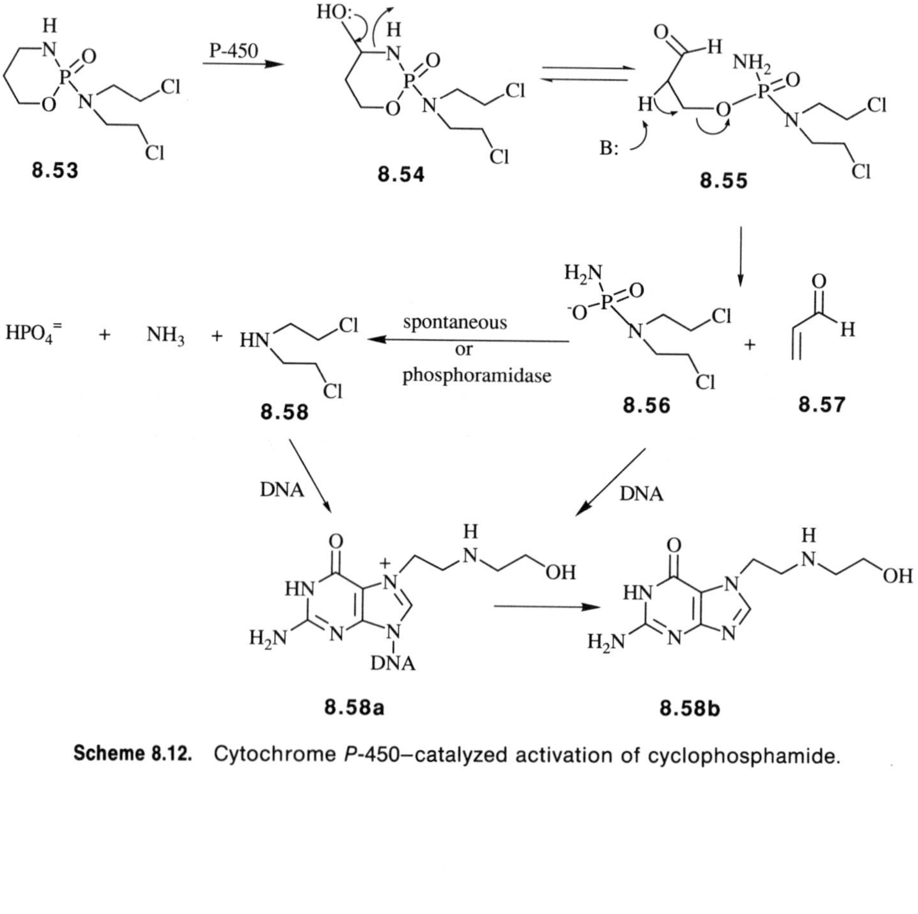

8.53 —P-450→ 8.54 ⇌ 8.55

8.55 → 8.56 + 8.57

$HPO_4^{=}$ + $NH_3$ + 8.58 ←spontaneous or phosphoramidase— 8.56

8.58 —DNA→ 8.58a → 8.58b ←DNA— 8.56

**Scheme 8.12.** Cytochrome *P*-450–catalyzed activation of cyclophosphamide.

tion of sulfhydryl compounds, which react readily with acrolein, can prevent this side effect. Aldehyde dehydrogenase catalyzes the oxidation of **8.54** to the corresponding cyclic amide and the oxidation of **8.55** to the corresponding carboxylic acid; however, both of these metabolites are inactive. It has been suggested that these detoxification reactions occur to a greater extent in normal cells than in cancer cells, which may account for the selective toxicity of cyclophosphamide.[60]

***c. N-Oxidation.*** The antitumor drug used against advanced Hodgkin's disease, procarbazine (**8.59**), is believed to be activated by N-oxidation (Scheme 8.13); it is inert unless treated with liver homogenates or oxidized in neutral solution.[61] Those who have studied Chapter 7 are probably wondering why this circuitous mechanism to **8.61** and methylhydrazine, starting with an N-oxidation reaction, was written instead of a direct conversion of **8.59** to these same metabolites by an oxidative deamination mechanism. The reason is that azoprocarbazine (**8.60**) was identified as the initial metabolic product.[62] 7-Methylguanine (**8.62**) was identified in the urine of mice given procarbazine,[63] which suggests that an activated methylating agent such as methyl diazonium or methyl radical[64] is the reactive intermediate.

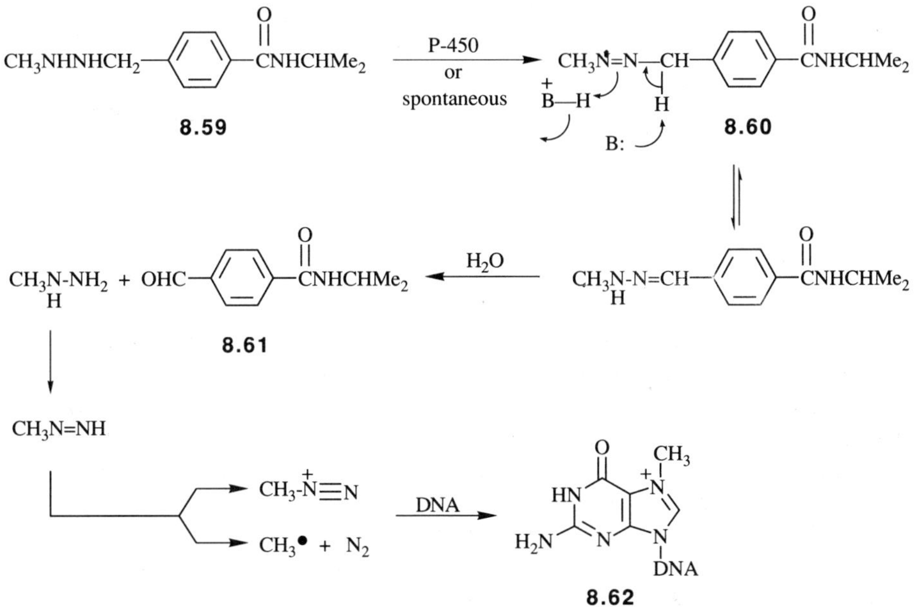

**Scheme 8.13.** Activation of procarbazine.

Another N-oxidation prodrug activation reaction is the reversible redox drug delivery strategy of Bodor and co-workers for getting drugs into the brain[22,23] (see Section II,A,2,c). In the case of pralidoxime chloride (**8.63**), an antidote for poisoning by organophosphorus pesticides and nerve toxins, the oxidation reaction converts the prodrug to the drug as well as prevents efflux of the drug from the brain. The neurotoxic organophosphorus compounds exert their effects by reacting with acetylcholinesterase, the enzyme found in nervous tissue of all species of animals that catalyzes the hydrolysis of the neurotransmitter acetylcholine after it has served its neurohumoral transmission function. The active site of the enzyme is believed to contain two important binding sites, the anionic site that binds the quaternary ammonium cation of acetylcholine and the ester site where the catalytic hydrolysis of the acetyl group occurs (Scheme 8.14).[65]

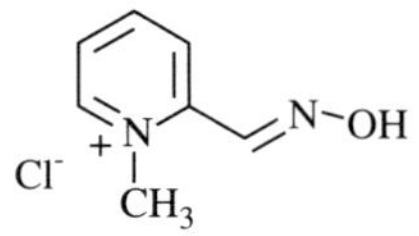


**8.63**

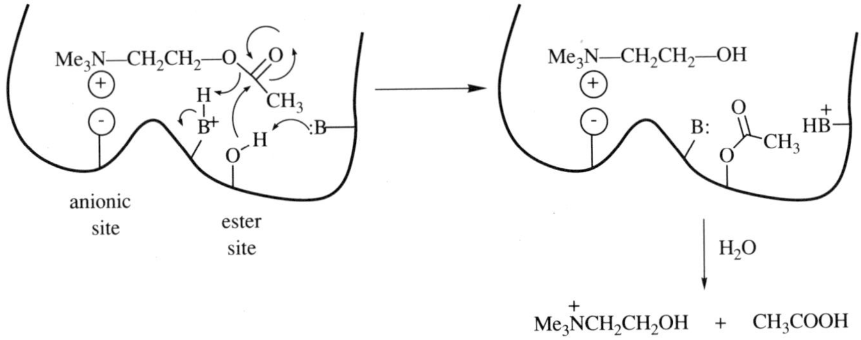


**Scheme 8.14.** Acetylcholinesterase-catalyzed hydrolysis of acetylcholine.

Organophosphorus compounds, such as the nerve poison diisopropyl phosphorofluoridate (**8.64**), phosphorylate acetylcholinesterase at the ester site[66] (Scheme 8.15). It was thought that a nucleophilic agent may be capable of dephosphorylating the ester site and reactivating the enzyme. Hydroxylamine appeared to be effective but also was quite toxic. Because acetylcholinesterase has an anionic binding site, quaternary amine analogs were designed, and 2-formyl-1-methylpyridinium chloride oxime (pralidoxime chloride, **8.63**) was found to be an effective reactivator of the enzyme (Scheme 8.16). However, **8.63** is very poorly soluble in lipids, so its generation is most likely restricted to the peripheral nervous system; little reactivation of brain acetylcholinesterase was observed *in vivo*.[67] Apparently, the effectiveness of **8.63** as an antidote for organophosphorus nerve poisons results from the fact that the primary damage done by these poisons is to the peripheral nervous system. To

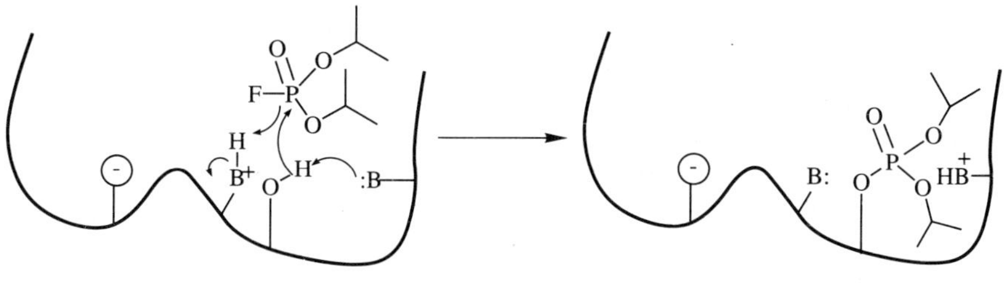

**8.64**

**Scheme 8.15.** Phosphorylation of acetylcholinesterase by diisopropyl phosphorofluoridate.

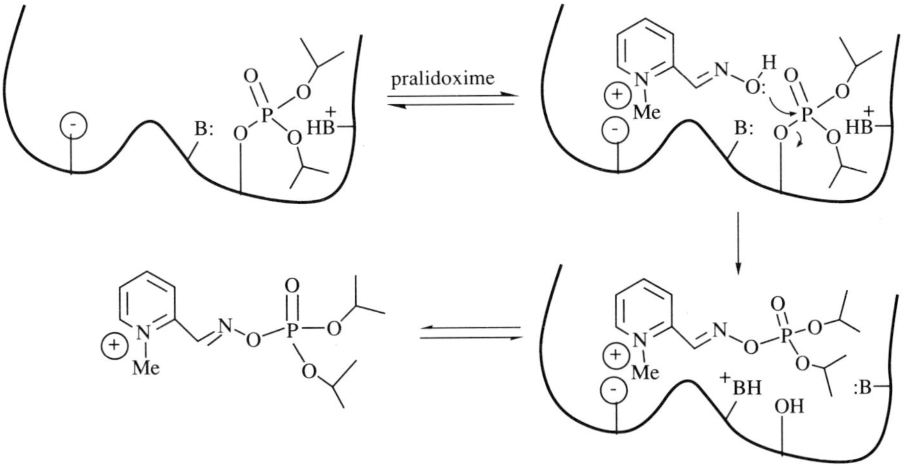

**Scheme 8.16.** Reactivation of phosphorylated acetylcholinesterase by pralidoxime chloride.

improve the permeability of **8.63** into the central nervous system Bodor and co-workers[22,68] prepared the 5,6-dihydropyridine analog **8.65**. As **8.65** is uncharged, its permeability through the blood–brain barrier was quite good. Once inside the brain it was oxidized to **8.63**.

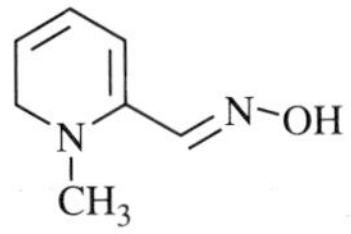

**8.65**

It is interesting to note that whereas irreversible inactivators of acetylcholinesterase, such as organophosphorus nerve gases, are highly toxic, compounds that form weakly stable covalent bonds to the serine residue in the ester binding site are useful therapeutic agents. This inhibition of acetylcholinesterase results in the enhancement of cholinergic action by facilitating the transmission of impulses across neuromuscular junctions, which has a cholinomimetic effect on skeletal muscle. An example of this is neostigmine (**8.66**), a drug used in the treatment of the neuromuscular disease myasthenia gravis. Neostigmine carbamylates the active site serine residue of acetylcholinesterase; the carbamate, however, hydrolyzes slowly so that, in effect, **8.66** acts as a reversible inhibitor of the enzyme (Scheme 8.17). Therefore, the difference in effects of the acetylcholinesterase substrates and inhibitors is derived from the stabilities of the covalent adducts. The acetylated serine formed from acetylcholine (a substrate) hydrolyzes readily, the carbamylated serine produced from neostigmine (an inhibitor) hydrolyzes slowly, and the phosphorylated serine from organophosphorus compounds (inactivators) is stable to hydrolysis.

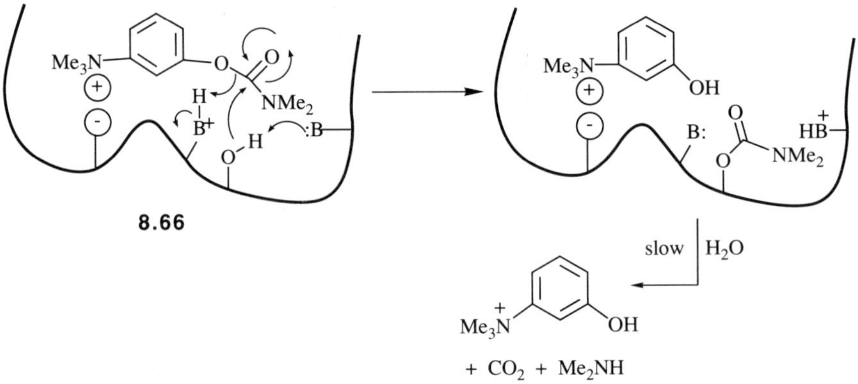

**8.66**

slow $H_2O$

$+ CO_2 + Me_2NH$

**Scheme 8.17.** Carbamylation of acetylcholinesterase by neostigmine.

***d. Other Oxidations.*** Carbamazepine (**8.67**) is an anticonvulsant drug that is the metabolic precursor of the active agent, carbamazepine 10,11-oxide (**8.68**).[69] Cysteine conjugates of the antitumor agent elliptinium acetate (**8.69**) found in the urine of patients taking the drug indicate that an oxidative mechanism is important (Scheme 8.18) and suggest that alkylation, as well as intercalation, of DNA by **8.69** may be a viable mechanism of action.[70]

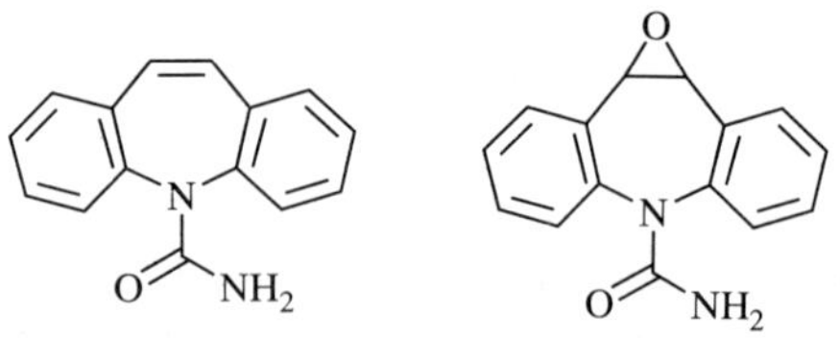

**8.67** **8.68**

Stimulation of pyruvate dehydrogenase results in a change of myocardial metabolism from fatty acid to glucose utilization. Because the latter requires

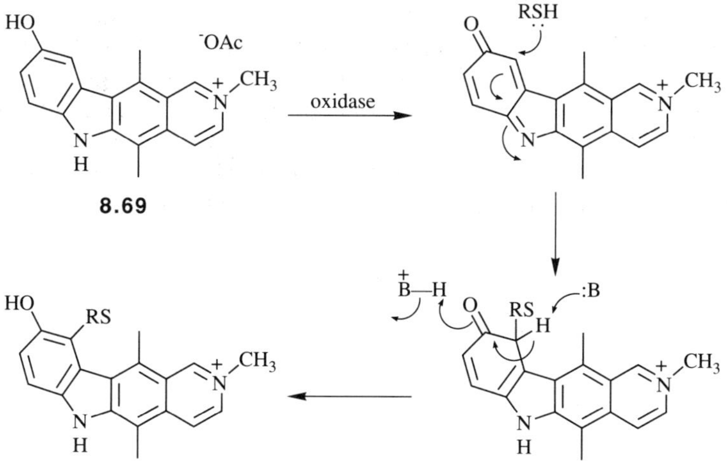

Scheme 8.18. Activation of elliptinium acetate.

less oxygen consumption, glucose utilization is beneficial to patients with ischemic heart disease in which arterial blood flow is blocked and therefore less oxygen is available.[71] Arylglyoxylic acids (**8.70**) are important stimulators of this enzyme, but they have short durations of action. L-(+)-2-(4-Hydroxyphenyl)glycine (oxfenicine; **8.71**, R = OH) is a stable amino acid that is actively transported across lipid membranes and is rapidly transaminated (see Section III,A,3 of Chapter 4) to 4-hydroxyphenylglyoxylic acid (**8.70**, R = OH).[72] This active transport system and rapid conversion of the prodrug to the drug allow a higher concentration of the active drug to persist at the desired site of action longer.

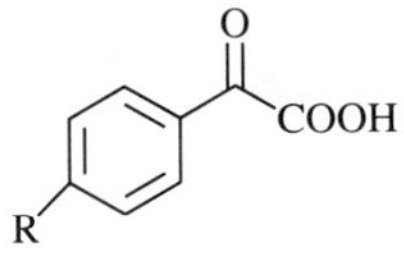

**8.70** 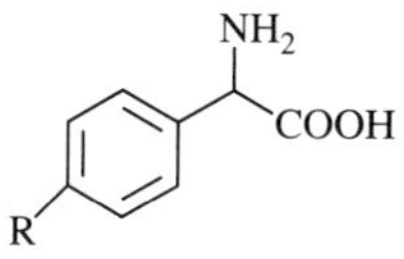 **8.71**

### 3. Reductive Activation

***a. Azo Reduction.*** As described in Section II,B,1 the paradigm for bioprecursor prodrugs, prontosil, is activated by reduction of its azo linkage to the true bacteriostatic agent, sulfanilamide.

Sulfasalazine (**8.72**), which is used in the treatment of inflammatory bowel disease (ulcerative colitis), is reductively cleaved by anaerobic bacteria in the lower bowel to 5-aminosalicylic acid (**8.73**) and sulfapyridine (**8.74**); **8.73** is the therapeutic agent, and **8.74** produces adverse side effects (Scheme 8.19).[73] A macromolecular drug delivery system was developed to improve the therapeutic index of this drug. The drug (**8.73**) was azo-linked at the 5 position through a spacer to poly(vinyl amine) (**8.75**).[74] The advantages of this polymeric drug delivery system are that it is not absorbed or metabolized in the small intestine, **8.73** can be released by reduction at the disease site, and the carrier polymer is not absorbed or metabolized. The water-soluble polymer-linked drug (**8.75**) was more active than **8.72** or **8.73** in the guinea pig ulcerative colitis model.

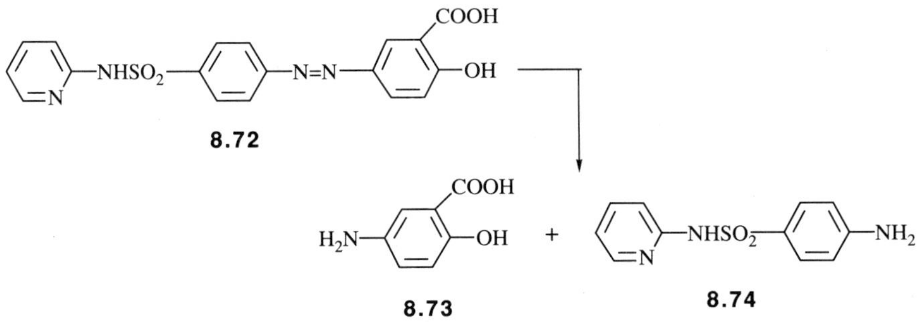

COOH
NHSO$_2$ N=N OH
N
**8.72**
COOH
$H_2N$ OH + NHSO$_2$ $NH_2$
N
**8.73** **8.74**

**Scheme 8.19.** Reductive activation of sulfasalazine.

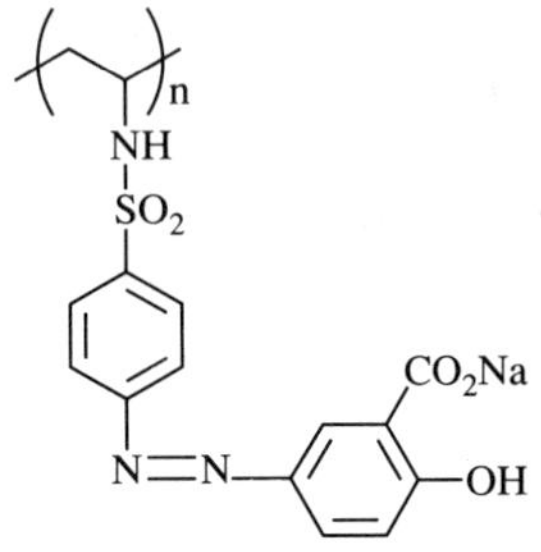

n
NH
SO$_2$
CO$_2$Na
N=N OH
**8.75**

***b. Sulfoxide Reduction.*** The antiarthritis drug sulindac (**8.76**) is an indene isostere (see Section II,D,4 of Chapter 2) of the nonsteroidal anti-inflammatory (antiarthritis) drug indomethacin (**8.77**), which originally was designed as a serotonin analog. Sulindac is less irritating to the gastrointestinal tract and produces many fewer and more mild central nervous system effects than does indomethacin.[75] The 5-fluoro group was substituted for the methoxyl group to improve the analgesic properties, and the *p*-methylsulfinyl group was substituted for the chlorine atom to increase the solubility. Sulindac is inactive *in vitro* but is highly active *in vivo*. The corresponding sulfide, however, is active *in vitro* and *in vivo*. Therefore, sulindac is a prodrug for the sulfide, the metabolic reduction product.

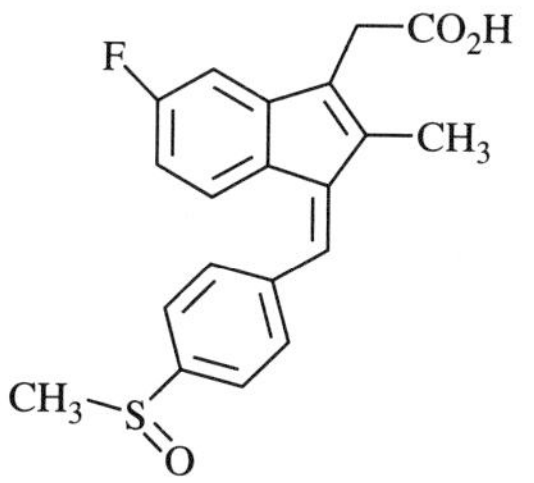

**8.76**

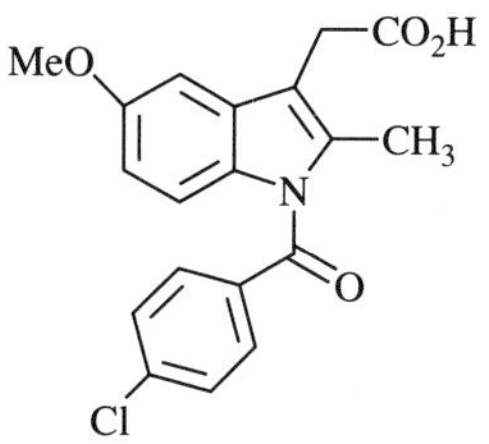

**8.77**

***c. Disulfide Reduction.*** Because thiamin (vitamin $B_1$; **8.79**) is a quaternary ammonium salt, it is poorly absorbed into the central nervous system and from the gastrointestinal tract. To increase the lipophilicity thiamin tetrahydrofurfuryl disulfide (**8.78**, Scheme 8.20) was designed as a lipid-soluble prodrug of thiamin.[76] The prodrug permeates rapidly through red blood cell membranes (as a model for other membranes) and reacts with glutathione to produce thiamin.[77]

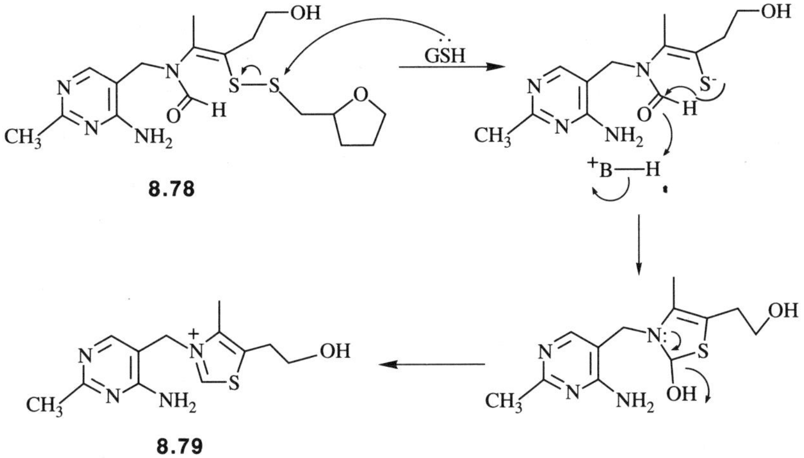

**8.78**

**8.79**

**Scheme 8.20.** Conversion of thiamin tetrahydrofurfuryl disulfide to thiamin.

To diminish the toxicity of the antimalarial drug primaquine (**8.80**) and target it for cells that contain the malaria parasite, a macromolecular drug delivery system was designed[78] (**8.81**). The lactose-linked albumin was used for improved uptake in the liver via the asialoglycoprotein receptor system. Because the concentration of free thiol in the blood is relatively low, but is high intracellularly, it was expected that thiol reduction of the disulfide linkage would occur mostly inside the cell. It is not known if after disulfide reduction the cysteinyl residue is detached by hydrolysis or remains attached to primaquine. The therapeutic index of **8.81**, however, was 12 times higher than that of the free drug in *Plasmodium*-infected mice.

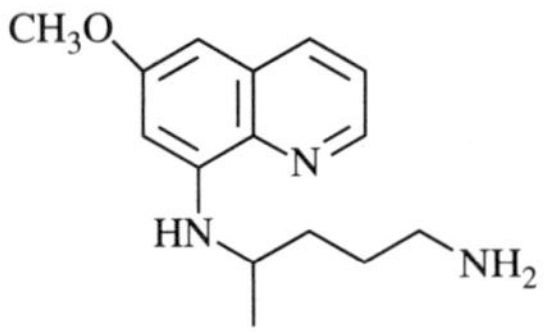

$CH_3O$ N HN $NH_2$

**8.80**

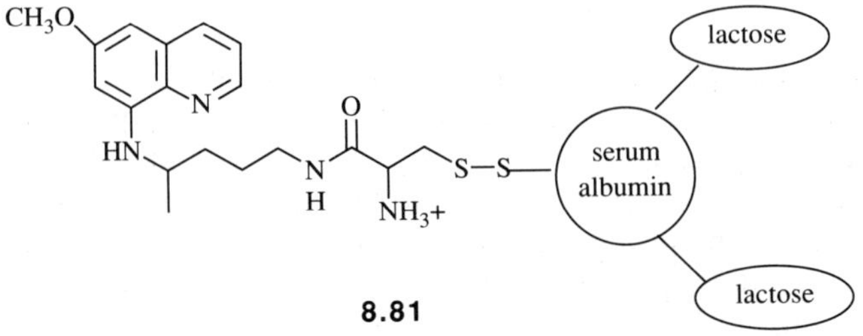

$CH_3O$ N HN N H O $NH_3$+ S—S— serum albumin lactose lactose

**8.81**

***d. Bioreductive Alkylation.*** *Bioreductive alkylation* is a prodrug strategy in which an inactive compound is metabolically reduced to an alkylating agent.[79] The prototype for antitumor antibiotics that act as bioreductive alkylating agents of DNA is mitomycin C (**8.82**, Scheme 8.21) which contains three important carcinostatic functional groups, the quinone, the aziridine, and the carbamate group.[79–81] The mechanism proposed by Iyer and Szybalski[80] as modified by Moore and Czerniak[79] is shown in Scheme 8.21. Reduction of the quinone by one electron to the semiquinone (**8.83**, R = electron) or by two electrons to the hydroquinone (**8.83**, R = H) converts the heterocyclic nitrogen from a vinylogous amide nitrogen (the nonbonded electrons of the nitrogen are in conjugation with the quinone carbonyl via the intermediate double bond), which is not nucleophilic, to an amine nitrogen, which can eliminate the $\beta$-methoxide ion (**8.83**). Tautomerization of the resultant immonium ion (**8.84**) gives **8.85**, which is set up for aziridine ring opening. This activates the drug by unmasking the electrophilic site at C-1 which alkylates the DNA (**8.86**). A subsequent reaction of DNA at C-10 (**8.87**) results in the cross-linking of the DNA (**8.88**). Bean and Kohn[82] showed in chemical models that nucleophiles react most rapidly at C-1; the reaction at C-10 to displace the carbamate also occurs, but at a slower rate. Reduction of the quinone is necessary for the covalent reaction of **8.82** to DNA, but controversy exists as to whether the semiquinone (**8.83**, R = electron) or hydroquinone (**8.83**, R = H) is the viable intermediate.[81] Chemical model studies on the mechanism of action of mitomycin C indicate that the conversion of **8.82** to **8.87** can occur at the semiquinone stage[83] and the conversion of **8.87** to **8.88** occurs at the hydroquinone oxidation state.[84] Both monoalkylated and bis-alkylated DNA adducts have been identified; the extent of mono- and bis-alkylation increases with increasing guanine base composition of the DNA.[85] The site of attachment of these adducts is at N-2 of the guanine bases[86a,b] with preferential interstrand cross-linking at 5′-CG rather than 5′-GC sequences.[86c]

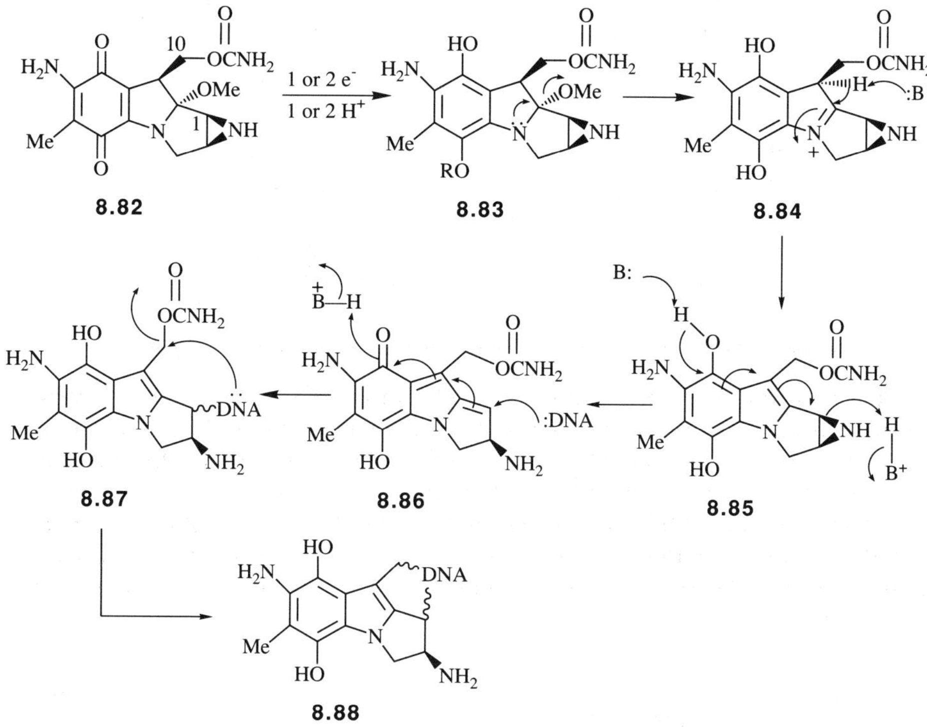

**Scheme 8.21.** Bioactivation of mitomycin C.

Several other naturally occurring antitumor quinones may be involved in this type of mechanism.[79] Anthracycline antitumor agents such as doxorubicin (**8.89**, R = OH) and daunorubicin (**8.89**, R = H) were suggested to act as one-electron bioreductive alkylating agents[87a] (Scheme 8.22); however, the hydroquinone has been synthesized, and it does not eliminate the sugar.[87b] Therefore, if the quinone methide mechanism is relevant to the anthracycline antibiotics, elimination must occur at the semiquinone oxidation state. Radical-induced reactions of anthracycline antitumor antibiotics were discussed in Section III,C,1 of Chapter 6.

The bioreductive alkylation approach was directed toward the design of new antineoplastic agents that may be selective for hypoxic ($O_2$-deficient) cells in solid tumors.[88] These cells are remote from blood vessels and are located at the center of the solid tumors. Hypoxia protects the tumor cells from radiation therapy, and because these cells are buried deep inside the

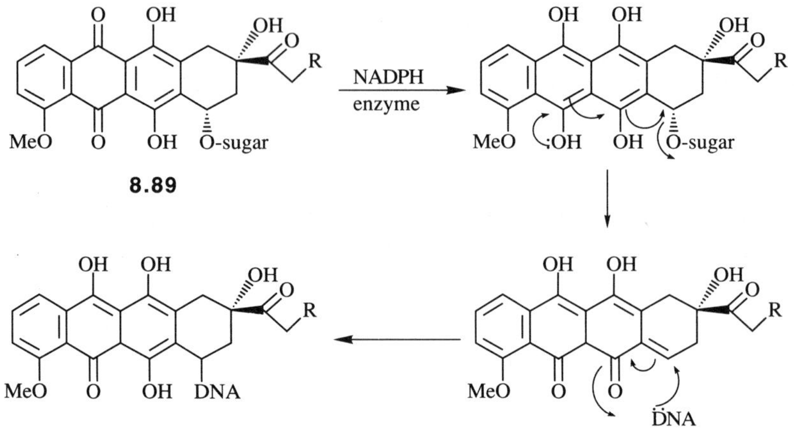

8.89

Scheme 8.22. Anthracycline antitumor agents as bioreductive alkylators.

tumor, appropriate concentrations of antitumor drugs may not reach them prior to drug metabolism. As these cells might have a more efficient reducing environment, bioreductive alkylation seemed to be well suited. The bioreductive alkylation approach based on reduction of a quinone to the corresponding hydroquinone was utilized in the design of the prodrugs. Both mono- (Scheme 8.23)[89] and bisalkylating agents (Scheme 8.24)[90] were developed. Electron-rich substituents lower the reduction potential of the quinones and make them more reactive.[91]

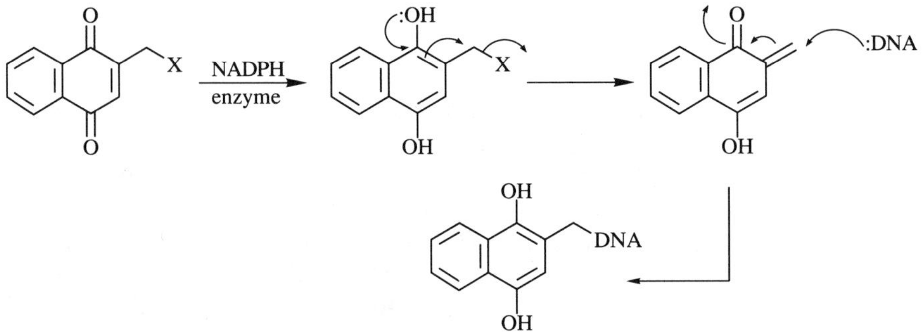

Scheme 8.23. Bioreductive monoalkylating agents.

***e. Nitro Reduction.*** The mechanism of action of the antiprotozoal agent ronidazole (**8.90**) is not known, but on the basis of metabolism studies using several radioactively labeled analogs, it was suggested that **8.90** is activated by initial four-electron reduction of the 5-nitro group to the corresponding hydroxylamine which can react with protein thiols by one of two mechanisms (Scheme 8.25).[92]

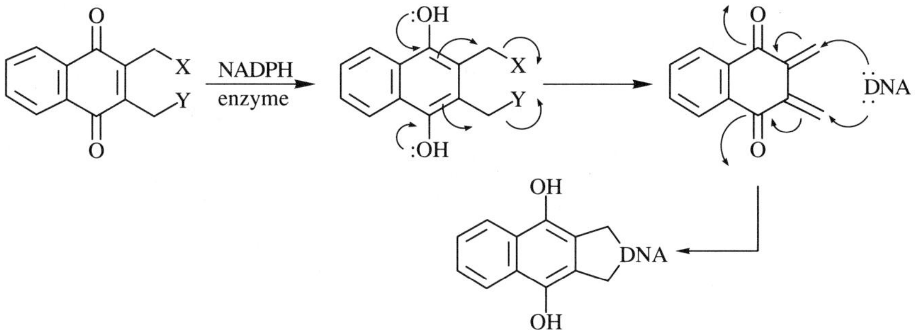

**Scheme 8.24.** Bioreductive bisalkylating agents.

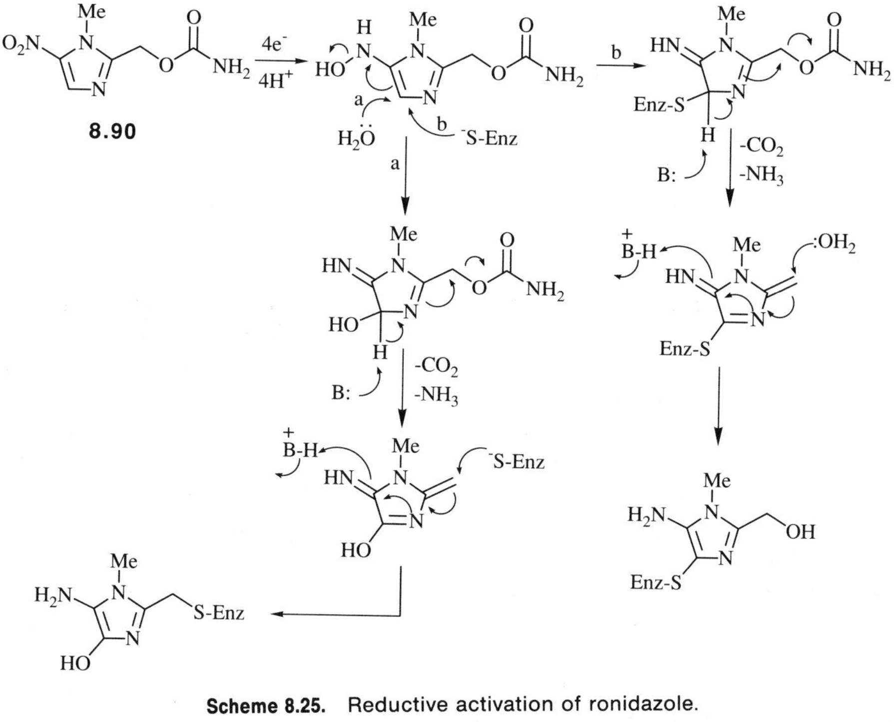

**8.90**

**Scheme 8.25.** Reductive activation of ronidazole.

## 4. Nucleotide Activation

The antineoplastic agent 6-mercaptopurine (**8.91**) produces a 50% remission rate for acute childhood leukemias. Although **8.91** inhibits several enzyme systems, these inhibitions are irrelevant to its anticancer activity. Only tumors that convert the drug to its nucleotide are affected. 6-Mercaptopurine is activated by a reaction with 5-phosphoribosylpyrophosphate, catalyzed by hypoxanthine–guanine phosphoribosyltransferase (hypoxanthine phosphoribosyltransferase) (Scheme 8.26). The nucleotide (**8.92**) inhibits several enzymes in the purine nucleotide biosynthetic pathway, but the most prominent site is one of the early enzymes in the *de novo* pathway, namely, phosphoribosylpyrophosphate amidotransferase (amidophosphoribosyltransferase), which catalyzes the conversion of phosphoribosylpyrophosphate to phosphoribosylamine.[93] 5-Fluorouracil (see Section V,C,3,e of Chapter 5) is similar to 6-mercaptopurine in the sense that it must first be converted to the corresponding deoxyribonucleotide in order for it to be active.

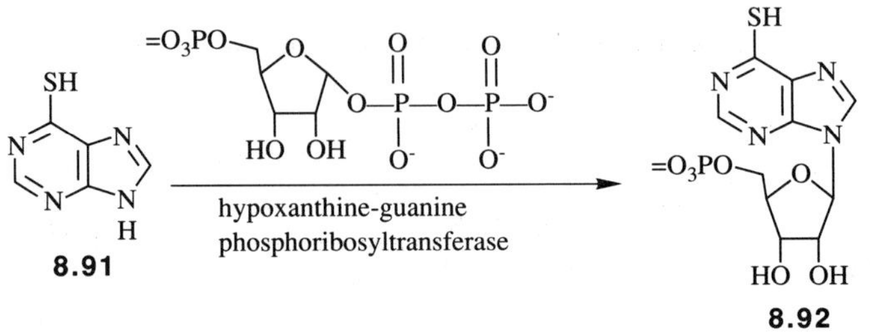

SH N N N N H **8.91** =O$_3$PO O O O O—P—O—P—O$^-$ HO OH O$^-$ O$^-$ hypoxanthine-guanine phosphoribosyltransferase → SH N N N N =O$_3$PO O HO OH **8.92**

**Scheme 8.26.** Nucleotide activation of 6-mercaptopurine.

## 5. Phosphorylation Activation

The antiviral drug acyclovir (**8.93**, R = H) is highly effective against genital herpes simplex virus and varicella zoster virus infections. Its structure can be drawn so that it closely resembles the structure of 2′-deoxyguanosine (**8.94**). Acyclovir itself is inactive, but it is selectively phosphorylated by a viral thymidine kinase to the corresponding monophosphate (**8.93**, R = $PO_3^{2-}$).[94] Uninfected cells do not phosphorylate acyclovir, and this accounts for the selective toxicity of acyclovir toward viral cells. The second step in the activation of acyclovir is the conversion of the monophosphate (**8.93**, R = $PO_3^{2-}$) to the diphosphate (**8.93**, R = $P_2O_6^{3-}$), catalyzed by guanylate kinase.[95] The final activation step is the conversion of the diphosphate to the triphosphate (**8.93**, R = $P_3O_9^{4-}$), which could be accomplished by a variety of enzymes, particularly phosphoglycerate kinase.[96] Further selective toxicity is derived from the fact that acyclovir triphosphate is selectively taken up by viral $\alpha$-DNA polymerases as its structure resembles that of the essential DNA precursor, deoxyguanosine triphosphate. The $K_i$ for viral $\alpha$-DNA polymerase is up to 40 times lower than that for normal cellular $\alpha$-DNA polymerase.[97] Acyclovir triphosphate is a substrate for the viral polymerase but not for the normal cellular polymerase; however, incorporation of acyclovir triphosphate into the viral DNA leads to the formation of a *dead-end complex* (an enzyme–substrate complex that is no longer active) after the next deoxynucleotide triphosphate unit is incorporated.[98] This disrupts the replication cycle of the virus and destroys it. Even if the phosphorylated acyclovir were released from the virus cell, it would be too polar to be taken up by normal cells, and, as indicated above, the triphosphate is a poor substrate for normal human α-DNA polymerase anyway. Therefore, this drug exhibits a high degree of selective toxicity against viral cells.

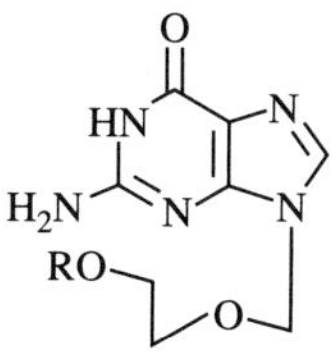

**8.93** 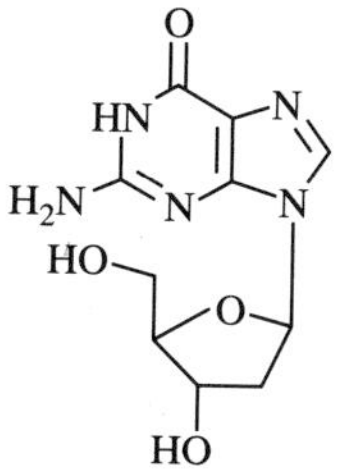 **8.94**

As might be predicted from knowledge of the mechanism of acyclovir, acquired resistance to the drug can occur by three different mechanisms. Because of the importance of the thymidine kinase to the activation of acyclovir, resistance could arise from a deletion of this enzyme[99] or a change in its substrate specificity.[100] The third mechanism could be an altered viral DNA polymerase.[99]

Ganciclovir (**8.95**) is an analog of acyclovir that has a conformation resembling the structure of 2′-deoxyguanosine even closer than does acyclovir. This compound is about as active as acyclovir against herpes simplex viruses and varicella zoster virus but is much more inhibitory than acyclovir against human cytomegalovirus,[101] an important pathogen in immunocompromised and acquired immune deficiency syndrome (AIDS) patients.

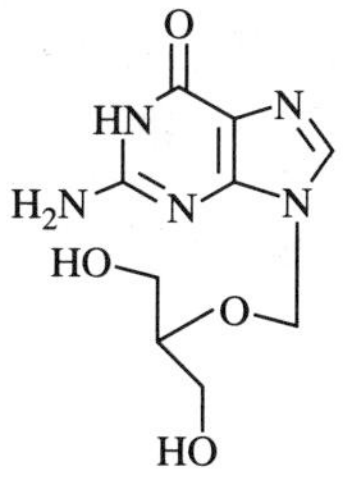

**8.95**

Only 15–20% of acyclovir is absorbed after oral administration. Consequently, two prodrugs for acyclovir have been designed to improve gastrointestinal absorption and to protect acyclovir against biotransformations to inactive metabolites. 2,6-Diamino-9-(2-hydroxyethoxymethyl)purine (**8.96**) is converted to acyclovir by the enzyme adenosine deaminase[102] (catalyzes the hydrolysis of adenosine to inosine), and 6-amino-9-(2-hydroxyethoxymethyl)purine (**8.97**, 6-deoxyacyclovir) is oxidized to acyclovir by xanthine oxidase.[103] The latter compound is 18 times more water soluble than acyclovir. In humans urinary excretion of acyclovir is 5–6 times greater when **8.97** is given than an equivalent dose of acyclovir.

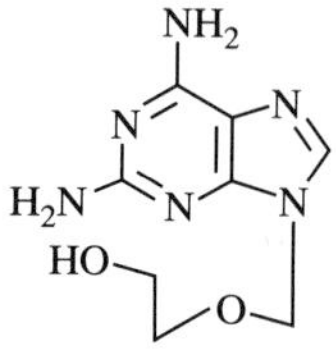

$NH_2$ N N $H_2N$ N N HO O

**8.96**

N N $H_2N$ N N HO O

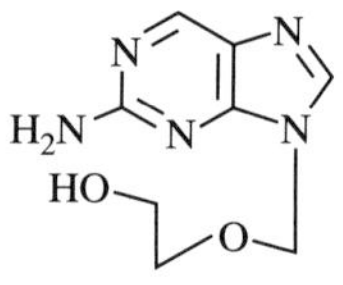

**8.97**

### 6. Decarboxylation Activation

The striatal tracts, which are important for the control of voluntary movements, contain a balance of the inhibitory neurotransmitter dopamine and the excitatory neurotransmitter acetylcholine. An imbalance in the dopaminergic and cholinergic components produces disorders of movement. In Parkinson's disease there is a marked deficiency in the dopaminergic component which is attributed to the loss of dopaminergic neurons in the substantia nigra. The obvious treatment for Parkinson's disease would be to give high doses of dopamine (**8.98**, R = H), but this does not work because dopamine does not cross the blood–brain barrier. However, there is an active transport system for L-amino acids; consequently, L-dopa (**8.98**, R = COOH) is transported into the brain where it is decarboxylated by the pyridoxal 5′-phosphate–dependent enzyme (see Section III,A,2 of Chapter 4) aromatic-L-amino acid decarboxylase (also called dopa decarboxylase) to dopamine. Since the D,L-mixture produces unwanted side effects, levodopa (L-dopa) is used as a prodrug for dopamine. Unfortunately, because dopaminergic neurons cannot be rejuvenated, levodopa does not reverse the course of the disease, it merely halts (actually only slows) its progression.[104]

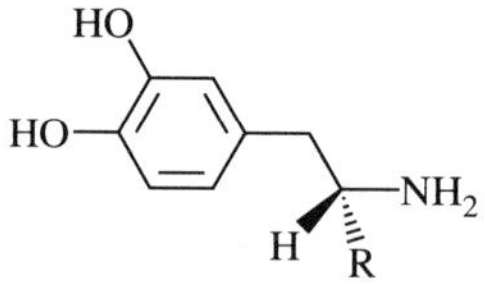

HO HO $NH_2$ H R

**8.98**

As discussed in Section V,C,3,d of Chapter 5, dopamine is a substrate for monoamine oxidase B; consequently, as levodopa is being converted to dopamine in the brain, monoamine oxidase B is degrading the dopamine. An inactivator of monoamine oxidase B, L-deprenyl, is now used in combination with levodopa to minimize the degradation of the dopamine generated by levodopa.[105]

One major complication with the use of levodopa therapy arises from the fact that aromatic-L-amino acid decarboxylase also exists in the periphery (outside of the central nervous system), and greater than 95% of the orally administered levodopa is decarboxylated in its first pass through the liver and kidneys. Possibly only 1% of the levodopa taken actually penetrates into the central nervous system. If the peripheral aromatic-L-amino acid decarboxylase could be inhibited without inhibition of the same enzyme in the brain, the levodopa would be protected from the undesired metabolism. This, in fact, is possible because inhibitors of aromatic-L-amino acid decarboxylase are charged molecules, and unless they are actively transported, they will not cross the blood–brain barrier. Carbidopa (**8.99**) is used in the United States, and benserazide (**8.100**) is used in Europe and Canada in combination with levodopa for the treatment of Parkinson's disease. With the combined use of a peripheral aromatic-L-amino acid decarboxylase inhibitor, the optimal effective dose of levodopa can be reduced by greater than 75%.

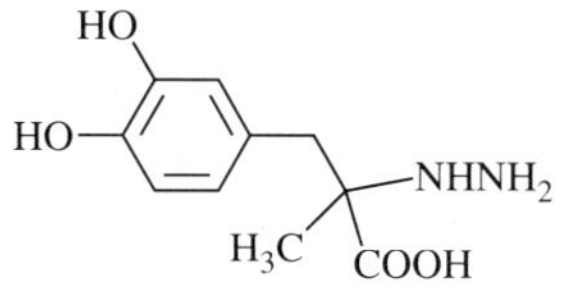

**8.99** 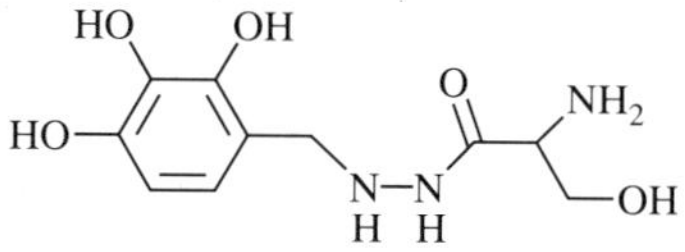 **8.100**

In Section V,C,3,c of Chapter 5, the application of inactivators of monoamine oxidase A (MAO A) as antidepressant agents was discussed. Although MAO inactivators are used in the treatment of depression, a severe cardiovascular side effect can result unless the diet is controlled to minimize the intake of tyramine-containing foods. This side effect results from the concurrent inactivation of the peripheral MAO A along with the brain MAO A. A brain-specific MAO A inactivator would give the desired antidepressant effect without the undesirable cardiovascular effect. A prodrug approach for the brain-selective delivery of a MAO A–selective inactivator was developed at Merrell Dow.[106,107] This particular type of prodrug was termed a *dual enzyme-activated inhibitor*[107] because the activating enzyme is, by design, part of the same metabolic pathway as the enzyme that is targeted for inhibition. In this case the activating enzyme is aromatic-L-amino acid decarboxylase and the target enzyme is MAO A. (*E*)-$\beta$-Fluoromethylene-*m*-tyramine (**8.101**, R = H) is a mechanism-based inactivator (see Section V,C of Chapter 5) of monoamine oxidase with selectivity for MAO A.[108] The corresponding amino acid, (*E*)-$\beta$-fluoromethylene-*m*-tyrosine (**8.101**, R = COOH)[107] is not an inhibitor of MAO, but it is a good substrate for aromatic-L-amino acid decarboxylase, which converts **8.101** (R = COOH) to **8.101** (R = H). The amino acid

(**8.101**, R = COOH) is actively transported into the central nervous system and is concentrated in the synaptosomes. Because brain aromatic-L-amino acid decarboxylase is located predominantly in monoamine nerve endings, **8.101** (R = COOH) is decarboxylated to **8.101** (R = H) at the desired site of action. To prevent inactivation of peripheral MAO A, **8.101** (R = COOH) is administered with carbidopa, which blocks peripheral aromatic-L-amino acid decarboxylase–catalyzed decarboxylation of **8.101** (R = COOH). This results in brain-selective MAO A inactivation with little or no peripheral MAO A inhibition and only a minimal tyramine effect.

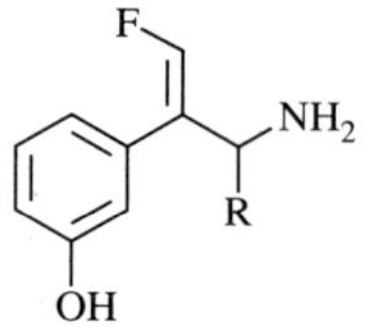

F
$NH_2$
R
OH

**8.101**

Not only is dopamine a major inhibitory neurotransmitter, but it also plays an important role in the kidneys. Dopamine increases systolic and pulse blood pressure and renal blood flow. If it is desired to have selective delivery of dopamine to the kidneys in order to attain renal vasodilation without a blood pressure effect, a prodrug for dopamine can be used. There is a high concentration of L-γ-glutamyltranspeptidase, the enzyme that catalyzes the transfer of the L-glutamyl group from the N-terminus of one peptide to another, in kidney cells. Consequently, an L-γ-glutamyl derivative of an amino acid or amine drug could be cleaved selectively in the kidneys.[109] L-γ-Glutamyl-L-dopa (**8.102**) is selectively accumulated in the kidneys, and the L-dopa released by L-glutamyltranspeptidase (γ-glutamyltransferase) is decarboxylated to dopamine by aromatic-L-amino acid decarboxylase, which also is abundant in kidneys (Scheme 8.27).[110] Even at high concentrations of this compound

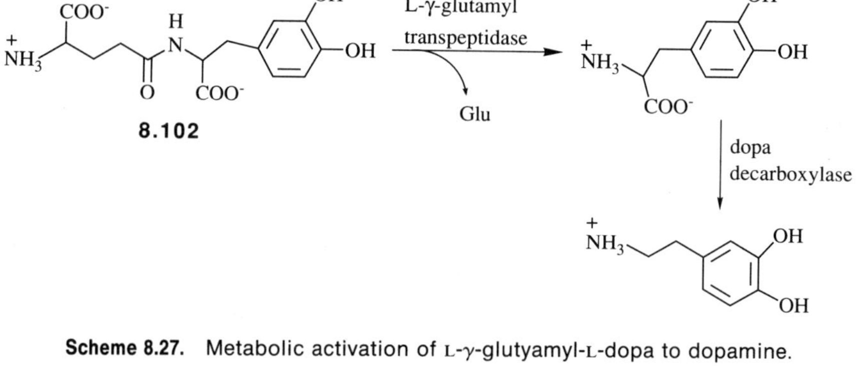

$COO^-$ H $\overset{+}{N}H_3$ N O $COO^-$ OH OH

**8.102**

L-γ-glutamyl transpeptidase

Glu

$\overset{+}{N}H_3$ $COO^-$ OH OH

dopa decarboxylase

$\overset{+}{N}H_3$ OH OH

**Scheme 8.27.** Metabolic activation of L-γ-glutyamyl-L-dopa to dopamine.

little central nervous system effect is apparent. This, then, is an example of a site-selective carrier-linked prodrug of a bioprecursor prodrug for dopamine.

Drug design is typically initiated with approaches to maximize the pharmacodynamic properties of molecules (increased binding to a receptor). A compound may be found that has the desired *in vitro* properties but has unfavorable *in vivo* properties. It should be apparent, then, from the discussion in this chapter that it may be possible to alter the structure of the compound to improve its pharmacokinetic properties and, thereby, transform it into a promising drug candidate.

## References

1. Albert, A. 1951. "Selective Toxicity." Chapman & Hall, London; Albert, A. 1958. *Nature (London)* **182**, 421.
2. Harper, N. J. 1959. *J. Med. Pharm. Chem.* **1**, 467
3. Wermuth, C. G. 1983. *In* "Drug Metabolism and Drug Design: *Quo Vadis*?" (Briot, M., Cautreels, W., and Roncucci, R., eds.), p. 253. Sanofi-Clin-Midy, Montpellier, France.
4. Greene, T. W. 1981. "Protective Groups in Organic Synthesis." Wiley, New York.
5. Long, L., Jr. 1940. *Chem. Rev.* **27**, 437.
6. Bundgaard, H. 1985. *In* "Design of Prodrugs" (Bundgaard, H., ed.), p. 1. Elsevier, Amsterdam.
7. Reynolds, W. F. 1983. *Prog. Phys. Org. Chem.* **14**, 165.
8. Bundgaard, H. 1987. *In* "Bioreversible Carriers in Drug Design" (Roche, E. B., ed.), p. 13. Pergamon, New York.
9. Johansen, M., and Bundgaard, H. 1982. *Arch. Pharm. Chem. Sci. Ed.* **10**, 111.
10. Johansen, M., and Bundgaard, H. 1980. *Int. J. Pharm.* **7**, 119; Bundgaard, H., and Johansen, M. 1981. *Int. J. Pharm.* **8**, 183.
11. Kaplan, J.-P., Raizon, B. M., Desarmenien, M., Feltz, P., Headley, P. M., Worms, P., Lloyd, K. G., and Bartholini, G. 1980. *J. Med. Chem.* **23**, 702.
12. Anderson, B. D., Conradi, R. A., and Knuth, K. E. 1985. *J. Pharm. Sci.* **74**, 365; Anderson, B. D., Conradi, R. A., Knuth, K. E., and Nail, S. L. 1985. *J. Pharm. Sci.* **74**, 375.
13. Anderson, B. D., Conradi, R. A., Spilman, C. H., and Forbes, A. D. 1985. *J. Pharm. Sci.* **74**, 382.
14. Slojkowska, Z., Krakuska, H. J., and Pachecka, J. 1982. *Xenobiotica* **12**, 359.
15. Hassall, C. H., Holmes, S. W., Johnson, W. H., Kröhn, A., Smithen, C. E., and Thomas, W. A. 1977. *Experientia* **33**, 1492.
16. Hirai, K., Ishiba, T., Sugimoto, H., Fujishita, T., Tsukinoki, Y., and Hirose, K. 1981. *J. Med. Chem.* **24**, 20.
17. Hirai, K., Fujishita, T., Ishiba, T., Sugimoto, H., Matsutani, S., Tsukinoki, Y., and Hirose, K. 1982. *J. Med. Chem.* **25**, 1466.
18. Cho, M. J., Sethy, V. H., and Haynes, L. C. 1986. *J. Med. Chem.* **29**, 1346.
19. Hadgraft, J. 1985. *In* "Design of Prodrugs" (Bundgaard, H., ed.), p. 271. Elsevier, Amsterdam.
20. Mandell, A. I., Stentz, F., and Kitabuchi, A. E. 1978. *Ophthalmology* **85**, 268.
21. Rappoport, S. I. 1967. "The Blood–Brain Barrier in Physiology and Medicine." Raven, New York; Pardridge, W. M., Conor, J. D., and Crawford, I. L. 1975. *Crit. Rev. Toxicol.* **3**, 159.
22. Bodor, N., and Brewster, M. 1983. *Pharmacol. Ther.* **19**, 337.
23. Bodor, N. 1987. *Ann. N.Y. Acad. Sci.* **507**, 289.

24. Pop, E., Wu, W.-M., Shek, E., and Bodor, N. 1989. *J. Med. Chem.* **32**, 1774; Wu, W.-M., Pop, E., Shek, E., and Bodor, N. 1989. *J. Med. Chem.* **32**, 1782. Pop, E., Wu, W.-M. and Bodor, N. 1989. *J. Med. Chem.* **32**, 1789.
25. Worms, P., Depoortere, H., Durand, A., Morselli, P. L., Lloyd, K. G., and Bartholini, G. 1982. *J. Pharmacol. Exp. Ther.* **220**, 660.
26. Jacob, J. N., Hesse, G. W., and Shashoua, V. E. 1990. *J. Med. Chem.* **33**, 733.
27. Brandes, D., and Bourne, G. H. 1955. *Lancet 1*, 481.
28. Garceau, Y., Davis, I., and Hasegawa, J. 1978. *J. Pharm. Sci.* **67**, 1360.
29. Hussain, M. A., Koval, C. A., Myers, M. J., Shami, E. G., and Shefter, E. 1987. *J. Pharm. Sci.* **76**, 356.
30. Deberdt, R., Elens, P., Berghmans, W., Heykants, J., Woestenborghs, R., Driesens, F., Reyntjens, A., and Van Wijngaarden, I. 1980. *Acta Psychiatr. Scand.* **62**, 356.
31. Chouinard, G., Annable, L., and Ross-Chouinard, A. 1982. *Am. J. Psychiatry* **139**, 312.
32. Persico, F. J., Pritchard, J. F., Fischer, M. C., Yorgey, K., Wong, S., and Carson, J. 1988. *J. Pharmacol. Exp. Ther.* **247**, 889.
33. Nielsen, N. M., and Bundgaard, H. 1989. *J. Med. Chem.* **32**, 727.
34. De Haan, R. M., Metzler, C. M., Schellenberg, D., and Vanderbosch, W. D. 1973. *J. Clin. Pharmacol.* **13**, 190.
35. Sinkula, A. A., Morozowich, W., and Rowe, E. L. 1973. *J. Pharm. Sci.* **62**, 1106.
36. Notari, R. E. 1973. *J. Pharm. Sci.* **62**, 865.
37. de Duve, C., de Barsy, T., Poole, B., Trouet, A., Tulkens, P., and Van Hoof, F. 1974. *Biochem. Pharmacol.* **23**, 2495.
38. Cecchi, R., Rusconi, L., Tanzi, M. C., Danusso, F., and Ferruti, P. 1981. *J. Med. Chem.* **24**, 622.
39. Shen, W.-C., and Ryser, H. J.-P. 1979. *Mol. Pharmacol.* **16**, 614.
40. Chu, B. C. F., and Howell, S. B. 1981. *Biochem. Pharmacol.* **30**, 2545.
41. Zupon, M. A., Fang, S. M., Christensen, J. M., and Petersen, R. V. 1983. *J. Pharm. Sci.* **72**, 1323.
42. Ringsdorf, H. 1975. *J. Polym. Sci. Part C: Polym. Symp.* **51**, 135.
43. Rowland, G. F., O'Neill, G. J., and Davies, D. A. L. 1975. *Nature (London)* **255**, 487.
44. Rowland, G. F. 1977. *Eur. J. Cancer* **13**, 593.
45. Balboni, P. G., Minia, A., Grossi, M. P., Barbarti-Brodano, G., Mattioli, A., and Fiume, L. 1976. *Nature (London)* **264**, 181.
46. Senter, P. D. 1990. *FASEB J.* **4**, 188.
47. Carl, P. L., Chakravarty, P. K., and Katzenellenbogen, J. A. 1981. *J. Med. Chem.* **24**, 479.
48. Jansen, A. B. A., and Russell, T. J. 1965. *J. Chem. Soc.*, 2127.
49. Bodin, N. D., Ekström, B., Forsgren, U., Jalar, L. P., Magni, L., Ramsey, C. H., and Sjöberg, B. 1975. *Antimicrob. Agents Chemother.* **9**, 518.
50. Daehne, W. V., Frederiksen, E., Gundersen, E., Lund, F., Mørch, P., Petersen, H. J., Roholt, K., Tybring, L., and Godtfredsen, W. O. 1970. *J. Med. Chem.* **13**, 607.
51. Møllgaard, B., Hoelgaard, A., and Bundgaard, H. 1982. *Int. J. Pharm.* **12**, 153.
52. Kingsbury, W. D., Boehm, J. C., Mehta, R. J., Grappel, S. F., and Gilvarg, C. 1984. *J. Med. Chem.* **27**, 1447.
53. Hartley, S., and Wise, R. 1982. *J. Antimicrob. Chemother.* **10**, 49; Baltzer, B., Binderup, E., Von Daehne, W., Godtfredsen, W. O., Hansen, K., Nielsen, B., Sørensen, H., and Vangedal, S. 1980. *J. Antibiot.* **33**, 1183.
54. Lahti, R. A., and Gall, M. 1976. *J. Med. Chem.* **19**, 1064; Gall, M., Hester, J. B., Jr., Rudzik, A. D., and Lahti, R. A. 1976. *J. Med. Chem.* **19**, 1057.
55. Preussmann, R., Druckrey, H., Ivankovic, S., von Hodenberg, A. 1969. *Ann. N.Y. Acad. Sci.* **163**, 697; Shealy, Y. F., O'Dell, C. A., and Krauth, C. A. 1975. *J. Pharm. Sci.* **64**, 177; Montgomery, J. A. 1976. *Cancer Treat. Rep.* **60**, 125.
56. Comis, R. L. 1976. *Cancer Treat. Rep.* **60**, 165.

57. Skibba, J. L., Ramirez, G., Beal, D. D., and Bryan, G. T. 1970. *Biochem. Pharmacol.* **19**, 2043; Mizuno, N. S., and Humphrey, E. W. 1972. *Cancer Chemother. Rep. (Part 1)* **56**, 465.
58. Brodie, B. B., and Axelrod, J. 1949. *J. Pharmacol. Exp. Ther.* **97**, 58.
59a. Colvin, M., and Chabner, B. A. 1990. *In* "Cancer Chemotherapy: Principles and Practice" (Chabner, B. A., and Collins, J. M., eds.), p. 276. Lippincott, Philadelphia, Pennsylvania.
59b. Cox, P. J., Farmer, P. B., and Jarman, M. 1976. *Cancer Treat. Rep.* **60**, 299.
59c. Hill, D. L. 1975. "A Review of Cyclophosphamide." Thomas, Springfield, Illinois.
59d. Benson, A. J., Martin, C. N., and Garner, R. C. 1988. *Biochem. Pharmacol.* **37**, 2979.
60. Connors, T. A., Cox, P. J., Farmer, P. B., Foster, A. B., and Jarman, M. 1974. *Biochem. Pharmacol.* **23**, 115.
61. Oliverio, V. T. 1982. *In* "Cancer Medicine" (Holland, J. F., and Frei III, E., eds.) 2nd Ed., p. 850. Lea & Febiger, Philadelphia, Pennsylvania; Weinkam, R. J., Shiba, D. A., and Chabner, B. A. 1982. *In* "Pharmacologic Principles of Cancer Treatment" (Chabner, B. E., ed.), p. 340. Saunders, Philadelphia, Pennsylvania.
62. Raaflaub, J., and Schwartz, D. E. 1965. *Experientia* **21**, 44.
63. Kreis, W., Piepho, S. B., and Bernhard, H. V. 1966. *Experientia* **22**, 431.
64. Tsuji, T., and Kosower, E. M. 1971. *J. Am. Chem. Soc.* **93**, 1992.
65. Froede, H. C., and Wilson, I. B. 1971. *In* "The Enzymes" (Boyer, P., ed.), 3rd Ed., Vol. 5, p. 87. Academic Press, New York.
66. Jansen, E. F., Nutting, M.-D. F., and Balls, A. K. 1949. *J. Biol. Chem.* **179**, 201.
67. Wilson, I. B. 1958. *Biochim. Biophys. Acta* **27**, 196.
68. Shek, E., Higuchi, T., and Bodor, N. 1976. *J. Med. Chem.* **19**, 113.
69. Frigerio, A., Fanelli, R., Biandrate, P., Passerini, G., Morselli, P. L., and Garattini, S. 1972. *J. Pharm. Sci.* **61**, 1144; Johannessen, S. I., Gerna, N. M., Bakke, J., Strandjord, R. E., and Morselli, P. L. 1976. *Br. J. Clin. Pharmacol.* **3**, 575.
70. Monsarrat, B., Maftouh, M., Meunier, G., Dugué, B., Bernadou, J., Armand, J.-P., Picard-Fraire, C., Meunier, B., and Paoletti, C. 1983. *Biochem. Pharmacol.* **32**, 3887.
71. Neely, J. R., and Morgan, H. E. 1974. *Annu. Rev. Physiol.* **36**, 413.
72. Barnish, I. T., Cross, P. E., Danilewicz, J. C., Dickinson, R. P., and Stopher, D. A. 1981. *J. Med. Chem.* **24**, 399.
73. Kirsner, J. B. 1980. *J. Am. Med. Assoc.* **243**, 557; Eastwood, M. A. 1980. *Ther. Drug Monit.* **2**, 149.
74. Brown, J. P., McGarraugh, G. V., Parkinson, T. M., Wingard, R. E., Jr., and Onderdonk, A. B. 1983. *J. Med. Chem.* **26**, 1300.
75. Shen, T. Y., and Winter, C. A. 1977. *Adv. Drug Res.* **12**, 90.
76. Matsukawa, T., Yurugi, S., and Oka, Y. 1962. *Ann. N.Y. Acad. Sci.* **98**, 430.
77. Stella, V. J., and Himmelstein, K. J. 1985. *In* "Design of Prodrugs" (Bundgaard, H., ed.), p. 177. Elsevier, Amsterdam.
78. Hofsteenge, J., Capuano, A., Altszuler, R., and Moore, S. 1986. *J. Med. Chem.* **29**, 1765.
79. Moore, H. W., and Czerniak, R. 1981. *Med. Res. Rev.* **1**, 249; Moore, H. W. 1977. *Science* **197**, 527.
80. Iyer, V. N., and Szybalski, W. 1964. *Science* **145**, 55.
81. Franck, R. W., and Tomasz, M. 1990. *In* "The Chemistry of Antitumor Agents" (Wilman, D. E. V., ed.), p. 379. Blackie and Son, Glasgow; Remers, W. A. 1979. "The Chemistry of Antitumor Antibiotics," Vol. 1, p. 271. Wiley, New York.
82. Bean, M., and Kohn, H. 1985. *J. Org. Chem.* **50**, 293.
83. Kohn, H., Zein, N., Lin, X. Q., Ding, J.-Q., and Kadish, K. M. 1987. *J. Am. Chem. Soc.* **109**, 1833; Danishefsky, S. J., and Egbertson, M. 1986. *J. Am. Chem. Soc.* **108**, 4648; Andrews, P. A., Pan, S.-S., and Bachur, N. R. 1986. *J. Am. Chem. Soc.* **108**, 4158.
84. Kohn, H., and Hong, Y. P. 1990. *J. Am. Chem. Soc.* **112**, 4596.
85. Borowy-Borowski, H., Lipman, R., Chowdary, D., and Tomasz, M. 1990 *Biochemistry* **29**, 2992; Borowy-Borowski, H., Lipman, R., and Tomasz, M. 1990. *Biochemistry* **29**, 2999.

86a. Tomasz, M., Lipman, R., Chowdary, D., Pawlak, J., Verdine, G. L., Nakanishi, K. 1987. *Science* **235**, 1204.
86b. Tomasz, M., Lipman, R., McGuinness, B. F., and Nakanishi, K. 1988. *J. Am. Chem. Soc.* **110**, 5892.
86c. Millard, J. T., Weidner, M. F., Raucher, S., and Hopkins, P. B. 1990. *J. Am. Chem. Soc.* **112**, 3637.
87a. Kleyer, D. L., and Koch, T. H. 1984. *J. Am. Chem. Soc.* **106**, 2380.
87b. Sulikowski, G. A., Turos, E., Danishefsky, S. J., and Shulte, G. M. 1991. *J. Am. Chem. Soc.* **113**, 1373.
88. Kennedy, K. A., Teicher, B. A., Rockwell, S., and Sartorelli, A. C. 1980. *Biochem. Pharmacol.* **29**, 1.
89. Antonini, I., Lin, T.-S., Cosby, L. A., Dai, Y.-R., and Sartorelli, A. C. 1982. *J. Med. Chem.* **25**, 730.
90. Lin, A. J., Lillis, B. J., and Sartorelli, A. C. 1975. *J. Med. Chem.* **18**, 917.
91. Lin, A. J., and Sartorelli, A. C. 1976. *Biochem. Pharmacol.* **25**, 206; Prakash, G., and Hodnett, E. M. 1978. *J. Med. Chem.* **21**, 369.
92. Miwa, G. T., Wang, R., Alvaro, R., Walsh, J. S., and Lu, A. Y. H. 1986. *Biochem. Pharmacol.* **35**, 33.
93. McCollister, R. J., Gilbert, W. R., Jr., Ashton, D. M., and Wyngaarden, J. B. 1964. *J. Biol. Chem.* **239**, 1560; Caskey, C. T., Ashton, D. M., and Wyngaarden, J. B. 1964. *J. Biol. Chem.* **239**, 2570; Henderson, J. F., and Khoo, M. K. Y. 1965. *J. Biol. Chem.* **240**, 3104.
94. Furman, P. A., McGuirt, P. V., Keller, P. M., Fyfe, J. A., and Elion, G. B. 1980. *Virology* **102**, 420.
95. Miller, W. H., and Miller, R. L. 1980. *J. Biol. Chem.* **255**, 7204.
96. Miller, W. H., and Miller, R. L. 1982. *Biochem. Pharmacol.* **31**, 3879.
97. Furman, P. A., St. Clair, M. H., Fyfe, J. A., Rideout, J. L., Keller, P. M., and Elion, G. B. 1979. *J. Virol.* **32**, 72.
98. Reardon, J. E., and Spector, T. 1989. *J. Biol. Chem.* **264**, 7405.
99. Coen, D. M., Schaffer, P. A., Furman, P. A., Keller, P. M., and St. Clair, M. H. 1982. *Am. J. Med.* **73**(1A), 351; Schnipper, L. E., and Crumpacker, C. S. 1980. *Proc. Natl. Acad. Sci. U.S.A.* **77**, 2270.
100. Larder, B. A., Cheng, Y.-C., and Darby, G. 1983. *J. Gen. Virol.* **64**, 523.
101. Elion, G. B. 1986. *In* "Antiviral Chemotherapy: New Directions for Clinical Application and Research" (Mills, J., and Corey, L., eds.), p. 118. Elsevier, New York.
102. Good, S. S., Krasny, H. C., Elion, G. B., and de Miranda, P. 1983. *J. Pharmacol. Exp. Ther.* **227**, 644.
103. Krenitsky, T. A., Hall, W. W., de Miranda, P., Beauchamp, L. M., Schaeffer, H. J., and Whiteman, P. D. 1984. *Proc. Natl. Acad. Sci. U.S.A.* **81**, 3209.
104. Bernheimer, H., Birkmayer, W., Hornykiewicz, O., Jellinger, K., and Seitelberger, F. 1973. *J. Neurol. Sci.* **20**, 415.
105. Birkmayer, W., and Riederer, P. 1984. *Adv. Neurol.* **40**, 475; Tetrud, J. W., and Langston, J. W. 1989. *Science* **245**, 519.
106. Palfreyman, M. G., McDonald, I. A., Fozard, J. R., Mely, Y., Sleight, A. J., Zreika, M., Wagner, J., Bey, P., and Lewis, P. J. 1985. *J. Neurochem.* **45**, 1850.
107. McDonald, I. A., Lacoste, J. M., Bey, P., Wagner, J., Zreika, M., and Palfreyman, M. G. 1986. *Bioorg. Chem.* **14**, 103.
108. McDonald, I. A., Lacoste, J. M., Bey, P., Palfreyman, M. G., and Zreika, M. 1985. *J. Med. Chem.* **28**, 186.
109. Magnan, S. D. J., Shirota, F. N., and Nagasawa, H. T. 1982. *J. Med. Chem.* **25**, 1018.
110. Wilk, S., Mizoguchi, H., and Orlowski, M. 1978. *J. Pharmacol. Exp. Ther.* **206**, 227; Kyncl, J. J., Minard, F. N., and Jones, P. H. 1979. *Adv. Biosci.* **20**, 369.

# General References

## Prodrugs

Bundgaard, H., ed. 1985. "Design of Prodrugs." Elsevier, Amsterdam.

## Macromolecular Drug Carrier Systems

Friend, D. R., and Pangburn, S. 1987. *Med. Res. Rev.* **7**, 53.
Goldberg, E. P., ed. 1983. "Targeted Drugs." Wiley, New York.
Gregoriadis, G., Senior, J., and Trouet, A., eds. 1982. "Targeting of Drugs." Plenum, New York.
Poznansky, M. J., and Juliano, K. L. 1984. *Pharmacol. Rev.* **36**, 277.
Roerdink, F. H. D., and Kroon, A. M., eds. 1989. "Drug Carrier Systems." Wiley, Chichester.

# Index

**B**

**D**

## E

## F



## G

## H

## I



## K



## L



## M

## Q



## R